Praise for *The Carbon Farming Solution*

"*The Carbon Farming Solution* is a book we will look back upon decades from now and wonder why something so critically relevant could have been so overlooked until that time. We are told we have a choice between chemical/GMO agriculture if we want to feed the world or we can see children starve and adopt organic agriculture as a romantic and sentimental pursuit. Really? Toensmeier describes a future that is in alignment with how life works, a scientific and sophisticated agricultural understanding of husbandry and biology that surpasses the productivity of industrial agriculture. What is phenomenal about these land-use solutions is that they are the only way we can bring carbon back home if we are to reverse climate change. The title is accurate but humble: *The Carbon Farming Solution* describes the foundation of the future of civilization."

—Paul Hawken, author of *Blessed Unrest*

"Dealing with climate change requires action on many fronts, and this book is the toolkit for making the soil itself a sponge for carbon. It's a powerful vision, one that I've seen playing out in enough places to make me very hopeful it can presage major changes in our species's use of the land."

—Bill McKibben, author of *Deep Economy*

"Eric Toensmeier presents a convincing argument that carbon farming is crucial to addressing global issues of the 21st century including climate change, food and nutritional insecurity, eutrophication and contamination of water, and dwindling of soil biodiversity. Implemented in a transparent manner and with payments of a just and fair price based on the true societal value, carbon farming is also pertinent to alleviating poverty and addressing several Sustainable Development Goals of the United Nations. Carbon farming as a strategy is in accord with the "4 pour 1000" initiative of the French Government presented during the COP21 Summit in Paris on December 1, 2015, and *The Carbon Farming Solution* is a befitting tribute to the 2015 International Year of Soils."

—Dr. Rattan Lal, Distinguished University Professor of Soil Science
and director of The Carbon Management and Sequestration Center,
The Ohio State University; President Elect, International Union of Soil Sciences

"Eric Toensmeier has done a hugely impressive job putting together this magnum opus. It is packed with an enormous amount of information about seven hundred plant species that have a role to play in saving the planet from land degradation and climate change while at the same time improving the lives of millions of poor farmers, especially in the tropics and sub-tropics. *The Carbon Farming Solution* covers species for every use and every situation that can be assembled in infinite agroecological combinations. On top of that, the cultivation of these crops can lead to new industries in the production of food, medicines, cosmetics, and materials—creating wealth and employment. This information should be absorbed by everyone engaged in agriculture; everyone concerned about the future of the world and the well-being and health of its people; and everyone interested in protecting biodiversity. Indeed, *The Carbon Farming Solution* offers a path to a bright new world!"

—Professor Roger Leakey, vice chairman, International Tree Foundation;
author of *Living with the Trees of Life*

"*The Carbon Farming Solution* is an excellent reference book that convincingly explains the potential of farming practices based on perennial crops for carbon sequestration and climate change mitigation and adaptation. The numerous photographs and charts included help illustrate the food-security and multi-functionality attributes of agroforestry and other such farming systems. In addition to professionals who work on food security and climate stabilization issues, undergraduate and graduate students of these topics will find the book useful."

—Dr. P. K. Ramachandran Nair, Distinguished Professor in the School of Forest Resources and Conservation, University of Florida

"In *The Carbon Farming Solution*, Eric Toensmeier admirably harnesses available data with traditional wisdom to propose a practical response to climate change. Toensmeier's solution-oriented ideas combine his clear understanding of ecology, agriculture, and the magnitude of the challenge we face with a set of agriculture-based solutions that are suited to various livelihoods, communities, and systems of production. This book will surely be a benchmark in policy-relevant knowledge."

—Dr. Cheikh Mbow, senior scientist on climate change and development, World Agroforestry Centre

"Scientific observations and models are building an increasingly dire picture of the obstacles that must be crossed on the road to achieving climate and ecological health and stability on a planet filled with humans. The relentlessly hopeful (but not naïvely optimistic) author of *The Carbon Farming Solution* reminds us that our planet is still rich in biological resources and that humanity is capable of astonishing feats of creativity and collaborative action; the picture painted here in word and image depicts both the barriers and paths through them. Eric Toensmeier draws upon both the scientific literature and the world's ethnobotanical knowledge bank to construct a logical and compelling road map for future research and investment to reinvent agriculture. But reason and facts alone are insufficient to sustain a global and long-term agenda; passion is required. In the end, it is the perennial plants (and their human and microbial partners) themselves—lovingly portrayed here in their glorious diversity and elegant functionality—that steal the show and our hearts. This 'Who's Who' of wild or orphaned potential crops can inspire a new generation of plant lovers and gardeners to become the convention-questioning, dedicated, passionate, hopeful scientists, farmers, and leaders that the movement requires."

—David Van Tassel, PhD, senior scientist, The Land Institute

"These are exciting times for soil carbon! What was once an obscure topic mainly of interest to agronomists and gardeners is now viewed by many people as a key to solving multiple challenges in the 21st century, including climate change, hunger, and drought. For urgent times, we need an urgent agriculture. That's exactly what we get in Eric Toensmeier's new book—a detailed, practical explanation of how to increase carbon in our soils, written with passion and skill by a leader in regenerative agriculture. We know what to do, and with *The Carbon Farming Solution* we know how to do it. Let's get going!"

—Courtney White, author of *Grass, Soil, Hope* and *Two Percent Solutions for the Planet*

"Agriculture is currently a major net producer of greenhouse gases with little prospect of improvement unless things change markedly. In *The Carbon Farming Solution*, Eric Toensmeier puts carbon sequestration at the forefront and shows how agriculture can be a net absorber of carbon. Improved forms of annual-based agriculture can help to a degree; however, to maximize carbon sequestration, it is perennial crops we must look at, whether it be perennial grains, other perennial staples, or agroforestry systems incorporating trees and other crops. In this impressive book, backed up with numerous tables and references, the author has assembled a toolkit that will be of great use to anybody involved in agriculture whether in the tropics or colder northern regions. For me the highlights are the chapters covering perennial crop species organized by use—staple crops, protein crops, oil crops, industrial crops, etc.—with some seven hundred species described. There are crops here for all climate types, with good information on cultivation and yields, so that wherever you are, you will be able to find suitable recommended perennial crops. This is an excellent book that gives great hope without being naïve and makes a clear reasoned argument for a more perennial-based agriculture to both feed people and take carbon out of the air."

—Martin Crawford, director, The Agroforestry Research Trust; author of
Creating a Forest Garden and *Trees for Gardens, Orchards, and Permaculture*

"*The Carbon Farming Solution* is a book whose time has come. This detailed documentation of regenerative practices from around the world, including principles and methods, provides a practical guide for others to follow and expand upon as humanity takes on the 'Great Work of Our Time'—to restore the Earth's natural systems to ecological health. *The Carbon Farming Solution* is of enormous importance."

—John D. Liu, founder and director, Environmental Education Media Project (EEMP)

"Eric Toensmeier has done it again! *The Carbon Farming Solution* is a detailed vision that will become the go-to reference guide for everyone who is interested in an accessible toolkit showcasing global agroecological carbon farming in action. This indispensable book needs to be put in the hands of all climate-change policy makers, agrarians, and people who eat food, drink water, and breathe air. Mr. Toensmeier's book is not ground-breaking—it is ground-healing!"

—Brock Dolman, director, Permaculture Program and
WATER Institute at Occidental Arts and Ecology Center

"If we seriously put our minds to it, we could easily provide ourselves with enough food forever, and do so in ecologically sound ways, and at the same time—a huge bonus!—trap enough carbon in the soil to tip the battle against global warming. The methods are those of agroecology—including organic farming in general and permaculture in particular; and as Eric Toensmeier excellently describes, farmers worldwide are already on the case. So this book offers what governments at present spectacularly do not: hope."

—Colin Tudge, author of *Good Food for Everyone Forever* and
Why Genes Are Not Selfish and People Are Nice

"Eric Toensmeier is one of North America's most inventive and scientifically minded permaculture experimenters. In this book, he offers nothing less than a new vision for world agriculture that is more resilient, supports traditional farmers, and also helps relieve the global climate crisis. *The Carbon Farming Solution* offers an encyclopedic but also highly readable view of new and old carbon-trapping farming methods that can be applied around the world and a profile of the highly adaptable, soil-enhancing perennial plant species that may just be the key to a livable human future."

—Brian Tokar, director, Institute for Social Ecology;
author of *Toward Climate Justice*

THE
CARBON FARMING
SOLUTION

A Global Toolkit of **Perennial Crops** and **Regenerative Agriculture**
Practices for **Climate Change Mitigation** and **Food Security**

ERIC TOENSMEIER

Foreword by Dr. Hans Herren

Chelsea Green Publishing
White River Junction, Vermont

Project Manager: Patricia Stone
Developmental Editor: Brianne Goodspeed
Copy Editor: Laura Jorstad
Proofreader: Eileen M. Clawson
Indexer: Shana Milkie
Designer: Melissa Jacobson
Page Layout: Abrah Griggs

Printed in the United States of America.
First printing January, 2016.
10 9 8 7 6 5 4 3   21 22 23 24

Our Commitment to Green Publishing
Chelsea Green sees publishing as a tool for cultural change and ecological stewardship. We strive to align our book manufacturing practices with our editorial mission and to reduce the impact of our business enterprise in the environment. We print our books and catalogs on chlorine-free recycled paper, using vegetable-based inks whenever possible. This book may cost slightly more because it was printed on paper from responsibly managed forests, and we hope you'll agree that it's worth it. *The Carbon Farming Solution* was printed on paper supplied by Versa Press that is certified by the Forest Stewardship Council.

Library of Congress Cataloging-in-Publication Data
Names: Toensmeier, Eric, author.
Title: The carbon farming solution : a global toolkit of perennial crops and regenerative agriculture practices for climate change mitigation and food security / Eric Toensmeier.
Description: White River Junction, Vermont : Chelsea Green Publishing, [2016]
 | Includes bibliographical references and index.
Identifiers: LCCN 2015034506| ISBN 9781603585712 (hardcover) | ISBN 9781603585729 (ebook)
Subjects: LCSH: Alternative agriculture. | Carbon sequestration. | Climate change mitigation.
Classification: LCC S494.5.A65 T64 2016 | DDC 338.1/62--dc23
LC record available at http://lccn.loc.gov/2015034506

Chelsea Green Publishing
85 North Main Street, Suite 120
White River Junction, VT 05001
(802) 295-6300
www.chelseagreen.com

This book is dedicated to my Kickstarter backers
who made it possible, and especially to my patrons:
Woodbine Ecology Center, Gordon G. Thorne, Scott Huffman,
and HRH Princess Basma bint Ali.

CONTENTS

TABLES

Foreword

Restoring carbon into the soil, where it belongs, and out of the atmosphere, where it is causing havoc, is one of the few win-win solutions to global (as well as local) problems and is the underlying theme of this book.

The timing of the publication of this book could not have been better planned. It arrives at a crucial point in time, just after the Paris Climate Summit, or COP21, held early in December, 2015. The summit was well prepared with a number of pre meetings. Of particular interest to the readers of this book is the "4 pour 1000" initiative launched by the French government in an effort to both reduce emissions from agriculture as well as to start the sequestration of surplus atmospheric carbon, originating both from agriculture and other industrial activities. The pressure from civil society on the Paris negotiators was loud and clear, articulating the need to deliver this time around, given the accelerated melting of the ice cap and CO_2 atmospheric concentrations increasing to dangerous and nonreversible levels.

Some of the very valuable and vast information that the book's author has collected, analyzed, and presented, in a format that is easy to read, understand, and set into practice, has been put forward in many side events and panel discussions in Paris and served as a basis for much of the progress made in pushing the agriculture and food system agenda in the climate discussions. Clearly, attention is now being paid to some workable alternative solutions to the usual energy transition ones in the discussion on how to deal with climate change. Is it enough? No, not yet, but with this book being available to both practitioner and policy constituencies, the excuse of lack of evidence to get started has been removed from the negotiation table. That there is need for more research, analysis, and discussion on the agriculture and food system transformation to live up to the needed results on climate change mitigation and actual reversal, while caring for the food nutrition, feed, and fiber needs of a growing population remains unchallenged. But here a strong scientific and field experience is documented, to build upon and stop the damage, while further solutions are being investigated.

The fact that we are navigating a very narrow range before reaching irreversible tipping points should tell us that the time to act on what we know is now. The safest option would be to restrict our atmospheric CO_2 to 350 ppm and end up with a 0- to .5-degree Celsius temperature change, which is without doubt a challenge. But as indicated in this book, this is an achievable goal if we get started now by implementing what is known already while we look for more and better options to sequester carbon under ground and also in biomass. Additional efforts in energy conservation and transitions to more sustainable energy sources will also be needed, but the evidence presented of a multifaceted and multifunctional agriculture linked to a less wasteful food system is already very convincing and worthy of inclusion in any policy discussion on the transformation of our food-production-to-consumption systems, as a major contribution to climate change mitigation and better, reversal.

The organization of this book will please all readers with a short but much needed introduction and introductory chapters on the basics of climate change and agricultural production systems, followed by practical examples, e.g., the "evidence," and closing with an implementation roadmap. The described agricultural system examples are universal; one can draw the basic principles and adapt them to any part of the world. The examples need to be seen as points of departure for the transformation, for the learning process that must accompany any implementation, given that agriculture is extremely location specific in its details, while remaining quite general on a larger scale. One can also draw the conclusion that the food systems in their wider sense will need to be more localized and respond to the agricultural and local cultural needs. Agroecology is what would come to mind as a description of what is expected as an end result of the transformation exercise. Carbon farming and regenerative/organic agriculture are variations on the same theme of sustainable agriculture and, more often now, the entire food system because clearly we can't separate one from the other. In fact, the main driver for a transformation toward carbon farming will come from the demand side, both for a climate change reversal and for quality and nutritious food.

Key to the transformation process is an enabling policy environment, addressing and including all actors in the system. The call is for a long-term, holistic, and systemic approach to planning as well as an adaptive implementation model where new knowledge, science, technology, and experience are activating a positive implementation spiral. The final section of the book admirably addresses these needed policy measures.

The future of agriculture and food systems is certainly not a world of corn, soybeans, and oil palms. As Eric Toensmeier's eloquent and extremely well-documented book demonstrates, the future lies in more diversity at all organizational levels in the three sustainable development dimensions (environment, society, and the economy). Smart use of nudging mechanisms, such as reassigned subsidies, ecosystem service payments (carbon sequestration being one of them), and true pricing, will help the transformation process, as well as a rethink of how we manage our agriculture and food systems now and how they should be redesigned to fulfill the ambitions set in the Sustainable Development Goals to which all governments have subscribed. Knowing that today we produce at global levels enough food for some 14 billion people, our primary concern needs to be how will a world of 9.5 billion people sustainably and for the long haul nourish itself. Again, Eric Toensmeier's brilliant book supports the case that this is well within the realm of the possible, if not the needed—to both nourish the world and reverse climate change.

DR. HANS HERREN

Introduction

High in the mountains of Veracruz, Mexico, a small cooperative is practicing agriculture in a way that fights climate change while simultaneously meeting human needs. Although millions of people around the world use these practices in some way, people in Western nations are largely unfamiliar with them, and there is little coordinated support to encourage farmers to adopt them. But if widely supported, implemented, and developed on a global scale in conjunction with a massive reduction in fossil fuel emissions, these "carbon farming" practices—a suite of crops and practices that sequester carbon while simultaneously meeting human needs—could play a critical role in preventing catastrophic climate change by removing carbon from the atmosphere and safely storing it in soils and perennial vegetation.

The cloud forest region of Veracruz, Mexico, is unique and beautiful. This humid tropical highland ecosystem combines a mostly temperate canopy of trees such as oaks and hickories encrusted with epiphytic ferns, orchids, and bromeliads with an understory of mostly tropical vegetation such as cannas, wild taros, passion fruits, and tree ferns. Mexico is one of the five most biodiverse countries in the world, and this cloud forest is home to 10 to 12 percent of the country's plant species on only 0.8 percent of its land. It is also home to 550 species of ferns and 750 plants found nowhere else in the world.[1]

But although it has a long history of human use, the cloud forest is disappearing. Between 70 and 90 percent of it has been deforested, and what remains is highly fragmented, with only tiny pockets of old growth. Much of the former forest is degraded pasture.[2] The clouds that give the region its name don't just bring rain. Moisture comes in the form of "horizontal precipitation"—fog from the coast that is captured by the epiphyte-covered

FIGURE I.1. At 5,636 meters (18,491 feet), Pico de Orizaba towers over the highland cloud forest landscape in Veracruz, Mexico. The humid tropical highland cloud forest of this region is home to 10 to 12 percent of the country's plant species on only 0.8 percent of its land.

trees. Water slowly makes its way through the spongy forest floor to streams and then to rivers. Intact cloud forest provides a year-round flow of water to drier regions downstream. In deforested areas, rain instead brings floods followed by dry riverbeds.[3]

Many people in this region are farmers. Cattle and coffee are the primary products. Neither provides much income, and cattle farming as practiced degrades the soil. But people are creative and resilient and are actively experimenting with alternatives. One model that provides income while preserving much of the forest and its functions is the *cafetal*, an agroforestry system in which farmers grow coffee in the cloud forest understory. The *cafetales* help maintain the cloud forest's ecological function and integrity. For example, pasture has no frog species, while *cafetales* have 12 compared with cloud forest's 21.[4] Capture of horizontal precipitation is good as well. *Cafetales* also sequester a lot of carbon in their forest soils and the biomass of the trees and coffee shrubs.

To Ricardo Romero of Las Cañadas, the small cooperative described above, *cafetales* don't provide enough income to farmers, however. Nor do they provide farmers with anything close to a balanced diet. He is working to develop food production systems that provide a complete diet while incorporating as much of the ecosystem function of the cloud forest as possible. Such systems could also serve as corridors to reconnect fragments of intact forest. And it could do all this while sequestering impressive amounts of carbon, helping return our world to a livable climate.

In 1988 Romero began managing the site for pastured cattle. Over the ensuing seasons, he observed the continued degradation of the soils and ecosystem functions. Degraded soils give up much of their carbon to the atmosphere as carbon dioxide, a greenhouse gas. In 1995 he sold his cows and undertook an impressive ecological restoration effort, propagating and planting 50,000 native trees on 60 hectares (148 acres) while allowing another 40 hectares (99 acres) to regenerate naturally. This was

FIGURE I.2. An aerial view of Las Cañadas in Veracruz, Mexico. The many sustainable practices employed at Las Cañadas also sequester carbon, helping to mitigate climate change while producing food, fodder, materials, chemicals, and energy. Photograph courtesy of Ricardo Romero.

the beginning of an ecotourism enterprise that included tours of an awe-inspiring old-growth cloud forest.

Romero also planted native trees on 22 hectares (54 acres) of the remaining pasture and carefully reintroduced cattle. This system, called silvopasture, combines livestock production with the ecological benefits of trees, including soil regeneration and capture of horizontal precipitation. Silvopasture sequesters carbon in the soil and the biomass of the trees. From there the team at Las Cañadas developed a successful organic dairy business.[5]

In 2006 Ricardo and his team formed a cooperative, which today has 22 worker-members. In 2007 they hosted a workshop with permaculture co-founder David Holmgren and decided to change direction toward even greater self-sufficiency. Instead of exporting nutrients from the farm in the form of cheese while buying organic fertilizers to replace them, they began to focus on raising as much of their own food, firewood, and building materials as possible, and generating income from training others in these techniques. Permaculture provided a design framework and principles with which to do so.[6] Today a tour of their farm is like walking through an encyclopedia of sustainable practices and crop diversity. But Romero and his team are doing something very important beyond practicing small-scale sustainable agriculture, fostering community self-reliance, creating jobs, improving biodiversity, and bringing degraded land back to life. These same practices sequester carbon, making Las Cañadas a showcase of some of the world's best climate mitigation techniques.

In this book the term *carbon farming* is used to describe a suite of crops and agricultural practices that sequester carbon in the soil and in perennial biomass. If widely implemented, these practices have the capacity to sequester hundreds of billions of tons of carbon from the atmosphere in the coming decades. (Throughout this book, "tons" refers to metric tons.) And if we combine carbon farming with a massive global reduction in fossil fuel emissions, it can bring us back from the brink of disaster and return our atmosphere to the "magic number" of 350 parts per million of carbon dioxide. Unlike high-tech geoengineering strategies, these practices can also feed people, build more fertile soils, and contribute to ecosystem health.

FIGURE I.3. Las Cañadas practices annual cropping methods that sequester carbon. In this photo, maize and beans are intercropped in a polyculture system. Management of this field uses carbon-friendly practices, including cover cropping, crop rotation, and compost application.

This may seem like a bold claim—and it is—but as we scramble for solutions to our climate catastrophe, the incredible sequestration potential of the crops and practices I describe in this book has been largely ignored. In chapter 3 we'll look more specifically at the available data on sequestration rates, as well as some of the challenges of quantifying it. Despite the challenges and the need for additional research, the evidence is already clear: these crops and practices have the potential to contribute mightily to what is perhaps the most pressing issue of our time.

Carbon farming can take many forms. First and simplest are modifications to annual crop production to reverse the loss of soil carbon from tillage. For example, Las Cañadas practices biointensive crop production with very high yields in small spaces through sophisticated organic techniques. Organic practices like this have been found to sequester more carbon than even the best conventional annual cropping systems. Their larger *milpas*, or crop fields, demonstrate carbon-sequestering agroecological approaches to production of maize, beans, and soybeans, including crop rotation, cover crops, and contour hedgerows. Although these practices have a fairly low carbon sequestration rate,

FIGURE I.4. Livestock can also be raised using techniques that sequester carbon. At Las Cañadas dairy cows graze under native alder trees in a carbon farming silvopasture system. Photograph courtesy of Ricardo Romero.

FIGURE I.5. Perennial crops have high carbon-sequestering ability. Ricardo Romero of Las Cañadas with perennial staple crop plantings that provide protein (perennial beans), carbohydrates (banana, peach palm, air potato), and fats (macadamia).

FIGURE I.6. Bamboo is an outstanding building material and a powerful tool for sequestering carbon. Ricardo Romero with edible shoots of giant timber bamboos in a bamboo grove at Las Cañadas.

they are widely applicable and easily adopted and thus have great global mitigation potential. They also allow us to continue growing the crops we know and love.

Certain livestock systems also constitute carbon farming, which is especially significant because both extensive grazing and confined livestock production (which depends on annual crop production for feed) have been identified as part of the climate change problem. These carbon farming livestock production systems are climate-friendly even when we account for methane releases. For example, Las Cañadas practices managed grazing, fodder banks, and silvopasture—all of which have been shown to sequester carbon in addition to their other benefits. Improved livestock production models typically have a low to moderate carbon sequestration on a per-area basis, but like improved annual cropping systems, they don't require people to change their diets. Given that more than two-thirds of global farmland is pasture, there is great potential to scale up these practices to mitigate climate change.

It is perennial crops, however, that offer the highest potential of any food production system to sequester carbon, especially when they are grown in diverse multilayered systems. (On the other hand, these systems can be challenging to establish and manage, and many people are also not interested in changing their diet to new and unfamiliar perennial crops.) With their plant nursery and seed company, Romero and Karla Arroyo have assembled a world-class collection of perennial crops for their climate with a special focus on perennial staple crops, analogs to maize and beans that grow on trees, vines, palms, and herbaceous perennials. The cooperative has also planted a highly diverse *bosque comestible*, or edible forest, of these species in a system called multistrata agroforestry—the gold standard of biodiversity and carbon sequestration in agriculture.

In their quest for self-sufficiency, resilience, and livelihoods, Las Cañadas is also interested in more than food. It produces many of its own materials, chemicals, and energy. One of their emphases is a 2-hectare (5-acre) planting of clumping bamboos sufficient to provide building materials for all members of the cooperative for the next 100 years. Bamboo is a powerful climate mitigation tool, as it stores lots of carbon in the soil and in its woody parts. Because few resources cover "industrial crops" such as bamboo that not only provide the material, chemicals, and energy that communities need, but also sequester significant amounts of carbon, I devote the entirety of part 4 to these incredible and typically overlooked crops.

All that said, producing food, growing industrial materials, and sequestering carbon is not enough for a 21st-century farmer. Agriculture must also *adapt* to a changing climate. Las Cañadas has a stated goal to "establish production systems that are resilient to prolonged droughts, excessive rains, floods, or abnormal frosts . . ."[7] Although carbon farming practices aren't necessarily, by definition, adaptive, in practice almost all of them are. This is a great co-benefit of carbon farming: Not only do they, by definition, help mitigate climate change, but they also help ecosystems and communities adapt to it. Among many agricultural adaptation techniques on display at Las Cañadas are increases in soil organic matter, crop diversification, and livestock integration. Many carbon farming practices and crops also yield as well as or better than conventional agriculture.

Many of the crops and practices I describe in the chapters ahead are already implemented on a scale of hundreds of millions of hectares globally, although they are still a small fraction of the nearly 5 billion hectares (12 billion acres) of world farmland. These are not minor or marginal efforts, but win–win solutions that also provide food, fodder, and feedstocks while building soils and preserving a climate amenable to civilization. At present, the tropics have stronger carbon farming options than colder climates; many of the agroforestry techniques that have the highest sequestration rates are largely confined to the tropics, at least at present, and most of the best perennial crops available today are also native to, or grown best in, the tropics. The head start the tropics have on carbon farming provides an excellent opportunity for wealthy countries to repay climate debt by bankrolling mitigation, adaptation, and development projects in the Global South and to take lessons from the endeavors already under way there.

This book doesn't offer a prescription for a percentage of cropland that should be used in a particular way. (The many factors that go into selecting appropriate strategies for any given region or farm are touched on in part 5.) Nor is this book a how-to manual, although following up on the references to a given section will frequently provide such information. Nor does it focus on strategies for agriculture-related emissions reduction or adaptation to a changing climate. Likewise, reforestation and timber plantations—both excellent and essential climate mitigation strategies—are outside of its agricultural purview. And you will not find much information on the economics and profitability of these practices here.

What this book does offer is a *toolkit* for communities, governments, and farmers. It is a starting place for selecting appropriate crops and practices for your home region. It provides the rationale behind carbon farming and discusses strategies for global implementation. And it delves into improved annual cropping and pasture systems, two sets of mitigation strategies that have gotten a lot of attention lately. (You will learn that both annual crops and pastures sequester much more

DEFINING CARBON FARMING

There are several, sometimes conflicting, definitions of carbon farming currently in circulation. Most definitions agree that the term refers to farming practices that sequester carbon. Some stop there. For example, here's such a definition from the Marin Carbon Project: "Carbon farming involves implementing practices that are known to improve the rate at which CO_2 is removed from the atmosphere and converted to plant material and/or soil organic matter. Carbon farming is successful when carbon gains resulting from enhanced land management and/or conservation practices exceed carbon losses."[8]

Some other definitions explicitly link carbon farming to carbon offsets.[9] Offsets are a strategy wherein entities that release greenhouse gas emissions pay other entities to sequester equivalent carbon or reduce equivalent emissions. These credits are typically traded on markets. There are two primary problems with carbon offsets. One is that even when they're functioning optimally, they don't reduce the total amount of greenhouse gases and instead just maintain their current dangerous level (unless a "shrinking cap" is built in, which has not yet happened anywhere to my knowledge).[10] Two, and more important, is that they have largely failed to work and have also been vulnerable to corruption.[11] For the record, I'm opposed to the use of offsets as a climate change mitigation strategy and don't include them in the definition I use in this book.

For an expert perspective on the subject, I wrote to Dr. Rattan Lal, the director of the Carbon Management and Sequestration Center at Ohio State University, to ask if carbon farming explicitly requires a link to offsets, or to financing more generally. Dr. Lal is the author of several key articles that serve as a theoretical underpinning for this book. Dr. Lal provided me with his definition of carbon farming. First, he said that *farming* implies products or ecosystem services. For example, hog farming produces pork, and organic farming provides the service of cleaner water downstream.[12] He went on to define carbon farming as "a system of increasing carbon in terrestrial ecosystem[s] for adaptation and mitigation of climate change, [to] enhance ecosystem goods and services, and trade carbon credits for economic gains."[13] In response to my question, Dr. Lal said that while paying the farmer for the service of carbon sequestration is essential to the definition, offsets were not an obligatory mechanism for doing so as long as the farmers were remunerated "in one way or the other."[14]

Does this mean you can't say you are carbon farming if your farm or ranch sequesters carbon, but you don't get paid for it? It depends whose definition you like. What's clear is that it is impossible to scale up the use of carbon-sequestering agricultural practices to the level that could provide serious mitigation without major financing efforts in place.[15] We'll discuss the world of carbon finance options in chapter 27.

carbon when trees are added to them.) The core of this book unpacks the impressive and neglected climate mitigation potential of perennial crops and perennial cropping systems. This includes first-of-their-kind comprehensive profiles of perennial staple crops that provide protein, carbohydrates, and fats, and perennial industrial crops for materials, chemicals, and energy.

Ultimately the goals of this book are to place carbon farming firmly in the center of the climate solutions platform, steer mitigation funds to the millions of people around the world who are already doing the work, and help ignite a massive movement to transform global agriculture.

Carbon farming alone is not enough to avoid catastrophic climate change, even if it were practiced on every square meter of farmland. But it does belong at the center of our transformation as a civilization. Along with new economic priorities, a massive switch to clean energy, and big changes to much of the rest of the way our societies work, carbon farming offers a pathway out of destruction and a route to hope. Along the way it can help address food insecurity, injustice, environmental degradation, and some of the core problems with the global food system. In the pages to come we'll explore the promise and pitfalls of this timely climate change solution.

PART ONE

The Big Idea

CHAPTER ONE

Climate Realities

This is not a book about bad news, but before we discuss agricultural climate change solutions we need to take a good, hard look at the facts. Some of them are grim indeed, not just for polar bears and golden frogs, but for my favorite species of all: human beings. The looming impact of climate change on humanity, especially our most vulnerable members, is what makes it the central issue of our century. As you read the bad news, however, keep in mind that catastrophic climate change is preventable. I promise we'll spend the rest of the book discussing the very real and very hopeful solutions agriculture can offer.

Climate Change Is Real and Is Caused by Human Activity

Between 97 and 98 percent of climate scientists agree that climate change is real and is caused by human activity; namely, through land management practices and burning fossil fuels.[1] Climate scientist James Hansen writes, "Humanity today, collectively, must face the uncomfortable fact that industrial civilization itself has become the principal driver of global climate."[2]

When this book refers to humans causing climate change, it's important to note that not all humans have contributed equally. Wealthy nations and wealthier people disproportionately drive the process through excessive use of energy and consumption of goods and services. In a cruel irony, it is those who have done the least to cause climate change who suffer the most and will suffer much more in years to come unless a global movement arises to reverse the trend. These wrongs have led to the movement for climate justice. Climate

justice activists argue that climate change is hardest not only on the poorest countries but also on the poorest within countries.[3]

The Planetary Carbon Cycle

The earth is endowed with a natural planetary carbon cycle that has moved carbon back and forth among five major pools for billions of years.[4]

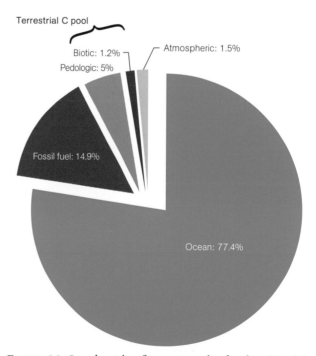

FIGURE 1.1. Our planet has five great pools of carbon. Burning fossil fuels and farming have moved carbon from the fossil and terrestrial pools to the atmospheric pool. Adapted from Lal, "Managing Soils and Ecosystems for Mitigating Anthropogenic Carbon Emissions and Advancing Global Food Security."

- The ocean currently holds 39 trillion tons of carbon (77.4 percent of the global carbon pool). The oceanic pool is growing at 2.3 billion tons per year because it is absorbing excess atmospheric carbon.[5] This in turn is causing harm as it acidifies the ocean.
- Fossil carbon currently accounts for 5 to 10 trillion tons (14.9 percent). This is shrinking, as we are burning it up at about 8 billion tons of carbon per year.[6]
- The soil currently contains 2.5 trillion tons (5 percent).[7]
- The biotic pool contains living biomass (plants, animals, et cetera) and dead detritus such as leaf litter. It currently accounts for 560 billion tons (1.2 percent).[8]
- The atmospheric pool currently holds 780 billion tons (1.5 percent) and is growing at 4 billion tons per year due to fossil fuel burning and land use.[9]

The terrestrial pool, which includes soil and biotic carbon, is roughly four times larger than the atmospheric pool.[10] Deforestation and agriculture have caused the loss of 320 billion tons of carbon from the terrestrial pool in the last 10,000 years, almost half of which has been since 1850. Burning fossil carbon for energy (in the form of coal, petroleum, and natural gas) has moved 292 billion tons of carbon to the atmospheric pool—most of it in the form of carbon dioxide—and is projected to emit another 200 billion tons in the first three decades of the 21st century.[11] What's the effect of all that carbon in the atmosphere?

The Greenhouse Effect

In the winter my garden can reach daytime temperatures of −18°C (0°F) or below. But if it is sunny, it can be as high as 27°C (80°F) inside my greenhouse because the plastic lets sunlight in and traps the heat inside. Gases such as carbon dioxide are called greenhouse gases because they work like the plastic on my greenhouse. They let light from the sun pass through them on its way to the planetary surface but capture some of the heat, or thermal radiation, that "bounces back." This means that heat that would otherwise escape into space is trapped in the atmosphere. We need some of that heat to have a functioning planet, but the dramatic increase of these gases in our atmosphere means the planet is heating up.

SKEPTICAL ABOUT CLIMATE CHANGE?

Fossil fuel billionaires and others who stand to lose profits from climate change mitigation efforts have funded some quite successful disinformation campaigns.[12] Climate skeptics often point to particular studies or arguments that they say disprove anthropogenic climate change. The website Skeptical Science addresses and debunks almost 200 such arguments with links to relevant scientific papers.[13]

Distrust of climate science (in the United States, at least) also appears to be driven by a perception that the issue belongs to the political left. But opinions don't change facts. The political right *and* the political left both need to do a better job of accepting and prioritizing climate change—and propose and implement solutions rapidly.

Carbon dioxide is responsible, by far, for the majority of global warming.[14] Other greenhouse gases, including methane and nitrous oxide, are also significant. Today all are at levels unprecedented in the last 800,000 years.[15] None of these gases, including carbon dioxide, is "bad"; they are all essential parts of natural planetary cycles. The problem is that we have supercharged the atmosphere with them, with serious consequences.

How Warm?

The Intergovernmental Panel on Climate Change (IPCC) states that we are most likely headed toward a rise of 3.7 to 4.8°C (6.7 to 8.6°F) by 2100 unless we undertake aggressive mitigation measures. When climate uncertainty is included in the projections, they range from 2.5 to 7.8°C (4.5 to 14.0°F).[16] Other sources estimate 4 to 6°C (7.2 to 10.8°F).[17] No one knows exactly how fast these changes might take place. We could hit 2°C (3.6°F) by 2030 or 2100. The faster the change, the harder it will be to adapt and the less time there will be to respond.[18]

Keep in mind that the IPCC is fairly conservative and actual climate changes keep ending up at the high end of their estimates or above.[19] Also remember that carbon dioxide takes decades to begin its warming effect, so we have not yet started to feel the impact of much of the carbon we've already emitted. We are already committed to a temperature increase of 1 to 1.5°C (1.8 to 2.7°F) even if we stop all emissions tomorrow. This time lag makes it hard to appreciate how serious the problem already is.[20]

Climate Change Is Already Damaging Our World

We are already seeing the impacts of climate change on all continents and throughout the oceans. Permafrost is melting, and glaciers are shrinking. There is less fresh water available, and the fresh water that is available is declining in quality. Organisms are altering their range, seasonality, populations, and behavior. Coral reefs and Arctic ecosystems are declining rapidly.[21] The number of Category 4 and 5 hurricanes almost doubled between 1970 and 2004.[22] And farmers around the world report that the rains are coming at different times, throwing off the farming season.[23] We are already seeing lower crop yields except in some high-latitude regions.[24]

Oxfam International's landmark publication *Suffering the Science: Climate Change, People, and Poverty* reports that hundreds of millions of people living in poverty around the world are already dealing with climate challenges today. This includes 26 million people who have been displaced from their homes.[25] The World Health Organization estimates 150,000 lives are lost every year already due to climate change.[26]

Is a Couple of Degrees So Bad? Hasn't Climate Change Happened Before?

The world's climate has changed before many times, but human-caused climate change is different from previous natural climate changes. First, it is much faster by orders of magnitude. Second, it is happening in a context of tremendous habitat fragmentation and already-damaged natural systems due to destructive agricultural and settlement practices.[27] Third, we now

have a global civilization brimming with vulnerable people, farms, and infrastructure.

What's the big deal with a couple of degrees warmer? It's actually a very big deal. At the coldest part of the last ice age, the planet was only 6°C colder than today.[28] The last time it was 6°C warmer was 55 million years ago—when there were rain forests in Greenland.[29] It was also 6°C warmer than today during the great Permian extinction 251 million years ago.[30]

Tipping Points and Point of No Return

The American Association for the Advancement of Science writes, "Pushing global temperatures past certain thresholds could trigger abrupt, unpredictable and potentially irreversible changes that have massively disruptive and large-scale impacts. At that point, even if we do not add any additional carbon dioxide to the atmosphere, potentially unstoppable processes are set in motion."[31]

Tipping points are points at which positive feedback loops would trigger additional warming. For example, when Arctic sea ice melts, we're left with more ocean water whose darkness absorbs more of the sun's heat, further contributing to warming.[32] Theoretically, we could reach such a tipping point without triggering this kind of feedback loop if the excess atmospheric carbon was removed quickly enough. A point of no return, which is harder to pinpoint than a tipping point, is the level at which these feedback results are irreversible.[33] Currently, we have already crossed a global tipping point and are on the way to a point of no return. We have a narrow window of time, perhaps a few decades, in which to lower the concentration of greenhouse gases enough to prevent catastrophe.[34]

Many climate tipping points are related to vast carbon sinks, or natural stores of ancient carbon. When temperatures rise, some of these sinks start to give off their carbon, dramatically speeding up the process of climate change. For example, when it gets too hot, stressed plants actually emit carbon dioxide instead of sequestering it. At 2°C, an estimated 15 percent of vegetation and soils may start to off-gas their carbon,

AGRICULTURE IS A LEADING CAUSE OF CLIMATE CHANGE

The production, processing, and distribution of food is the world's largest economic activity. The food system (including production of food, fiber, and other products) is responsible for roughly half of all greenhouse gas emissions worldwide (see figure 1.2). Emissions that result directly from agricultural production account for 11 to 15 percent of total greenhouse gas emissions worldwide. Processing, packing, refrigerating, and retailing food is responsible for 15 to 20 percent, and food waste is another 2 to 4 percent. Land clearing and deforestation for agriculture accounts for 15 to 18 percent. In fact, 70 to 90 percent of all deforestation globally is for agriculture, most of which is for industrial sugarcane, oil palm, and plantations of annual crops such as soy, maize, and rapeseed. From farm to plate, our food system is responsible for a staggering 44 to 57 percent of anthropogenic greenhouse gas emissions.[35]

Let's look more closely at the greenhouse gas emissions resulting from agricultural production per se, not including, for the moment, processing, distribution, waste, et cetera. The Food and Agriculture Organization

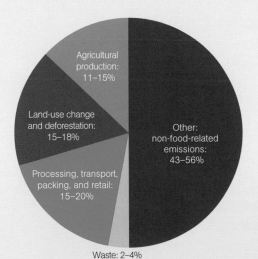

FIGURE 1.2. Emissions from agriculture, land clearing for agriculture, the food system, and food waste amount to roughly half of *all* anthropogenic emissions. Adapted from GRAIN, "Food, climate change, and healthy soils: The forgotten link," in *Wake Up Before It Is Too Late: Make Agriculture Truly Sustainable Now for Food Security in a Changing Climate*, UNCTAD, 20.

of the United Nations (FAO) estimates that methane and nitrous oxide from rice paddies, ruminant livestock, and poorly managed manure accounted for a whopping 76 percent of emissions from agriculture from 1990 through 2012.[36] And then are the fossil fuels that power our farms; the worst culprits are agrochemicals, plowing and subsoiling, harvesting with combines, and pumping water for irrigation.[37] We burn fossil fuels to remove nitrogen from the atmosphere to manufacture fertilizer, much of which then off-gases as nitrous oxide from runoff and leaching.[38]

But agriculture's most damning contribution to climate change is the release of carbon held in the soil, primarily from deforestation and land clearing. This is euphemistically referred to as "changes in land use" and exposes soil organic carbon to oxygen in the air, converting it to carbon dioxide. Since the dawn of agriculture 10,000 years ago or more, land clearing and degradation have resulted in 320 billion tons of emissions, 155 billion tons of which were released between 1850 and 2010.[39] Two and a half trillion tons of carbon are currently held in the top meter of soils around the world, with an additional 560 billion in living aboveground biomass and detritus. Together these amount to six times the amount of carbon currently in the atmosphere.[40] As of 2012, the rate of emissions from land clearing was 0.9 to 1.7 billion tons per year.[41]

Plowing, or tillage, in particular, is highly destructive of soil organic carbon. Most agricultural soils have lost between 30 and 40 tons of carbon per hectare, or 25 to 75 percent of what existed before the land was cleared.[42] This varies by ecosystem, soil type, and farming practice. For example, soil loses 30 to 50 percent of its organic carbon by the time it has been cultivated for 50 years in temperate climates, but only 10 years in tropical climates.[43] It loses more from eroded or degraded soils.[44] In fact, accelerated erosion from poor farming practices accounts for 1.1 billion tons of carbon emissions annually.[45] All of this degradation of soils has another impact—reduced productivity of the farmland itself.[46]

The good news, which I'll keep coming back to, is that agriculture can not only reduce emissions but actually sequester billions of tons of atmospheric carbon, trapping it in the soil and aboveground biomass.

becoming a net contributor to climate change instead of a bulwark against it. This may increase to 40 percent at 4°C (7.2°F).[47] Sinks also harbor points of no return. At 4°C, carbon-rich Arctic soils could thaw, resulting in vast emissions; perhaps 900 billion tons are stored there today. Sometime in the 22nd century, we may see the release of vast amounts of carbon stored as methane hydrates (also called clathrates) at the bottom of the sea.[48]

The Projected Impacts of Climate Change

It's hard to forecast exactly what the impact of climate change will be. There are many variables and unknowns, including how humanity will respond. What we do know with some certainty is that some people will suffer the consequences of climate change to a much greater degree than others. In a 2014 report the IPCC stated that "risks are unevenly distributed and are generally greater for disadvantaged people and communities in countries at all levels of development."[49]

The projections covered here are not good news, but I've chosen to present them in some detail for several reasons. First, it catches you up with the grim reality of climate change, which is rarely given its due in the media. Second, it makes a strong case for the need for carbon farming. Finally, most policy efforts are aiming at 2°C (and policy makers are failing to come close even to this goal).[50] The research presented here shows that 2°C would be disastrous for much of the world. Indeed, African delegates at the UN climate summit in Copenhagen cried out "2 degrees is suicide" for their countries.[51] The better news is that IPCC scenarios that are aimed at attaining 2°C rely on carbon farming with improved annual crops and pasture management for much of their sequestration despite their low per-hectare impact. But reaching the safer ground of 350ppm (less than 1°C, or 1.8°F) means a more vigorous effort—that's where the extraordinary sequestration rates of agroforestry and perennial crops, the strategies at the heart of this book, come in.

I have divided the projected impacts of climate change that follow into three parts—environmental impacts, agricultural impacts, and human impacts—for easier digestion. But it's important to acknowledge that these categories are more or less artificial. You can't have an environmental impact without having an impact on humans, for example. In covering this material I am indebted to Mark Lynas's *Six Degrees: Our Future on a Hotter Planet*, for which he reviewed tens of thousands of scientific papers in order to conduct an analysis of which climate impacts we can expect with every increase of 1°C.[52]

ENVIRONMENTAL IMPACTS

The frequency of climate-related disasters is projected to triple from 2009 to 2030.[53] These events would impact 660 million people each year, mostly in less developed countries with less infrastructure. According to a report from Oxfam International, disasters of equivalent strength kill between 12 and 45 times more people in poorer countries than in wealthy ones. Women and children are also disproportionately killed in disasters.[54] Storms that we have thought of as once in a century may come every 30 years by the 2050s and every 4 years in the 2080s and beyond.[55]

Because warmer air holds more moisture, there will be not only more storms, but stronger storms.[56] The intensity index of storms has already doubled as ocean temperatures have risen.[57] At 3°C (5.4°F), hurricane intensity increases so much we may need to add a new category—Category 6—to reflect these new superstorms.[58] At 6°C, hurricanes of "staggering ferocity" could sweep the globe.[59]

Storms will also affect regions that have not previously experienced them. For example, the first recorded hurricane in the South Atlantic hit Brazil in 2004, and scientists project increasing hurricanes in South America, which has not been subjected to them historically. The Mediterranean may also see the arrival of historically unprecedented hurricanes. Hurricanes are devastating enough to begin with, but when they strike regions with no experience of them, the results can be horrific. This trend of wider-ranging hurricanes is projected to continue.[60]

Some regions of the world are projected to become much drier, while others will become more humid.

CLIMATE JUSTICE

The climate justice movement "embodies the fundamental understanding that those who contribute the least to the excess of carbon dioxide and other greenhouse gases in the earth's atmosphere consistently and disproportionately experience the most severe and disruptive consequences of global warming, and are often the least prepared to cope with its consequences."[61] Today the wealthiest 7 percent of the world's people are responsible for 50 percent of all emissions.[62]

Climate change is already disrupting lives in the Global South and is projected to have devastating impacts throughout the tropics and subtropics, particularly South Asia, Africa, and Central America. Wealthy countries have determined to set 2°C as an "acceptable" level of warming (although their emissions commitments have us headed to 4 to 6°C). Climate scientists are clear that 2°C would be an unprecedented disaster, submerging entire island nations and wide swaths of Bangladesh and other countries.[63]

It is a vicious kind of racism that is willing to tolerate disaster in other people's countries rather than change your way of life. Pope Francis reflects extensively on climate justice in the encyclical *Laudato Si'*. He writes, "We have to realize that a true ecological approach *always* becomes a social approach; it must integrate questions of justice in debates on the environment, so as to hear *both the cry of the earth and the cry of the poor*"[64] (italics in original).

Racism doesn't discount the suffering just of poor countries but also of poor individuals within rich countries. Brian Tokar's *Toward Climate Justice* points out, "The six U.S. states with the highest African American populations are all in the Atlantic hurricane zone, and African Americans also have the highest historic rates of heat death."[65] Naomi Klein draws a direct connection between the shootings of unarmed black men by police (in response to which the Black Lives Matter movement arose) in the United States and the discounting of climate impacts on the world's tropical countries. She writes, "If black lives matter—and they do—then global warming is already a five-alarm fire, and the lives it has taken are already too many."[66] In this sense, climate change mitigation may be one of the most important anti-racism projects of the 21st century.

Rainfall will likely come in the form of intense storms, causing flash floods and poor infiltration.[67] At 1°C of warming, regions such as western North America will see stronger droughts.[68] Indeed, the devastating California drought of 2015 is "very likely" linked to climate change and is in line with projections for the region.[69] Southern Africa will become drier, while northern Africa will become more humid, if only in the short term.[70] At 3°C of warming, southern Africa would experience permanent drought; Botswana, for example, would shift from savanna to parched sand dunes by 2070.[71] "Super El Niño" effects could bring an end to the Indian monsoon rains, which provide critical rainfall for billions of people.[72] And the Sahara desert would move north into Southern Europe.[73] Meanwhile, other regions of the world, including central and eastern Africa, will see excessive rainfall and flash flooding.[74]

We are already living through the sixth mass extinction in our planet's history—an extinction caused by human activities. The International Union for Conservation of Nature (IUCN) Red List documents 830 organism extinctions, with 29 more extinct in the wild and 4,735 critically endangered.[75] (Organisms that are extinct in the wild may still exist in captivity, such as in zoos, botanical gardens, or—rarely—on farms.) Runaway climate change will only further the destruction.[76] The IPCC reports that up to 30 percent of species are at increasing risk of extinction at 1°C, which we have already exceeded.[77]

Even at 1°C of warming, sensitive habitats such as the Australian Queensland Wet Tropics and the South African Cape Floristic Region begin to disappear.[78] Depending on the speed at which changes happen, many species will be unable to keep up with the poleward move of their habitats.[79] Above 2°C, there is a "moderate" chance that the Amazon will dry up and become a desert, resulting in a tremendous loss of biodiversity, stored carbon, and land inhabited by indigenous people.[80]

At 3°C a shocking 33 to 50 percent of all living species will be on their way to extinction.[81] Between 10 and 50 percent of the world's habitats will simply be gone. These include many of our most diverse ecosystems. Many of the organisms adapted to these conditions will have nowhere to go and will perish; others will adapt. Other habitats will move hundreds of miles poleward or up in elevation. Species adapted to them will need to follow, adapt to new conditions elsewhere, or perish.[82] At 6°C we will have raised temperatures as high as they were during the Permian extinction event—a mass extinction 251 million years ago in which the world lost as much as 95 percent of all land and sea life, but over the course of a century instead of 10,000 years.[83]

Coral reefs protect coastlines from storm surge and are bastions of biodiversity. A third of all ocean life relies on them, including the young of many fish species that humanity relies on for food.[84] Heat can kill the symbiotic algal partners of corals, a phenomenon known as coral bleaching. It takes 30 years to recover, but even with 1°C warming, annual bleaching will wipe out much of the world's coral.[85] At 2°C warming most corals will be bleached, and at 3°C the IPCC expects "widespread mortality" of corals.[86]

Carbon dioxide taken up by the oceans makes the water more acid. At 2°C warming, parts of the Southern Ocean and Pacific will become toxic to organisms such as plankton, crabs, and shellfish that use calcium carbonate to make shells. The effect on plankton is especially worrisome, because they are critical to oceanic carbon sequestration (remember that the ocean is the largest of our global carbon pools). Their loss could lead to the spread of "marine deserts" devoid of the base of their food chain.[87] A global rise in temperatures of 6°C could eventually cause the loss of oxygen in the deeper waters of the ocean, causing a massive die-off of ocean life, such as happened in previous mass extinctions.[88]

Glaciers are already melting at rates not seen in more than 11,000 years.[89] The loss of steady glacial meltwater would mean water shortages for one to two billion Chinese people this century.[90] The melting of glaciers and loss of year-round flow of rivers will affect hundreds of millions of people on the Indian subcontinent. Cities such as Lima, Quito, La Paz, and Kathmandu whose water comes from glacier runoff will face serious water shortages.[91]

The Greenland ice cap and Antarctic ice are beginning to melt. Greenland alone contains enough ice to raise the sea level by 7 meters (23 feet). Scientists think this will happen once the global temperature rise hits 1.2°C (2.2°F). This melting will probably take centuries or millennia, although climate expert James Hansen warns we could be facing "explosively rapid" melting and sea rise.[92] He asserts that business-as-usual emissions will almost certainly result in multimeter sea-level rise by 2100.[93] The IPCC predicts a much more comfortable 0.25- to 1-meter (0.8- to 3.3-foot) rise for this century.[94] Whether it happens this century or not, melting of Greenland ice from a 1.2°C temperature rise would flood Miami, Manhattan, London, Bangkok, Bombay, and Shanghai. Half of humanity would have to relocate.[95]

Sea-level rise is essentially irreversible once begun. At temperatures above 2°C, the great ice sheets of both Greenland and east Antarctica will eventually melt. Their contribution will raise sea level by 25 meters (82 feet). Fortunately, under most projections this melting will take thousands of years. On the other hand, if seawater penetrates the foundation of the west Antarctic ice sheet, we could see 5 meters (16 feet) in a matter of decades. The last time the planet was 4°C warmer the poles were ice-free.[96]

AGRICULTURAL IMPACTS

At present, most climate change impacts on agriculture are undesirable, except in some high-latitude regions.[97] Food prices are projected to rise by as much as 84 percent this century.[98] Declines of agricultural productivity of 15 to 30 percent are projected by 2080 in Africa, South Asia, and Central America.[99] Chinese staple crop production suffers at 40 percent loss at 4°C.[100] In some countries this could reach 50 percent loss of agricultural productivity, and in some regions agriculture will likely become impossible.[101]

Scientists anticipate yield reductions in maize of 3 to 15 percent, wheat 2 to 14 percent, rice 1 to 3 percent, and soybean 2 to 7 percent, although some of these may be offset 1 to 2 percent by increases due to more available carbon dioxide. These will be much more severe in poorer regions where food security is already problematic.[102] The International Food Policy Research Institute predicts that the result will be price increases of 131 percent for maize, 78 percent for rice, and 67 percent for wheat by 2050.[103] Year-to-year yields are also projected to vary more widely due to erratic and intense weather.[104] Increased carbon dioxide will have little net effect on crops, although it could impact nutrition by changing the protein, mineral, and amino acid content of cereal grains.[105] Fruits and vegetables will also suffer serious declines.[106]

Heat and water stress are projected to reduce productivity of livestock, although pastures in colder regions will benefit from longer grazing seasons.[107] Forage quality is expected to decline in many regions.[108] Food security means more than agriculture. Food processing, transportation, and other aspects are also at risk, if little investigated.[109]

Increasing aridity is projected to increase irrigation needs by 7 to 40 percent this century.[110] At 3°C half of Australia's wheat land dries up, and Central America dries up enough to devastate agricultural production.[111] In Pakistan the loss of glacial water will reduce river flows to agricultural areas by 90 percent. This could displace tens of millions of Pakistani people.[112] The loss of snowmelt in the western United States may well create a permanent Dust Bowl drought and greatly increase wildfires.[113]

In a 4°C world terrible permanent drought strikes the southwestern United States, Central America, the Mediterranean, South Africa, Australia. Seasonal droughts affect Southeast Asia, Siberia, the Amazon, and West Africa.[114] Good news for Canada, Siberia, and other high-latitude areas, as they are projected to become warmer and wetter, perhaps serving as new breadbaskets for humanity.[115] This option has limits in regards to areas north of the existing boreal agricultural zones where much soil is poorly suited to agriculture. Note that many tropical crops are well adapted to heat, and agriculture in the humid tropical belt might simply shift more to such crops as breadfruit, cassava, and pigeon pea.

At 5°C (9°F) most of sub-Saharan Africa dries out. This would result in decreased crop and pasture yields in most of the continent. Crop failures on rain-fed farms are projected to happen every other year in much of southern Africa.[116] Ulrich Hoffmann, senior economic affairs officer for the United Nations Conference on Trade and Development, writes, "In some locations, a combination of temperature and precipitation changes might result in complete loss of agricultural activity; in a few locations, agriculture might become impossible."[117]

Some temperatures are just too high for our common food crops. Rice, wheat, and maize all cease to yield in temperatures above 40°C (104°F). Their yields go down 10 percent for every degree above 30°C (86°F).[118] At 4°C almost all of Australia will be too dry and hot for the annual crops we rely on. India at 4°C is too hot for most crops.[119] (Some of the tropical tree crops profiled in part 3 could become suitable since many of them can survive in extremely hot climates.) Tropical crop pests may move into temperate and tropical highland growing regions.[120] Some of the world's worst weeds are also projected to spread more widely and become more competitive with crops.[121] Mycotoxins, deadly fungi that contaminate stored grain, are projected to spread their range as well.[122]

While it is possible at 1 to 2°C to shift to other crops, changing the staple food of whole regions and countries is difficult.[123] Some perennial crops such as apples and walnuts require chilling hours—a period of cold—to produce the following year. As warming increases, plantings of these crops will have to move poleward.[124] It is the poorest and most food insecure countries that face the worst impacts of climate change to their farming systems.[125] At 5°C the tropics are too hot for most crops, while the subtropics are too dry. The belt of habitability where farming as we know it is possible moves closer to the poles.[126]

Sub-Saharan Africa will see a $2 billion US loss in maize alone at 1 to 2°C.[127] The same temperature increase is likely to improve food production in humid temperate and boreal areas. The United States might see an overall farm revenue increase of $1.3 billion despite the serious losses in California and the western half of the country. Productivity may increase in tropical highlands and extend growing seasons in temperate zones.[128] While wealthy people can "buy" food security, poor families and countries will be hit hard by climate change.[129]

HUMAN IMPACTS

The range of diseases such as malaria will change, with projections from a decrease in 150 million to an increase of 400 million additional people exposed. In Africa it only stands to increase, moving up the mountains to reach 21 to 67 million more people at 3°C.[130] Overall hundreds of millions of people are expected to be exposed to new diseases at 2°C.[131] The IPCC warns of a "substantial burden on health services" at 4°C.[132] Water access and quality will impact drinking water and other human health issues.[133] "Increased water stress" will impact hundreds of millions at 3°C and above.[134] During the 2003 heat wave in Europe 22,000 to 35,000 people died from heat stress. This kind of heat will become increasingly normal.[135] By 2050 London is projected to see five times more deaths due to heat. This is particularly important for farmers and others who work outside.[136]

At 3°C millions will be forced to flee permanent droughts in Central America and southern Africa.[137] At 4°C hundreds of millions of people will have to move inland, upslope, and toward the poles.[138] At 2°C scientists anticipate 200 million refugees on the move annually by 2050. With migration comes much suffering, including human trafficking.[139] At 4°C sea-level rise would displace 72 to 182 million people living at low elevations by 2100, especially in South and Southeast Asia.[140]

The IPCC warns, "Human security will be progressively threatened as the climate changes . . . Climate change is an important factor threatening human security through 1) undermining livelihoods, 2) compromising culture and identity, 3) increasing migration that people would rather have avoided, and 4) challenging the ability of states to provide the conditions necessary for human security . . . Human rights to life, health, shelter and food are fundamentally breached by the impacts of climate change."[141]

Massive migrations, devastating storm disasters and droughts, famines, and loss of coastal and low-lying land will put tremendous stress on governments and citizens. Some of our governments don't function terribly well today! Mass migration and strain on resources provide fertile ground for conflict. Researchers estimate that at 2°C, climate change will endanger 2.7 billion people in 46 countries by fueling violent conflict.[142] The IPCC states that "climate change can indirectly increase risks of violent conflict in the form of civil war and inter-group violence by amplifying well-documented drivers of these conflicts such as poverty and economic shocks."[143]

A leading climate researcher bluntly states that 4°C warming "is incompatible with any reasonable characterization of an organized, equitable and civilized global community."[144] At 5°C drought and sea-level rise would corral humans into shrinking habitats. Economic depression would be inevitable but might be the least of our worries.[145] Millions or billions of people may die in a 5°C world.[146] In fact, a 5°C world may only be capable of supporting a billion human beings.[147]

It's Not Too Late to Act

There is still time to act, although the window is closing. If we reduce greenhouse gas emissions, we'll stop making the problem worse. Even better, we can actually remove vast amounts of carbon dioxide from the atmosphere. Remember the terrestrial pool of carbon, made up of soil and plants? It has lost 320 billion tons of carbon since the dawn of agriculture, but it can reabsorb much of that. (If we didn't need any farms or living spaces, it could reabsorb it all.)[148]

Climate scientist James Hansen writes: "If humanity wishes to preserve a planet similar to that on which civilization developed and to which life on Earth is adapted . . . CO_2 will need to be reduced . . . to at most 350ppm, but likely less than that."[149] He continues: "An initial 350ppm target may be achievable by phasing out coal use . . . and *adopting agricultural and forestry practices that sequester carbon* [my emphasis]."[150] These changes must be achieved in a few short decades in order to step back from tipping points and avoid catastrophic changes.[151]

That's right. Some kinds of farming sequester carbon and fight climate change, all while providing the food and other products of agriculture. As of 2009, the total excess carbon dioxide in the atmosphere was 200 billion tons.[152] Climate change will continue for centuries even if we stop emissions now because of the persistence of the greenhouse gases we've already released. This overshoot is why sequestration is necessary to return to 350ppm even if we reduce emissions.[153]

Two-hundred-plus billion tons is a lot to draw back down, but the good news is that agriculture can shift from being a contributor to climate change to a major part of the solution. Soil carbon that has been lost due to land clearing and poor farming practices can be recaptured and held in the soil for a long time. By changing the way we farm, we can reduce emissions, sequester carbon, and be part of adapting to the new less stable world climate change brings us. The remainder of this book lays out these carbon farming solutions.

CHAPTER TWO

Agricultural Climate Change Mitigation and Adaptation

When it comes to agriculture, climate change mitigation involves three broad areas: reduction of agriculture-related greenhouse gas emissions, agroecological intensification of existing farmland to prevent additional land clearing for agriculture, and carbon sequestration in soil and aboveground biomass. Although the focus of this book is on carbon sequestration, it's not possible to entirely separate that out from emissions reduction and agroecological intensification. Making progress in one area has an impact on the others and vice versa. For example, according to a 2013 report issued by the international non-profit organization GRAIN, we could cut agricultural emissions in half in several decades if we transition to an agroecological food system.[1]

Mitigation is not the only issue we need to grapple with when it comes to climate change. Because climate change is already impacting agriculture, farming must adapt by making itself more resilient in the face of extreme and changing weather. Luckily, most of the carbon-sequestering practices profiled in this book are also adaptation strategies.

Reducing Agricultural Emissions

We can reduce carbon emissions from soil by slowing or ending land clearing and wetland drainage for agriculture, preventing erosion and reversing the degradation of agricultural soils, and reducing tillage. We can reduce fossil fuel emissions by reducing the use of mechanized equipment and cutting back on chemical nitrogen fertilizers, which are energy-intensive to manufacture. In 2011 the University of Minnesota constructed a facility that uses wind power instead of fossil fuels to power the Haber-Bosch process, which converts atmospheric nitrogen (N_2) into conventional nitrogen fertilizers. Their vision is of windmills located directly in the fields to produce fertilizer for use right there. This kind of innovation would greatly reduce the carbon footprint of nitrogen fertilizers (although off-gassing can still be an issue). It remains to be seen if this kind of small-scale, decentralized nitrogen fertilizer production can be economically competitive with industrial Haber-Bosch facilities.

Likewise, we can reduce methane emissions by changing the way we farm. Rice paddies emit substantial amounts of methane; letting paddies dry out (among other strategies) can greatly reduce the impact. Emissions from livestock manure and the digestion of ruminant livestock such as cattle can also be reduced. We can reduce nitrous oxide emissions from chemical fertilizer by using fertilizer more efficiently with better timing and applying appropriate amounts, or by replacing it with manure or nitrogen-fixing plants such as legume cover crops or agroforestry support trees.[2]

We know that shortening supply lines in industry reduces emissions. Is localizing agricultural production a viable emissions reduction strategy?[3] This is not as simple a question as it appears. One reason is that there are many times more farms than factories, and they are highly dispersed compared with industrial production. And transportation is a small fraction of food system emissions, only 3 percent in the United States.[4] Food from supermarkets often actually has a smaller carbon footprint than food from local farms.[5] This is due in part to more efficient transportation and processing in the global food system.[6] Most of the difference comes from the extra individual customers driving to farms for on-farm sales. It is more energy-efficient for local farms to deliver produce to the customer or to central pickup sites than for every customer to drive to the farm.[7] Redesigning cities and improving transportation systems can greatly improve the carbon efficiency of locally produced foods.[8] Food production is also more carbon-efficient in some climates. For example, it is more carbon-efficient to raise beef on large pasture operations in Argentina and ship it to Germany than it is to produce it in smaller feedlots in Germany.[9] And international trade is critical to the livelihoods of millions of farmers in the tropics. None of this is to suggest that we should stop supporting our local farmers and instead consume mass-produced foods from far away; it's to suggest that issues of transportation and distribution are complex and we have to conduct some hard and honest accounting if we're going to improve those systems.

Agroecological Intensification

As we saw in chapter 1, clearing land for agriculture is responsible for 15 to 18 percent of human emissions today.[10] Yet scientists and policy makers are concerned that we will need to increase food production substantially to feed a population of nine billion by 2050.[11] In order to meet our needs without sacrificing more carbon-storing natural ecosystems, we'll have to improve our yields on the farmland we already have—a challenge called agricultural intensification.[12]

There's disagreement on the best way to go about this. Some people argue for more industrial agriculture —synthetic fertilizers, improved irrigation, mechanization, GMO and hybrid seeds, and so on—a high-input, high-output kind of farming. Admittedly, even when we factor in emissions, improved yields from industrial agriculture have prevented 161 billion tons of emissions since 1961 due to land that wasn't deforested. We would have had to clear and convert 1.7 billion more hectares

to reach the same yield using the "baseline" (previously widespread) agricultural practices.[13]

Sadly, industrial agriculture's gains in yield have come at a huge social and ecological cost.[14] This includes climate costs such as emissions from the manufacture and use of synthetic nitrogen fertilizers. For decades farmers, activists, and researchers have advocated for alternative agroecological production systems that can maintain these impressive yields without sacrificing people or ecosystems. Their arguments are being taken more and more seriously. In its 2013 report, *Wake Up Before It Is Too Late: Make Agriculture Truly Sustainable Now for Food Security in a Changing Climate*, the United Nations Conference on Trade and Development came out strongly in favor of agroecological intensification and strongly opposed to the industrial food system.[15]

Can agroecological farming provide high enough yields to prevent further land clearing for agriculture? The answer varies considerably from system to system and site to site. Agroecological systems, once established (which can take several years), often yield better than industrial agriculture.[16] Many studies show that agroecological farms have reduced emissions and sequester more carbon than industrial agriculture.[17] One study reviewed 286 sustainable agriculture projects spanning 37 million hectares (91 million acres), 12.6 million farms, and 3 percent of all cropland in developing countries. The projects included integrated pest management, integrated nutrient management, conservation tillage, agroforestry, aquaculture, water harvesting, and livestock integration. At a conservative rate of less than 1 t/ha/yr, these projects were estimated to sequester 11.4 million tons of carbon annually.[18] The average increase in yield was 79 percent.[19]

Polycultures, which are the integration of multiple crops in the same area, are a great example of agroecological intensification. Well-designed polycultures can improve yields. For example, researchers in France have found that the yield of 100 hectares (247 acres) of timber trees intercropped with cereals would require 130 to 140 hectares (321 to 346 acres) of separate monocultures of timber and cereals. In a Brazilian study, oil palm yields increased by roughly 125 to 175 percent when grown in polyculture—and the yields of other products

IT'S NOT JUST ABOUT CARBON DIOXIDE

For simplicity in this book I'm focused on carbon sequestration and emissions of carbon dioxide from agricultural soils. There are several other issues of concern, however.[20]

Methane (CH₄). Methane is another greenhouse gas emitted by agricultural sources, including ruminant livestock digestion, manure, and flooded rice paddies. It is 21 times stronger than carbon dioxide but only remains in the atmosphere for 9 to 15 years.[21]

Nitrous oxide (N₂O). Nitrous oxide is much more powerful than methane or CO_2, with 296 times the global warming power of carbon dioxide. It stays in the atmosphere about 114 years, which makes it a serious concern. Agricultural sources include synthetic fertilizer, manure, and to some extent nitrogen-fixing plants.[22]

Volatile organic compounds (VOCs). Volatile organic compounds are hydrocarbons emitted into the atmosphere by trees. They are part of the natural carbon cycle and help to destroy ozone and methane when nitrous oxide concentration is normal. However, when nitrous oxide concentrations in the atmosphere are high (as they are today due to human-induced pollution), VOCs actually make more ozone and methane. Some have used this as an argument for deforestation, but it is not only more logical but also more cost-effective to stop polluting with nitrous oxide instead of cutting down forests.[23]

(fruit, timber, spices, cacao) grown in the system must be added to this. These are just a few of many examples of agroecological intensification working to increasing food production and carbon sequestration while reducing emissions from tillage and nitrogen fertilizers.

So in many cases agroecological production can yield as well as or better than industrial agriculture,

but we need to be careful not to assume it always will. And although yield is a critical factor in selecting carbon farming techniques, so are social implications, economics, ecosystem services, and of course carbon sequestration capacity.

The Basics of Carbon Sequestration

Agricultural carbon sequestration involves removing excess carbon dioxide from the atmosphere and storing it in soil organic matter and in the aboveground biomass of long-lived plants and trees (perennials). This natural part of the carbon cycle provides us with a powerful tool for climate mitigation. Let's take a closer look at this phenomenon, so we can better understand its potential and limitations.

HOW SEQUESTRATION WORKS

Phase one of agricultural carbon sequestration is photosynthesis. Plants convert sunlight, water, and atmospheric carbon dioxide into carbohydrates. Carbohydrates are molecules made up exclusively of carbon, hydrogen, and oxygen. These include sugars, starches, and cellulose. In plants the basic unit is glucose, a short-chain sugar molecule. Polysaccharides are a group of molecules built out of long chains of sugars. Collectively polysaccharides account for 75 percent of all living and dead organic matter on the planet—all the product of photosynthesis![24] Starch and fiber are polysaccharides, as is cellulose—which alone makes up 40 percent of all organic matter.[25] Many plants also produce hydrocarbons (molecules with only hydrogen and carbon) as a product of photosynthesis.

If you were to dehydrate plant biomass, you would find that, on average, 50 percent of the weight is carbon.[26] Over time, some or all of the plant's carbon-rich aboveground biomass dies and falls to the ground as leaf litter or other residues. About two-thirds of this material is released into the atmosphere as carbon dioxide as part of the global carbon cycle. The remaining third becomes long-lived soil organic matter, which we will discuss in a moment.[27] Meanwhile, roots are equivalent to 25 to 40 percent of the weight of the aboveground

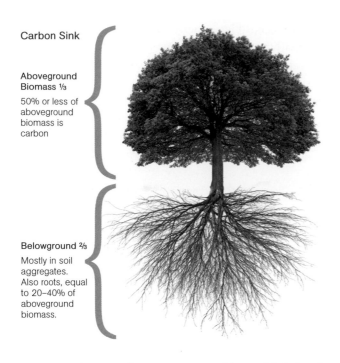

Carbon Sink

Aboveground
Biomass ⅓
50% or less of
aboveground
biomass is
carbon

Belowground ⅔
Mostly in soil
aggregates.
Also roots, equal
to 20–40% of
aboveground
biomass.

FIGURE 2.1. In an agroforestry system, roughly one-third of total carbon is held in aboveground biomass, and two-thirds below ground. Most soil carbon is organic matter, with some held in roots. Roots are roughly 20 to 40 percent the weight of aboveground biomass.[29]

CARBON VERSUS CARBON DIOXIDE

Sometimes people confuse carbon and carbon dioxide. We are looking to sequester carbon—200 to 250 billion tons of it. That carbon is captured by oxygen to form carbon dioxide, which weighs 3.67 times more.[33] Thus the weight of the carbon dioxide we want to remove from the atmosphere is 734 to 917 billion tons. However, we usually speak of atmospheric carbon dioxide in parts per million. Every part per million of carbon dioxide is equivalent to roughly 2 billion tons of carbon.[34]

We also look at annual carbon sequestration rate, which is expressed here (and usually) in metric tons per hectare. This is equivalent to 0.4 English ton per acre. We sometimes look at the lifetime sequestration rate, which typically looks at carbon sequestered over a 20- to 50-year period until saturation is more or less reached. Finally, the term *carbon stocks* refers to the total amount of carbon currently held in soil and/or aboveground biomass at a particular time.

biomass.[28] Every year some root hairs die even on healthy plants. Some of the carbon contained in these roots becomes long-lived soil carbon as well.

There is also a faster way that carbon from photosynthesis gets into the soil. Plant roots exude a complex blend of more than 200 compounds, many rich in carbon. These exudates in turn assist the plants by feeding soil organisms, which help with nutrient cycling, suppression of pests and diseases, and other benefits.[30] Between 10 and 40 percent of all photosynthesized carbon passes through the roots within an hour![31] Australian soil scientist Dr. Christine Jones calls this rapid transfer of carbon through root exudates the "liquid carbon pathway."[32]

In other words, a lot of the photosynthesized carbon ends up in the soil. Several physical, chemical, and biological processes then combine to bind soil carbon into clumpy aggregates, including glues from roots, fungi, and bacteria. The addition of chemical bonds

with inorganic soil particles creates a long-lived, stable form of carbon—perfect for long-term sequestration.[35] These aggregates are more or less the same as humus.[36] Soil carbon is not only in the shallow layers of the soil, but a meter or more deep. These deeper soils typically contain about half of the total soil organic carbon, and keep it secure for long periods of time.[37]

Soil organic matter is about 58 percent carbon; soil organic carbon can be multiplied by 1.7 to give estimated total organic matter.[38] Every ton of soil organic carbon (which is half the weight of the organic matter) is equivalent to 3.67 tons of atmospheric carbon dioxide.[39] By one writer's rough calculation, an average 1 percent of organic matter in the soil weighs 36.5 tons, meaning 21.2 tons of carbon. This would mean every 1 percent increase of soil organic matter indicates roughly 21 tons of carbon sequestered per hectare.[40]

Overall an average of a third of carbon in agroforestry systems is held in aboveground biomass, though this varies by ecosystem and farming practices. The remaining carbon is held in the soil. Some soil carbon is in living roots, which are estimated at about 25 to 40 percent of the total weight of aboveground biomass.[41]

MEAN RESIDENCE TIME

How long does sequestered carbon remain in the soil? Scientists call this the mean residence time (MRT). Some soils have inherently better or worse MRT based on their composition. For instance, an abundance of clay particles provides long-term binding sites for humus (which is half carbon). Humus persists 100 to 5,000 years in soil in the absence of disturbances such as tillage.[42]

The IPCC describes soil and biomass carbon reserves as non-permanent and reversible. Trees can be cut down or burned, and soil can be tilled again.[43] There is also a point at which soil and biomass carbon sinks may become saturated and no longer able to absorb carbon. Some scientists think that sequestration slows down to zero, although others report that even in mature old-growth forests carbon stocks in soil and dead organic matter still grow even when living biomass virtually stops increasing. Biochar in soils is likely to persist for 100 to 1,000 years.[44]

As we will see in chapter 3, many farming practices sequester carbon, some much more than others. Agricultural practices can have a profound impact on the mean residence time of carbon in soil. Farming practices and crops that sequester and increase the MRT of carbon in the soil include perennial crops, mulching, non-flooded rice, reduced tillage, managed grazing, crop–livestock integration, and continuous cover through green manures or cover crops. In contrast, annual crops, tillage, and bare soils can quickly release soil carbon and return it to the atmosphere.[45]

Mean residence time in aboveground biomass and roots depends on the life span of the plant. Although some leaves and roots are shed every year, carbon in tree trunks, for example, can be sequestered for a long time. Some of the tree crops profiled in this book can live hundreds of years or longer.

CARBON-HOLDING CAPACITY OF SOILS AND BIOMASS

How much carbon can be held by soils and biomass? We know that since the dawn of agriculture, our soils and biomass have lost 320 billion tons of carbon from land clearing.[46] If all of that land was restored to its original pre-agricultural state, we might be able to reabsorb it all.[47] Our use of the land for farming, living, and working makes this effectively impossible. There may also be only so much carbon any soil can hold.[48] In grasslands this saturation point is probably reached after 10 to 25 years of implementing carbon farming practices.[49] For new forest plantations and regenerating forests it takes 20 to 50 years for the period of active sequestration to end.[50] On the other hand, until recently it was believed that old-growth forests hit a carbon sequestration ceiling, then enter carbon "stasis" and simply continue to hold the carbon. Scientists now know that forests can continue to sequester some additional carbon annually even when they are 800 years old.[51]

Many soil scientists, and the IPCC, estimate that agricultural soils have a maximum storage capacity of 30 to 50 tons per hectare.[52] Others have estimated much higher storage for agroforestry systems (see chapter 3). Some ecosystems hold much more than this in their soils, as do many of the regenerative farming practices we will touch on in part 2 and perennial crops from parts 3 and 4. Some of these practices and species also sequester significant additional carbon in their aboveground biomass. Chapter 3 goes into detail on this, and appendix C provides specific examples.

TABLE 2.1. Productive Life Spans of Selected Perennial Staple Crops

Productive Life Span	Perennial Staple Crop Species
3–9 years	Lima bean, winged bean, wattleseed acacias, sugarcane, perennial pigeon pea
10–24 years	Yellowhorn, sugar palm
25–49 years	Breadfruit, Mayan breadnut, runner bean, standard hazelnut, buffalo gourd, date palm
50–99 years	Peach palm, carob, coconut, honey locust, Tahitian chestnut, karuka, avocado, mesquite
100–499 years	Nara melon, pecan, Chinese chestnut
500–999 years	Coppiced hazelnut, many nut pines
1,000–2,000 years	Brazil nut, olive

ADDITIONAL BENEFITS OF SOIL ORGANIC CARBON ON THE FARM

Soil organic matter does much more than seques-ter carbon. It improves soil fertility, resulting in better crop yields, pH buffering, disease prevention, water-holding capacity, and more. Soils that are high in soil organic carbon (more than 1.5 to 2.0 percent carbon or 2.5 to 3.5 percent organic matter) produce higher crop yields and are more resistant to drought.[53] An increase of 1 ton of soil organic carbon per hect-are to the croplands of developing countries could increase food production there by 30 to 50 million tons per year.[54] Indeed, as Courtney White writes in *Grass, Soil, Hope: A Journey Through Carbon Country*, "You can't increase soil carbon with a practice that degrades the land."[55] That's not to say that all carbon-sequestering techniques yield as well as or better than the alternative, but a lot of them do, as you will find in part 2 of this book.

It is this synergy of fighting climate change while improving farm health and productivity that makes carbon farming such an appealing idea. It is somewhat remarkable that farming practices are so rarely men-tioned as climate solutions. One goal for this book is to bring carbon farming into the center of the climate conversation and to emphasize the potential of exist-ing perennial crops and tree-based farming systems, which sequester the most carbon of any food produc-tion model.

AN IMPORTANT CAVEAT: ALBEDO

Have you ever noticed that a dark surface is hotter than a light one on a sunny day? Dark surfaces absorb more light and heat up as a result. A surface's ability to reflect light is called its albedo. Albedo is measured on a scale from no reflectivity (a perfectly black surface) with a score of 0 to high reflectivity (a white surface) with a score of 1. Agriculture and other land uses have increased the planet's albedo.[56]

Different land uses have different albedo values. For example, forests and dark wet soils have low albedo (.05 to .15), grasslands' ranges are fairly low (.10 to .20), annual crops are intermediate (.15 to .25), savanna is .17 to .18, and bare agricultural soil is about .20 to .35.[57] Deciduous forest ranges from 0.15 to 0.18, with ever-green forest .09 to .15.[58] Many agroforestry systems are, or would be, similar to savanna.

Albedo, like the other effects described here, is a neu-tral natural phenomena, but in the context of climate change a higher albedo is desirable. In tropical climates, there is little impact, but in colder climates the lower albedo of tree plantings can somewhat counteract the benefit of carbon sequestration. In fact, in high-latitude boreal climates (above 50 degrees latitude) such as Siberia and Canada, dense plantings of evergreen trees actually cause net warming. This is because the low-albedo trees soak up heat even in winter, especially compared with bare snow on level ground. Even in temperate climates of 20 to 50 degrees latitude, albedo mitigates a significant amount of the carbon seques-tered by trees, although it is still a net benefit.[59] One study found that in boreal climates, evergreen forest plantations caused a net equivalent stock change (in terms of loss of tons of carbon per hectare equivalent) reduction of 50 to 130 percent. In temperate climates this impact from evergreen plantations caused the effec-tive loss of 10 to 70 percent of the carbon sequestered.[60] Thus, ironically, clearing forests in high latitudes could be beneficial to climate mitigation (which, I hope goes without saying, doesn't mean we should).

In the end, this may be of minimal importance because there are few evergreen perennial crop trees for boreal and temperate climates. However, it's important to note that this is the trade-off in colder climates: The practices that sequester the most carbon are also those that have the least desirable albedo effects. The use of managed grazing and prospect of perennial grains are albedo-friendly practices for high latitudes. Of course, trees still have important ecological and agroecological functions regardless.[61]

Adaptation

Climate change is having an impact on agriculture already, and this impact is likely to increase in the decades to come. Intense droughts and floods, unpredictable weather, and powerful storms will require us to change the way we farm. These changes are referred to as adaptation strategies. Although adaptation is distinct from mitigation, it's difficult to talk about one without talking about the other. Sometimes there are trade-offs between adaptation and mitigation; for example, tree plantings for carbon sequestration might compete for water or land with food production.[62] Fortunately, however, many strategies serve dual purposes of adaptation and mitigation. For example, soils richer in carbon hold moisture longer during droughts and may survive floods better as well.[63] According to a 2014 report by the IPCC, "Most climate change adaptation measures also have positive impacts on mitigation."[64]

Adaptation can mean switching crops to species that are more adapted to the new climate. For example, an ideal mitigation–adaptation annual cropping system is diversified, uses soil conservation practices, incorporates trees, and reduces energy use. The ideal perennial system has diverse species, multiple layers of vegetation, and sustainable soil and fertility management. The ideal livestock system is a silvopasture with high density of trees and shrubs providing fodder, over shade-tolerant pasture, with managed grazing.[65] Some adaptation strategies, such as flood mitigation, only

FIGURE 2.2. Unseasonable weather is bad for agriculture. This flood at Nuestras Raíces farm in Massachusetts occurred in September following a hurricane.

work at the whole-landscape level.[66] And most take time to implement and see benefits; they need to be in place before disaster strikes.[67]

WATER

Water is becoming increasingly unpredictable. Adaptation means being ready for both droughts and floods. At the most basic level, increasing soil organic matter (and therefore carbon) builds the water-holding capacity of the soil, increasing rainwater capture and reducing runoff and flooding. Rainwater-harvesting earthworks (zai holes, contour infiltration basins, and more) and tanks store water for later use. Drip irrigation and other improved irrigation systems allow conservation of this limited resource. In rice paddy areas, water use can be reduced through improved paddy management and the System of Rice Intensification (SRI). Finally, conversion to dryland-adapted crops and varieties reduces water needs.[68]

A landscape-level approach to flood control must begin at the top of the watershed, with erosion control and perennial vegetation to slow and infiltrate rainwater before it can cause destructive flooding downstream.[69] Downstream, farms in floodplains can slow and infiltrate floodwater with perennial riparian buffer plantings and vegetated perennial water-breaks in the fields.[70] Some water management strategies are covered in chapter 9.

SOIL

Improving soil carbon is a critical adaptation strategy because it is useful against almost any climate challenge.[71] Soils high in organic matter (which, again, is 58 percent carbon dry weight) can hold more water. This means they both capture and store more rainfall while reducing runoff and the resulting flooding. High-carbon soils are less vulnerable to erosion. They also are less likely to lose nutrients through leaching and have better fertility.[72]

Soil-based strategies for adaptation include conservation agriculture (no-till, cover cropping, mulching), agroforestry and perennial crops, managed grazing, and silvopasture. Erosion control strategies such as terracing and contour planting are also adaptation tools.[73]

FIGURE 2.3. Guatemalan farmers Doña Corina and Don Victoriano use contour hedgerows and other techniques to build soils and increase yields. These practices also sequester carbon to mitigate climate change and make their farms more resilient in the face of a changing climate.

CROP DIVERSITY AND PRODUCTION

Crop diversification means that even if some crop yields are poor in a given year, other crops may be producing well. In this way, farmers manage the risk and food and income keep coming.[74] Biodiversity on the farm can improve adaptation, even in response to hurricanes. Farmers practicing cover cropping, intercropping, and/or agroforestry were less impacted than their monoculture-practicing neighbors by Hurricane Mitch in Central America. Biodiverse coffee plantings were less damaged than simpler coffee farms in Mexico by Hurricane Stan. In Cuba diversified farms had 50 percent

TABLE 2.2. Strategies That Both Sequester Carbon and Adapt to Climate Change

Issue	Adaptive Challenge	Sequestration and Adaptation Strategies
Water	Unpredictable and reduced/increased rainfall	Rainwater harvesting, drip irrigation, arid-land crops, SRI
Soil	Need to build organic matter to improve resilience	No-till, SRI, mulching, rotations and cover crops, perennial crops, agroforestry, silvopasture, managed grazing
Crop production	Utilize efficient and resilient production techniques	Nitrogen fixation, SRI, diversification, agroecology, agroforestry
Livestock	Utilize efficient and resilient techniques	Managed grazing, agroforestry, silvopasture, fodder banks

Source: Adapted from the FAO Climate-Smart Agriculture Handbook.

losses during Hurricane Ike, compared with 90 to 100 percent for their monoculture neighbors. These farms also recuperated more quickly.[75] Yields in polycultures are less affected by droughts than monocultures.[76] At the landscape level, a mosaic of farm and natural habitats provides production and ecosystem services.[77] For example, biodiversity can create reservoirs for beneficial insects and other organisms that help with pest control.[78]

Improved annual cropping systems, including conservation farming and agroecological approaches, help to buffer farmers against unpredictable and intense climate. These include no-till, SRI, mulching, crop rotations and cover crops, livestock integration, and on-farm nitrogen fixation.[79] This group of strategies is covered in chapter 6.

Some perennial crops are more resilient than annuals.[80] Trees help stabilize slopes to prevent landslides during storms.[81] Perennials often survive floods that destroy annual crops.[82] Trees have a significant impact on the local microclimate, cooling crops, pasture, people, and livestock in their shade and in the vicinity.[83] The deep root systems of trees move water from deep to shallower soils and vice versa in a process called hydraulic lift. This can make water available for adjacent shallow-rooted annual crops in agroforestry systems, though trees also compete for water.[84] Like trees, perennial crops are critical to erosion control, rainwater infiltration, and flood mitigation. As with annual crops, switching to or breeding climate-hardy varieties reduces climate risks. Perennial cropping systems are described in chapter 8.

LIVESTOCK

As with crops, breeding climate-resilient livestock or switching to hardier breeds or species is an important adaptation strategy. Integrating livestock and crop production is also important, including feeding livestock with crop residues and utilization of manure as fertilizer.[85] Managed grazing increases vegetation cover and improves soil organic matter content and water infiltration. Silvopasture has many adaptation benefits, from cooling pastures and livestock to windbreak and erosion control.[86] These and other practices are covered in chapter 7.

CHAPTER THREE

Carbon Sequestration Potentials

Are all carbon farming practices equal in terms of their sequestration capacity? What are the variables? Where are these techniques applicable, and how widely are they likely to be adopted? How much carbon can agriculture sequester? Definitive answers are hard to come by, but in this chapter we'll catch up with the research.

The IPCC rates mitigation practices by several criteria (see table 3.1). First it looks at potential mitigation impact. This is a combination of the annual sequestration rate per hectare and the amount of eligible farmland it is suited to. Next, they assess the ease of adoption by farmers. Is it easy to learn? Does it require new equipment or training? Is it affordable? Finally they look at readiness, dividing practices into those ready for wide implementation, those that require 5 to 10 years before scaling up, and those that are still under development. While I don't agree with all their conclusions and I'd certainly add some more practices, it is an important analysis and I refer to it throughout part 2.

Carbon Sequestration Capacity Compared

When it comes to the annual rate of carbon sequestration, all carbon farming practices are not created equal, as shown in figure 3.1. The more trees, the higher the level of carbon sequestration. Yet agroforestry practices are given little attention in most mitigation literature, and tree crops are virtually absent. The impressive sequestration power of perennials—along with this gap

in the literature—is the reason parts 3 and 4 of this book explore perennial crops so extensively. All of the figures discussed in this chapter are in tons per hectare per year. Throughout this chapter I've ranked sequestration rates as follows:

"Very low" is 0–0.5 t/ha/yr.
"Low" is 0.5–1 t/ha/yr.
"Medium" is 1–5 t/ha/yr.
"High" is 5–10 t/ha/yr.
"Very high" is 10–20 t/ha/yr.
"Extremely high" is 20 t/ha/yr or more.

The lifetime carbon accumulation ability of soils also varies widely from practice to practice. This is illustrated in table 3.2. Again the superior ability of systems with woody perennials can be seen. Note that this table does not include lifetime accumulation in aboveground biomass. In this chapter soil carbon stocks are ranked:

"Modest" is 0–50 t/ha.
"Large" is 50–150 t/ha
"Very large" is 150–300 t/ha.

The practices described in this book are not the only carbon farming practices. Others exist, and more will be discovered. As more information becomes available, the information in this chapter, and the book as a whole, will shift. I suspect that the general pattern present in figure 3.1 will not shift, though I would welcome news

TABLE 3.1. IPCC Summary of Mitigation, Implementation, and Time-Scale Potential for Selected Carbon Farming Techniques

Category	Practices and Impacts	Potential Global Mitigation Impact[a]	Ease of Adoption by Farmers	Readiness of Practice
Afforestation of farmland	Monocultures or mixed species, including tree crops and multipurpose trees	Medium	Easy	Ready
Crop management	Rotations, cover crops, perennial crops, improved varieties	Medium	Easy	5–10 years
Tillage and residue management	Reduced tillage, crop residue retention	High	Easy	Ready
Water management	Rainwater harvesting and other strategies	Medium	Moderate	5–10 years
Rice paddy management	Straw retention, reduced flooding, nutrient management	Medium to high	Moderate to easy	Ready
Biochar application	Application of biochar for fertility and carbon sequestration	High	Moderate	Still under development
Pasture management	Improved pasture species, fodder banks, etc.	Low	Easy	Ready
Managed grazing	Stocking densities, improved grazing management, fodder production and diversification	Low	Moderate to easy	Ready
Manure application	Application of manure to cropland for fertility; livestock–crop integration	High	Easy	Ready
Livestock feeding	Methane-reducing feed and forage	Medium	Moderate	5–10 years
Manure management	Modified bedding, changed feeds, biodigestion, etc.	High	Moderate to easy	Ready
Agroforestry	Integration of trees with crops and/or livestock	Medium	Moderate	Ready
Mixed biomass production	Productive shelterbelts and riparian buffers, biomass crop integration	Medium	Easy	Ready
Perennial protein and biomass coproduction	Concentrated protein as a by-product of perennial biomass processing	Medium	Easy	Ready

Source: Adapted from IPCC, Climate Change 2014: Mitigation of Climate Change, 830–32.

Note: Although I would add many practices and change some of the emphasis, it is interesting to see what the IPCC reports regarding the potential impact, acceptability, and readiness of carbon farming techniques. Most of the strategies profiled in this book are not yet on IPCC's radar.

[a] This does not refer to carbon sequestration per hectare but to the impact it could potentially have if implemented at the global scale.

of higher sequestration rates for any of the practices described here.

Throughout this chapter, the emphasis is on carbon sequestration without looking at methane, nitrous oxide, and albedo impacts. This reflects a lack of available data. Ideally we could use a full life-cycle analysis (LCA), looking at the climate pros and cons of each practice from cradle to grave. Hopefully this information will become available as mitigation efforts ramp up.

Annual Cropping Systems

Annual crops are grown on 22 percent of world farmland, 1.3 billion hectares (3.2 billion acres).[1] They are the primary source of human food and the primary source of concentrated feed for livestock. Annual industrial crops such as cotton are also important sources of materials, chemicals, and energy. A number of annual cropping practices have emerged that sequester carbon, including agroforestry practices that integrate trees in annual crop fields. Sequestration rates generally range from very low to medium (in the case of tree–annual intercropping). These practices are reviewed in chapter 6.

The first cluster is a group known as conservation agriculture, including no-till, cover cropping, and crop rotation. These practices have very low to low rates of sequestration.[2] Lifetime soil carbon stocks in these systems are modest at an estimated 30 to 50 t/ha.[3] They are practiced on roughly 124 million hectares (306 million acres) and are continuing to spread.[4]

Organic and agroecological approaches to annual cropping sequester carbon at low to medium rates.

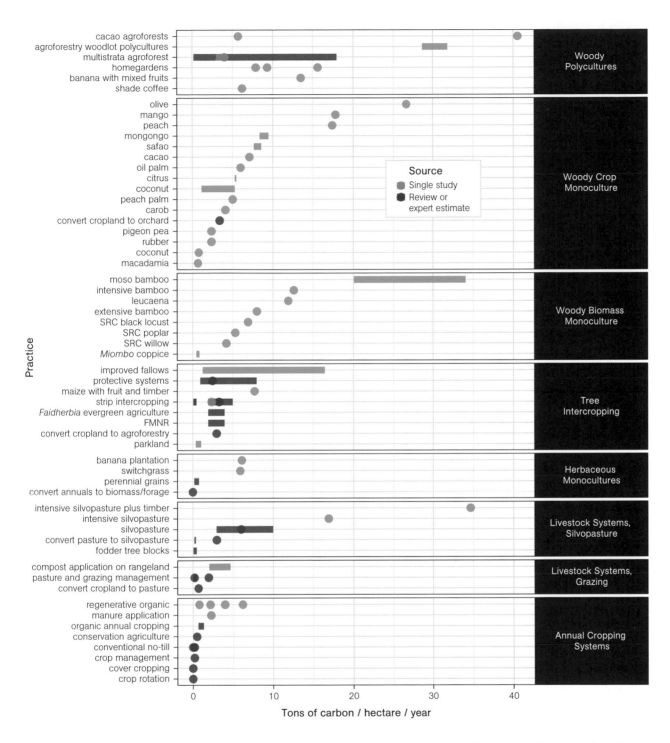

FIGURE 3.1. The data and citations on which these rates are based are in appendix C. Blue indicates results from a single study; red indicates scientific reviews or estimations by experts in the field. Circles represent single points of data; rectangles represent ranges. The data is not adjusted to show increases or decreases of methane or other greenhouse gases. Based on this data, woody plant-based systems generally have far higher sequestration rates than improved annual cropping or grazing strategies alone. Data visualization by Rafter Ferguson.

TABLE 3.2. Lifetime Soil Carbon Stocks of Carbon Farming Systems Compared

Category	Practice	Location	Lifetime Soil Organic Carbon Gains per Hectare	Study Type
Annual cropping	Conservation agriculture	India, USA	-4 to +20	Review[a]
	Improved annual cropping	Global	30–50	Expert estimation[b]
Silvoarable agroforestry	Tree intercropping	Tropical humid	"Very low"–150	Expert estimation[c]
	Tree intercropping	Temperate	Up to 200	Expert estimation[c]
	Protective systems (e.g., windbreaks)	Arid and semi-arid	Up to 100	Expert estimation[c]
Grazing systems	Improved grazing and pasture management	Global	30–50	Expert estimation[b]
	Improved grazing and pasture management	Global	Up to 16	Review[d]
Livestock with trees	Silvopasture	Global	Up to 250	Expert estimation[c]
	Fodder banks	Tropical	Up to 140	Expert estimation[c]
Woody biomass	Coppiced firewood lots	Tropical	Up to 140	Expert estimation[c]
Woody polyculture	Multistrata agroforestry	Humid tropics	Up to 300	Expert estimation[c]

Note: Carbon stocks represent the total carbon gained and stored over the active lifetime of a system. The ranges shown in this table represent the net total gains (or losses) from these practices over decades, until the saturation point is reached and sequestration slows considerably. There is much less data available on carbon stocks compared with annual sequestration rates.
Many of these estimates are from table 3 of P. K. Nair's "Climate change mitigation: A low-hanging fruit of agroforestry." The article as published has mislabeled headings, but I confirmed this interpretation with Nair himself via email.

[a] Srinivasarao et al., "Conservation agriculture and soil carbon sequestration," 492–99.
[b] Lal, "Abating climate change and feeding the world through soil carbon sequestration," 447.
[c] Nair, "Climate change mitigation," 53.
[d] McSherry and Ritchie, "Effects of grazing on grassland soil carbon: a global review."

Organic production systems generally have higher soil organic carbon than conventional (chemical) systems.[5] The Rodale Institute has proposed the term *regenerative organic* production, which includes cover crops, compost application, and crop rotation. They have put forward a few impressive studies with medium sequestration rates, including one from their long-term on-farm research site. Organic annual crops occupy 6.3 million hectares (15.6 million acres) of cropland.[6]

Silvoarable agroforestry systems integrate trees with annual crops. This includes strip intercropping, alley cropping, evergreen agriculture, and other practices. Sequestration rates are generally low to medium, with exceptions. Soil organic carbon stocks are modest to very large, ranging from very low to 200 t/ha lifetime.[7] These practices represent a significant improvement over annual cropping alone and don't require major shifts in diets, farm mechanization, or the food system. Tree intercropping of one kind of another is practiced on an estimated 700 million hectares (1.7 billion acres) of cropland.[8]

Livestock Systems

Pasture occupies 3.3 billion hectares (8.2 billion acres), or 67 percent of the world's farmland.[9] Livestock is a major contributor to the food supply, and often the only available option in dry, steep, or otherwise untillable lands. Sequestration rates of livestock-based carbon farming practices range from very low to extremely high. The drawback of livestock, particularly ruminant livestock such as cattle, sheep, and goats, is their production of methane, a potent greenhouse gas. This issue is discussed extensively in chapter 7, as are the following practices.

Pasture and grazing management have very wide potential applicability. Rates of sequestration are generally very low, with some estimates at the low end of medium. The Marin Carbon Project has given us a new tool for sequestering carbon on these lands, the application of compost. This kicks carbon sequestration into overdrive, with solid "medium" carbon sequestration rates continuing for years after application.[10] Lifetime soil organic carbon stocks of grazing systems are modest at 30 to 50 t/ha.[11]

GRASS VERSUS TREES

Some grazing enthusiasts argue that grasslands are superior to tree-based agriculture. Why? They claim that grasslands have more soil carbon than forests and that aboveground biomass "doesn't count" because it is vulnerable to burning or being cut for firewood. Even the excellent book *Grass, Soil, Hope* recommends removing trees from pastures to permit better carbon sequestration.[12]

How do the soil carbon levels of grasslands stack up when compared with those of forests? Soil organic carbon stocks (lifetime accumulations) range as follows: forest: 80 to 471 t/ha; savanna: 21 to 264 t/ha; grasslands: 42 to 295 t/ha.[13] Some types of grasslands have more soil organic carbon than some types of forest, but this is the exception, not the rule.

So is carbon in aboveground biomass more vulnerable to loss than carbon in the soil? The IPCC considers both to be non-permanent and reversible.[14] All trees eventually die, but many crop trees live for centuries (see table 2.1). The wood of many crop trees can be used in construction after their bearing years, sequestering carbon in buildings for additional decades. Aboveground biomass can indeed be burned in fires, although many trees are quite fire-resistant; indeed, some require fire to thrive. Trees can be cut for firewood or killed by storms, errant livestock, wild animals, and diseases. Carbon in soils can likewise be quickly lost by a return to tillage. The FAO Grasslands Carbon Working Group notes that carbon gains in grasslands are also subject to rapid reversals from ecological disturbances or poor livestock management.[15]

In addition to aboveground biomass, soil organic carbon stocks are estimated to be much higher in silvopasture. Total lifetime accumulation from improved pasture and grazing is estimated at 30 to 50 t/ha, while silvopasture figures are estimated at up to 250 t/ha.[16] These numbers, however, are based on estimates and not on side-by-side comparisons of pasture and silvopasture on the same site.

Adding trees to pasture ("silvopasture") increases sequestration substantially. Silvopasture generally receives a medium to high rating for its sequestration rates. Soil organic carbon stocks are very large in silvopasture systems at an estimated 250 t/ha.[17] Most remarkable is the practice of intensive silvopasture, in which fodder trees are planted every square meter, to be browsed repeatedly by livestock. When timber and other trees are added, intensive silvopasture sequestration rates are extremely high.

Perennial Cropping Systems

Perennial crops are grown on at least 153 million hectares (378 million acres), 11 percent of world cropland.[18] Their carbon sequestration rates range from very low to extremely high. Perennial replacements are available for staple crops in some climates, with others under development. Perennial sources of materials, chemicals, and energy are also available. These cropping systems include both herbaceous and woody perennials, and both monocultures and mixed-species systems. Along with intensive silvopasture, these represent the highest levels of carbon of any food production system.

Herbaceous perennials tend to sequester less carbon than woody plants, at very low to high rates. Bananas and biomass grasses demonstrate high rates of sequestration, while others are far lower. More data is needed on the potential impact of herbaceous perennials.

Woody perennial crops such as fruits and nuts are powerful sequesterers of carbon, with most rated medium to very high. Bamboos sequester carbon at high to extremely high rates. Coppiced woody plants have low to very high rates. Soil carbon stocks in woody biomass systems are large at an estimated 140 tons/ha.[19]

Multistrata agroforestry systems feature multiple perennial crops, often growing in layers, with shrubs and small trees growing under a canopy of tall timber or crop trees. These systems sequester carbon at medium to extremely high rates. In multistrata systems, the majority of carbon is sequestered by the woody overstory. Within the understory, shrubs sequester more than herbs and grasses.[20] Lifetime soil carbon

SEQUESTERING MORE CARBON THAN NATURAL FOREST

A study in the Philippines found that both homegardens and multistrata agroforests (which include timber, fruit, nut, coffee, banana, and native trees) outcompete natural forest in annual carbon sequestration and aboveground carbon stocks, and outcompete preserved old-growth forest in annual carbon sequestration.[21] In the Philippines, a country of 99 million people, 70 percent of all households have homegardens.[22]

TABLE 3.3. Carbon Sequestration by Multistrata Agroforestry Systems and Natural Forest in the Philippines

System	Average Age of Three Sampling Sites	Annual Sequestration Rates	Aboveground Carbon Stocks
Preserved forest	100+	2.3	234.5
Natural forest	100+	1.5	147.5
Multistrata agroforest	38	4.1	155.8
Homegarden	21	9.4	159.7

Source: Brakas and Aune, "Biomass and carbon accumulation in land use systems of Claveria, the Philippines," 170.

accumulation in these systems is very large at an estimated 300 t/ha.

Generally speaking, agroforestry systems sequester less carbon than natural forest, but long-term agroforestry systems can sequester equal or higher amounts of carbon than adjacent natural forests.[23] This is likely due to factors such as the high proportion of nitrogen-fixing trees, the presence of livestock manure, and the active participation of human managers.

Variables in Carbon Sequestration Rates

The challenge in understanding and assessing all these tools is that there is a huge variation in the carbon sequestration potential of various practices, as well as variation within the same practices, depending on other important factors such as climate and soil type. For example, differences in regional climate affect carbon sequestration in both natural and agricultural systems. Temperate-climate systems generally sequester less carbon in aboveground biomass annually than tropical-climate systems.[24] Instead they store it belowground, where organic matter breaks down more slowly in colder climates, giving these regions more potential for soil carbon storage compared with the tropics.[25]

Arid, semi-arid, and degraded lands all have less potential to sequester carbon.[26] That said, because there is so much arid, semi-arid, and degraded land around the world, we shouldn't discount its potential for carbon storage. Based on scale alone, these lands represent significant potential.

There are also challenges in understanding precise rates of carbon sequestration of various practices. For one, there is a great range of practices that people consider to be carbon farming, and in how those practices are, well, practiced. I define *carbon farming* as "the suite of agricultural practices that sequester carbon." And as a framework for understanding and implementing those practices, I focus on agroforestry and perennial crops, as two aspects of carbon farming that are the most effective way to focus our efforts.

There is also great variability in the techniques researchers use for measuring and estimating carbon content in aboveground biomass and in soils. For example, some sample carbon at 10cm (4-inch) depth in the soil, while others go 30cm, 60cm, 100cm, or even 200cm. It's difficult to compare these accurately. Some studies adjust for emissions of methane or nitrous oxide while others don't. None adjusts for albedo. The lack of a standardized system and comprehensive global data is a major limitation to carbon farming awareness and funding.[27]

What this all means is that, as of 2015, we still lack a comprehensive understanding of (and ability to predict) the carbon sequestration impact of various practices and crops in different regions, with different soils, and with different climates. What we do have, however, is a fair amount of data from individual studies that collectively point to some clear trends for core practices. There is data available to compare against natural

10 years, but slower-growing ones sequester more after that.[28] Slow-growing trees often have denser wood, and denser wood sequesters more carbon. Denser wood is also used for longer-lived, durable products after timber harvest, leading to long-term sequestration.[29]

Deciduous and semi-evergreen trees compete less for light with understory crops and may therefore allow a higher density of trees with higher carbon capture as a result.[30] Species with higher root densities are more effective, and trees sequester more carbon than shallow-rooted warm-season grasses.[31] Combining cool-season and warm-season plants (and other types of resource partitioning among plants) increases carbon sequestration.[32]

Comparisons of soil carbon consistently find 20 to 100 percent more carbon in soils under nitrogen-fixing plants.[33] On the other hand, nitrogen-fixing plants give off extremely potent greenhouse gases such as nitrous oxide; scientists have not reached conclusions about how significant this is yet.[34] There is no evidence that native species, as a class, sequester more or less carbon than non-native species in any given region.[35] Different cultivars within a species can vary in carbon sequestration potential.[36]

Design and management can also affect carbon sequestration in complex ways. Design issues such as density of trees and species selection, and management practices such as timing and type of pruning, tree life span, and harvesting, all can have positive and negative impacts.[37] Burning wood on the forest floor after timber harvest releases carbon but can also lead to long-term increases in soil carbon in some cases.[38] Emissions from activities such as the use of machinery and water pumps can counteract gains from tree planting.[39]

Managed bamboo forests sequester more carbon than natural bamboo forests. This is because harvesting stimulates more growth, and treatment of harvested culms allows for long-term sequestration in bamboo products.[40] One general trend is that natural forests sequester the most carbon, followed by multistrata systems, followed by tree plantations, followed by annual agriculture.[41] The more diversity, higher density, and longer life span of trees, the more carbon is sequestered.[42]

FIGURE 3.2. Measuring carbon as part of the Standard Assessment of Mitigation Potential and Livelihoods in Smallholder Systems (SAMPLES), a global program of the CGIAR research program on climate change, agriculture, and food security. Photograph courtesy of the World Agroforestry Centre.

ecosystems, improved annual cropping systems, annual systems that incorporate perennial elements, livestock systems with perennial elements, and fully perennial systems. Some data is also available on different classes of perennial crops. We don't have every kind of data for every practice—far from it—but it is certainly possible to make some comparisons and draw initial conclusions.

We know that plant species differ in their ability to sequester carbon. Comprehensive lists of species and their carbon storage capacities are not yet available, but researchers have determined some general guidelines. Fast-growing trees sequester more carbon in the first

If forest is cleared for agroforestry systems (or for new plantations), there is a net loss of carbon to the atmosphere. When planted on farmland, degraded land, or other treeless areas, agroforestry and reforestation are net gains for sequestration.[43] Management with fire can actually increase biomass and soil carbon by stimulating vigorous plant growth.

Global Estimates of Agricultural Carbon Sequestration Potential

So can carbon farming alone solve our climate change problem? Not even close. Carbon farming doesn't work without dramatic emissions reductions (including clean energy and reduced consumption in wealthy countries), as even a small fraction of the remaining 5 to 10 trillion tons of carbon in the fossil pool would far overwhelm the theoretical maximum sequestration capacity of soils and biomass, estimated at 320 billion tons.[44] Emissions reduction also doesn't work without carbon farming since even if emissions stop today, we're already over the tipping point with no way to return without sequestration. These strategies are two halves of a whole, and each cannot succeed without the other.

Of the 200-plus billion tons of carbon that must be sequestered, how much can come through agriculture? These calculations are outside the scope of this book, although I am currently developing such scenarios with Project Drawdown. (Making such calculations is complex for many reasons. One is double accounting, in which the same practice is counted more than once. For example, the same land may be included under grazing, silvopasture, and/or forestry, depending on who is counting. Other farms may fall through the cracks and not be counted at all for similar reasons.) Scientists have come up with various proposals, some of which are quite hopeful. The challenge lies in scaling up these practices to a global scale, which we will discuss in part 5.

This is where the second half of the carbon farming definition comes in—the need to remunerate farmers for their carbon sequestration work. Studies have shown that the more incentives are available for farmers, the more carbon they can drawn down. One way farmers can be paid for sequestration is through carbon offsets. (Prices for "carbon" actually reflect carbon dioxide and should be read as such throughout this book.) In this model, the amount of money received by farmers for their mitigation impact depends on the price of carbon dioxide in international markets. (I explain my distaste for carbon offsets in the introduction and chapters 27 and 28. Nonetheless it is of interest here to point out that higher payments to farmers result in higher levels of mitigation.) The IPCC has developed several scenarios to predict the impact of agricultural mitigation, assuming different prices on carbon. The higher the price on carbon, the more money gets to farmers (in theory). The IPCC's most optimistic scenario assumes a carbon price of $100/ton US of carbon dioxide equivalent. Under this scenario they estimate global sequestration of 81 million to 1.2 billion tons annually by 2030.[45] As the price of carbon drops to $50 and $20, the mitigation impact of carbon farming also drops dramatically.

Rattan Lal of the Ohio State University Carbon Management and Sequestration Center estimates that global

TABLE 3.4. Estimates of Global Sequestration Impact of Carbon Farming

Scenario	Global Annual Rate in Tons	Lifetime Impact	Details
Global agricultural impact	81 million–1.2 billion		By 2030
Conservation agriculture and forestry[b]	400 million–1.2 billion	20–120 billion tons or 50ppm	Over 50–100 years
Pasture and grazing management[c]	54–216 million		By 2030
Global agricultural impact [d]	1.5–1.6 billion		By 2030
Agroforestry[e]		1.8 million–135 billion	
Woody staple crop conversion[f]	394 million	10.2 billion	10% of tropical annual staple cropland

[a] IPCC, *Climate Change 2014: Mitigation of Climate Change*, 851.
[b] Lal, "Abating climate change and feeding the world through soil carbon sequestration," 444.
[c] FAO, *Challenges and Opportunities for Carbon Sequestration in Grassland Systems*, 13.
[d] Smith et al., "Greenhouse gas mitigation in agriculture," 789.
[e] Leakey, *Living with the Trees of Life*, 56.
[f] See text.

adoption of conservation agriculture, managed grazing, and improved forest management, world soils could sequester 400 million to 1.2 billion tons of carbon annually over a period of 50 to 100 years. He equates this with a drawdown of 50ppm of carbon dioxide, although by then we are likely to need far more than that to return to 350ppm.[46] Grasslands alone could sequester 54 to 216 million tons annually by 2030 according to the IPCC. This assumes a price on carbon of $20 to $50/ton US.

An impressive study reviewing the global potential of multiple practices estimated a theoretical capacity of 1.5 to 1.6 billion tons per year. The authors did not think this would be achieved and estimated 409 to 436 million, 681 to 736 million, and 1.0 to 1.2 billion tons respectively at carbon prices of $20, $50, and $100/ton US.[47] Reviewing these scenarios it is clear that financing is necessary to achieve high impact. Whether this comes through a price on carbon or through the many other options remains to be seen. We'll discuss financing carbon farming in chapter 27.

It is also noteworthy that few of these studies consider much impact for agroforestry. For example, the IPCC scenario includes virtually no agroforestry, and no mention at all of perennial crops.[48] What scenarios have been run to determine the potential of agroforestry and perennial crops? The World Agroforestry Centre estimates that 900 million hectares (2.2 billion acres) are suitable for conversion to agroforestry. Their sequestration rate estimate was 2 to 50t/ha lifetime.[49] This gives the wide range of 1.8 million to 135 billion tons. (Both the IPCC and World Agroforestry Centre are looking primarily at tropical climates in these estimates, although agroforestry is practiced in temperate climates as well.)

In the absence of published estimates on conversion to woody staple crops, I've developed a sketch of a scenario to show their impressive sequestration capacity. There are many tropical woody staples that yield as well as annuals, and for purposes of this exercise we'll assume equivalent yields. Annual staples are grown on about 789 million hectares in the tropics.[50] Let's assume a conservative 5 t/ha/yr sequestration rate (some tree staples exceed 20 t/ha/yr). What if 10 percent of tropical annual staple cropland was converted to perennials? I'll assume this is primarily to feed livestock to avoid the need for market development. This scenario sequesters 394 million tons of carbon annually. Forest and timber trees typically continue to sequester for 25 to 50 years, so let's go with a conservative 25 years for our smaller food trees.[51] During that period, our 78.9 million hectares of staples (10% of total) would sequester 10.2 billion tons of carbon—equivalent to a reduction of 5 parts per million of atmospheric carbon dioxide.

What are the prospects for shifting a sizable percentage of staple crop production to perennials? To answer this we must know more about these crops. Part 3 of this book contains what I believe to be the first comprehensive global inventory of cultivated perennial staple crops. Some of the many social, financial, and political aspects of this kind of perennialization of agriculture are discussed in part 5.

CHAPTER FOUR

Agroforestry and Perennial Crops

We've learned in chapter 3 that the presence of perennial crops, especially woody crops, greatly increases carbon sequestration, in terms of both annual rates and lifetime storage. This includes perennial crops, polycultures of perennial crops, and the incorporation of trees in pastures and annual crop fields. Yet these strategies are given little attention, even within the small world that is carbon farming. Let's explore them in more detail.

The World Agroforestry Centre defines *agroforestry* as "a dynamic, ecologically-based, natural resources management system that, through the integration of trees on farms and in the agricultural landscape, diversifies and sustains production for increased social,

FIGURE 4.1. Agroforestry is the intentional, intensive, integrated, and interactive integration of trees with crops, often annuals. Here grain is being harvested in the rows of a cherry–walnut tree intercropping agroforestry system in France. Photograph courtesy of AGROOF, © AGROOF scop/www.agroof.net.

economic, and environmental benefits for land users at all levels."[1] Agroforestry is more than the random incorporation of trees on farms; it is "intentional, intensive, integrated, and interactive."[2] Agroforestry systems sequester significant carbon and have multiple social and environmental benefits. Along with perennial crops, it is one of the two most powerful carbon farming strategies in terms of per-hectare sequestration rates.

Some agroforestry systems are ancient. Tropical homegardens go back 13,000 years in some regions, while the *dehesa* silvopasture system in Spain and Portugal has a 4,500-plus-year history. Modern agroforestry systems have about three decades of history, having begun in the late 1970s and early 1980s as a response to the environmental and food crises.[3]

The amount of land worldwide in agroforestry systems is approximately 1 billion hectares (2.5 billion acres). An estimated 1.5 billion farmers practice some degree of agroforestry. Most are smallholders in the tropics.[4] Agroforestry is making gains in temperate climates as well, with more than 2,700 farmers in the United States practicing alley cropping or silvopasture.[5]

A lot of emphasis in agroforestry is on making annual crop production more sustainable by incorporating woody plants to help reduce erosion, fix nitrogen, and build biomass and tilth. In most agroforestry systems, the trees themselves do not provide the staple crops. Important exceptions include homegardens, and other complex multistrata agroforests, in which staple tree crops such as breadfruit, coconut, peach palm, and avocado are grown. There are some perennial farming systems that are not agroforestry, including established (monocultures, orchards, plantations, and bamboo production), new (pasture cropping), and hypothetical (perennial grains) systems. Anything without woody plants is not agroforestry, even if it has a strong herbaceous perennial component. These practices and more are covered in more detail in part 2.

For the purposes of this discussion, I've divided agroforestry into three broad classifications. The first is tree intercropping systems, which integrate trees or other woody plants with annual crops. This includes alley cropping and contour hedgerows, windbreaks, evergreen agriculture, and other practices covered in chapter 6. Tree intercropping systems of this kind are grown on an estimated 700 million hectares (1.7 billion acres) worldwide.[6] These practices generally have better sequestration than improved annual techniques alone.

The second classification is silvopastoral systems, which integrate trees or other woody plants with livestock. Practices include silvopasture, tree crop cultivation with pasture understory, fodder banks, and other practices covered in chapter 7. Silvopastures occupy an estimated 450 million hectares (1.1 billion acres) globally.[7]

And the third classification is multistrata agroforestry systems, which consist solely of perennials or integrate trees or other woody plants with annual crops and livestock. These include tropical homegardens and other new and traditional systems such as woody agriculture. Multistrata systems are grown on an estimated 100 million hectares (247 million acres), mostly in the tropics.[8] These practices are described in chapter 8.

In agroforestry there are many ways to integrate trees with the other components of the system. Tree plantings can be dense or dispersed, in rows that alternate with other crops, or along the boundaries of a field. Likewise, the trees and other components of a system might share space throughout the entire life of that system (as is common in silvopasture and plantation crop combinations), only in the early stages (as in successional intercropping), intermittently (as with intensive silvopasture or *Faidherbia* evergreen agriculture), sequentially or in rotation (as in improved fallows), and in other more complex combinations.[9]

In addition to carbon sequestration, agroforestry systems provide people with food, fodder, fuel, and other important products. They also provide a range of important protective functions, such as windbreaks, erosion control, soil and moisture conservation, and shade for crops, livestock, and people.[10] For example, my own temperate homegarden provides food, fodder for silkworms and poultry, medicinal plants, cut flowers, shade, privacy, and habitat for beneficial insects and birds that control pests.

Agroforestry systems may be low, medium, or high input (labor, fertilizers, financial capital, irrigation).

FIGURE 4.2. Silvopasture is the deliberate integration of certain beneficial trees on pasturelands. Here sheep graze under encina oaks in a traditional Spanish *dehesa* silvopasture system. Photograph courtesy of Brock Dolman.

FIGURE 4.3. Multistrata systems involve different layers of vegetation. In this example from Veracruz, Mexico, avocados and macadamias are grown above bananas, which in turn provide shade for coffee.

This depends in part on whether they are intended for subsistence, commercial production, or even educational and recreational use. There are options from small-scale intensive to broad-scale extensive systems.[11] My friend Roberto Muj operates an intensive 1-hectare (2.5-acre) perennial market garden in the Guatemalan highlands. It is a strip intercropping system with rows of such trees as avocado, macadamia, peach, citrus, and mulberry. The understory comprises rows of perennial crops, including cut flowers, beans, vegetables, medicinal plants, and poultry forage. All are direct-marketed in the nearby city of Antigua. Extensive systems are larger and require less labor per hectare. An example is the *espinal* silvopasture system in Chile, in which ruminants graze on large tracts of savanna with pasture and nitrogen-fixing acacia trees.

Although there are hundreds of agroforestry systems around the world, most are developed from a basic palette of roughly 20 practices. With this palette of practices as a foundation, individuals and communities can tailor an agroforestry system to reflect regional socioeconomic and environmental conditions, such that appropriate agroforestry systems have been developed for most of the world's climates. The most successful examples build from the basic palette and patterns of agroforestry and tailor them for locally adapted perennial crops.[12]

Tropical Homegardens

To many of us from outside the world of agroecological science, the term *homegarden* just sounds like a "home garden." In the Western world this may evoke lawns, flowers, and perhaps annual vegetables, but as used here it signifies the tropical multistrata tradition—an ancient and highly sophisticated form of agroforestry. Indeed, if a single practice exemplifies agroforestry's climate potential, it would be the tropical homegarden. Homegardens have been defined as "intimate, multi-story combinations of various trees and crops, sometimes in association with domestic animals, around the homestead."[13] They are a small-scale, intensive type of multistrata agroforestry that typically occupies the size of the home yard.

Homegardens are widespread in the tropics and unremarkable there; my mother-in-law in Guatemala has one, and nobody thinks it in any way unusual. In the permaculture movement they are known as food forests, forest gardens, and edible forest gardens.

Homegardens are often difficult for the Western eye to even appreciate as agriculture, so closely do they replicate structures and patterns found in nature, but they are one of the oldest human land uses and in Southeast Asia may go back as far as 13,000 years.[14] It is a kind of agricultural production that has been a way of life for centuries in parts of both the Old and New World.

The majority of homegardens are in the humid tropics. There are also some homegarden traditions in tropical highlands, as well as Mediterranean regions of Spain and South Africa.[15] Some arid tropical regions have homegardening traditions, including the island of Soqotra in Yemen.[16] No one is certain exactly how many homegardens there are or how much of the world they cover. We do know that there are 5.1 million

FIGURE 4.4. A tropical homegarden at Maya Mountain Research Farm in Belize. It incorporates mango, breadnut, coffee, cacao, coconut, "golden plum" (*Spondias dulcis*), cocoyam, chi'kai, banana, peach palm, erthyrina, caesalpinia, heliconia, and *Calathea lutea*, an important traditional Maya bush food. This homegarden, which was formerly a citrus grove, now provides a diversity of food and marketable products. Photograph courtesy of Christopher Nesbit.

hectares (12.6 million acres) in Indonesia, half a million in Bangladesh, a million in Sri Lanka, and 1.4 million hectares (3.5 million acres) in Kerala, India. In the Philippines more than 70 percent of all households have homegardens. They are common in Mesoamerica, the Caribbean, South India, Southeast Asia, and the Pacific. There are also homegarden traditions in tropical South America and sub-Saharan Africa.[17]

Homegardens have been characterized by leading agroforestry scientist P. K. Nair as "the epitome of sustainability"[18] for their social and environmental impact, in the areas of "biodiversity conservation, gender equity, social justice, environmental integrity, appreciation of indigenous knowledge, preservation of cultural knowledge, and so on."[19] Nair goes on to state, specifically in terms of biodiversity: "A classification based on the production systems and species diversity ranked home gardens top with its highest biological diversity among all man-made ecosystems."[20]

These remarkable systems also demonstrate some of the highest levels of carbon sequestration in agriculture. In some cases carbon sequestration in homegardens is as good as or better than that of natural forests.[21] According to agroforestry researcher B. M. Kumar, "Homegardens resemble young secondary forests in structure and biomass accumulation and may be considered as a human-made forest . . . with considerable productive potential."[22]

Interestingly, tropical homegardens around the world have a shared set of species, be they in Africa, Asia, the Americas, or the Pacific. This shared palette is shown in table 4.1. These species are supplemented by regional crops and useful native plants.[23] Food plants are the most important products, but homegardens also produce cash crops, medicinals, and other crops for materials, chemicals, and energy.[24]

Today many homegardens are threatened by urbanization, industrialization of agriculture, and other challenges.[25] In some regions, market pressures are causing people to strip their homegardens down to fewer species in order to meet consumer demand for specific products. On the other hand, more and more homegardens are turning up in urban areas, an exciting new development. P. K. Nair writes in "Whither Homegardens?" that cash crops have been part of homegardens for a long time, and he is confident that homegardens will stay with us in some form even as they evolve and change.[26]

Why have there been no organized institutional efforts to spread this "epitome of sustainability"? The complexity and the uniqueness of each garden has slowed adoption of homegardens outside their core homelands. According to Nair, this complexity and uniqueness has also hindered development efforts from promoting their spread and has made it relatively difficult for researchers to study them.[27] Although homegardens have thus far been ignored by carbon finance efforts and development agencies, researchers have identified three priority areas to help realize the potential of homegardens to sequester carbon and meet human needs sustainably. First, we can prioritize the

TABLE 4.1. Common Species of Humid Tropical Homegardens

Category	Common Homegarden Species	
Nuts	Cashew	*Anacardium occidentale*
	Jakfruit	*Artocarpus heterophyllus*
	Coconut	*Cocos nucifera*
Staple fruits	Breadfruit	*Artocarpus altilis*
	Bananas and plantains	*Musa* spp.
	Avocado	*Persea americana*
Other tree fruits	Annonas	*Annona* spp.
	Carambola	*Averrhoa carambola*
	Citrus	*Citrus* spp.
	Figs	*Ficus* spp.
	Mango	*Mangifera* spp.
	Guava	*Psidium guajava*
	Golden apple	*Spondias dulcis*
	Malay apple	*Syzygium malaccense*
	Tamarind	*Tamarindus indica*
Other food crops	Pineapple	*Ananas comosus*
	Peanut	*Arachis hypogaea*
	Pigeon pea	*Cajanus cajan*
	Passionfruit	*Passiflora edulis*
	Beans	*Phaseolus, Psophocarpus,* and *Vigna* spp.
	Sugarcane	*Saccharum officinarum*
	Maize	*Zea mays*
	Vegetables	Various
Root crops	Taro	*Colocasia esculenta*
	Yams	*Dioscorea* spp.
	Sweet potato	*Ipomoea batatas*
	Cassava, yuca	*Manihot esculenta*
	Tannier, cocoyam	*Xanthosoma* spp.
Spices, social beverages, and stimulants	Betel nut	*Areca catechu*
	Cinnamon	*Cinnamomum verum*
	Turmeric	*Curcuma longa*
	Lemongrass	*Cymbopogon citratus*
	Betel vine	*Piper betle*
	Kava	*Piper methysticum*
	Ginger	*Zingiber officinale*

Source: Adapted from Nair, "Whither homegardens?," 359.

conservation of existing homegardens. Second, we can intensify homegardens by optimizing design, management, and species selection. Finally, we can bring more land into homegardens.[28]

COLD-CLIMATE HOMEGARDENS

China has temperate as well as tropical homegardening traditions. These temperate homegardens include species such as poplars and willows, nitrogen-fixing tree species such as *Styphnolobium japonicum*, and fruits such as apple, pear, peach, and grape. They also often include bamboo and firewood; mushrooms and medicinal plants in shady areas; and the incorporation of livestock, worms, eels, and fish.[29] Studies in China have found that Chinese homegardens generate six times the income of farmland of the same size.[30]

In Western temperate and even boreal regions such as the United States, Canada, Europe, and Australia, a movement to develop homegardens emerged in the 1970s. One of the driving forces behind this has been the permaculture movement. These gardens, variously known as food forests, forest gardens, or edible forest gardens, involve many of the same elements as tropical homegardens: tree crops, fungi, annual and perennial herbs, microlivestock. I myself have a temperate homegarden and recognize the great potential and present limitations of the practice in cold climates—notably the lack of domestication of sufficient perennial crops besides fruits. The permaculture movement as a whole has not acknowledged its debt to the tropical homegarden and could greatly benefit from studying the design and management of this ancient practice.[31] Though temperate homegardens are small in scale (like their tropical counterparts), they are growing in number. Permaculture-inspired homegardens have been successful not only in humid temperate and boreal regions but also in arid and semi-arid temperate (such as Nevada and Colorado in the United States), in Mediterranean climates, and arid tropics and subtropics (such as Jordan and Arizona).

FIGURE 4.5. The author's cold temperate homegarden system with 300 species of useful perennial plants, as well as annual crops, mushrooms, and small livestock.

Perennial Crops

Perennial crops are those that live for three years or more, although many of the crops profiled in this book can live for hundreds of years and some for thousands. (The longer they live, the longer they sequester carbon.) They include trees, palms, bamboos, shrubs, vines, succulents, and cacti, as well as non-woody species such as grasses, herbaceous perennials, and ground covers. All perennials sequester carbon, but not all equally.

Some perennials particularly lend themselves to carbon farming. Ideally, carbon farming systems are no-till, meaning the soil is not turned over with shovels or plows after establishment. This rules out perennial root crops, which require digging for harvest. Other perennial crops are grown for many years but are harvested in a way that kills them. This ends their carbon-sequestering life span, although in the case of timber trees, for example, the lumber used in buildings can last for decades or centuries, preserving the carbon long after the tree has died.

This book is focused on perennials that are non-destructively harvested, meaning they are suitable for no-till once established, and are not killed by harvest. "Non-destructively harvested" means just what it sounds like: Plants are not killed in the process of harvesting. All annual crops by definition die within a year of planting whether harvested or not, but not all perennial crops survive the harvest process. For example, cassava plants are perennial shrubs that can live for several years, but when the roots are harvested the plants die. In contrast, breadfruit and chestnut trees will continue to bear starchy crops year after year. Pine trees that are harvested for timber are killed in the process, whereas timber bamboos and coast redwoods continue to resprout after cutting.

When you combine perennial crops with no-till and non-destructive harvest practices, you get plants that yield for many years and hold the soil (and its carbon) in place. The palette of over 600 such species presented in this book—never before compiled in one place to my knowledge—is a critical tool for restoring degraded land and sequestering carbon.

While the perennial crops that probably come to mind for most people are those that provide fruits and perhaps vegetables, many non-destructively harvested perennial (NDHP) crops can provide staple foods such as protein, oil, and carbohydrates. Others can provide a basis for replacing petroleum for materials, chemicals, and energy, although not remotely enough energy to meet our current use.

Managing and Harvesting NDHP Crops

Despite the great diversity of non-destructively harvested perennial crops, there are only five basic strategies used to manage them. All maintain the soil organic carbon, though only standard woody, standard

FIGURE 4.6. The five basic management strategies of NDHP crops, *from left to right,* standard woody, managed multistem woody, coppiced woody, standard herbaceous, and herbaceous hay.

TABLE 4.2. Management Strategies for NDHP Crops

Management Strategy	Description	Sample Crops
Standard woody crops	Crops grow normally Typically fruits or seeds are harvested Includes basic orchard- and plantation-type management	Coconut, chestnut, castor bean
Standard herbaceous crops	Crops grow normally Products are harvested without cutting of whole-plant biomass	Air potato, perennial beans, oyster nut
Coppiced woody crops	Whole plants are cut on a rotation (typically 3–10 years) Resprout vigorously after cutting	Hazelnut, willow, moringa leaf
Managed multistemmed woody crops	Crop produces multiple trunks or stems from the roots Some are cut while others remain in place New stems typically emerge annually	Bamboo, banana, sago
Herbaceous hay crops	Whole plant is harvested, typically annually or more often Resprout vigorously May include harvest of seeds or other parts as well Processing may involve extraction of fiber or chemicals	Sugarcane, ramie, giant miscanthus

herbaceous, and managed multistem systems maintain consistent carbon in aboveground biomass.

Standard woody crops—which include trees, shrubs, most vines, cacti, succulents such as agaves, and palms—are managed by allowing them to grow to their normal height, then harvesting the fruit, seeds, or other products. This includes familiar orchard crops such as coconuts and some tapped crops such as rubber.

Standard herbaceous crops are similar—the plant is allowed to grow to maturity; then you harvest seeds or other products while leaving the rest of the plant intact. Perennial vining beans are a good example. Perennial grains such as rice are another.

Coppiced woody crops are species that can be cut completely to the ground and resprout vigorously. They are cut repeatedly on a rotation. This includes many biomass crops such as willows, as well as woody leaf crops for human consumption and livestock fodder.

Herbaceous hay crops are similar, in that the whole plant is cut to the ground and harvested annually. Hay from perennial hay fields is the most common example. Other herbaceous hay crops include nettles and kendyr for fiber, and potential hydrocarbon crops such as leafy spurge.

Managed multistem crops produce multiple stems or trunks, which are regularly removed, while others are allowed to remain. This includes biomass crops such as bamboo, where most stems remain but some are harvested every year. Bananas are another example—each

stem fruits once, then dies, but new stems continually emerge from the rootstock.

Harvest of NDHP crops includes many familiar practices as well as some unusual ones. For example, fruits, pods, nuts, and seeds, can be harvested by hand or with machinery. These may be used as staple foods (protein, carbohydrates, oil), produce (fruits, vegetables), or spices. Some fruits, pods, nuts, and seeds provide industrial products such as materials (natural building materials, plastics), chemicals (industrial lubricants, soaps, waxes, solvents, medicines), and energy (biodiesel, ethanol).

Most of us are less familiar with trees that are tapped for food or industrial products. In some cases, such as when palms are felled for their sugary sap, this damages or eventually kills the tree. In other cases, as with rubber trees or true sugar palms, tapping is non-destructive. Tapping pines for turpentine can be done either destructively or so as to ensure a long, healthy life for the tree.[32] Tree saps can provide foods such as sugars and, remarkably, edible tree milks. They can also provide industrial products, including materials (rubber), chemicals (gums, turpentine), and energy (diesel, ethanol).

Some trees can be non-destructively harvested for their bark. Cork is the most familiar example, although other barks can provide tannins and fibers. For example, paper mulberry bark was historically an important fiber crop throughout the Pacific.

TABLE 4.3. Harvested Parts of NDHP Crops

Harvested Part	Food and Fodder Products	Materials, Chemicals, Energy	Sample Crops
Nuts, fruits, seeds, pods	Protein, starch, oil, fruits, vegetables, spices	Industrial lubricants, biodiesel, plastics, fibers	Coconut, breadfruit, mesquite, banana, chestnut, olive, castor bean
Saps, latexes, and resins	Sugar, tree milks	Rubber, turpentine, diesel	Sugar palm, para rubber, turpentine pines, cow trees, diesel tree
Barks		Cork, tannins, medicine, fiber	Cork oak, baobab, linden
Trunk starch (from multistemmed or coppiced only)	Starch	Plastics, ethanol	Sago, enset, agave
Aerial tubers	Starch	Plastics, ethanol	Air potato
Leaves	Food		Baobab, alfalfa, moringa, mulberry
Biomass		Paper, natural building materials, firewood	Coast redwood, willows, rattan, sugarcane, ramie fiber
Biomass extracts	Sugar	Stem fibers, ethanol, hydrocarbons	Sugarcane, ramie fiber, petroleum plant
Crop residues (wastes)	Livestock feed	Composite materials such as plywood, energy, livestock feed	Sugarcane bagasse, nut shells, presscake from seed oil extraction

Note: To my knowledge there are no cultivated NDHP root crops besides air potato, although there are some promising undomesticated species.

TABLE 4.4. Perennial Crops in World Production

Crop	Million hectares in production	Rank by # of hectares in production
Sugarcane	26.0	11
Oil palm	17.2	17
Coconuts	12.1	19
Bananas and plantains	10.3	21
Olives	10.2	22
Coffee	10.0	23
Cacao	9.9	24
Rubber	9.8	25

Source: FAO Statistical Service, 2012.
Note: This table shows 2012 global production of the world's most important non-destructively harvested perennial crops. Their "rank" is based on their order within all crops, annual and perennial, edible and industrial. Thus sugarcane is the 11th most widely grown of all the world's crops, although it is a short-lived perennial grown at great environmental cost.

Remarkably from my cold-climate perspective, some plants have trunks filled with starch and fiber. They are grown as managed multistem or coppiced crops. Sago is a daily staple to a million people, enset has prevented famines in parts of Ethiopia, and although starch agaves today are mainly grown for tequila, they were once an important staple. Many of these starchy foods require processing to remove fibers, after which they are made into tapioca, fermented breads, and other staple foods.

There are some less common kinds of crops, such as the aerial tubers produced by air potato and other members of the genus *Dioscorea*, that grow on vines aboveground like fruits. These tubers are important sources of NDHP starch for food and industrial uses. Many NDHP crops produce edible leaves or other vegetative parts for human and livestock food; typically these are trees and shrubs that are coppiced for ease of harvest. Biomass crops are growing in importance. Bamboos, coppiced woody plants, and giant perennial grasses are used for paper, building materials, and other uses. Fibers, sugar, and hydrocarbons are extracted from some of these biomass crops. Crop residues (wastes) are also used as a source of biomass for purposes that include energy, plywoods, and livestock feed.

The Global Status of NDHP Crops

These crops are already playing an important global role in meeting human needs. NDHP crops were grown on 174 million hectares (430 million acres) worldwide in 2012, accounting for 11 percent of world cropland. Although this is impressive, it's worth noting that we devoted more than 215 million hectares (531 million acres) to the world's number one crop, annual wheat, the same year.[33]

BAOBAB: A REMARKABLE PERENNIAL CROP

The baobab tree (*Adansonia digitata*) is an icon of the vast semi-arid African savanna. With their enormous swollen trunks and often bare branches, baobabs are said to look as though they were planted upside down, with their roots coming out into the air and their leaves somehow underground. Baobabs offer far more than beauty in a landscape, however; they are one of the most important food plants across a wide swath of Africa, and one of the most useful trees in the world. To quote *Lost Crops of Africa*, volume 2: "A tree as productive and important to people as this one is worthy of massive and pan-African research. Programs dealing with food, nutrition, forestry, agriculture, agroforestry, rural development, home economics, horticulture, and other subjects should embrace this species as a potential tool for helping achieve their individual goals."[34]

Uses. The most important product of the baobab tree is probably its edible leaves. They are high in protein (5 to 17 percent dry weight) and very high in vitamin A and other important vitamins and minerals. Interestingly, baobab leaves are a popular vegetable only in West Africa. The baobabs that grow in southern and eastern Africa (a variation on the same species) have fuzzy leaves that are less enjoyable to eat.[35]

Baobab produces large fruits in a woody shell. Inside is a dry, chalky flesh that is 5 percent protein—which is quite high for a fruit—and very high in vitamins. A sour, gingery traditional African beverage from the fruits of the baobab tree has recently broken into the global market; the fruit can also be dried in cakes and can be stored for years as a safeguard against hard times. The woody shells enclose small nuts, which are roughly 33 percent protein and 30 percent oil.[36]

Food is only one of many uses for this remarkable tree. The leaves are used for livestock fodder. The bark is used for fiber, the roots produce a red dye, and the shade the trees provide is welcome to people and animals. The frequently hollow trunks are used as housing, bus stops, and even jails and bars. These cavities can store as much as 38,000 liters (10,000 gallons) of water, which through some property of the trees stays clean indefinitely.[37]

Harvest and processing. Most baobab leaves are wild-harvested, but increasingly the trees are cultivated for leaf production. Dried in the shade, the leaves preserve their vitamins and offer sustenance through the dry season. The seeds are roasted and eaten like peanuts, ground and used in soups, or pressed for oil.[38]

Carbon farming applications. Baobabs are a popular component of the parkland agroforestry system in West and central Africa, where trees are scattered through annual croplands. There they are found alongside such fruits as tamarind, the nitrogen-fixing apple-ring acacia, and perennial staples such as shea, African locust bean, and balanites. The trees are increasingly planted as fodder banks. They could also conceivably be grown for the fruits and seeds as a woody agriculture crop.[39]

Baobabs are at home in the semi-arid lowland tropics. In more humid climates they produce leaves very well, but fruit production suffers. It takes 8 to 10 years for fruit to set, and 30 until they bear fully. It even takes as much as six years until significant leaf harvests are possible. After that, however, they frequently live for hundreds of years. In recent decades there have been efforts to identify superior fruiting and vegetable forms. Five thousand grafted trees with superior fruit were recently planted in Mali.

Coppice blocks (also known as human fodder banks) are increasingly planted for commercial leaf production. Almost all wild baobab trees are claimed as the "tree tenure" of a nearby family who harvests their fruit, leaves, or other products.[40] That is to say, though these families do not own the land, they have

FIGURE 4.7. The remarkable, multifunctional baobab tree. Photograph courtesy of Brock Dolman.

the more or less exclusive right to harvest and manage those particular trees.

Crop development. There are seven other species in the genus *Adansonia*, mostly in Madagascar, with one in Australia. Researchers know little about their uses. Surely a tree with such wide potential human impact should be a major focus of carbon farming efforts. Indeed, momentum is growing for a baobab movement across Africa, to plant this remarkable tree far and wide. With the development and distribution of improved fruit, nut, and leaf varieties, this remarkable tree on which humanity has depended for centuries may take its rightful place in Africa and semi-arid tropics around the world.

CHAPTER FIVE

A Multifunctional Solution

The fact that carbon farming has the potential to sequester more than 100 billion tons of carbon is evidence enough of its merit for global implementation and funding for new projects and research. However, the beauty of carbon farming, unlike many other so-called solutions to the climate crisis—which would surely have potentially catastrophic unintended consequences—is that it also offers elegant and proven solutions to many of the other 21st-century challenges we face. This isn't to say that carbon farming in practice will always be without unintended consequences of its own. But by design, definition, and necessity, carbon farming is—and should be—multifunctional. It draws from agricultural traditions and frameworks—such as permaculture—that recognize the value of multifunctionality.

Permaculture arose in Australia in the 1970s as a movement with a distinctive worldview, utilizing a design system and a best-practice framework.[1] It was based on an early-20th-century movement for permanent agriculture, perhaps best exemplified by J. Russell Smith's seminal *Tree Crops: A Permanent Agriculture*, which remains an important work on temperate-climate perennial farming systems.[2] Though permaculture itself has developed few if any new practices, it uses the criteria of ecosystem mimicry and system optimization (agricultural ergonomics) to evaluate practices and assemble a toolkit that includes many of the species and practices profiled in this book.[3]

Permaculture researchers Rafter Ferguson and Sarah Taylor Lovell propose that "if it were possible to distill the agroecological content of the permaculture literature into a single thesis, it might appear in this way: with systematic site design, emphasizing diversity at multiple scales, integrated water management, and access to global germplasm, we can increase the productivity demonstrated by heritage agroecosystems [traditional ecological farming systems such as homegardens]—especially labor productivity—while retaining their most desirable attributes of sustainability and multifunctionality."[4]

Permaculture is not without flaws, including a tendency toward distrust of and disconnection from science, an overemphasis on simple solutions, an underestimation of the difficulty of managing complex integrated systems, and an overemphasis on itinerant "permaculture celebrity" teachers (including myself). But I also find permaculture to be an essential framework for approaching the world's challenges. The idea that "humans can meet their needs while increasing ecosystem health"[5] is a remarkably hopeful one, especially in contrast with the bad-news approach of many environmental movements. And it offers a helpful framework and insight into carbon farming and how we might most effectively focus our energies and attention.

Permaculture is based on a set of design principles, including two that are related to multifunctionality. The first is that "each element performs many functions."[6] In the context of carbon farming, this means that the strategies we use to sequester carbon should also perform other functions, such as producing food and stabilizing slopes. Indeed, as long as humanity is investing a titanic effort in mitigating climate change (which we are not doing yet, but must), we should derive as many benefits from it as possible. The second principle is that "each important function is supported by many elements."[7] This means the critical function of carbon sequestration should be supported by a diversity of tactics, including a diverse and productive mosaic of perennial farming systems and perennial crops. This principle builds resilience into a system with a network of interwoven relationships, rather than a putting-all-your-eggs-in-one-basket approach to complex problem solving.

Since the late 1990s, the wider world outside the permaculture community has also become interested in multifunctional agriculture, although rather than the complex and integrated vision of permaculture, "multifunctional agriculture" has a more general goal of valuing the positive effects of ecological agriculture beyond food production. This acknowledges "that agricultural activity beyond its role of producing food and fibre may also have several other functions such as renewable natural resources management, landscape and biodiversity conservation, and contribution to the socio-economic viability of rural areas."[8] Multifunctional agriculture has attracted attention from scientists and policy makers and become part of the international discussion about climate, agriculture, and land use.

Regardless of how we arrive at an appreciation of multifunctionality, it is an essential idea for addressing massive 21st-century challenges such as climate change, food production, social justice, and restoring ecosystem health. From a multifunctional perspective, carbon farming is a home run (for reasons we'll explore later in this chapter). Unfortunately, agricultural solutions are a small part of the climate change mitigation conversation today, much of which is dominated by clean energy conversion, which I consider essential, and "geoengineering," which is more problematic.

Geoengineering Compared

Geoengineering, or climate engineering, is a still-theoretical set of practices that aim to mitigate climate change through massive interventions in planetary systems at the global scale. Some scientists and policy makers are looking seriously at climate engineering techniques because governments have so thoroughly failed to reduce emissions in a significant way.[9]

Some climate engineering strategies aim to manipulate albedo and reflect light away from the earth to reduce heating. These "solar radiation management" strategies include giant orbiting solar mirrors to reflect the sun's light away from the earth and increasing reflective cloud cover through aerial sulfur application or volatilizing seawater on a massive scale.[10] Other strategies seek to harness the power of ocean algae to sequester carbon. These include alkalinizing the oceans to reverse acidification, massive application of iron filings, or installing gargantuan pumps to bring up nutrient-rich bottom waters to encourage growth of carbon-sequestering ocean algae.[11] Scientists are also looking at capturing carbon from coal-burning power plants and burying it underground or in the ocean floor.[12]

Climate leader Al Gore's opinion of geoengineering is that it is "insane, utterly mad and delusional in the extreme."[13] A recent study found that even were the strategies to be done at as large a scale as possible and run for years, their mitigation impacts would be small. Some would likely have extreme consequences and, once initiated, could not be stopped without triggering catastrophic climate change.[14] Worst of all, some geoengineering strategies have terrible impacts on the world's most vulnerable people. Aerial spraying of sulfur dioxide, for example, could fundamentally alter the summer monsoons that billions of people in Africa and Asia depend on for food.[15]

Geoengineering fails to pass the multifunctionality test. It does not feed people, does not increase habitat or rainwater infiltration, and does not shrink the gap between rich and poor. It would be hard to come up with a better example of a strategy that does not address climate justice. Instead of taking mitigation funds supplied by the wealthy nations that are most responsible for climate change and spending them in the developing nations that are the least responsible but suffer the most from it, geoengineering projects would enrich the already-wealthy companies that design, implement, and manage them.

That said, we might end up needing one or two of these strategies in order to keep a livable climate. But they should be a distant seventh or eighth line of defense after reducing emissions, switching to clean energy, overhauling transportation, reducing consumption, reforestation, and, of course, carbon farming. Meanwhile, geoengineering attracts far more media attention than the humble task of planting useful trees.

In a 2010 article on climate change and food security, Rattan Lal concludes: "In comparison with the engineering techniques of geologic and oceanic sequestration, C sequestration in terrestrial ecosystems has numerous co-benefits, such as increasing net primary productivity, advancing food security, improving the quality and quantity of water resources, enhancing biodiversity, and others."[16] With the right funding, it might also reduce the gap between rich and poor nations and address climate justice.

Ecosystem Services

When people talk about "ecosystem services," it is usually in the context of assigning a value to the many benefits ecosystems provide to human "well-being, health, livelihoods, and survival."[17] These benefits include such things as the oxygen that plants produce and that we breathe, and the soil that plants, fungi, and bacteria help to build and that we grow our food in. A recent study estimated the value of 2011 global ecosystem services as $142 trillion 2014 US dollars, twice 2012's gross world product of $71 trillion.[18] From the first ecosystem service assessment in 1997 until the 2011 assessment, environmental damage resulted in a loss of $23 trillion in ecosystem services.[19]

The idea of ecosystem services is controversial. Critiques include anthropocentrism, reduction of nature to purely economic value, commodification of nature, and an assumption that all nature's services are positive—as opposed to habitat for disease-bearing mosquitoes, for example.[20] Nonetheless, the idea has gained traction and is currently an important driver of policy and international discussion.

We already know that carbon farming systems provide the critical ecosystem service of carbon sequestration, but they also provide a number of other "services." Complex perennial vegetation such as multistrata agroforestry systems, for example, appear to do a better job than herbaceous systems at transpiring water to create clouds to bring rain downwind.[21]

BIODIVERSITY

P. K. Nair considers homegardens to have the highest standard of biodiversity of any human-created ecosystem.[22] Most agroforestry systems are less diverse than homegardens, but they nevertheless provide an important anchor for biodiversity. Agroforestry systems can directly provide habitat for many organisms, from plants and animals to fungi and bacteria. These plantings can provide critical corridors to create connection between intact natural habitat fragments. They also reduce the clearing of nearby habitats for firewood and fresh farmland. By maintaining good soil and water quality they ensure that habitats downstream remain in good health.[23] Converting annual cropland to perennial agriculture can result in a marked increase in habitat for wildlife.[24]

Biodiversity benefits are not limited to agroforestry. Many practices that sequester carbon in pastures also increase biodiversity.[25] So do hedgerows in annual crop landscapes.[26] Biodiversity is not just useful for its own sake. Habitat can bring pollinators (essential for production of most crops) and beneficial insects and other organisms that control pests.

WATER QUALITY

Agroforestry systems can help to reduce flooding by increasing rainwater infiltration. This process also recharges groundwater.[27] (Trees can also compete with adjacent crops for water and can have other negative effects on water.)

One of the worst impacts conventional agriculture has is its impact on downstream water quality. This is often the result of soil erosion and nutrient leaching from annual cropping systems and poor livestock management. Perennial farming systems result in less nutrient leaching.[28] In fact, erosion and nitrogen leaching in perennial crops are less than 5 percent of those in annual crops.[29] One reason for this is that perennial crops such as trees are present and using nutrients year-round, while annuals are only actively using nutrients for a small fraction of the year.

PRESERVATION OF NATURAL FOREST AND COMBATING DESERTIFICATION

As we discussed in chapter 4, agroforestry in rural tropical regions can reduce pressure on natural forests for firewood and land clearing. In such cases, every hectare of agroforestry can prevent the deforestation of 5 to 20 hectares of natural forest.[30]

Desertification is "the process of loss of soil fertility in the semi-arid and in the dry sub-humid regions of the world."[31] Many perennial crops are well adapted to these conditions and can play a role in combating desertification while providing the basis for human livelihood.[32] Large-scale anti-desertification agroforestry projects have been undertaken, particularly in western China.

Agroecosystem Benefits

Carbon farming also has multiple beneficial impacts on farms where agroforestry practices and perennial crops are integrated into the farm system. These benefits can help maintain and improve the productivity of the farm over time.

I'm aware of almost no carbon farming techniques that were developed for the purpose of carbon sequestration. They have emerged for reasons such as building soil fertility, erosion-proofing farms, and providing diverse crops as insurance in case the one crop fails.

SOIL IMPROVEMENT

All carbon farming practices increase soil organic matter. This includes improved annual cropping,

RESTORING DEGRADED LANDS THROUGH REGENERATIVE AGRICULTURE

"The potential of agroforestry to reduce the hazards of soil erosion . . . and desertification, as well as to rehabilitate salt-affected lands . . . has been well recognized."
—*P. K. Nair*[37]

Land degradation is a "decline in land quality caused by human activity."[38] Degraded land has diminished carbon above- and belowground, which has off-gassed as carbon dioxide. Unsustainable farming practices such as tillage, poorly managed grazing, and poorly managed dryland irrigation can result in problems such as compaction, erosion, desertification, acidification, fertility depletion, loss of organic matter, salinization, and crusting. "Under stress, fragile lands degrade to a new steady-state and the altered state is unfavorable to plant growth and less capable of performing environmental regulatory functions."[39]

The amount of degraded land worldwide is between 1.9 and 3.6 billion hectares (4.7 and 8.9 billion acres), which is 32 to 73 percent of the world's 4.9 billion hectares (12.1 billion acres) of farmland (including pasture).[40] Roughly 70 percent of grasslands are degraded.[41] Because arid and semi-arid lands are the most vulnerable, it means that 33 percent of the planet's land surface is vulnerable to desertification, or the degradation of drylands. Sixty-seven percent of Australia's cropland is degraded.[42]

Globally, we lose 75 billion tons of farm soil to erosion annually, at an estimated cost of US $400 billion.[43] About 950 million hectares (2.4 billion acres) of farmland worldwide have become salty due to poor irrigation practices in arid and semi-arid regions. A remarkable 33 percent of all potentially arable lands in the world have been affected by salinization.[44]

What is the impact of degradation on the yield of food and other products we depend on from our farms? In Africa reduced yields due to land degradation range from 2 to 40 percent, with an average of 8.2 percent across the continent. In South Asia reduced yields due to land degradation are estimated at 36 million tons of grain per year.[45] Soil compaction has resulted in a 20 to 50 percent reduction in yield in parts of Europe and North America.[46]

Degraded land has lost much of its carbon—and some carbon farming systems can restore that carbon while bringing the land back into productivity. Restoring degraded land therefore represents a tremendous opportunity to sequester and store carbon while also

livestock, and perennial systems. More soil organic matter means higher fertility, better drought resistance, less vulnerability to erosion, and less loss of nutrients via leaching.[33] Carbon farming systems from conservation agriculture to perennial crops also reduce or eliminate tillage of the soil. The elimination or reduction of tillage reduces erosion and increases the complexity and diversity of life in the soil, notably of the highly beneficial arbuscular mycorrhizal (AM) fungi. These mycorrhizae help plants use nutrients more efficiently by capturing and transferring nutrients from plant to plant.[34]

Perennial crops have many positive impacts on soils. Overall they can improve the structure of the soil, reduce compaction, improve water cycling and capture, and boost plant growth. Deep-rooted perennial

crops can increase the water storage capacity of soils, which helps to reduce salinity.[35] It's also thought that perennial crops and agroforestry systems utilize water and nutrients more efficiently than annual crops.[36] As I mentioned in the passage above on water quality, perennial root systems are dramatically better at preventing the loss of nutrients through leaching than are the root systems of annual crops.

Agroforestry systems benefit soils through the addition of nitrogen from nitrogen-fixing plants and the addition of organic matter from leaves that fall from woody plants and "chop-and-drop" coppicing and mulching. Even strips of perennials intercropped with annuals can have a positive effect on preventing nutrient losses through leaching. Agroforestry systems can reduce soil acidity and salinity. Like perennial crops,

producing food and meeting human needs, though some lands may be irreversibly degraded.[47]

Regenerative agriculture is a set of practices that are intended to restore degraded lands and improve the health of soils and ecosystems while producing a yield.

FIGURE 5.1. Farmers in North Korea work to restore degraded hillsides used for annual cropping by adding contour strips of perennial legumes and tree crops such as Asian pears and chestnuts. Photograph courtesy of the World Agroforestry Centre.

Many practices can accomplish this goal, including no-till and organic annual cropping, managed grazing, agroforestry, and perennial crops, although not all of them would be appropriate on any given site, with variations in soil and climate.[48] Farmers need the freedom to determine which practices are best for them, although as Courtney White remarks in his excellent book *Grass, Soil, Hope,* "You can't increase soil carbon with a practice that degrades the land."[49]

Perennial crops and agroforestry may be especially well suited to address the unique challenge of soil salinization. Perennial crops increase soil water storage and can thus help to flush salt out of soils.[50] Many perennial crops—including staple crops such as mesquite—are halophytes, or salt-loving species, and can produce well in salty areas where it is impossible to grow conventional crops. (See table 10.2 in chapter 10 for a list of perennial halophyte crops.) In "The coming of age of agroforestry," P. K. Nair states that "decades of research in India and other places where soil salinization is a major problem has shown that planting salt-tolerant trees on salt-affected soils is one of the best ways of rehabilitating such degraded lands."[51] Nair lists a number of agroforestry tree species that can be used to reverse desertification and restore saline soils.

agroforestry has a positive impact on the beneficial life in the soil.[52]

Even in relatively level soils, tillage can cause erosion. Carbon farming systems help to reduce the loss of soil nutrients from erosion. For example, agroforestry trees bring nutrients from deep in the soil, including those that might otherwise be lost from leaching, and make them available to crops.[53]

PREVENTING EROSION AND STABILIZING SLOPES

Not all carbon farming systems prevent erosion, but many do. One study found that multistory tree gardens actually had less erosion than a natural rain forest.[54] Annual cropping systems such as cover cropping and no-till were developed to prevent erosion.

Agroforestry systems that prevent or reduce erosion on slopes include plantation polycultures, homegardens and multistrata systems, alley cropping and contour hedgerows, and silvopasture. In addition, windbreaks can reduce wind erosion on annual cropland regardless of slope.[55]

REDUCED NEED FOR EXTERNAL INPUTS

The crop diversity of agroforestry polycultures provides pest protection.[56] The protection increases with greater diversity; homegarden and multistrata agroforests typically have few pest problems. Perennial crop monocultures require fewer fossil fuel inputs than annual cropping systems, and agroforestry systems fewer still, with multistrata systems needing least of all.[57] The US National Research Council notes that perennial crops have reduced needs for fuel, fertilizer, and pesticides.[58] And as we discussed in chapter 2, carbon farming systems include important tools for reducing agricultural fossil fuel emissions.

WATER RETENTION AND MICROCLIMATE CREATION

Carbon farming systems help to reduce runoff and infiltrate groundwater, making more available for use on-farm. Practices noted for this effect include managed grazing and agroforestry.[59] The increased organic matter content that characterizes carbon farming practices helps to improve the water-holding capacity of agricultural soils.[60] In drier climates competition between agroforestry trees and annual crops can be a concern.

Shade from crop and agroforestry support trees is important for livestock, humans, and shade crops, particularly as the planet grows warmer. Windbreaks, shelterbelts, and green corrals can protect crops, people, and livestock from the effects of winds and dust.[61]

RESILIENCE AND RISK MANAGEMENT

Merriam-Webster's Dictionary defines *resilience* as "an ability to recover from or easily adjust to misfortune or change; the ability to become strong, healthy, or successful again after something bad happens."[62] Resilient agriculture can bounce back to some degree after drought, storms, pest outbreaks, and other disturbances. Resilience is at the heart of climate change adaptation strategies. Carbon-friendly grazing and pasture management, for example, generally makes grasslands more resilient to climate change.[63] Increases in soil organic matter, the backbone of all carbon farming systems, make soils more drought-resistant.[64]

Some perennial crops can continue to produce during drought, for example, when shallow-rooted annual crops fail; some trees will go dormant during a drought but will return to full production the following year if and when the rains resume. Many perennial crops also produce food or fodder during the "hungry season," which occurs at the end of the dry season but before annual crops begin yielding.[65] Similarly, fodder bank species can provide fodder far into the dry season.

Growth that accumulates on trees during healthy and productive seasons provides a resource of fuel and timber that can be drawn on in drought years. Agroforestry can also provide surpluses and diversified sources of income to draw on during hard times.[66] In general, crop and livestock diversification provides a buffer against crop failure of any individual product in any given year, whether due to drought, storms, or pest outbreaks. Diversification also protects farmers against sudden drops in market prices for specific products.[67]

YIELDS

Increasing organic matter carbon increases crop yields.[68] Chapter 6 profiles many annual cropping systems with improved yields. Many improved grazing systems sequester carbon while increasing livestock production.[69] Intensive silvopasture increases livestock production by an incredible factor of 2 to 10 times on a per-hectare basis.[70]

You may have heard some talk about the trade-off between perennial crops and their ecosystem benefits versus their lower yields.[71] Many tropical perennial staples yield as well as or better than annual staples. Outside of the tropics, yields are typically lower than annuals. These lower yields are not necessarily a permanent limitation of cold-climate crops because breeding work is ongoing.[72] I've included yield comparisons at the beginning of most chapters in parts 3 and 4, and a lengthy discussion of yield issues in chapter 11. Some agroforestry systems increase yields of the adjacent annual crops or grazing livestock, while others reduce them. See the profiles of individual systems in part 2 for details.

Socioeconomic Benefits

Perennial crops and agroforestry systems also provide benefits to people—in many cases to the people who are most in need. The social impacts are a powerful argument for the use of climate mitigation funds to support the development of carbon farming systems. I discussed climate justice in chapter 1, but there are also many other important—and related—socioeconomic benefits to carbon farming.

PRODUCTS AND REVENUE

Carbon farming systems can provide an impressive portion of the material needs of humanity. Perennial crops can provide fruits and vegetables, herbs and spices, protein, carbohydrates, fats, and other foods for human sustenance.

The same crops can be fed to livestock to supplement pasture and perennial fodder species. Livestock in turn can provide food, fiber, transportation, and traction, as well as performing useful functions in integrated farming systems.

Carbon farming can provide much more than food. Bamboo alone has more than 1,500 uses, and many other species provide materials for uses ranging from basketry and lumber to biobased plastics and composites. Plants can provide medicines, solvents, feedstocks for the chemical industry, and more. They can also provide energy, if not nearly enough to replace the fossil fuels we use today.

Agroforestry systems do more than produce these important products; they help improve the long-term productivity of farms by building soil, making the production of annual crops more sustainable. Agroforestry practices can reduce reliance on fertilizers and pesticides and reduce carbon emissions.

All of these products and services add up to increased income for farmers. This comes not just from sales but greater self-sufficiency. Regional-scale processing facilities can also drive economic development through localized eco-industry.

THE EMPOWERMENT OF FEMALE FARMERS

Women produce half the world's food. In the developing world, the number is even higher—between 60 and 80 percent. Somehow they manage this while receiving only 7 percent of investment and aid.[73] A 2011 report by Evelyne Kiptot and Steven Franzel titled *Gender and Agroforestry in Africa: Are Women Participating?* states, "Women . . . remain disadvantaged in the agricultural sector due to cultural, socio-economic and sociological factors. Such factors include ownership and access to resources, land tenure systems, access to education and extension services, among many others."[74]

The report concludes that "if women had access to the same resources (i.e., education, farm inputs and labor) as men, food production would be boosted by 10 to 20 percent."[75]

Agroforestry follows the same pattern as the rest of agriculture: Most of the people who manage homegardens around the world are women.[76] And women perform most agroforestry labor in Africa.[77] However, as in the rest of agriculture, female agroforestry farmers have less access to decision-making power, land, agroforestry information, and markets.[78]

FIGURE 5.2. Women carrying firewood in Africa. Photograph courtesy of the World Agroforestry Centre.

That said, agroforestry offers many benefits to the world's female farmers. It requires fewer inputs, is less expensive, requires less labor, and provides multiple products and services on the farm.[79] Table 5.1 shows the labor, land, and capital benefits of agroforestry for African women who have little of each to spare.

Firewood provides an excellent example of the benefit to women from agroforestry. In regions where firewood is the primary fuel for cooking, it is collected by women. Many agroforestry practices produce fuelwood on the farm, whether from intentional firewood plantings or agroforestry practices such as hedgerow intercropping, fodder banks, and improved fallows. *Gender and Agroforestry in Africa* determined that "women who collect fuel wood for cooking away from the farm spend on average 130 hours per year, as

compared to only 36 hours spent by those who harvest firewood from their own farms."[80]

Agroforestry policy makers and institutions need to focus particularly on the needs of women. Development of some new appropriate technologies could dramatically reduce women's labor in agroforestry operations. New-crop domestication efforts should include projects led by female farmers to domesticate the indigenous crops important to women.

FOOD SECURITY

The World Health Organization defines *food security* as the condition that exists "when all people at all times have access to sufficient, safe, nutritious food to maintain a healthy and active life."[81] Carbon farming can have many positive impacts on food security. The

TABLE 5.1. Agroforestry as a Low-Cost, Low-Labor, Limited-Land Option for Women in Africa

	Land	Labor	Capital
Wild-collect off-farm	None	High	None
Purchase	None	None	High
Grow annual crops	High	High	Low or none
Practice agroforestry	Low	Low[a]	Low or none

Source: Adapted from Kiptot and Franzel, *Gender and Agroforestry in Africa*, 3.
Note: Female farmers in Africa have less land, time, and capital than men.
Agroforestry provides women with a good alternative to wild-harvesting, purchasing, or growing annual crops for meeting a family's needs.
[a] Agroforestry labor is medium in cases where a nursery is required.

improved yields associated with increased soil organic matter can have a significant food security impact.

At issue is not only overall production, but seasonal availability. Many perennial crops have a longer or complementary season compared with annuals. Higher diversity of crops offers not only protection in case of crop failure but a more diverse and nutritious diet and additional income opportunities.[82] Agroforestry systems are also important in providing firewood and charcoal, which roughly two billion people depend on for cooking their daily meals.[83]

Carbon-sequestering grazing practices are also essential for food security. First, they often improve livestock yields.[84] This is crucial to the 880 million rural people who live on less than $1 a day, 70 percent of whom depend on livestock for their livelihood.[85] The ability of carbon farming systems to adapt to a changing climate may prove to be an important contributor to food security in coming years. Carbon farming offers the opportunity for a win–win solution that addresses food security and climate change at the same time.

FOOD SOVEREIGNTY

Food is about more than having enough calories and nutrition. Our food system, indeed our entire civilization and global economy, is fraught with social and environmental problems. Food sovereignty is "the right of each nation or region to maintain and develop their capacity to produce basic food crops with the corresponding productive and cultural diversity . . . [It] emphasizes farmers' access to land, seeds, and water while focusing on local autonomy, local markets, local

production–consumption cycles, energy and technological sovereignty, and farmer-to-farmer networks."[86]

Via Campesina, an international food sovereignty movement, has roughly 300 million members.[87] Mostly small farmers in the tropics, they embody the remarkable success of agroecology and the strength of traditional (non-chemical) farming practices. In fact, food sovereignty is already a movement that encourages and practices carbon-sequestering agriculture. For millennia, agroforestry and perennial crops have been part of the traditional practices the movement fights for.[88]

These farmers and their allies would benefit, however, from policy changes that would incentivize their sustainable practices instead of the reverse. Funds for research, breeding, education, and establishment would help Via Campesina and other food sovereignty efforts expand their important work further.

CLIMATE JUSTICE AND SHRINKING THE WEALTH GAP

Climate justice is the notion that wealthy countries have done the most to cause climate change, but poorer countries are the most affected by its impacts. Thus mitigation strategies that help to level the playing field between rich and poor nations help to address climate justice. Using funds from wealthy nations to support development of perennial farming systems in the tropics is an application of this idea.

Drawbacks, Trade-Offs, and Risks

Carbon farming is not a silver bullet. It brings its own costs. As decisions are made about changing land use for mitigation, adaptation, and development, there are some potentially problematic consequences that must be considered and addressed.

Perhaps the most serious risk is that carbon farming may be a strategy with an expiration date. Climate change itself is an enemy of carbon farming. The IPCC projects that by the time we've reached 2.5°C (4.5°F), perhaps 15 percent of the world's soil and biomass will become net emitters of carbon. By the time 4°C (7.2°F) of warming has occurred, this could increase to 40 percent![89] This would surely limit, and perhaps reverse, the

PART TWO

A Global Toolkit of Practices and Species

CHAPTER SIX

Annual Cropping Systems

Annual crops are grown on roughly 1.2 billion hectares (3 billion acres) globally, accounting for 89 percent of all cropland.[1] Thus it is of utmost importance to transition annual cropping from a net emitter of carbon (as it is today) to a net sequesterer of carbon even though the ideal scenario, in my mind, would be a global transition to multifunctional, agroforestry-based perennial staple and industrial cropping systems. Since that's still a long way off, we're fortunate that annual crop farmers have a diverse palette of carbon farming practices to choose from in the meantime. Some of these are already widely practiced, such as conservation agriculture, which is practiced on an impressive 124 million hectares (306 million acres), 10 percent of world annual cropland today.[2] These practices tend to sequester carbon at very low to low rates, with some promising exceptions, such as regenerative organic. Soil carbon stocks, representing lifetime accumulation of carbon, are modest at 50 t/ha or less (sometimes negative!).[3] This chapter does not attempt to comprehensively review all carbon-friendly annual cropping practices; instead, it gives a broad overview of a place to start.

Annual crops are also intercropped with trees and other perennials. This integration can be randomly scattered, consist of rows or strips of woody plants, be seasonal in nature, or involve intercropping perennials with annuals in other ways. Perennial–annual integration includes agroforestry, but it also includes practices that use non-woody perennials (such as perennial grasses) and are therefore not technically agroforestry. (By definition agroforestry must include a woody plant, and some perennial farming systems use herbaceous perennials instead of woody ones. Examples include vetiver contour hedgerows and pasture cropping.) Perennial–annual integration systems of one form or another are currently practiced on an estimated 700 million hectares (1.7 billion acres) of cropland worldwide.[4] Sequestration rates are typically low to medium, with exceptions on either end. Soil carbon stocks are more impressive, up to 150 t/ha in the tropics and 200 t/ha in temperate climates.[5] Perennial grains will likely be excellent options for integrated perennial–annual systems once perennial grains are ready for market.

The practices profiled in this chapter allow us to continue growing the annual crops we know and love while sequestering decent amounts of carbon. This translates to only minimal changes in production, harvest, processing, and diet. Although the sequestration rates are lower compared with some other practices, such as the use of staple tree crops, the ease of adoption by farmers and the vast scale of annual cropland worldwide means the impact could be profound—and certainly far better than the carbon losses associated with standard tillage agriculture.

With the exception of pasture cropping, all of the practices I describe in this chapter allow you to substitute perennial crops for the annual crops. This would result in fully perennial systems, which I describe in chapter 8. For example, bananas are as suitable as a main crop for alley cropping with nitrogen-fixing hedgerows as annual crops are.

The IPCC rates planting of trees, including useful trees on farms, as having a medium potential global impact on carbon. (I propose it could be much higher.)

FIGURE 6.1. Polycultures are cultivated systems in which two or more crops are grown in a given area at the same time. This Guatemalan polyculture features mango, papaya, and moringa as overstory crops with chipilín, sweet potato, and lemongrass as understory crops.

They felt that it would have easy adoption by farmers and is ready for implementation immediately. Agroforestry (meaning the integration of tree with annual crops) was rated as medium mitigation potential, moderate ease of adoption, and ready for widespread implementation.[6] It is also noted as an important adaptation strategy.[7]

Many of these practices defy easy categorization. That said, although there is some overlap among various practices, I've tried to distill out important differences and distinctions as a way for carbon farmers, researchers, and policy makers to make sense of the options that may be most suitable in a given situation and region.

Each profile that I've included in this chapter—and in chapters 7, 8, and 9 as well—includes an overview of the system, along with its historical, cultural, geographic context. Then I evaluate the impact a given system has, or could have, on carbon sequestration, as well as other important factors such as the impact the system has on the yield of annual crops. The profiles in this chapter (and in those to come) are clustered based on similarities. In this chapter the order is: fully annual systems; modern agroforestry systems; protective systems; traditional agroforestry systems; and systems under development.

POLYCULTURES

Polycultures are cultivated systems in which two or more crops are grown in a given area at the same time. Well-designed polycultures maximize mutual benefit, minimize competition, and minimize the amount of management and harvest labor required. For example, in the annual crop polyculture of corn, beans, and squash, it would take almost twice as much land to grow the three crops in monoculture to yield as much as the polyculture yields. When a polyculture yields more than the equivalent land in a monoculture of the same species yields, it is referred to as overyielding. This is an important concept. Yields of individual crops may be higher in the monoculture, but the total system yield may be higher in the overyielding polyculture.[8]

Overyielding polycultures are not limited to the tropics. The société coopérative et participative spécialisée en Agroforesterie (AGROOF) in France has studied yields when rows of timber trees are planted in strip intercropping systems within annual crop fields. They found no immediate reduction in crop yields when the trees were planted, but a gradual decrease of up to 20 percent by year 30 when the timber was harvested. The researchers determined that the optimal spacing was 50 mature trees per hectare (2.5 acres) to allow minimum impact on the crops. The trees benefited from less competition than they would have in plantations or forests. They also benefited from the fertilizers that were applied to adjacent annual crops as well as the tillage of adjacent annual crop fields, which resulted in the trees having deeper roots and thereby improved drought resistance. There were also challenges, including the need to work around the widths of annual crop machinery, potential damage to the trees from equipment, and determining the most appropriate spacing and density of trees. Despite these challenges, the researchers and participating farmers found that 100 hectares of agroforestry yielded as much as 130 to 140 hectares of monoculture trees and annual crops.[9]

Systems with higher diversity and more layers of crops sequester more carbon than monocultures.[10] A Brazilian study reported the remarkable potential of overyielding oil palm polycultures. Oil palm was intercropped with fruit and timber trees, cacao, nitrogen-fixing trees and ground covers, and productive vines such as passion fruit and black pepper. The oil palm polyculture sequestered more carbon than adjacent natural forest regrowth. And yields of oil were 6.4 to 8.7 t/ha/yr of oil compared with a regional average for 5 t/ha/yr for monoculture oil palm.[11]

Overyielding polycultures are also a critical strategy for agroecological intensification. They allow increased production on existing farmland, thereby reducing the need to clear additional land. This is also an important adaptation strategy since diverse crops will provide at least some yield in all but the worst conditions.[12] In fact, polycultures have been shown to outyield monocultures in droughts.[13]

System Classifications

Carbon farming systems fall into several key categories: Traditional systems go back hundreds or thousands of years. Recent systems have mostly been developed since the 1970s as a response to the environmental crisis. Both traditional and recent systems can be intensive, with high yields in a relatively small space, more inputs, more labor and/or management, and often on a smaller scale; or they can be extensive, with relatively lower yields, fewer inputs, less labor and/or management, and often on a larger scale. Most traditional systems are extensive, although some traditional systems such as homegardens are highly intensive. Among recent systems, there is a stronger tendency toward intensivity. Although carbon farming systems tend to defy easy classification and these system classifications are more or less subjective, these classifications still nonetheless represent a useful way to evaluate a given carbon farming strategy for a given situation. It is also of interest to note the scale at which these techniques are currently practiced, as this often indicates a level of readiness for wider implementation.

TABLE 6.1. Status and Scale of Selected Improved Annual and Integrated Perennial–Annual Systems

System	Example	Recent or Traditional	Scale	Region
All tree intercropping systems	Irregular intercropping, FMNR, strip intercropping, alley cropping, protective systems, etc.	Mixed	700 million ha[a]	Global
Protective systems	Windbreaks, shelterbelts, hedgerows, riparian buffers, etc.	Mixed	300 million ha protected by these systems[b]	Global
Improved annual cropping systems	Conservation agriculture	Since the 1970s	124 million ha[c]	Global
Improved annual cropping systems	No-till	Since the 1970s	100 million ha[d]	Global
Successional intercropping	*Taungya*: timber trees with annuals at establishment	Since the 1800s	"Most of" 54 million ha[e]	Asia and Africa
Complex swidden		Traditional	25.2–39.6 million ha[f]	Southeast Asia: Thailand, Laos, Cambodia, Vietnam, Myanmar, Malaysia, Indonesia
Irregular intercropping	Shea nut parkland	Traditional	22.9 million ha[g]	Semi-arid West Africa
Improved annual cropping systems	System of Rice Intensification	Since the 1960s	4–5 million farmers[h]	Global
Improved annual cropping systems	Organic	Since the 1970s	6.3 million ha[i]	Global
Irregular intercropping	Parkland	Traditional	5.1 million ha[j]	Mali
Farmer-managed natural regeneration	"Wild" trees permitted to sprout in annual crop fields, mostly *Faidherbia*	Recent decades	4.8 million ha[k]	Niger
Strip intercropping	*Paulownia* intercropping	Recent	3 million ha[l]	China
Irregular intercropping	*Streuobst*: fruit and nut trees irregularly intercropped with annual crops	Traditional (and at risk)	1 million ha[m]	Europe
Windbreaks	Windbreaks protecting annual crops	20th century	1 million ha[n]	Canada and United States
Strip intercropping	Olive intercropping	Traditional	650,000 ha[o]	Greece
Faidherbia evergreen agriculture	Nitrogen-fixing trees with annual crops	Recent decades	300,000 ha[p]	Zambia
Strip intercropping	Poplar intercropping	Since the 1970s	280,000 ha[q]	Northern India
Strip intercropping	Jujube intercropping	Since the 1960s	200,000 ha[r]	Semi-arid temperate China
Strip intercropping	Intercropped vineyards	Traditional	153,000 ha[s]	Sicily
Irregular intercropping	*Quesungual*: irregular intercropping of legume and timber trees with annual crops	Recent decades	60,000 ha[t]	Central American dry-season tropics
Living terrace edges	Woody crops at edges of constructed terraces	Recent decades	52,200 ha[u]	Cold, dry regions of western China

Note: These are just a few systems—not necessarily the largest or most representative—for which data on the scale at which they are practiced is available. Some practices are listed more than once here; an individual practice such as "Windbreaks" is also included under "Protective systems" and "All tree intercropping systems." The systems in this table are presented in descending order of scale. "Region" does not necessarily indicate the only place where a system is practiced, but rather the region described in the study that is quoted.

[a] Nair, "Climate change mitigation," 47.

[b] Ibid., 31.

[c] Farooq and Siddique, "Conservation agriculture," 8.

[d] Lal, "Managing soils and ecosystems for mitigating anthropogenic carbon emissions and advancing global food security," 710.

[e] Cubbage, "Global timber investments, wood costs, regulation, and risk"; Nair, *An Introduction to Agroforestry*, 78.

[f] Schmidt-Vogt et al., "An assessment of trends in the extent of swidden in Southeast Asia," 272.

[g] Boffa, *Agroforestry Parklands in Sub-Saharan Africa*.

[h] Vidal, "India's rice revolution."

[i] IFOAM, *The World of Organic Agriculture*, 61.

[j] Zomer et al., *Trees on Farm*, 3.

[k] Garrity et al., "Evergreen agriculture," 205.

[l] Li, *Agro-Ecological Farming Systems in China*, 107.

[m] Mosquera-Losada et al., "Definitions and components of agroforestry practices in Europe," 9.

[n] Williams et al., "Agroforestry in North America and its role in farming systems," 21–22.

[o] Crawford, "Silvoarable systems in Europe," 7.

[p] Garrity et al., "Evergreen agriculture," 201.

[q] Zomer et al., *Trees and Water*, 29.

[r] Li, *Agro-Ecological Farming Systems in China*, 150.

[s] McAdam et al., "Classifications and functions of agroforestry systems in Europe," 28.

[t] Cherrett, "Quesungual agroforestry."

[u] Li, *Agro-Ecological Farming Systems in China*, 283–88.

Conservation Agriculture

Conservation agriculture was developed in the 1970s in response to an erosion crisis resulting from tillage-based farming. It is a suite of practices including reduced tillage ("no-till"), cover cropping, and crop rotation. Other practices such as nutrient management are often included in definitions of conservation agriculture as well.

Conservation agriculture has been very widely adopted by conventional farmers, accounting for more than 124 million hectares (306 million acres) worldwide. This represents a very impressive adoption rate.[14]

Unfortunately, in the Americas and Europe, conservation agriculture tends to be highly chemical intensive. A preceding crop or cover crop is typically "burned down" with herbicide, leaving crop residues and undisturbed soil. The farmer then uses a special no-till planter that digs narrow furrows into which he or she sows seeds or transplants, often herbicide-resistant GMO crops such as cotton or corn. In this way, soils can go without tilling for many years.

This is a challenging issue for me as someone coming from an organic farming background. Herbicide-based farming is not particularly ecologically sound, and I do not, by any means, support it, nor am I in favor of using GMO crops. However, if the choice is between conventional conservation agriculture and conventional tillage-based farming, I'd opt for the option that at least sequesters carbon, especially if it is a transitional step toward a more agroecological approach.

Fortunately, there are no-till practices that don't use herbicides, and this is the direction we need to be moving in. No-till practices are widely used by African smallholders with non-mechanized operations, including as part of the evergreen agriculture system described on pages 75–77.[15] The Rodale Institute has spent several decades developing an organic no-till production system at their research farm in Pennsylvania. In 2013 they began trialing their system on other farms. Many farmers found the precise timing requirements a challenge.[16] This is a promising effort, however, because once it is successful, it will combine the benefits of no-till with the superior soil carbon content of organic production for mechanized farms, as well as reduced reliance on herbicides and GMO crops.

Initially no-till was widely touted as "the" carbon farming solution. However, its climate mitigation impact was overstated at the time. Multiple international review papers show extremely modest sequestration rates from 0 to 0.4 ton per hectare per year. A recent review estimated 0.3 t/ha/yr.[17] The more fully integrated approach of conservation agriculture sequesters 0.6 t/ha according to a recent global review.[18]

Conservation agriculture does, however, have other benefits. It is promising as an adaptation strategy because it increases the water-holding capacity of most soils, allowing for better drought resistance. It reduces erosion

FIGURE 6.2. Conservation agriculture. No-till soybeans growing in the residue of a harvested maize crop. Photograph courtesy of the National Resource Conservation Service of the US Department of Agriculture.

FIGURE 6.3. Jeff Moyer using Rodale's organic no-till roller-crimper. Organic no-till is still under development but holds great promise. Photograph courtesy of Rodale Institute.

and emissions that result from plowing; the resulting healthier, more stable soil can sometimes support greater yields.[19] The IPCC rates it as having high global mitigation potential, being easily adoptable by farmers, and ready for rapid implementation.[20] This is because even though the amount of carbon fixed is low, it is easy for farmers to transition their farms, and it is applicable on hundreds of millions of hectares of cropland globally (as are most of the practices discussed in this chapter).

System of Rice Intensification

The System of Rice Intensification (SRI) was developed in Madagascar in the 1980s and is today practiced by ten million farmers, mostly in Asia.[21] (That's likely more land than is in organic annual crops globally, although SRI is less well known.) It involves transplanting young seedlings, individually spacing them in wide rows, avoiding continuous flooding, using a rotating hoe for weeding, and applying compost.[22] Among other benefits, it requires no new seed varieties, chemical fertilizers, or pesticides; it uses only 10 to 20 percent as much seed for planting as paddy rice and 25 to 50 percent as much water.[23] Its goal is to unleash the genetic potential of rice by providing optimal growing conditions rather than the industrial agriculture approach of fancy seeds, fertilizers, and pesticides.

SRI uses few off-farm inputs but has high requirements for labor, organic material, and water management. It is usually practiced on a smallholder scale.[24] Farmers and researchers are working on adapting the SRI approach to other annual grain crops. Collectively these approaches are known as the System of Crop Intensification.[25]

FIGURE 6.4. Cambodian farmer Sin Chhukrath harvesting SRI rice. Few people outside Asia have heard of this agroecological practice that is extremely popular among small-scale rice farmers. Photograph courtesy of Oxfam International.

A comparison of SRI and conventional rice fields in Madagascar found the mean lifetime soil carbon value at 30 centimeters (1 foot) soil depth in SRI fields was more than 150 percent higher than the lifetime soil carbon value in conventional fields.[26] Flooded rice paddies are a major source of methane emissions, and SRI reduces these emissions in addition to sequestering carbon.[27] Because SRI uses less water than paddy rice, it is also an important climate change adaptation practice.[28]

Typical SRI rice yields are 7 to 8 t/ha, which is twice the world average for rice yields. Some growers have apparently attained 15 t/ha, the theoretical maximum potential yield of rice.[29] Yield impacts on other crops are also impressive. A recent publication reports increases of 60 percent for sugarcane, 72 percent for wheat, 56 percent for pulses, 50 percent for oilseeds, and 20 percent for vegetables. The costs of production go up about 50 percent per hectare, but with these increased yields the costs per unit of crop are 40 percent less. Farm incomes typically double.[30] Recent studies have called some of these reported into question, alleging that the studies fail to account for better initial soil fertility and other methodological concerns. By the time you read this, this controversy may be resolved.[31]

Regenerative Organic Annual Crop Production

The FAO defines *organic agriculture* as "a holistic production management system which promotes and enhances agro-ecosystem health, including biodiversity, biological cycles, and soil biological activity. It emphasizes the use of management practices in preference to the use of off-farm inputs, taking into account that regional conditions require locally adapted systems. This is accomplished by using, where possible, agronomic, biological, and mechanical methods, as opposed to using synthetic materials, to fulfil any specific function within the system."[32] Organic farming generally avoids the use of synthetic inputs such as pesticides and fertilizers. Worldwide 1.9 million farmers grow organic annual crops on 6.3 million hectares (15.6 million acres) of farmland.

There are different approaches to organic annual crop production. Some large-scale organic farms operate like industrial agriculture but substitute "natural" inputs for synthetic agrochemicals. Others, classified as "regenerative organic farming systems" by the Rodale Institute, have a complex agroecological approach incorporating cover crops, rotations, intercropping, composting, and so on.[33]

A 2007 paper published in *Renewable Agriculture and Food Systems* found that organic yields are somewhat lower than conventional yields in wealthy countries, but somewhat higher in the Global South.[34] A 2015 meta-analysis of 115 studies comparing more than 1,000 observations (three times larger than the meta-data sets of previously published comparable analyses) found that organic yields are 19.2 percent lower than conventional yields (a smaller gap than previous estimates)—but the incorporation of multi-cropping (including both polycultures and sequential cropping) and crop rotation can cut this gap in half or more. It also found "entirely different effects of crop types and management practices on the yield gap compared with previous studies." The authors concluded that investment into agroecological research could greatly reduce or potentially eliminate this yield gap between conventional and organic farming at least for some crops or regions.[35]

Sequestration rates for organic annual cropland worldwide have been estimated at 0.7 to 1.4 t/ha/yr.[36] However, there have also been some individual reports of incredible sequestration rates for regenerative organic systems—those that employ a diversity of regenerative practices including cover cropping, no-till, the application of manure, and more, in addition to the absence of synthetic pesticides, herbicides, and fertilizers—that are higher than any other annual crop systems I am aware of. These rates range from 2.3 to 6.8 t/ha/yr. In one case, rates were high at 2.7 to 4.1 t/ha/yr the first five years, but over the lifetime of the system averaged 0.9 t/ha as the rate quickly slowed down.[37] Organic and regenerative organic yields can be higher in drought years due to high soil organic matter content, making organic an important adaptation strategy.[38]

ADDITIONAL PRACTICES FOR CARBON SEQUESTRATION WITH ANNUAL CROPS

Although annual crop production is not the ideal carbon farming strategy—in a maximum carbon scenario we'd transition much of our farmland to fully perennial, integrated, multistrata systems—there are many carbon farming techniques that annual crop producers can turn to as a way to improve their existing farming systems without transitioning to something entirely new. According to Rattan Lal in "Abating Climate Change and Feeding the World Through Soil Carbon Sequestration," in general, anything that improves soil organic matter is also desirable in terms of carbon sequestration.[39] The system of conservation agriculture combines many of these practices, such as reduced tillage, cover cropping, and crop rotation. The carbon sequestration rate of conservation agriculture has been estimated at 0.6 t/ha/yr.[40] (Source: Most practices below are from Rattan Lal's "Abating Climate Change and Feeding the World Through Carbon Sequestration.")

Cover cropping. Planting crops for erosion control, weed suppression, nitrogen fixation, and other benefits rather than as a marketable commodity.

Crop rotations. Alternating crops, usually from different families, rather than growing the same crop season after season. Crop rotations that include a period in perennial grassland for grazing or hay are especially beneficial for carbon sequestration.[41]

Nutrient management. Using slow-release fertilizers, applying the optimal amount rather than an excessive amount, and carefully timing the application. Conventional nitrogen fertilizers are energy-intensive to manufacture and emit nitrous oxides when they are over-applied. Most farmers are only 50 percent efficient in their nitrogen applications.[42]

Biological nitrogen fixation. Providing nitrogen to crops with nitrogen-fixing cover crops, intercrops, undersown crops, and other strategies—including agroforestry with nitrogen-fixing trees, which provides higher carbon sequestration than annual cover crops—as an alternative to synthetic fertilizers.

Mulching and residue retention. Leaving crop residues such as stalks and stubble at the soil surface rather

Strip Intercropping

In strip intercropping, tree crops are grown in rows, usually alternating with annual crops. The tree crops typically produce timber or food, although many products are possible. This is often described as a form of alley cropping, although many people distinguish between rows of trees that are crops in their own right (strip intercropping) and rows of coppiced, nitrogen-fixing trees primarily present to support the annual crops (alley cropping). Strip intercropping is the most widely practiced temperate-climate agroforestry system involving annual crops.

In China, *Paulownia* timber trees are intercropped in alternating rows with cereals and other annuals, fruit trees, tea, mushrooms, and medicinal herbs.[46] The cultivated *Paulownia* species are deep-rooted, minimizing competition with adjacent crops, and cast light shade, allowing good growth of annual crops. Their leaves emerge late

FIGURE 6.5. A strip intercropping system of timber trees with young annual grain crops growing between them. Photograph courtesy of société coopérative et participative spécialisée en Agroforesterie (AGROOF).

than burning or tilling them into the soil. This not only improves soil organic matter but also sequesters more carbon than burning or tilling the residues. Mulching with biomass has similar benefits.

Amendments. Using manure and compost (as well as other more controversial amendments such as zeolites and biosolids) to improve soil organic matter and build soil carbon.

Inoculations. Inoculating soil with beneficial organisms such as mycorrhizal fungi.

Improved rice paddy management. In addition to the SRI techniques profiled earlier in this chapter, composting crop residues and draining rice paddies several times per season to reduce methane emissions from fermenting bacteria in standing paddy water. Draining and flooding several times a season does, however, require more water than flooding paddies once.

Reduced tillage. Besides conventional no-till, a diversity of reduced tillage systems exists for large- and small-scale operations. For example, zai holes are small pits dug for a clump of crops as an alternative to tilling an entire field.

Minimizing bare fallows. Not leaving soils bare—sometimes used as a weed control strategy—for extended periods. Mulches and cover crops are good alternatives in many cases.

Improved irrigation. Poor irrigation wastes water, increases erosion, lowers soil organic matter, and can increase soil salt levels in dry regions. Improved irrigation includes proper types and timing of irrigation, such as well-managed drip irrigation.

Conversion to these practices can mean reduced yields for three to five years during the transition. After this period, yields commonly increase by 20 to 120 percent.[43] The IPCC rates rotations and cover crops as having medium potential global impact, being easily adoptable by farmers, and ready for wide implementation in 5 to 10 years; tillage and crop residue retention as high potential, easy adoption, and ready to go; improved rice paddy management as medium to high impact, moderate to easy adoption, and ready for wide use.[44] Improving soil organic matter, cover cropping, and mulching are also important adaptation strategies.[45]

TABLE 6.2. Crop and Timber Yields of 10-Year-Old *Paulownia* Intercrop Systems in China

	Trees/ha	Crop yields/ha compared with monoculture	Timber yields/ha in cubic meters
Paulownia-focused system	200–400	First 3 years normal, years 4–6 summer crop yield normal but fall crop significantly reduced; years 6–10, 20% reduction of summer crops	80–140
Crop-focused system	40–67, 50% thinned in years 6–7	57–70% increase	Unstated
Equal priority	80–133	Years 1–5 yields increase; years 6–10 summer crops higher and fall crops lower; overall yield same as control	36–53

Source: Li, *Agro-Ecological Farming Systems in China*, 152.

in the spring, which provides minimal light competition even with winter annual grains directly beneath them.[47] *Paulownia* timber trees are intercropped on a total of 3 million hectares (7.4 million acres) in China.[48]

Jujube (*Ziziphus jujuba*) is a fruit tree that is tolerant of cold and dry conditions and has become a popular intercrop in western China, where people began growing it in the 1960s to combat damage to crops from wind and natural disasters. Today jujubes are intercropped with vegetables and other annual crops on about 200,000 hectares (494,000 acres). This has improved yields of annual crops and today provides 65 percent of the country's jujube fruits.[49]

In Greece, olives have been intercropped with annual crops for ages with about 650,000 hectares (1.6 million acres) of intercropped olives in production today.[50] Other common combinations include poplars intercropped with annual crops; in the United States black walnut and pecan trees are often intercropped with annuals.

Estimates of the sequestration rates of strip inter-cropping vary, in part because of different definitions used. One study estimated a rate of 3.4 t/ha/yr for strip intercropping in temperate North America.[51] Other estimates are more conservative, like the USDA's esti-mate of 0.2 to 0.5 t/ha.[52] P. K. Nair estimates 2 to 5 t/ha/yr for tropical intercropping systems.[53]

Yield impacts vary depending on tree species, tree spacing, and the understory crop. With the right combi-nation of species, annual crop yields can be higher than monocultures of the same crop due to benefits from trees such as wind protection and increased organic matter. In Chinese jujube systems, shade-tolerant crops are grown beneath the tree rows while sun-loving crops can be grown in the alleys.[54] One drawback of strip intercropping is that in mechanized systems, tree row spacing must be set to allow access for tractors and implements.

Farmer-Managed Natural Regeneration

Farmer-managed natural regeneration (FMNR) is one of the most extraordinary success stories in regenerative agriculture. FMNR was developed in the Sahel, where many on-farm tree-planting projects have failed. The key to FMNR is that rather than planting new trees, the "underground forest" of tree stumps and roots already present in farmers' fields, traditionally suppressed by tillage, is permitted to grow. Some are cut annually, some are heavily pruned each year, and others are allowed to grow large. The trees provide products and income from firewood, building materials, and fodder, while acting as windbreaks and improving the soil. The most common resprouting tree in much of the region is the amaz-ing *Faidherbia albida* (see the Evergreen Agriculture section on page 75), which farmers prune for single or multiple stems depending on the products they desire.[55] Farmers and researchers in Niger have also developed the Farmer-Managed Agroforestry Farming System (FMAFS). This system builds on FMNR but adds wind-breaks of edible-seeded, nitrogen-fixing acacia shrubs, mulching with crop residues, and plantings of multipur-pose trees such as moringa and baobab.[56]

FIGURE 6.6. A farmer practices farmer-managed natural regen-eration (FMNR) by pruning spontaneous, resprouting trees in her pearl millet field. She will use the biomass for mulch and fuel. Photograph courtesy of the World Agroforestry Centre.

While most tree-planting projects in the region have failed, FMNR has largely succeeded. Tony Rinaudo of World Vision, one of the developers of FMNR, describes it as "a social, rather than a technical breakthrough . . . The greatest barriers to reforestation were neither the absence of an exotic super tree nor ignorance of best practice nursery and tree husbandry techniques. The greatest barriers were the collective mindset which saw trees on farmland as 'weeds' needing to be cleared and inappropriate laws which put responsibility for and ownership of trees in the hands of the government and not in the hands of the people."[57]

FMNR may represent "the single largest environ-mental transformation in Africa."[58] From its beginnings in 1983 in Niger, FMNR spread to cover 4.8 million hectares (11.9 million acres) of land by 2010. It has been estimated that Niger today has between 10 and 20 times more trees then it did in the 1980s thanks to FMNR.

FIGURE 6.7. A Farmer-Managed Agroforestry Farming System (FMAFS) during Niger's dry season with rows of edible-seeded acacia shrubs at the field's edge and FMNR *Faidherbia albida* in irregular overstory. Photograph courtesy of Peter Cunningham.

Today it is a standard practice in Niger and is spreading across semi-arid Africa and beyond.[59]

FMNR has been estimated to sequester roughly 2 to 4 t/ha/yr of carbon,[60] although both sequestration and yields probably vary depending on the species and the density at which they are planted. Although data on the impacts that FMNR has on yields is not currently available, farmers report better yields with FMNR than without it. In the Sahel, farmers often must replant their crops three or four times due to damage from sand-blasting winds. FMNR farmers report no need to replant their crops, thanks to the windbreak benefits of their trees.[61]

Evergreen Agriculture

Evergreen agriculture is a subset of agroforestry intended to provide green cover on cropland year-round. It involves the "direct and intimate intercropping of trees within annual crop fields," including inter-cropping, FMNR, and other systems.[62] One unique and fascinating element of evergreen agriculture is the frequent use of the nitrogen-fixing apple-ring acacia (*Faidherbia albida*) as a companion to annual crops. This truly remarkable tree exhibits "reverse leaf phenology," meaning it leafs out in the dry season when crops are not present beneath it and loses its leaves for most of the rainy season. Thus it does not shade out annual crops and can be planted at high densities, avoiding some of the pitfalls of competition between agroforestry trees and their companion crops. Competition for water is also minimized.[63] This ingenious practice is really just a systematization of the widespread and ancient agroforestry system of *Faidherbia* parkland.[64] As such, it is a great example of building on traditional knowledge, which may be part of why it has spread rapidly since its development just a few decades ago.

Although attention has rightfully been focused on *Faidherbia*, there are other appropriate species that express reverse phenology and can be good candidates for evergreen agriculture. For example, the nitrogen-fixing African oil bean (*Pentaclethra macrophylla*), a large tree that produces edible, protein- and oil-rich

FIGURE 6.8. *Faidherbia albida* with annual crops growing below in Zambia. Photograph courtesy of the World Agroforestry Centre.

beans, sometimes expresses reverse phenology.[65] A good candidate for domestication as a staple crop, African oil bean could become the *Faidherbia* of the humid African tropics. (See African Oil Bean in chapter 15.) Each region of the world should search for the most useful native species that could be incorporated into evergreen agriculture systems.

Evergreen agriculture is estimated to sequester carbon above- and belowground at a rate of 2 to 4 t/ha/yr.[66] This is measurably better than the less densely treed parkland systems from which it developed. The impact *Faidherbia* evergreen agriculture has on yields is also impressive. In Africa average yield for maize is 1.25 t/ha and 1 t/ha for cotton.[67] But one study in Zambia found that farmers produced 1.5 tons more maize per hectare and 0.46 ton more cotton per hectare in fields densely planted with *Faidherbia albida*. Another Zambian study found that maize under *Faidherbia* trees yielded 4.1 tons per hectare, while maize beyond the tree canopy yielded only 1.3 t/ha. In Malawi one study found 280 percent greater yields under the canopy; another found 100 to 400 percent increases, with farmers reporting

APPLE-RING ACACIA (*FAIDHERBIA ALBIDA*)

Among the world's many remarkable multipurpose workhorse species, *Faidherbia albida*, commonly referred to as apple-ring (or winterthorn) acacia, stands out as the anchor of the exciting practice of evergreen agriculture. Native to Africa and the Middle East, *F. albida* can grow well in dry areas and in tropical lowlands and highlands.[68] In some regions it is so ubiquitous that it's simply allowed to sprout up wild in the fields through FMNR. In others regions it is planted, typically at a density of about 100 trees per hectare, then thinned to about 25 per hectare once they are mature.[69] With FMNR, the tree density is typically similar.[70]

In Niger 4.8 million hectares of farmland are dominated with *Faidherbia*, mostly through FMNR.[71] In Zambia, there are about 300,000 hectares (741,000 acres), most of which were planted.[72] Already 13 percent of Africa's millet and sorghum are grown in

Faidherbia systems, although still less than 2 percent of maize is.[73] Annual crops are managed with minimal tillage, retention of crop residue, legume rotations, rainwater harvesting, and other carbon-friendly practices.[74]

Faidherbia fixes nitrogen, coppices, and grows quickly. It is grown as shade for coffee and sometimes along the edges of rice paddies. The foliage can be fed to livestock, although excessive pruning initiates leafy growth in the rainy season, which can be a drawback. *Faidherbia* trees drop their pods, which are also eaten by livestock at the end of the dry season; *Faidherbia* spacing in a mature evergreen agriculture system is 25 trees/ha, which could yield 1.3 tons of pods for livestock. One tree yielded 40 to 339 kg in different years—a substantial variation. Another trial showed an average pod yield of 590 kg (1,300 pounds) per hectare per year with 11 trees/ha. People sometimes eat the seeds when other foods are scarce, such as at the end of a long dry season and during famines.[75]

that it took between one and six years for effects to become clear.[76]

The downside of evergreen systems with species such as *Faidherbia* is that wide enough plantings of a single species could potentially invite pests and diseases to attack the tree. Likewise, the usefulness of a single species such as *Faidherbia* should not prevent the incorporation of other useful trees on African farms.

Alley Cropping

Alley cropping, which is based on traditional practices in tropical Africa and Southeast Asia, integrates annual crops with rows of perennials.[77] It is currently practiced by thousands of farmers in Southeast Asia but has not been rapidly adopted in other tropical regions.[78]

FIGURE 6.9. Alley cropping with coppiced legume trees between strips of maize in the Philippines. Photograph courtesy of Henrylito D. Tacio, Mindanao Baptist Rural Life Center.

VETIVER (*CHRYSOPOGON ZIZANIOIDES*)

Vetiver shows us what an agroforestry support species can become with some breeding work and an international network behind it. Native to India, the root systems of this clumping perennial grass are enormous and deep. It is widely adaptable, ranging from warm temperate to tropical climates and from the fairly arid (300 millimeters, or 12 inches, of rainfall per year) through very humid regions, even tolerating more than a month of flooding.[79] It has been planted around the world for contour hedgerows, erosion control, wastewater treatment, dune stabilization, rice paddy borders, and more. Considering its use as a "living nail" for erosion control and contour farming, vetiver represents a species that has fully achieved its potential.[80]

Vetiver is also used for biomass products including mulch, thatch, and basketry. Its tender new growth is used as fodder. An essential oil can be distilled for use in perfumes and insect repellents.[81] These plants are long-lived, and a set of cultivars has been developed, each with sterile seeds and no invasive potential.[82] To promote vetiver and its many benefits, The Vetiver Network International publishes guides and organizes courses and conferences.[83]

FIGURE 6.10. Vetiver, known as the living nail of slope stabilization in Ethiopia. Photograph courtesy of The Vetiver Network International.

Typically, although not always, the perennials are coppiced, nitrogen-fixing woody species. In some cases perennial crops such as bananas are substituted for the annual crops.[84] Sometimes people refer to strip intercropping as alley cropping, but for clarity I refer to alley cropping systems as those whose perennials primarily support annual crop production, and strip intercropping systems as those that are equally focused on both the tree crop production and annual crop production.

The impact on yield can be either positive or negative, depending on the perennial crop, the annual crop, soils, and rainfall. Benefits include increased organic matter and mulch, nitrogen fixation, and erosion control. The downside of alley cropping is that the perennials can compete for light, nutrients, and water. While the competition for light can be managed with timely pruning, competition for water typically makes alley cropping a poor choice in semi-arid and arid environments.[85]

Contour Hedgerows and Living Terrace Edges

Contour hedgerows can be thought of as alley cropping systems that are laid out on the contour of slopes. Typically farmers alternate annual crop contour strips with coppiced woody legume strips or other perennial erosion control species. The perennial strips are coppiced or cut back annually for mulch or livestock fodder. Over time, the annual crop soils upslope of the strips tend to slump, leveling out the soil and creating "living terraces."

Although they have been slow to spread beyond Asia, contour hedgerows are fairly widespread in tropical Southeast Asia, particularly in Indonesia, as well as in China.[86] In the cold, dry north and west of China, they are relatively less successful because of limited water, but researchers developed a practice called living terrace edges, in which terraces are constructed and the edges are planted to productive perennial vegetation to provide long-term stabilization. (See appendix A for a global list of species used in contour hedgerows.) The Chinese province of Shaanxi has constructed 52,200 hectares of living terrace edges.[87]

A similar system known as *metepantli* was popular in the Mexican highlands at the time of conquest. *Metepantli* are terraces on moderately sloping land, stabilized with agaves, mesquites, nopal cacti, or fruit trees; they remain fairly widespread today in Mexico.[88] A sophisticated integrated farming system known as the Sloping Agricultural Land Technology (SALT) builds on contour hedgerows, alternating a strip of perennial crops instead of annuals in one out of every three contour

FIGURE 6.11. In Ethiopia these vetiver hedgerows are planted just off contour for ease of plowing. Photograph courtesy of The Vetiver Network International.

FIGURE 6.12. An aerial view of *metepantli*, vegetated terraces in the Mexican highlands.

CARBON FARMING BUILDING BLOCKS

All of the practices profiled in part 2 should be looked at as building blocks. They can be combined and integrated in creative, productive, and increasingly sophisticated ways. For example, Farmer-Managed Agroforestry Farming Systems (FMAFS) builds on FMNR, adding hedgerows of nitrogen-fixing Australian edible-seeded acacias and tree crops such as moringa and baobab. These perennial elements are integrated with improved annual cropping practices such as crop rotation and retention of crop residues as mulch.[89]

The Sloping Agricultural Land Technology (SALT) system builds on contour hedgerows and incorporates them into a landscape-level farm planning system. Forests are managed at the top of the slopes, with rainwater harvesting and contour hedgerows on the middle and lower slopes. Intensive gardens and living spaces are sited in the valley bottoms. Between every third contour hedgerow, perennial crops such as bananas and avocados are grown for improved erosion control.

bands. SALT originated in the Philippines in the 1970s. Contour hedgerows, SALT systems, and living terrace edges are labor-intensive to establish and do not provide immediate results, although they are critical for the long-term sustainability of hillside annual crop production.[90] Their high labor demands and competition between the hedgerow and annual crops have been cited as barriers to wider implementation.[91] The US Department of Agriculture promotes a similar practice, with contour strips of perennial grasses replacing the woody hedgerows.

Carbon sequestration data is unavailable but probably similar to alley cropping and strip intercropping systems. The impact on yield varies and can include the adverse impacts noted above. Soil conservation impacts are very positive, however; without some form of erosion control, yields of annual crops on slopes will decline over time.[92]

Perennial Crop Rotations and Improved Fallows

Another way to build carbon in annual cropping systems is to plant fields to a perennial crop for a few years. Annual crop farmers often convert fields to hay for several years to rebuild organic matter, a practice found to sequester carbon.[93]

A practice that has emerged in Africa in the last few decades is the improved fallow. This is a sort of substitution for the traditional swidden systems described on pages 82 and 83, in which fields are fallowed for decades and allowed to return to forest. In improved fallows, fast-growing woody plants, typically nitrogen fixers, are planted at high density. In one to three years they have accumulated tremendous biomass, which is typically used as firewood. This practice also builds impressive soil carbon: 1.3 to 16.5 t/ha/yr in a study in Kenya.[94]

The soil carbon gained from hay and woody fallow periods is eventually burned back up by tillage once the land is returned to annual crops. Some African farmers have gone a step further and do not remove the fallow woody species. Instead they select coppicing woody fallow species that resprout vigorously after cutting. Annuals are planted between the fresh-cut stumps. The woody plants are cut multiple times each year for mulch, turning this into an intercropping system.[95] The scale of this practice is unknown.

The average maize yield in Africa is 1.25 t/ha.[96] The use of permanent interplanted coppiced woody legumes has resulted in an average yield increase of 1.6 t/ha, which in 67 percent of cases was a doubling or tripling of maize yields. This dramatic increase was only seen in degraded soils where yields were already below 2 t/ha. These results were higher than both annual legume cover crops and non-coppicing improved fallows.[97]

Windbreaks

Windbreaks, or shelterbelts, are an ancient and widespread practice. Trees, or sometimes shrubs or other perennials, are planted in long strips against prevailing winds to protect crops, livestock, and people from potentially catastrophic winds. Windbreaks also help minimize wind erosion of bare tilled soils and can shelter annual crops, pastures, and tree crops from elements other than the wind. (See appendix A for a global inventory of windbreak species.) In Canada 43,000 linear kilometers (26,700 miles) of windbreaks protect 700,000 hectares (1.7 million acres) of farmland. In the United States, 95,000 linear kilometers (59,000 miles) shelter 300,000 hectares of farmland.[98]

Windbreaks can sequester 6.4 tons of carbon per linear kilometer in aboveground biomass. The lifetime aboveground biomass carbon of windbreaks can be as high as 105 t/ha at 20 years of age.[99] P. K. Nair estimates that in arid and semi-arid regions, protective systems such as windbreaks sequester 1 to 8 t/ha/yr.[100]

Well-designed windbreaks can also increase overall yields of field and forage crops.[101] Crops that are planted right next to the windbreak strips often compete for light, water, and nutrients, resulting in a reduction of yield for those areas, but overall field yields are generally still higher. In colder regions, the benefit of melting snow trapped in the windbreaks can increase yields to make up for the competition for light and nutrients.[102] A review of 78 studies found that windbreaks increased wheat yields in the former Soviet Union by 24 percent.[103] Positive yield effects are felt in many climates, soils, and regions, especially dry climates or drought

years.[104] The yield impacts are especially high for vegetables crops, tree crops, and vineyards.[105]

Living Fences and Hedgerows

Living fences and hedgerows are simply managed rows of shrubs and trees. Their use is an ancient practice intended to delineate the borders and boundaries of farms and fields while providing products and benefits. There are living fence and hedgerow traditions in Europe, Asia, Africa, and Latin America.[106] In the United States, Osage orange living fences were the essential farm fence before the invention of barbed wire. The number of living fences has declined since the spread of conventional agriculture in the 1950s and 1960s, however, and many have now been removed.[107]

When there are many living fences in an area they form a hedgerow network landscape or *bocage*.[108] Typically the woody plants are coppiced or heavily pruned, and they are sometimes woven together intricately. In other systems, live stakes are planted to serve as living fence posts for barbed wire. Living fence and hedgerows allow farmers to choose when and where livestock have, or do not have, access to fields, and can keep pest herbivores such as deer out of crops. (See appendix A for a global inventory of living fence and hedgerow species.)

FIGURE 6.13. Windbreaks protecting annual cereal crops in North Dakota, USA. Photograph courtesy of the National Resource Conservation Service of the US Department of Agriculture.

FIGURE 6.14. A living fence of moringa with spiny *Euphorbia lactea* in Cuba.

FIGURE 6.15. Living fence posts of *Gliricidia sepium* in Cuba.

The carbon sequestration impact of living fences and hedgerows is probably similar to that of windbreaks. P. K. Nair estimates that in arid and semi-arid regions, protective systems such as hedgerows and living fences sequester 1 to 8 t/ha/yr.[109] Products from living fences and hedgerows include woodcrafts and semi-wild foods and medicines. These systems, along with the ditches and earthworks with which they are often associated, help to slow and infiltrate water, minimize erosion, and serve as windbreaks. They also provide important habitat and corridors for wildlife and help to buffer pest population explosions by harboring beneficial insects and other predators.[110]

Riparian Buffers and Water-Breaks

Riparian buffers are strips of perennial vegetation planted or managed where annual cropping fields abut streams and rivers. They are widely implemented in the United States, Canada, and Europe, with increasing utilization elsewhere in the world. Riparian buffers help catch leached nutrients from farming operations to maintain water quality. They also reduce flooding, both on-farm and downstream. Because they take land completely out of annual crop production, they reduce overall farm yields—although increasingly, productive riparian buffers known as "agriculturally productive buffers," or multifunctional

FIGURE 6.16. Riparian buffers protect water quality and can include productive perennial crops. Photograph courtesy of the National Resource Conservation Service of the US Department of Agriculture.

buffers, are being used in the United States. These productive buffers include perennial food crops, hay, timber, and biomass crops and help justify taking annual cropland out of production.[111]

Increased flooding, correlated with climate change, is driving development of new agroforestry practices. One such system is the water-break, a system of vegetated perennial buffers on farmed floodplains. Water-breaks, in concert with riparian buffers, help to slow floodwater and reduce its impact.[112] Vegetated water-breaks are also considered an important climate adaptation strategy.[113] Riparian buffers are estimated to sequester 2.6 t/ha/yr in the United States.[114] P. K. Nair estimates that in arid and semi-arid regions, protective systems such as riparian buffers sequester 1 to 8 t/ha/yr.[115]

Irregular Intercropping: Parkland, *Streuobst*, and More

Irregular intercropping systems, which are simply characterized as trees scattered throughout cropland, are traditional systems that appear to have developed independently in different parts of the world. In China the practice of intercropping trees and annual crops goes back to 1600 BC.[116] Today at least 120 woody species are intercropped with annual crops in China.[117] Intercropping also has a strong tradition in Europe, where farmers incorporate tree crops such as walnuts, olives, poplars, and fruit trees with grapes and annual crops. Although the practice is waning, the German *streuobst* model of fruit trees interspersed with annual crops still occupies 400,000 hectares (988,000 acres) of land.[118]

In Africa irregular intercropping is referred to as parkland, and most cultivated land in the Sahel is in one kind of parkland or another.[119] Parkland is almost certainly the most widely practiced form of agroforestry in the world, with almost 30 million hectares (74 million acres) of shea nut parkland alone in West Africa, and another 5 million hectares (12 million acres) of mixed-species parkland in Mali.[120]

Although traditional irregular intercropping systems are not highly efficient in terms of carbon farming—they weren't developed with that in mind—they do form the foundation of many modern high-yielding, carbon-sequestering agroforestry systems. Parkland can sequester 0.4 to 1.1 tons of carbon per hectare per year in aboveground biomass—a low rate that reflects low tree density and a dry climate with poor soils.[121]

The impact on annual crop yields varies depending on tree and crop species and the distance from the tree that the annual crops are planted. Some crop trees such as shea and locust bean can cause yield reductions under the tree, while others such as apple-ring acacia improve grain yields dramatically.[122] Of course, although some parkland trees lower yields of the crops beneath them, the trees themselves can also produce crops, including fruits, nuts, fodder, firewood, and more. Parkland trees also help to build and maintain soil fertility.

Swidden

In swidden, or shifting cultivation, the landscape is managed as a mosaic of different successional stages. A patch of forest is cleared, usually by burning, and annual and short-lived perennial crops are grown for one to three years. After this the land is allowed to revert to forest for 5 to 20 fallow years, rebuilding soil fertility. Then the cycle begins again.[123]

FIGURE 6.17. A baobab parkland in the Sahel during the dry season. Photograph courtesy of the World Agroforestry Centre.

FIGURE 6.18. A swidden mosaic of rice, bananas, abandoned cropland, and forest in Yunnan, China. Photograph courtesy of Desmanthus4food/Creative Commons 3.0.

TABLE 6.3. The Carbon Cost of Converting Swidden Landscapes to Other Farming Systems

Land Use Transformation from Swidden to:	Loss of Carbon in Aboveground Biomass	Loss of Soil Organic Carbon in Topsoil (Unspecified Depth)
Shorter fallow periods	88–90%	0–27%
Continuous annual cropping	95–99%	13–40%
Rubber plantations	-10 to 40% (a gain)	0–30%
Oil palm plantations	60%	0–40%

Source: Adapted from Bruun et al., "Environmental consequences of the demise in swidden cultivation in Southeast Asia," 383.

Swidden is an ancient practice in much of the world, still widely used in some areas, although it is losing ground to annual cropland and oil palm plantations. In some cases fallow periods have been shortened to as little as four years, removing many of the benefits.[124] In Southeast Asia swidden is currently practiced on 25 to 39 million hectares (62 to 96 million acres) in Thailand, Laos, Cambodia, Vietnam, Myanmar, Malaysia, and Indonesia, as well as other countries for which data is lacking.[125]

For many years this traditional practice was derided as a "backward" slash-and-burn technique that contributed to deforestation. But swidden is distinct from the permanent clearing of forest (with fire or otherwise) for farmland, which is terribly destructive of soil carbon. Although it's true that forest is cut and burned for permanent conversion to annual cropland, swidden today is being praised by scientists for its environmental benefits, including biodiversity and water quality protection.[126] The fallow period is hardly a period of abandonment; some useful trees are never cut during the clearing phase. And after the annual cropping period, species that produce timber, food, and other products are often planted and/or selected for from the natural regeneration. Over time this leads many forest fallows to be "guided" to a productive state.[127]

Despite the cycle of burning, clearing, and tilling, swidden has impressive carbon benefits. Carbon sequestration in swidden systems needs to be measured at the landscape level to account for the different stages of succession, from freshly cleared to fully fallow. When swidden landscapes are converted to other farming systems, there is usually a substantial loss of sequestered carbon. There is also an unwelcome short-term but significant climate impact of "black carbon" from

such burning.[128] As with homegardens, it is difficult to measure the yields of swidden systems. They are characterized by extremely high crop diversity, although yields are generally lower than with intensive industrial production.[129]

Successional Intercropping

In its simplest form, successional intercropping is the cultivation of sun-loving short-lived crops such as annuals or short-lived perennials into new tree crop plantings. Successional intercropping takes advantage of available sunlight before trees grow large and provides an early yield as well. One example of a successional intercropping system is *taungya*, which was developed in colonial Burma with annual crops grown during establishment of timber plantations.[130] Most

FIGURE 6.19. Nigerian farmer Ben Egbune with young rubber trees successionally intercropped with plantain. Photograph courtesy of the World Agroforestry Centre.

tropical timber plantations in Asia and Africa today began as *taungya* systems.[131] Asia and Africa had 44 and 10 million hectares (108 million and 24.7 million acres) of timber plantations respectively in 2005, so this represents a large area of *taungya* production.[132]

There is little information available about carbon sequestration potential and the impact on yields of successional intercropping. What is clear, however, is that the economic benefit of revenue in establishment years can be substantial, such as with rubber intercropping systems in China.[133] Such systems provide cash flow while farmers wait years or decades for revenue from tree products such as rubber, nuts, or timber.

Pasture Cropping

Pasture cropping is an innovative practice developed in Australia, in which the perennial element is pasture, specifically warm-season pasture. (In Australia there are many warm-season pastures.) Warm-season grasses use the C_4 respiratory pathway, which allows them to photosynthesize efficiently during hot dry seasons. Cool-season grasses, which use the standard C_3 respiratory pathway, grow when it is cooler out. In the early 1990s Australian farmer Darryl Cluff found a way to produce warm-season pasture and cool-season annual grains on the same piece of land. When the warm-season grass goes dormant in the fall, a no-till plow opens narrow furrows; these are planted to cool-season greens such as rye, oats, or wheat. These greens mature during the mild Australian winter and are harvested just as the warm-season pasture is coming back out of dormancy. This practice allows farmers to graze animals in the summer and grow grains in the winter on the same piece of land, all while getting the carbon sequestration, soil building, and erosion control benefits of a diverse perennial native pasture.[134]

Pasture cropping has been spreading rapidly. During the years between its development in 1993 and 2013, more than 2,000 Australian farms practiced pasture cropping, with interest beyond Australia as well.[135] It has proven especially popular on marginal and degraded lands.[136] (Reverse pasture cropping might also be possible, with cool-season C_3 pastures and warm-season C_4 grains and pseudocereals such as sorghum, maize, or amaranth; I grow a short-season dwarf sorghum that could be an excellent candidate.)

In one study, soil carbon under pasture cropping doubled over the course of 10 years.[137] A comparison of pasture cropping with conventional grazing and annual cropping found that "rotational grazing and pasture cropping . . . can increase perennial vegetative groundcover and litter inputs . . . [and] can improve landscape function while sustaining similar or higher stocking rates over the year compared to the conventional system."[138] In another study, grain yields were 65 percent or less than seen with conventional no-till, but given the lower inputs, both the grazing and cropping elements were found to be profitable.[139] (Of course, the livestock yields must be factored in to determine total yield per hectare.) Still another study found an overall farm profitability gain of 10 percent.[140]

FIGURE 6.20. Colin Seis and Darren Doherty in Seis's pasture-cropped oat field in Australia. Photograph courtesy of Kirsten Bradley.

CHAPTER SEVEN

Livestock Systems

About 3.5 billion hectares (8.6 billion acres) of the world's roughly 5 billion hectares (12 billion acres) of farmland—or 70 percent—is devoted to pasture. (This includes rangeland, shrubland, pastures, and cropland currently growing fodder or pasture.) These grasslands are often too dry, steep, or rocky to be suited to annual crop production, although some could support arid-land tree crops.[1] In total, 30 percent of the planet's land is used for livestock production, which provides 33 percent of the protein consumed by humans.[2] In fact, one-quarter of all terrestrial animal biomass consists of humans and our livestock.[3] And if we include the annual crops that are grown to feed them, livestock production is responsible for 18 percent of all human emissions. Within agriculture, livestock production accounts for a whopping 80 percent of emissions.[4] And global meat production is on the rise, tripling since the early 1970s.[5]

Livestock provide more than protein from meat, dairy, and eggs. Wool, hair, and silk are important fibers. Hides, skins, and feathers are also important materials from livestock. Manure and urine are essential fertilizers, and dung is used as a fuel and building material in many regions. Blood, bone, and hooves are also used as organic fertilizers. Some livestock, such as silkworm pupae and worms, are used as feed for other livestock. Several genera of domesticated bees provide honey and wax. Industrial products from livestock include dyes, shellac, pharmaceuticals, bioplastics, soaps, creams, and gelatin.[6]

Livestock animals also provide services, substituting for human labor or fossil-fuel-powered machinery. They provide transportation and draft power to pull tractors and power mills. Pigs and chickens can turn compost piles, while soldier flies and worms decompose food waste and produce quality soil amendments. Livestock can assist with control of pests and weeds. They have many important cultural aspects, provide income, and serve as "walking capital."[7]

Carbon sequestration rates of improved livestock production practices are generally very low to medium, although systems with trees generally sequester more carbon than pastures alone—and intensive silvopasture, a new practice, has very high sequestration rates. But even a small increase in carbon sequestration can have a huge impact when practiced on 70 percent of the world's farmland. Simply because pasture is humanity's largest use, improved pasture management may have more potential to sequester carbon than any other carbon farming practice.[8]

The IPCC rates managed grazing and improved pasture management as having low global climate change mitigation impact potential, being easily adopted by ranchers, and ready for implementation.[9] The IPCC also considers it to be one of the most cost-effective options for the amount of carbon sequestered.[10] The IPCC rates manure application on cropland as having a high global mitigation potential, being easily adopted by farmers, and ready for implementation. Manure management was rated as high mitigation, moderate to easy adoption, and ready to go. Production of protein concentrate for livestock feed as a by-product of biomass processing was rated as having medium mitigation potential, easy farmer adoption, and ready to go.[11]

The Controversy Around Livestock's Impact on Climate

But the topic of livestock has also become incredibly controversial among environmentalists and activists.[12] Livestock-related climate change mitigation proposals run the gamut from greatly increasing grazing to a global conversion to a vegetarian diet. Travis McKnight's August 2014 opinion piece in the *Guardian*, "Want to Have a Real Impact on Climate Change? Then Become a Vegetarian," lays out this latter argument: Since livestock as currently raised are such a huge part of the problem, we should reduce their numbers substantially. McKnight fails to recognize carbon-friendly livestock practices, as do many such arguments.[13]

Admittedly, I myself came to this subject thinking that managed grazing was a high-impact mitigation strategy and that confined livestock must be a huge part of the problem. What I've learned in my research resulted in a more nuanced understanding. Much of the controversy around livestock is rooted in the genuine complexity of accounting for emissions and sequestration in livestock production. Is pasture or confined livestock a better choice? Should we raise and eat less livestock or should we scale up carbon-sequestering practices? Do methane emissions erase the benefits of carbon sequestration? Farmers, scientists, and activists often find little common ground among and even within their communities. What follows is a brief tour through the complicated and controversial issues related to livestock production and its impact on climate, with my best attempt at a fair and objective assessment of the available science. My guiding principle on a subject like this—complicated and controversial—is to "teach the controversy."

Let's start with methane. Although this greenhouse gas only persists in the atmosphere for between 9 and 15 years, methane is 21 times stronger than carbon dioxide.[14] Ruminant livestock are the largest anthropogenic methane emitters due to fermenting bacteria in their digestive systems that produce methane as a waste produce.[15] Manure from all classes of livestock is also responsible for methane emissions.[16] All told, ruminant digestion and the manure of all livestock account for 80 percent of agricultural methane emissions and 35 to 40 percent of anthropogenic methane emissions.[17]

As a sustainable agriculture enthusiast I wanted to believe that pasture-raised livestock would emit substantially less methane than grain-fed feedlot and confined livestock. This is not the case, although there are few simple answers in this area. (And there are, of course, still strong animal welfare arguments against concentrated animal feeding operations.) Livestock in high-tech confinement systems that are fed concentrates from annual grains generally have lower methane emissions per productive unit *from digestion* than ruminants, at least on poor-quality pastures. This is because the annual grain concentrate feed is easier to digest due to its lower fiber content and spends less time in the digestive tract, giving less opportunity for fermentation by methane-producing bacteria.[18] A 2014 study published in the *International Journal of Biodiversity* found that cattle finished on pasture produced 30 percent more methane emissions than grain-finished cattle.[19]

But the kind of pasture and fodder also matters. Better-quality pastures improve productivity and reduce the percentage of methane per ruminant animal.[20] (This could still increase total methane emissions per hectare with higher stocking rates.[21]) High-tannin forages from woody legumes such as those used in fodder banks, intensive silvopasture, and other agroforestry systems described below also reduce methane emissions from ruminant digestion.[22] Farmers and scientists are working to find other options for reducing methane emissions from ruminants. Strategies that include low-fiber feeds, switching or crossing livestock breeds, and the addition of some feed supplements, medications, and inoculants, including oils and oilseeds, all show promise.[23] I've even heard farmers at conferences talking about feeding their livestock powdered biochar.

Manure of both ruminant and monogastric livestock also emits methane (as well as nitrous oxide compounds). However, in this case pasture-based systems are better. Manure in pastures is not a significant source of methane emissions, whereas the manure from confined livestock, including manure piles and lagoons, is.[24] Stored manure is also a significant source of nitrous oxide, as is manure of pasture.[25] Composting manure

TABLE 7.1. Ruminant Versus Non-Ruminant Livestock

	Ruminants	Non-Ruminants (Monogastrics)
Diet	Can subsist entirely on pasture and leaves. No grain required in diet.	Require grain, feed concentrates, crop residues, or (sometimes) food waste. Cannot subsist only on pasture and leaves.
Advantages	Can utilize world's vast grasslands, can consume perennial foliage.	Many (e.g., pigs, poultry) can consume food waste. Some (e.g., fish, insects) are extremely efficient converters of feed to meat.
Methane emissions	Digestion emits methane (varies some with diet). Manure can emit methane (varies with management).	Manure can emit methane (varies with management).
Species	Cattle, sheep, goats, water buffalo, camels, llamas and alpacas, yaks, microdeer.	Pigs, poultry, rabbits, rodents (e.g., guinea pig), fish, crustaceans (e.g., crayfish), edible insects, iguanas.

reduces methane but can increase nitrogen emissions. Non-aerated compost has significant methane emissions.[26] Biogas digesters have been identified as a key appropriate technology to capture methane from manure and convert it to energy. They are considered a best practice for reducing manure-related emissions from confinement operations.[27]

Although methane emissions from ruminants eating feed concentrates are lower than those of pasture-fed animals, the primary ingredients of these feeds are annual grains. In fact, much of livestock's current climate impact is actually from the annual grains (cereals, soybeans, and so on) grown to feed them. About 33 percent of global annual cropland produces feed for livestock.[28] All the challenges of land clearing, fossil fuel use in equipment, nitrogen fertilizer use, and carbon loss from tillage apply to this cropland—just as they do when the land is used for human food. The difference is that while humans need grain to live, livestock do not. Ruminant livestock (cattle, water buffalo, sheep, goats, camels, llamas, and the like) don't need to eat grain at all, although it can increase their productivity. Monogastric livestock don't need to eat food-grade annual grains.

The controversy is not just about manure and methane, however. It's also about livestock's impact on the ecosystem. Allan Savory, the developer of the Holistic Management school of managed grazing, ignited a global firestorm of controversy during a 2013 TED Talk during which he stated that, using his grazing techniques, "we can take enough carbon out of the atmosphere and safely store it in the grassland soils for thousands of years, and if we just do that on about half the world's grasslands that I've shown you, we can take us back to

pre-industrial [carbon] levels, while feeding people."[29] Although I respect Savory's work, I have trouble with this kind of rhetoric, because it suggests that Holistic Management (or carbon farming or geoengineering or anything else) can allow us to continue emitting all the fossil fuels we want; in fact, the total potential storage of carbon in soil and vegetation is much lower than the total amount of carbon remaining in unburned fossil fuels.[30] Carbon farming, whether that includes managed grazing or not, only works as part of a larger strategy that includes greatly reduced fossil carbon use.

Many scientists leapt to dispute Savory's claims. They argued that studies have found that while the adaptive management aspects (planning, goal setting, monitoring, and financial rigor) of Holistic Management are powerful tools for farmers, the purported ecological benefits don't always bear out. Intensive rotational grazing, a similar though less sophisticated practice than Holistic Management, has likewise shown few ecological benefits in numerous studies, at least in arid and semi-arid landscapes. Most evidence shows that grazing strategy (rotation, holistic management, et cetera) does not have a strong carbon sequestering practice impact, especially in the dry climates Savory spoke about.[31]

On the other hand, a global review of 300 studies on managed grazing found a carbon sequestration mean of 2.1 t/ha/yr, with high variation, including losses of carbon in some cases.[32] Another global review found a mean of 0.5 t/ha/yr.[33] Many farmers using Holistic Grazing have found increases in soil organic carbon of 3 percent (roughly 6 percent increase in organic matter).[34] The IPCC notes low rates of carbon sequestration from improved management of pastures and grazing. The real impact comes from the great scale of the world's grasslands.[35]

What the controversy indicates to me is the need to dig deeper for better answers. What grazing and pasture practices sequester carbon? How much? In what climates and on what scales? We need a clear body of evidence, and we can't get there without additional research and funding. This needs to be a top priority.

When it comes to managed grazing, and Holistic Grazing in particular, there is a disparity (and a tension) between what scientists and practitioners report. I've seen this in my own experience: While I've read many scientific papers that cast legitimate doubt on the efficacy of managed and Holistic Grazing (both carbon sequestering and other ecosystem benefits), I've also seen hard-to-refute results of managed grazing on farms and ranches—results that, even if they don't measure soil organic carbon per se, indicate the kind of ecosystem health that tends to go hand-in-hand with carbon sequestration. This apparent dichotomy has made this chapter the most difficult chapter of the book to write because it's required me to grapple with and attempt to reconcile these two kinds of "knowing" – science and the lived experience of farmers and ranchers who work the land every day—while also examining my own pre-conceived notions about "what seems right." Through the course of my own research, I've come to the conclusion that we really need to engage with, and respect, both kinds of "knowing." Rather than seeing these forms of evidence in opposition to one another, I believe we'll get further in understanding the carbon sequestering potential of various agricultural practices if we are willing to acknowledge the importance and legitimacy of *both*, engaging with the science and honoring the real-world experience of farmers and ranchers who are so well-positioned to understand what is happening on the land they work everyday.

To this end, I was excited to learn that there are scientists who have acknowledged and are working to better understand this dichotomy. In "Multi-paddock grazing on rangelands: Why the perceptual dichotomy between research results and rancher experience?", Richard Teague and his colleagues point out flaws in the methods some reviews and studies have been designed, such as conflating stocking rates with grazing treatments.[36] They lay out guidelines for future experiments that, with luck, will reconcile rancher experience with science and offer better grazing guidelines. There is a critical need for data to quantify the impact of Holistic Grazing in order to provide this method with the legitimacy I believe it deserves.

Can livestock systems sequester enough carbon to cool our climate even when the warming impact of methane is taken into account? Some interesting answers are emerging. One study of nine grazing systems in Europe found that methane emissions represented the equivalent of a 25 percent loss of the carbon sequestered in the pasture.[37] Another recent study compared top-ranked grass-based dairy farms in Ireland with confinement dairy operations in the United States and UK. The carbon emissions were roughly equal until researchers factored in carbon sequestration rates, which resulted in 5 percent and 7 percent lower net emissions for the Irish grass-based dairy farms than the UK and US confinement operations, respectively. On the other hand, the confinement systems

TABLE 7.2. Carbon Sequestration Balanced Against Methane Emissions

	Degraded Pasture	Improved Pasture	Intensive Silvopasture	Intensive Silvopasture plus Timber Trees
Carbon sequestered	-0.3 t/ha	0.9 t/ha	4.6 t/ha	9.4 t/ha
Emissions (methane and nitrous oxide, in carbon equivalent)	0.6 t/ha	1.8 t/ha	2.2 t/ha	2.2 t/ha
Balance	-0.9 t/ha	-0.9 t/ha	2.4 t/ha	7.2 t/ha

Source: Naranjo et al., "Balance de gases de efecto invernadero en sistemas silvopastoriles intensivos con *Leucaena leucocephala* en Colombia," 6.
Note: In this 2012 study, degraded pasture and improved pasture (seeded to quality grasses and legumes) without managed grazing both lost more than three tons of carbon dioxide equivalent to the atmosphere per hectare every year. In contrast, intensive silvopasture had sequestration as good as other silvopasture systems, and the addition of timber trees was equivalent to some of the best agroforestry systems.

produced more milk per cow, which helped balance the higher emissions resulting from annual crop production for feed.[38] Still, 7 percent less emission from grassfed systems doesn't suggest to me that improved grazing per se is a climate change mitigation silver bullet, despite its many other potential benefits.

A fascinating study compared the carbon sequestration and methane (and nitrogen) emissions of four livestock systems in Colombia. These were degraded pasture (the baseline practice in the area), improved pasture (seeded with high-quality forage species), and intensive silvopasture with and without extra timber trees. The results are summarized in table 7.2. The nitrous oxide and methane are converted to carbon dioxide equivalents for simplicity. The degraded pasture gave off emissions of soil carbon as well as methane and nitrous oxides, with total emissions of 0.9 t/ha carbon equivalent (not good). The improved pasture sequestered 0.9 t/ha, but permitted higher stocking rates (more cows per hectare). This meant higher methane and nitrous oxide emissions—so much that the net emissions were as bad as degraded pasture at 0.9 t/ha of carbon equivalent lost to the atmosphere. (Note that though the methane and nitrous oxide emissions increased on a per-hectare basis, it might in fact decrease on a per-animal basis due to higher quality pasture. Some view this as a net decrease.) When *Leucaena* fodder trees were added at one per square meter (again, see the detailed system description below), net

sequestration was finally achieved at 2.4 t/ha. The addition of timber trees permitted an extremely impressive 7.2 t/ha net sequestration. Meat production was also increased dramatically.[39] This shows that methane and nitrogen emissions can negate or even overcome carbon sequestration in ruminant systems.

What does all of this mean for individuals trying to decide what to eat? How much meat, dairy, and eggs are "okay" to eat as far as the climate is concerned? The answer depends on how much carbon-friendly livestock can be raised in your region and how its emissions compare with crops (annual or perennial) that could be raised on the same land. Most plant-based foods have lower emissions than animal products, with notable exceptions, such as produce transported by air or grown in heated greenhouses. Projections show that changing our diets, including reduced meat consumption, can have a profound impact on overall emissions reduction. In one scenario looking at emissions of methane and nitrous oxides (but not carbon dioxide), adding diet change to carbon farming practices more than tripled the potential reduction.[40] Still, the IPCC notes that "substituting plant-based diets for animal-based diets is complex since, in many circumstances, livestock can be fed on plants not suitable for human consumption or growing on land with high soil carbon stocks not suitable for cropping [for instance, grassy slopes unsuited to tillage] . . ."[41] There is also considerable variation in emissions of animal products based on their diet and

TABLE 7.3. Status and Scale of Selected Livestock Systems

System	Example	Region	Status	Scale
Silvopasture		Global	Mixed	450 million ha[a]
Managed grazing	Holistic Management	Global	Recent	12 million ha[b]
Fodder silvopasture	*Dehesa* and *montado*; acorn-bearing oaks and cork with hog and sheep grazing	Spain, Portugal	Traditional	5.5 million ha[c]
Fodder banks	Coppiced mulberry for silkworm production	Global	Traditional	220,000ha[d]
Intensive silvopasture	10,000 *Leucaena* trees per hectare of pasture	Australia, Colombia, Mexico	Recent	210,000ha[e]

Note: These are just a few systems—not necessarily the largest or most representative—for which data on the scale at which they are practiced is available. Some practices are listed more than once here; an individual practice such as "Fodder silvopasture" is also included under "Silvopasture." The systems in this table are presented in descending order of scale. "Region" does not necessarily indicate the only place where a system is practiced, but rather the region described in the study that is quoted.

[a] Nair, "Climate change mitigation,"47.
[b] Neely and De Leeuw, "Home on the range," 337.
[c] Boffa, *Agroforestry Parklands in Sub-Saharan Africa*.

[d] See the Insect Fodder Trees sidebar on page 94 for the calculation.
[e] Cuartas Cardona et al., "Contribution of intensive silvopastoral systems to animal performance and to adaptation and mitigation of climate change," 7.

the climate and production system in which they were raised.[42] Vegan and vegetarian diets rely heavily on annual grains, which are themselves part of the problem when raised with tillage, nitrogen fertilizers, and so on.

The notion of leaving livestock completely behind is also unrealistic, not just because it is unlikely but because much of the world's farmland is suited to little else. Abandoning livestock would also displace millions of the world's poorest people from their livelihoods. Of the 880 million people who live on less than $1 a day, 70 percent are dependent on livestock for part or all of their income.[43] An FAO report notes that "pastoralism is considered the most economically, culturally, and socially appropriate strategy for maintaining the well-being of communities in dryland landscapes, because it is the only one that can simultaneously provide secure livelihoods, conserve ecosystem services, promote wildlife conservation and honor cultural values and traditions."[44] Globally roughly one billion people depend on livestock for their income—most of these depend on pasture-raised animals.[45] Additionally, it is difficult to be a vegetarian on a dry rangeland where few annual crops will grow (though mesquites, perennial sorghum, and others offer promise).

I hope you can see that one-size-fits-all answers do not do justice to the complexities of livestock and climate. The power of managed grazing to sequester carbon may be slight, especially when methane is accounted for, but the great scale of our planet's grasslands could still multiply small gains by enormous factors. Some other practices sequester impressive carbon even when balanced against methane. Silvopasture, a well-established agroforestry practice integrating trees with pasture, has received far less press than grazing but has much more impressive carbon numbers. Converting pasture to silvopasture is estimated to result in sequestration rates of 3 t/ha/yr according to the IPCC (much more impressive numbers than most managed grazing estimates).[46] The new practice of intensive silvopasture is one of the best climate impacts of any livestock practice and should be widely promoted and tested outside its proven range.

A powerful step would be to stop feeding annual grains to livestock. Let them eat pasture, fodder (leaves), and feed (nuts, woody pods, fruits) from perennials and consume crop residues and food wastes. Insects such as soldier flies and silkworms can provide food for monogastric livestock. The great advantage of livestock is that they can convert plants we cannot eat into livestock products that we can consume. Converting the full third of annual cropland that grows food for livestock to perennial crops could result in serious carbon sequestration. This allows perennialization of substantial farmland without asking for a difficult transition to a perennial diet for humans.

The livestock-based carbon farming strategies in this chapter are presented in clusters representing different livestock production models. They are: livestock integration; improved grazing and pasture management; tree-based grazing systems; zero-grazing systems; and systems under development.

Livestock Integration

Livestock have been integrated with crops for thousands of years. It is really only in the last 60 years or so that synthetic fertilizers have permitted the separation of crops and livestock.[47] In many parts of the world, this

FIGURE 7.1. Livestock integration: trials grazing sheep in apple orchards in Ireland. Photograph courtesy of AgForward.

separation never occurred and integrated production continues to this day.[48]

There are many opportunities to integrate livestock and crops within or between farms. There are also many benefits. Integrating crops and farm animals can reduce the need for farm inputs and replace or reduce hand labor and machinery use. Practices include:

- Feeding livestock on crop residues and by-products, which converts it to manure.
- Providing manure for crop and field fertility.
- Adding pasture and perennial fodder crops to annual crop rotations.
- Grazing under tree crops.
- Consumption of household wastes (especially by poultry and pigs).
- Using geese for weeding grasses; turning hogs onto crop residues for cleanup; using goats to clear land and manage brush.
- Draft power; "tillage" by hogs and scratching chickens; compost turning.
- Pest control, such as direct consumption of pests by ranging poultry or cleanup of dropped fruit containing pest and disease propagules by hogs and poultry.
- Pollination by the many species of honeybees.
- Grazing of pasture cropping and perennial grain systems at appropriate times.

Livestock integration also has drawbacks. Poorly designed or managed systems can result in the livestock damaging or destroying crops or being poisoned by toxic plants. Livestock can also damage young trees, consume crops, and dig holes in fields and orchards.[49] Sheep have damaged drip irrigation systems in nut groves.[50] Brief livestock rotations, the use of portable electric fencing and/or protective tree tubes, matching crop and livestock species, and other strategies can mitigate some of these challenges.[51]

The FAO considers livestock reintegration as among the most promising climate change mitigation and adaptation strategies.[52] The IPCC also rates crop–livestock integration as having high potential carbon impact because it is easily adopted by farmers and ready for wide implementation.[53] Ruminant livestock

can obtain up to 50 percent of their diet from crop residues.[54] And adding manure from grazing cattle to legume–grain crop rotations has been shown to double carbon sequestration.[55]

Across Asia and the Pacific, grazing under coconuts is common. This typically increases coconut yields and reduces pasture productivity, although livestock manage the understory and reduce mowing and weeding labor. This is a form of silvopasture in that it involves grazing under trees, but it is covered here under livestock integration as in this case the livestock are primarily there to reduce labor, with meat a secondary, minor project.[56]

Managed Grazing and Improved Pasture Management

Managed grazing is a broad category that includes many practices, including managing stocking rates (the density of animals per unit area), controlling the intensity and timing of grazing, enclosure of grassland to encourage resting, and various kinds of planned and adaptive grazing. The effects of the same practices are sometimes inconsistent among different regions and sites.[57] Most improved grazing practices have emerged since the 1970s; in the United States more than 288,000 farmers currently practice rotational or management-intensive grazing.[58] One form of managed grazing, Holistic Management, is practiced on an estimated 12 million hectares (30 million acres) worldwide.[59] There are also more traditional managed grazing systems, including al-Hima grazing, which dates back 1,400 years on the Arabian peninsula.[60]

Improved pasture management techniques include planting improved pasture species such as deeper-rooted grasses and nitrogen-fixing legumes, fertilization, managed burning, irrigation, and fire management.[61] (The line between pasture management and grazing can be blurry; I treat them together here, but the IPCC treats them individually.) Breeding to improve pasture crops is also important.

Thousands of ranchers swear by managed grazing techniques such as Holistic Management, with some reporting gains of 6 percent in soil organic matter.[62]

Figure 7.2. Managed grazing in the western United States. Photograph courtesy of Owen Hablutzel.

Yet I have found little scientific evidence that Holistic Grazing increases livestock productivity or ecosystem services. Rotational grazing has likewise been found to have few ecological benefits when human variables of goal setting, experience, and decision making are removed from the experiments; what seems distinct about managed grazing is the human manager, not the rotations per se. Graziers and researchers are working on ways to account for this to better practice, study, and teach managed grazing.[63]

There is disagreement among scientists about the carbon-sequestering potential of grazing and pasture practices. The IPCC notes that "optimally grazed lands" often have better carbon sequestration than overgrazed and ungrazed lands.[64] One review of 115 studies found annual sequestration rates ranging from 0.1 to 3.0 t/ha, with a global mean of 0.5 t/ha/yr.[65] Another global review of managed grazing found an average sequestration rate of 2.1 t/ha/yr. Examples in this study ranged from a staggering gain of 33.4 t/ha/yr to a serious net loss of 12.4 t/ha yr. This variation is due to the impacts of climate, land-use history, vegetation, and soil types.[66] (Most of these sequestration rates are not adjusted to account for methane emissions.) Additional reviews and estimations offer rates of 0.2 to 0.5, 0.1 to 0.2, and

0.3 t/ha/yr.[67] Meanwhile other studies have found little or no carbon impact, especially in dry grasslands.[68] The application of compost to grazed rangeland has been shown to result in an average sequestration rate of 0.3 to 1.6 t/ha per year for three years (not including the addition of carbon from the compost itself), while pasture yields increased by 50 percent.[69] This very promising technique is deservedly getting a lot of attention in the carbon farming world at the moment. Carbon stocks are likely to be in the 30 to 50 t/ha range.[70]

The IPCC has come down cautiously on the side of managed grazing and improved pasture management. Though the carbon impacts are fairly small on a per-hectare basis, the vast extent of the world's grasslands means these strategies can provide a significant opportunity for climate mitigation. The IPCC rates both managed grazing and improved pasture management as having a low potential global mitigation impact but being easily adoptable by farmers and in a state of readiness for implementation.[71] It has been estimated that improving management on just 10 percent of Africa's grasslands would sequester 1.3 million tons per year for 25 years.[72] Pasture and grazing practices that sequester carbon often also improve productivity.[73] Managed grazing is noted as an important adaptation strategy as well.[74] A study of European grazing systems found that methane emissions offset about 25 percent of the carbon sequestered in the pastures.[75]

Silvopasture

Simply put, silvopasture systems integrate pasture with trees. Agroforestry expert P. K. Nair estimates that silvopasture is practiced on an impressive 450 million hectares (1.1 billion acres) worldwide, including 9.2 million hectares (22.7 million acres) in Central America, 1.5 million hectares (3.7 million acres) in Chile, and 1 million hectares (2.5 million acres) in Greece.[76] It is, by far, the most widely practiced type of agroforestry in the United States and Canada.[77] In some silvopasture systems, animals graze existing woodlands and savannas; in others the animals graze beneath tree crops or plantation forestry trees. A third option allows for spontaneous growth of woody plants.

NEW RESEARCH ON CARBON SEQUESTRATION AND GRAZING

Scientists are working hard to understand why grazing can sequester carbon under some circumstances but seems to result in a net loss of carbon under other circumstances. Two recent papers dig into this question in great detail and represent the frontier of our current understanding. Although the researchers use different approaches, they arrive at a similar conclusion: under certain circumstances, grazing can sequester carbon in the soil, but it depends on a multitude of factors, and not all grasslands are suitable candidates for climate change mitigation through grazing and pasture management.

A 2013 paper entitled "Effects of grazing on grassland soil carbon: a global review" published in the journal *Global Change Biology* reported the results of a meta-analysis of 47 independent experimental contrasts from 17 different studies in an effort to isolate the factors that result in soil carbon gains or losses from grazing. The paper's authors determined that grazing tends to increase soil carbon in sandy soils, especially in more humid environments, but they also determined that the opposite is true in clay soils where grazing leads to the loss of soil carbon, again particularly in more humid areas. In grasslands dominated by warm-season grasses, the authors determined that light grazing causes a loss of soil carbon, while moderate and heavy grazing causes an increase in soil carbon. In cool season grass-dominated grasslands, light grazing tends to sequester carbon while moderate and heavy grazing causes soil carbon losses.[78] The paper's authors concluded that "Soils of grasslands represent a large potential reservoir for storing CO2, but this potential likely depends on how grasslands are managed for large mammal grazing," and "Our results, which suggest a future focus on why C3- vs. C4-dominated grasslands differ so strongly in their response of SOC [soil organic carbon] to grazing, show that grazer effects on SOC are highly context-specific and imply that grazers in different regions might be managed differently to help mitigate greenhouse gas emissions."[79] The authors estimated a maximum annual sequestration rate of 1.5 t/ha/yr and a lifetime carbon stock maximum of 16 t/ha as a best case scenario for soil carbon sequestration through managed grazing.[80]

The authors of a 2015 paper entitled "Greenhouse gas mitigation potential of the world's grazing lands: Modeling soil carbon and nitrogen fluxes of mitigation practices" that was published in the journal *Agriculture, Ecosystems, and Environment* developed models to better understand the circumstances under which select mitigation strategies lead to increased soil carbon sequestration and the circumstances under which those same strategies might actually lead to a net loss. They found that pasture improvement through nitrogen fertilization always led to net greenhouse gas emissions.[81] They also determined that only 28 percent of global grazing lands—still an impressive 711 million hectares—are suitable to carbon sequestration through improved grazing management.[82] If managed grazing were limited to these lands, the sequestration potential would be 30 million tons this century. If, however, managed grazing were applied to all grazing lands including the 72 percent that is not suitable for sequestration via grazing, it would result in net greenhouse gas emissions even taking into account the suitable 28 percent.[83] The sowing of pasture legumes leads to similar results. The authors calculated that 71 million hectares of grassland are amenable to net sequestration from legume sowing. The net sequestration of sowing only suitable land is an impressive 55 million tons of carbon. (39 million tons if you account for the nitrous oxide emissions the pasture legumes would produce.)[84] As with grazing, if legume sowing was practiced on all grasslands and not just the suitable grasslands, it results in a net emission of greenhouse gases.[85]

The bottom line is that carbon sequestration in grasslands is complex and scientists and researchers such as those who authored these articles are on the leading edge of looking for global patterns to help guide our grazing and pasture management practices. In the near future, we might have a set of guidelines to determine the best management practices for any given grazing area based on factors such as soil, rainfall, vegetation, and management practices.

FIGURE 7.3. A traditional restored Ngitili silvopasture system with cattle grazing under trees in Tanzania. Photograph courtesy of the World Agroforestry Centre.

FIGURE 7.4. Cattle and pigs grazing under encina oaks in *dehesa*, a traditional fodder silvopasture system in Spain and Portugal. Photograph courtesy of AgForward.

In silvopasture systems the spacing between trees is typically much wider than it is in forests or timber plantations. In some systems, animals graze between double hedgerows in wide, sunny alleys. Denser plantings can only be grazed profitably when they are young, although sometimes animals can still be used as mowers. Shade-tolerant pasture grasses and legumes are an important component of these densely planted systems.

Silvopasture and other pastures with woody plants sequester up to three times as much carbon as ordinary pastures.[86] Silvopasture systems are estimated to sequester 3 to 10 t/ha/yr globally, and 6.1 t/ha in temperate North America.[87] Lifetime carbon stock accumulation could reach 250 t/ha/yr.[88] Productivity of the livestock, pasture, and tree crops varies widely depending on species selection, spacing, and other factors. While dense shade is detrimental to pasture productivity, moderate shade often increases pasture yields compared with full sun, especially in the tropics. This is why most silvopasture systems have savanna-type spacing.[89] The carbon-rich leaf litter of trees is also beneficial to pasture crops. Shading of livestock in hot weather can reduce stress and increase weight gain.[90] Silvopasture trees moderate temperature extremes

for pasture below them, extending the pasture season. Shaded cattle can reach their target weight 20 days before those in sun.[91] Silvopasture has been identified as an important adaptation strategy. The role of trees in providing shade and moderating the local microclimate is part of their adaptive role.[92]

Fodder Tree Silvopasture

In this modified silvopasture system, the trees themselves produce food for the livestock in the form of fodder or various nuts, fruits, acorns, or woody pods. This fodder may drop down to the pasture when ripe, or in some cases must be knocked off the trees with poles. Some fodder trees are suitable for ruminant livestock, some for non-ruminants, and some for both, a topic that is thoroughly explored for temperate climates in J. Russell Smith's 1927 classic *Tree Crops: A Permanent Agriculture*. Of the 450 million hectares of silvopasture worldwide, the number of hectares devoted specifically to fodder tree silvopasture is unknown, but likely substantial.

Common tree species in this system include oaks, guasimo (*Guazuma ulmifolia*), and many species of pod-bearing legume trees, including mesquites and

INSECT FODDER TREES

A surprising number of tree crop species around the world are grown for the production of useful insects. These insects can be used for food, fodder, silk, dye, shellac, and more. There is increasing recognition of the energy efficiency of insects as food and fodder. One study found that edible crickets, for example, were 2 times as efficient as chickens at converting fodder to human food, 4 times as efficient as pigs, and 12 times more efficient than cattle.[93] That means that more people can be fed from a much smaller space—often from perennial, carbon-sequestering fodder trees.

Unlike honeybees, most cultivated insects require specific host plants to feed them. The best known of these insects is the silkworm (*Bombyx mori*), which can only be fed the leaves of mulberries and closely related species. Silkworms produce silk, while the worms and pupae can be eaten by humans or fed to hogs, fish, or poultry. (My wife, Marikler, raises silkworms at home as a supplemental feed for our poultry.) Several other species of silkworms are raised or wild-harvested around the world, each with its own favored host plants. Global silk production in 2013 was 159,000 tons.[94] In China, 0.7 ton of silk is produced per hectare,[95] which translates to roughly 222,000 hectares (549,000 acres) in silkworm fodder mulberry production.

Mopane worms (*Gonimbrasia belina*) and palm grubs (*Rhynchophorus* spp.) are other important food insects that depend on host trees. Mopane worms are wild-harvested and sun-dried for later consumption. Their nutritional value is astounding; they are 48 to 61 percent protein, 16 to 20 percent fat, and high in iron.[96] Over nine billion mopane worms are consumed annually, representing a US $85 million/year industry and

FIGURE 7.5. Nutritious mopane worms feed on mopane tree foliage. Photograph courtesy of Brock Dolman.

a major portion of the daily diet of people in southern Africa.[97] There are efforts underway to cultivate the mopane tree and domesticate the mopane worm as an improved source of food and livestock feed, but early efforts have shown that intensive monoculture leaves the worms vulnerable to viruses.[98]

Shellac was originally made from the exudate of a group of scale insects called lac insects (*Kerria lacca* and related species). In fact, today a large percentage of shellac is still made from lac insects and serves as a feedstock for many varnishes and coatings, which lengthen the life of wood products. Insect dyes remain important in the food and textile industries, especially cochineal, which is also made from a scale insect (*Dactylopius coccus*); a range of prickly pear cacti (*Opuntia* spp.) are cultivated as hosts for cochineal.[99] A listing of insect fodder trees is found in appendix A.

carob. (See appendix A for a list of fodder pod legume trees.) In semi-arid regions of Indonesia, multistrata fodder silvopasture systems provide fodder year-round. The pasture is productive from the early wet season to the early dry season. Mid-layer tree legumes offer fodder from the early dry season to the mid-dry season. The canopy trees produce fodder from the mid-dry season

to the mid-wet season. The total fodder production is 90 percent more than pasture alone. with a 46 percent higher stocking rate in the dry season.[100]

Perhaps the most famous fodder tree silvopasture system is the *dehesa* or *montado* of Spain and Portugal. This is an extensive system of dry Mediterranean areas covering 5.5 million hectares (13.6 million acres) today, which is less than its historic coverage.[101] In it, high-yielding acorn oaks (*Quercus ilex*) are interspersed with cork oak (*Q. suber*) and other crop trees at a spacing of about 20 to 50 trees/ha. Livestock, including sheep, pigs, and cattle, are grazed on pasture and benefit from the acorns dropped into the pasture. This is the source of the famous Iberian acorn-fed pork. Occasional annual crops on 5- to 10-year rotations keep brush under control.[102]

Other fodder tree systems include fodder foliage trees grown in pastures or as living fences at field edges. These trees are lopped for fodder and thrown down to livestock, often during dry seasons when pasture is unproductive. Typically these trees are pollarded, a form of coppicing where shoots are cut from higher trunks out of browsing reach of livestock. This is a traditional practice in Europe and elsewhere.[103]

Although data is not available for the carbon sequestration rates of fodder silvopasture specifically, it's reasonable to assume that it would be similar to comparable non-fodder tree silvopasture systems. In terms of yields, a multistrata silvopasture system that provides fodder year-round can have a beneficial impact on stocking rates. Likewise, the addition of *Faidherbia albida* trees for pods and fodder leaves permits doubling of cattle stocking rates in sub-Saharan Africa.[104]

Intensive Silvopasture

Intensive silvopasture was developed in Australia in the 1970s, where it is currently practiced on 200,000 hectares (494,000 acres). It combines improved pasture with extremely high densities of woody nitrogen-fixing legumes, typically *Leucaena leucocephala*. These trees are planted at a remarkable 8,000 to 10,000 per hectare. Grazing is under a planned rotation regime with electric fencing. Livestock graze the trees along with the pasture, with *Leucaena* resprouting rapidly in the resting period when livestock are rotated out of the paddock. In some cases useful woody overstory species are also incorporated.[105]

There are more than 5,000 hectares (12,000 acres) in Colombia and 3,000 hectares (7,400 acres) in Mexico under intensive silvopasture management with ambitious

FIGURE 7.6. Intensive silvopasture systems feature *Leucaena* fodder trees planted at 8,000 to 10,000 per hectare, as well as rows of timber trees in this case, in Colombia. The fodder trees are browsed to the ground in a managed grazing rotation. Photograph courtesy of CIPAV.

plans for large-scale expansion in both Latin American nations. Colombian and Mexican producers have adapted the system by adding more timber, palm, and fruit trees.[106] Additional research is needed to determine the suitability of intensive silvopasture outside of humid tropical regions, where plentiful water and sunlight doubtless contribute to its productivity. Species with potential in colder regions include *Albizia julibrissin*, *Cytisus proliferus*, *Lespedeza bicolor*, *Amorpha fruticosa*, *Elaeagnus angustifolia*, and *Atriplex canescens*.[107]

In terms of carbon sequestration, intensive silvopasture has high potential compared with silvopasture in general. One Colombian study reported 4.6 t/ha/yr. When timber trees are incorporated, this soared to 9.4 t/ha, an extremely impressive number.[108] In addition, the abundance of *Leucaena* leaves in the ruminant diet results in a lower methane emission per cow.[109]

Yield impacts are equally astounding. Intensive silvopasture permits the stocking of two to four times more livestock per hectare and produces 2 to 10 times more meat per hectare.[110] Intensive silvopasture can help reduce the effects of parasites and diseases on livestock by providing habitat for beneficial organisms; it also improves water quality and biodiversity.[111]

Fodder Banks and Pollarded Species

Not all sustainable livestock systems involve grazing. On smaller farms, livestock can be confined in shelters to maximize manure collection. These livestock are fed with fodder banks, which are coppiced plantings of woody fodder shrubs (and some giant grasses and other herbaceous species). These plants are cut and carried to the livestock enclosure, where they are fed to animals. See appendix A for a list of fodder bank species.

Fodder banks are a recent formalization of a traditional system and are widely practiced on smallholder farms around the tropics. Sequestration rates for tropical fodder banks are estimated at 0.1 to 0.5 t/ha/yr, with

FIGURE 7.7. Kenyan farmer Rose Koech cutting woody legume foliage from a fodder bank for her dairy operation. Photograph courtesy of the World Agroforestry Centre.

lifetime stocks of up to 140 t/ha.[112] In terms of yields, the integration of fodder legume shrubs and productive grasses for slope stabilization on marginal land in Indonesia sustainably increased stocking rates by a factor of four to six.[113]

Outdoor Living Barns and Green Corrals

Outdoor living barns and green corrals are livestock enclosures made of living woody plants. Cold, windy conditions can impact livestock health. Living barns are essentially windbreaks with small patches of forest to provide shelter from the elements.[114] Green corrals are hedgerows or living fences of livestock-proof woody plants—spiny, poisonous, dense, and/or unpalatable. They are an affordable strategy for confining livestock, and especially important during the stage of establishing vulnerable young tree crops on smallholder farms.

There is no data available on the status of these practices, although both have become more widely promoted in recent decades. Carbon sequestration is probably comparable to that seen in living fences and windbreaks, with a high rate over a fairly small area. Outdoor living barns have a positive impact on winter health of livestock. Green corrals work similarly to or in tandem with fodder banks for excluding livestock from crop areas, with concomitant impacts on tree survival and manure capture.

Perennial Feed and Fodder for Storage

Perennial crops can also be harvested and carried to confined livestock or stored for later use in feeding, an excellent alternative to feeding livestock with annual crops. Crops include hay from well-managed hay fields; tree hay (the dried and stored foliage of fodder trees); perennial staple crops such as carob for use in feed concentrates; and fermented silages from perennial fodders. Hay can presumably be managed in a carbon-sequestering fashion, although hay fields are often degraded due to poor management, such as repeated harvest without fertilization to make up for nutrients lost in removed hay.

The IPCC has stated that the use of perennial livestock feeds is important for reducing the need for annual grains and grazing land. They are particularly interested in the potential for using leaf protein concentrate, a potential by-product of processing perennial biomass for biofuels, as a feed source. They rate the global mitigation potential of this strategy as medium, consider adoption by farmers easy, and declare the practice ready for immediate implementation.[115] See chapter 14 for details.

Little data is available on the current extent of these practices, but fruits such as carob, acorns, chestnuts, breadfruit, mesquite, and avocado are among those used to feed livestock. Tree hay was once much more widespread in Europe.[116]

Restoration Agriculture

Restoration agriculture is a hybrid of silvopasture and multistrata agroforestry for cold climates. It is under development by Mark Shepard at New Forest Farm in Wisconsin, using the historic savanna of the region as the model. Shepard grows fruit and nut woody crops in multistrata systems, some with fruiting vines trained on them, and berries underneath. These polycultures are generally laid out on keyline for rainwater harvest. Between these polyculture hedgerows, rotations of multiple livestock species are grazed using leader–follower, mob-stocked managed grazing techniques. Some annual crops are also grown in strip intercropping systems.[117]

Shepard has been developing this system for decades, as well as breeding tree crops that are suited to his management system and very cold climate. With the publication of his book *Restoration Agriculture* there has been much interest in the system, with a small wave of early-innovator farms springing up across the United States.[118] In theory, restoration agriculture could be practiced in other climates as well.

Currently, there is no reliable data available on yields or carbon sequestration, although the latter is likely to

FIGURE 7.8. Restoration agriculture: cattle, keyline hazelnuts, and oaks at New Forest Farm in Wisconsin. Photograph courtesy of Peter Allen.

be at or above the high end of silvopasture, based on the high density of woody plants. Shepard's book *Restoration Agriculture* features a comparison of the yield of calories in an acre of his system versus the yield of calories in an acre of maize. He claims that his system produces as many calories per acre as maize, although on closer examination this is because he discounts 79 percent of the maize calories, as they are, statistically (on a national basis), not used for human food but rather for livestock feed, industrial feedstocks, and so on. Regardless of their potential end use, an accurate comparison must compare calories with calories. Shepard also counts calories from livestock in his system but discounts the maize used for livestock feed. By his own numbers, restoration agriculture produces a good bit less than half the calories of an equivalent area of maize.[119] This is nothing to be ashamed of, as it has fewer inputs, has a far better carbon balance, and provides many other agroecosystem benefits.

Despite these somewhat misleading comparisons, restoration agriculture is a promising new technique worthy of investigation and implementation. A new NGO, the Savanna Institute, is dedicated to testing and promoting restoration agriculture with the goal of offering solid data on this under-development practice.[120]

CHAPTER EIGHT

Perennial Cropping Systems

Fully perennial and integrated systems typically sequester far more carbon than any other type of farming—a notable exception being silvopasture. Tree-based systems do have undesirable albedo effects, somewhat offsetting their sequestration impact in temperate climates and canceling them out entirely in boreal climates, at least with evergreen trees. Fully perennial and integrated systems also tend to be the ones that require the most change to human diets and to our farm and food system, although feeding perennials to livestock is a good solution to that problem. In parts 3 and 4, I profile many promising perennial crops that could replace many annual crops or serve as industrial raw materials.

TABLE 8.1. Status and Scale of Selected Perennial Systems

System	Example	Region	Status	Scale
Perennial crops	Orchard, plantation	Global	Traditional	153 million ha[a]
Multistrata systems		Global	Mixed	100 million ha[b]
Managed bamboo		Global	Traditional	22 million ha[c]
Multistrata agroforests	Cacao agroforests	Humid tropics	Traditional	7.8 million ha[d]
Coppice	Coppiced eucalyptus plantations	Global (outside Australia)	Traditional (19th and 20th century)	"Most of[e]" 7.2 million ha[f]
Multistrata agroforests	Shade-grown coffee	Global	Traditional	6.12 million ha[g]
Multistrata agroforests	Tropical homegardens	Indonesia	Traditional	5.13 million ha[h]
Multistrata agroforests	Jungle rubber	Indonesia	Traditional (19th and 20th century)	2.8 million ha[i]
Multistrata agroforests	Tropical homegardens	Kerala, India	Traditional	1.3 million ha[h]
Multistrata agroforests	Tropical homegardens	Philippines	Traditional	70% of all households[h]
Multistrata agroforests	Tropical homegardens	Sri Lanka	Traditional	1 million ha[h]
Multistrata agroforests	Tropical homegardens	Bangaladesh	Traditional	540,000 ha[h]
Short-rotation coppice	SRC willows for energy	Sweden	Recent	16,000 ha[j]
Aquaforestry	Chinampas: raised beds in wetlands stabilized with willows	Mexican highlands	Historic, mostly gone	12,000 ha in 1491[k]

Note: In addition, a category called "tree woodlots and specialty crops" occupies an estimated 50 million hectares, according to Nair's "Climate change mitigation," p. 47. This category includes border tree plantings, coppiced and non-coppiced woodlots, improved fallows, fodder banks, and woody biomass plantings. "Region" does not necessarily indicate the only place where a system is practiced but rather the region described in the study that is quoted. Some practices are listed more than once here; an individual practice such as "Tropical homegardens" is also included under "Multistrata systems."

[a] FAO Statistics Division online.
[b] Nair, "Climate change mitigation," 47.
[c] Henley and Yiping, "The climate change challenge and bamboo."
[d] Zomer et al., *Trees on Farm*, 3.
[e] El Bassam, *Handbook of Bioenergy Crops*, 186.
[f] Clay, *World Agriculture and the Environment*, 309.

[g] Jha et al. "Shade coffee," 4.
[h] Nair and Kumar, "Introduction," 3.
[i] Wibawa et al., "Rubber based Agroforestry Systems (RAS) as alternatives for rubber monoculture system," 2.
[j] Weih, "Intensive short rotation forestry in boreal climates," 1370.
[k] Whitmore and Turner, *Cultivated Landscapes of Middle America on the Eve of Conquest*, 221.

The IPCC determined that perennial crops in general have medium potential impact, are easily adoptable, and are ready for wide implementation in 5 to 10 years.[1] It reported that the integration of biomass crops into farm operations has medium global mitigation potential, is easily adoptable by farmers, and is ready for global implementation.[2] Perennial crops as a group are also an important adaptation strategy.[3]

Fully integrated perennial cropping systems can be clustered in the same way that annual and livestock systems are. Here is the organizational structure: multistrata systems; perennial monocultures; aquaforestry; systems under development.

Multistrata Agroforests

Multistrata agroforests are forest-like systems that feature multiple layers of trees, often with the incorporation of herbaceous perennials, annual crops, and livestock. They range from very simple, such as coffee grown under a single species of shade tree, to extremely biodiverse and complex, such as Mesoamerican homegardens that average 348 plant species per hectare.[4] The tropical homegarden, an ancient multistrata agroforest system, dates back more than 13,000 years in Southeast Asia and today accounts for more than 5 million hectares (12 million acres) in Indonesia, 1.3 million hectares (3.2 million acres) in Kerala, India; 1 million hectares (2.5 million acres) in Sri Lanka; and 0.5 million hectares (1.2 million acres) in Bangladesh.[5] Seventy percent of Filipino households have homegardens.[6] At present such systems are rare in temperate climates. Given the very high carbon and other benefits, development of such systems should be a priority.

Although homegardens focus on highly diverse products for home consumption, multistrata agroforests also produce global commodities. Cacao agroforests are one of the world's most widely practiced multistrata systems, with an estimated 7.8 million hectares (19.3 million acres) in the humid tropics, mostly West Africa.[7] Coffee under traditional diverse shade canopy accounts for 24 percent of the world's crop at 2.4 million hectares (5.9 million acres), and coffee with less diverse, sparse shade trees another 35 percent at 3.6 million hectares (8.9

million acres).[8] "Jungle rubber" polycultures in Indonesia occupy 2.8 million hectares (6.9 million acres).[9] On inhabited Micronesian coral atoll islands, multistrata agroforests dominated by coconuts and breadfruit cover 50 to 70 percent of the land surface. Other species within the agroforests include native timber trees, fruits such as banana, mango, and papaya, vegetables such as peppers and *Pisonia grandis*, and starchy root crops, including taro and its relatives.[10] Between homegardens and other multistrata systems, P. K. Nair estimates that there are about 100 million hectares (247 million acres) globally.[11] Only four crops occupy more land than that: wheat, maize, rice, and soybeans.[12]

There are many plantation-scale agroforests that feature only two or three species. For example, in Brazil peach palms are underplanted with cacao, and black pepper vines are trained on the peach palm trunks. In tropical Africa, passion fruit vines are trained on living trellises of nitrogen-fixing *Sesbania* trees. And in Hawaii, coffee is grown under macadamias.[13]

Multistrata agroforestry has by far the best carbon sequestration rates of any food production system, between 10 and 40 times higher than typical improved annual crop production and managed grazing.[14] Annual aboveground biomass sequestration ranges from 3 to 13

FIGURE 8.1. Jungle rubber plantations, such as this one in Jambi, Indonesia, are a widespread multistrata agroforestry system. Photograph courtesy of the World Agroforestry Centre.

tons per hectare and total annual sequestration between 13 and 40 tons per hectare. Lifetime aboveground biomass rates of 72.9 t/ha (at age 4 years) and 155.8 (at age 38) have been recorded in the Philippines. Sequestration rates for tropical multistrata systems are estimated at 0.2 to 18.0 t/ha/yr. Lifetime soil organic carbon stocks are estimated at up to 300 t/ha, the highest of any agricultural system I've seen.[15]

The impact on yields can be difficult to measure given the complexity of these systems. The yield of primary commodities in a complex multistrata agroforest system can be higher or lower than yields of the same primary commodity grown in monoculture. The oil yield from oil palms in complex agroforestry polycultures has been measured at 6.4 to 8.7 t/ha/yr compared with 5 t/ha/yr in baseline monocultures. On the other hand, shade-grown coffee yields are lower than those of coffee grown in a full-sun monoculture (in part because shade-grown coffee is planted less densely).[16] But shade-grown coffee plants are also productive for 18 to 24 years as opposed to 6 or 8.[17] It's also very difficult to account for all of the products of a complex multistrata system, such as fruits, macadamias, firewood, and timber, as well as the impact of each of these products on the others and on the primary commodity.[18]

Perennial Monocultures

Although we have a global food system largely geared toward annual crops, there are many well-known perennial crops, including trees such as apples, palms such as coconut, shrubs such as blueberries, succulents such as agaves, vines such as grapes, and giant herbs such as bananas. Many perennial crops have been grown for thousands of years; they currently cover 153 million hectares (378 million acres) worldwide.[19] These perennial crops are frequently grown in monoculture orchards or plantations that don't feature understory crops, polyculture companions, or livestock integration, although all of these variations are possible, and this chapter includes many such examples. As with some other cropping systems featured in this book, I felt it was important to include perennial monocultures because, in terms of carbon sequestration, they're a great strategy

FIGURE 8.2. A walnut orchard, a woody perennial staple crop monoculture, in California. Photograph courtesy of Robert Couse-Baker/Creative Commons 2.0.

for carbon sequestration. Still, they're not the best-case scenario because they lack the biodiversity that makes for strong ecosystems (and local economies).

The IPCC estimates that conversion from annual cropland to orchard sequesters an average of 3.5 t/ha/yr.[20] Annual sequestration rates of perennial crops are also reported in individual studies. They range from 0.8 to 17.5 t/ha (see appendix C). Some monoculture plantations maintain bare soil between perennial crop rows, a practice that surely reduces carbon sequestration. In some cases the yields of these perennial crops are competitive with annual crops.

Managed Bamboo

Bamboo cultivation and management is an ancient practice in Asia, the Americas, Africa, and North America. Managed bamboo is currently grown on more than 22 million hectares (54 million acres) worldwide, in addition to wild bamboo.[21] It is cultivated for 1,500 uses, including construction materials, paper feedstock, and many more.

Carbon sequestration ranges from 8 to 34 t/ha/yr, with lifetime aboveground and soil totals of 60 to 288 t/ha.[22] The world's 22 million-plus hectares

FIGURE 8.3. A managed bamboo planting at Las Cañadas, Mexico. Bamboo offers 1,500 uses and high carbon sequestration levels.

FIGURE 8.4. Chestnut coppice at Ben Law's site in West Sussex, England. Photograph courtesy of Mark Krawczyk.

of bamboo are estimated to store a total of 727 million tons of carbon.[23] As a feedstock for paper and cardboard production, bamboo can yield six times as well as pine plantations.[24] Bamboo biomass yields are up to 26 t/ha, which puts them about in the middle of the pack of perennial biomass crops.[25]

Coppice

Many species of woody plants resprout after being cut. Coppicing is a form of management in which these trees and shrubs are cut, and cut again, and cut again, on a rotation. Historically coppice has been widespread since ancient times in Europe and North America, notably in California. In the Netherlands, coppice once covered 130,000 hectares (321,000 acres).[26] In traditional European coppice, trees are cut on 7- to 25-year rotations, depending on what size wood is desired.[27] Its products include firewood, natural building and craft materials, wood pulp for paper, and other biomass products. In some ways coppiced trees can be thought of as filling bamboo's niche in regions that have historically had no bamboo.

Coppicing is used in many agroforestry systems, including alley cropping, contour hedgerows, fodder banks, FMNR, perennial improved fallows, and woody agriculture. Some plants thrive under coppicing—doubling or tripling their life spans to 500 or even 800

years.[28] Contemporary eucalyptus coppicing, on the other hand, has a much shorter cycle, with shorter intervals between cuttings and perhaps three to four cuttings before replanting.[29] Most eucalyptus plantations outside of Australia today are coppiced.[30] Collectively these plantations represent 7.2 million hectares, making this contemporary coppice system one of the world's most widespread woody perennial farming systems.[31]

Agroforestry woodlots in Kerala, India, have been shown to sequester 23.9 t/ha/yr—and a remarkable 55.8 t/ha/yr in Puerto Rico. (Of course, much of this ends up burned as fuel and returned to the atmosphere.) On the other hand, some coppiced wild Miombo woodlands showed only 0.5 to 0.9 t/ha/yr in semi-arid Zambia.[32] Lifetime soil carbon stocks are estimated at up to 140 t/ha.[33]

Biomass production in traditional and contemporary coppice systems can be very high. Eucalyptus can produce 45 cubic meters (1,589 cubic feet) per hectare per year, or about 22.5 tons of highly dense material for paper pulp and firewood. Sometimes the second cutting in eucalyptus systems yields more than the first, as mature root systems resprout vigorously.[34]

Short-Rotation Coppice

Short-rotation coppice (SRC) is a relatively new (since the 1970s) intensive monoculture system of coppicing

FIGURE 8.5. A short-rotation coppice (SRC) of willow with mechanized harvest in New York. Photograph courtesy of D. Angel, State University of New York College of Environmental Science and Forestry.

woody plants that represents the mechanization of traditional coppicing. The woody plants are typically cut on two- to three-year rotations and harvested with mechanized chipper-shredders for use as bioenergy. (As I discuss in chapter 19, this is not necessarily the best use of biomass feedstocks.) Sweden has pioneered SRC willow for energy production; there are more than 16,000 hectares (40,000 acres) devoted to SRC willow there today.[35] In some short-rotation coppice systems—willow being the notable example—plants are fed with nutrient-rich wastewater that they help to treat as it fertilizes them. Unlike traditional coppicing (where individual plants can live for centuries), these systems have a life span of 20 to 30 years. See appendix A for a list of species cultivated in SRC systems.

Total sequestration rates of both soil and aboveground biomass of 4.3 and 7.0 t/ha/yr have been found in SRC systems.[36] Sequestration in the soil alone was measured at 0.4 t/ha/yr. Generally SRC carbon totals are on the lower end of carbon sequestration. One tropical system in India had 45 t/ha/yr lifetime at age eight years.[37] Tropical SRC tends to yield 15 to 30 t/ha/yr of dry material, with 10 to 15 for temperate climates.[38] Yields of SRC willow range from 2 to 42 t/ha/year dry matter.[39]

Herbaceous Biomass Crops

Herbaceous biomass crops are large, robust perennial grasses—including *Pennisetum*, *Miscanthus*, *Phalaris*, and *Tripsacum*—that are grown as feedstock, typically (though not always) for energy production. They are usually grown in monoculture and mechanically harvested with extra-large hay equipment, although some are harvested with shredders, as with short-rotation coppice. A few herbaceous non-grasses, such as cardoon, are also grown for biomass. Although hay making and thatching are ancient practices, the large-scale

FIGURE 8.6. The author with the extremely impressive biomass grass *Gynerium sagittatum* in Cuba. Image courtesy of Rafter Ferguson.

mechanized production of biomass grasses is a recent development, and there is little data available on global hectares currently under production.

One study found carbon sequestration at 6 t/ha/yr total, while two others found 91 to 106 lifetime total carbon. Biomass production from grasses ranges from 10 to 100 t/ha/yr dry matter, as good as or better than the best woody biomass crops. More typical global yields are 10 to 20 t/ha/yr, or 3 to 10 on marginal or degraded land.[40]

Woody Agriculture

The Badgersett Research Corporation in Minnesota is at the forefront of woody agriculture and defines it as "the intensive production of agricultural staple commodities from highly domesticated woody plants."[41] Although this system is still relatively experimental, a small number of farmers have commercial-scale plantings of Badgersett hazelnuts in woody agriculture systems. Badgersett's vision is to convert the maize and soybean landscape of Iowa to a staple crop carbon-sequestering system. Their current system involves

FIGURE 8.7. Mechanical harvest of neohybrid hazel in a woody agriculture system. Photograph courtesy of Perry Rutter.

AQUAFORESTRY

Several complex and interesting farming systems integrate trees with fish production, and sometimes with livestock and annual crops. These systems are known as silvoaquaculture or aquaforestry; Chinese dike-pond farming and Mexican chinampas are among the best-known examples. Other silvoaquacultural systems include mangroves and seasonally flooded rain forests with aquaculture beneath; aquaculture under crops such as coconut and African oil palm; tree planting in aquaculture ponds (such as bald cypress in crayfish ponds in the southern United States); and homegardens.[42]

DIKE-POND SYSTEMS

Perhaps the most well-known silvoaquaculture model is the Chinese dike-pond system, in which crops and livestock are grown on the dikes of fishponds. Multiple species of fish are raised in the ponds (mostly species of carp), each with its own niche.[43] This practice goes back to the ninth century and remains popular in China

FIGURE 8.8. Harvesting mulberry leaves for silkworms in a traditional Chinese dike-pond aquaforestry system. Photograph by Elma Okic.

today due to its productivity and profitability, with the traditional version featuring coppiced mulberry trees on the dikes.[44] The leaves are fed to silkworms, and the silkworm waste is fed to the fish. The sludge from the pond bottom is periodically removed and spread beneath the mulberries.[45] In these complex systems the waste product of each component is used as food or fertilizer by another. Dike-pond systems are popular because they provide a "high and steady yield" throughout the year.[46]

Sugarcane, bananas, fodder grasses, fruit trees, bamboos, and water spinach are also cultivated on the dikes, depending on market needs.[47] Pigs, known in China as "walking fertilizer factories," are fed aquatic vegetation and fodder crops from the dikes. Their manure is used to fertilize the ponds. Ducks are also important to fertility and provide additional protein.[48]

CHINAMPAS

The chinampa system, in which long, narrow raised beds are built up in swamps, was developed in the Mexican highlands and practiced on about 12,000 hectares (30,000 acres) at the time of contact.[49] The edges are stabilized with willows (*Salix bonplandiana*). A variety of annual and perennial crops have been grown in chinampas.[50] Fish are harvested from (but apparently not cultivated in) the canals, which are also used to transport produce by boat.[51] Aquatic weeds are collected and used as compost, as was mud from the canals historically.[52] Chinampas remained important until the mid-20th century, but today are on the wane despite good productivity and ecological benefits, due to changing water levels, city encroachment, and global trade policies.[53] Many are being converted to greenhouses for flower production, given higher flower prices.[54]

A similar system is practiced in China, where poorly drained land is converted to a series of dikes and ditches. Fish, shrimp, turtles, and shellfish are raised in the dikes. The dikes are planted to timber trees, fruits, vegetables, and mushrooms.[55] Although no doubt labor-intensive to construct, Chinese dike-ditch systems produce between 5 and 10 times the income of cereal crop monocultures in the same region.[56]

SPECIES POTENTIALLY SUITED TO WOODY AGRICULTURE

The species listed in table 8.2 are suited to coppicing and produce seeds, pods, or other fruits durable enough for mechanical harvest and long-term storage, although there are many challenges—physiological, management, harvest—that could disqualify them as successful woody agriculture species. We need far more research, experimentation, and probably some substantial variety selection and breeding before any of these could realistically be ready for woody agriculture production. Woody agriculture could be a promising technique for producing perennial cotton in the tropics, with cotton plants annually mechanically harvested for a combined biomass–fiber–oilseed crop. (To my knowledge no one is working on this at present.)

neohybrid hazelnuts, chestnuts, and hickory-pecans grown in double rows and coppiced to the ground on roughly 10-year rotations. This kind of coppicing is an alternative to pruning for rejuvenation and removal of diseased or unproductive wood; the coppiced material is available as biomass feedstock. This mechanized biomass harvest is immensely less labor-intensive than hand pruning.

The staff at Badgersett is working on mechanical harvesting with machines that straddle the rows and had a successful trial of this equipment with hazelnuts in 2013. This improved harvesting system is designed to greatly reduce labor and machinery hours, as is replacing pruning with coppicing. At the farm, sheep and horses graze the grass between rows, poultry are used for insect control, and hogs are introduced after harvest to clean up fallen nuts and convert them to pork. Badgersett is also using interesting breeding techniques in conjunction with the development of this system. (Their breeding work is discussed in chapter 10, and woody agriculture is described in detail in *Growing Hybrid Hazelnuts: The New Resilient Crop for*

a Changing Climate.) A similar system is used for olive production in the Mediterranean, with coppicing on a 10-year rotation.[57] Likewise, in some regions farmers harvest coffee by mechanically cutting the entire plants to the ground every two years and separating out the fruit.[58] (The plants die after three or four cuttings.) Sea buckthorn is also mechanically harvested for a combined biomass–fruit crop in Europe.[59]

Jonah Adels, a student of mine, took roughly 1,000 soil samples at Badgersett to test for carbon sequestration in plantings of different ages. Tragically, he died on the return trip in a car accident. A team of his friends is working to analyze his data and publish the results in his honor. Yield data is not yet available from this system-under-development, but refer to chapter 10 for remarkable impacts of Badgersett breeding on the yields of hazel and chestnut.

Perennial Grains

Scientists have been working since the 1970s to develop perennial versions of the annual grains—cereals, legumes, and oilseeds—on which humanity relies. Today perennial grains are still largely hypothetical, although small amounts of some new and traditional perennial grains—some of which were regional crops historically—are being grown. In some cases researchers are crossing modern annual crops with perennial relatives; in others, they are domesticating wild perennials. As a group, these crops are still largely hypothetical or very low yielding. (Details can be found throughout part 3, particularly the discussion of perennial cereals in chapter 13. Perennial grain breeding issues are discussed in chapter 10.) But there is a ton of potential. A number of farming systems have been proposed:

- Monoculture production of perennial grains is feasible. Currently, some low-yielding perennial grains are commercially grown in monocultures for the gluten-free market.[60]
- Perennial grains could be grown in rotations with two to four years of cropping followed by seeding of annual crops or pasture, much as hay and pasture are included in crop rotations today.[61]

TABLE 8.2. Crops Potentially Suited to Woody Agriculture

Latin Name	Common Name	Product	Latin Name	Common Name	Product
Balanced Carbohydrate Crops			**Edible Oil Crops**		
Araucaria spp.	Bunya bunya, monkey puzzle	Nuts	*Carya* spp.	Hickory, pecan	Nuts
Barringtonia spp.	Cutnut	Nuts	*Elaeagnus rhamnoides*	Sea buckthorn	Edible oil
Brosimum spp.	Ramon, Maya breadnut	Nuts	*Gevuina avellana*	Chilean hazel	Nuts
Castanea spp.	Chestnut	Nuts	*Madhuca longifolia*	Butter tree	Nuts
Castanopsis spp.	Tropical chestnut	Nuts	*Moringa* spp.	Moringa	Oilseed
Cordeauxia edulis	Yeheb	Nuts	*Olea europaea*	Olive	Edible oil
Ginkgo biloba	Ginkgo	Nuts	*Prunus* spp.	Siberian and Manchurian apricot	
Gleditsia triacanthos	Honey locust	Pods	*Schleichera oleosa*	Macassar oil tree	Edible oil
Gnetum spp.	Jointfir	Nuts	*Virola sebifera*	Virola nut	Industrial oil
Inocarpus fagifer	Tahitian chestnut	Nuts	**Industrial Starch**		
Prosopis spp.	Mesquite	Pods	*Aesculus* spp.	Horse chestnut, buckeye	Industrial starch
Quercus spp.	Oaks	Acorns	*Corynocarpus* spp.	Karaka	Industrial starch
Protein Crops			**Industrial Oil (Inedible)**		
Acacia spp.	Australian wattleseed acacias	Beans	*Aleurites* spp.	Moluccana	Industrial oil
Albizia lucidior	Tapria siris	Beans	*Azadirachta indica*	Neem	Industrial oil
Cajanus cajan	Perennial pigeon pea	Beans	*Jatropha curcas*	Jatropha	Nuts, industrial oil
Caragana arborescens	Siberian pea shrub	Beans	*Melia azedarach*	Chinaberry	Industrial oil
Erythrina edulis	Chachafruto	Beans	*Pongamia pinnata*	Pongam	Industrial oil
Parkinsonia aculeata	Jerusalem thorn	Beans	*Ricinus communis*	Castor bean	Industrial oil
Protein & Oil Crops			*Sapindus saponaria*	Soapnut	Industrial oil
Adansonia digitata	Baobab	Fruit, nuts	*Simarouba amara*	Paradise tree	Industrial oil
Adenanthera pavonina	Condori wood	Beans, edible oil	*Trichilia* spp.	Mafura butter	Industrial oil
Anacardium occidentale	Cashew	Nuts	*Vernicia* spp.	Tung, mu oil	Industrial oil
Balanites aegyptiaca	Balanites	Nuts	*Virola sebifera*	Virola nut	Industrial oil
Chrysobalanus icaco	Cocoplum	Nuts	**Hydrocarbons**		
Corylus spp.	Hazel	Nuts	*Hymenaea courbaril*	Stinking toe	Hydrocarbons
Guibourtia coleosperma	African rosewood	Beans	**Fibers**		
Jatropha curcas	Jatropha (edible types)	Nuts, industrial oil	*Calotropis* spp.	French cotton	Fiber
Juglans spp.	Walnut	Nuts	*Ceiba* spp.	Kapok	Fiber
Pistacia vera	Pistachio	Nuts	*Gossypium* spp.	Perennial cotton	Fiber
Plukenetia volubilis	Sacha inchi		**Wax**		
Prunus armeniaca	Apricot	Nuts, edible oil	*Morella cerifera*	Wax myrtle	Wax
Prunus dulcis	Almond	Nuts	*Myrica pensylvanica*	Bayberry	Wax
Ricinodendron spp.	Djansang	Nuts	*Rhus succedanea*	Wax tree	Wax
Salvadora persica	Salvadora	Nuts	*Toxicodendron vernicifluum*	Lacquer tree	Wax
Terminalia catappa	Tropical almond	Nuts	**Gums**		
Edible Oil Crops			*Acacia senegal*	Gum arabic	Gum
Argania spinosa	Argan	Edible oil	*Prosopis* spp.	Mesquite	Gum
Canarium spp.	Pili, elemi	Nuts			

Note: Although we need more research, experimentation, breeding, and probably substantial variety selection before any of these could realistically be ready for woody agriculture production, these species are suited to coppicing and produce seeds, pods, or other fruits durable enough for mechanical harvest and long-term storage.

- Perennial grains might yield for a few years and then be grazed as pasture.[62] The Land Institute grazes the perennial grain Kernza to manage weeds, yielding both a cereal grain crop and seasonal livestock production.[63]
- Perennial cereal grains (grasses) and perennial legumes could be intercropped.[64] This could occur in alternate strips for ease of mechanical harvest.
- Perennial grains could be grown in diverse polycultures modeled on the prairie.[65] This was the Land Institute's original vision.
- Perennial grains could replace annual grain crops in agroforestry systems, leading to even better carbon sequestration and ecosystem services.[66]
- In a reversal of pasture cropping, annual green manure legumes could be sown into perennial cereal grain fields to provide nitrogen fixation.[67]
- The wild perennial grain nipa (*Distichlis palmeri*) is a salt grass and has potential for restoring degraded soils high in salt, as well as productive salt marshes for coastal storm surge protection.

Data on carbon sequestration from perennial grain systems is unavailable. Most sequestration would be in the soil since persistent aboveground biomass is minimal, such as in pasture systems. But researchers

FIGURE 8.9. A field of Kernza, a perennial grain under development at the Land Institute. Photograph courtesy of Lee Dehaan/ Creative Commons 3.0.

at the Land Institute estimate soil sequestration of 0.2 to 0.8 t/ha/yr for the perennial grains they are working on.[68] Although these sequestration rates are low, perennial grains would have a large impact on slopes and other vulnerable and degraded lands. Grain yields are generally low or still hypothetical, although perennial rice is yielding fairly competitively in research trials.[69]

CHAPTER NINE

Additional Tools

What the practices profiled in this chapter have in common is that they are not directly related to growing crops on the farm. Some are abiotic and involve constructed or manufactured elements to support crop production. Others look at the productive management of "wild" lands, which falls outside most definitions of agriculture. Many of the practices I include here are or were extremely widespread and permit us to continue eating foods we are familiar with, but the carbon sequestration rates are variable and frequently unknown. Afforestation, which is the establishment of trees in an area where previously there was no forest, and "pure" (non-productive) restoration of degraded land are also important sequestration tools, but fall outside the scope of this book, so I don't cover them here.

Rainwater Harvesting

Rainwater harvesting uses a suite of techniques to capture, store, or infiltrate rainfall to increase water availability for crops, livestock, and people during dry periods. It is an ancient practice; for example,

FIGURE 9.1. A Tarahumara Indian rainwater-harvesting system in Mexico. Photograph courtesy of Brad Lancaster.

earthworks in India's Thar desert are dated at 6,400 years old.[1] However, although it has been practiced in most or all arid environments, it is currently in decline worldwide, perhaps due to the increased availability of less sustainable water sources.[2] Rainwater-harvesting techniques include earthworks, mulching, vegetation, and water storage tanks to capture roof water and other rainwater sources. Earthworks include contour rock

TABLE 9.1. Status and Scale of Implementation of Selected Additional Tools

System	Example	Region	Status	Scale
Indigenous land management	Australia prior to colonization	Australia	Historic (until 1783)	769 million ha[a]
Indigenous land management	United States prior to colonization	United States	Historic (until 1492)	493 million ha[b]
Indigenous land management	Amazonia prior to colonization	Central Amazonia	Historic but some persists	21–200 million ha[c]
Biochar	Terra preta soils in Amazon	Amazonia	Historic (still present in soils)	55 million ha[d]

Note: "Region" does not necessarily indicate the only place where the tools are practiced, but rather the region described in the study that is quoted. Some practices are listed more than once here; an individual practice such as "Terra preta" is also included under "Amazonian indigenous land management."

[a] Gammage, *The Biggest Estate on Earth*, 173.
[b] Rough estimate based on half of area of contemporary United States.
[c] Mann, *1491*.
[d] Mann, "The real dirt on rainforest fertility," 920.

walls, berms, and basins; terraces; snow traps; French drains; infiltration basins; tractor-drawn imprinting rollers to create small pits; zai holes and waffle gardens (small hand-dug planting pits); and many others.[3] All of these techniques catch rainwater and store it or infiltrate it into the soil on-farm, rather than letting it wash away downstream.

The IPCC rates rainwater harvesting as having a medium global mitigation potential, with moderate ease of farmer adoption and 5 to 10 years required for wide implementation.[4] Not surprisingly, it is an important adaptation strategy in regions that will be receiving erratic or reduced rainfall.[5] Rainwater harvesting increases yields and biomass production by making more water available to crops. A study on sites in sub-Saharan Africa showed a yield increase of 150 percent.[6]

Terraces

Terraces are walls or other structures that create level planting areas on slopes. They are an ancient practice that developed independently in agricultural regions around the world. They're also long-lived; some terraces are still functioning 1,000 years after construction.[7] Although terraces are labor- and energy-intensive to install, they harvest rainwater and prevent erosion effectively. They are considered a best management practice for soil and water conservation.[8]

The erosion control benefits of terracing enable sustained yields on steep slopes for centuries and beyond. The water-harvesting impacts are also important. Terracing is considered an important adaptation strategy.[9]

Keyline

Keyline design was developed in Australia in the 1950s by P. A. Yeomans as a way to maximize water resources on farms through landform analysis. It combines soil improvement, water harvesting, and farm planning and management. Keyline was one of the inspirations for permaculture. Practically speaking, keyline can include construction of ponds in strategic locations, slightly off-contour subsoil plowing and perennial plantings, and managed grazing.[10] One important goal is the movement of water from gullies to drier ridges through subsoiling and off-contour vegetation rows.

FIGURE 9.2. Rice terraces in Sa Pa, Vietnam. Photograph courtesy of Topas Ecolodge/Creative Commons 2.0.

FIGURE 9.3. Keyline farming at Rancho San Ricardo in Oaxaca, Mexico. Keyline is an integrated farm planning and management tool. Photograph courtesy of Rodrigo Quiros.

FIGURE 9.4. Biochar is a special kind of charcoal with potential to improve yields and sequester carbon for centuries when used as a soil amendment. Photograph courtesy of Dale Hendricks.

Although it has been scarcely noticed by scientists and there is no information available on the global scale of adoption, keyline is popular with many farmers who credit it with impressive soil organic matter improvement.[11] It is applicable to large farms and ranches of hundreds of hectares or more and should be a priority of carbon farming research.

Biochar

Biochar is a type of charcoal that results from burning biomass in the absence of oxygen. Its manufacture produces energy that has a better life-cycle analysis than most biofuels (meaning it is more carbon-efficient), as well as the biochar itself, which is incorporated into agricultural soils for carbon sequestration and improved fertility.[12] More than 1,000 years ago, Amazonian farmers used biochar along with bone, compost, and other organic materials to improve infertile soils. Remarkably, these improvements have persisted to the present day in rich fertile soils as deep as 2 meters (7 feet).[13] Amazonian terra preta, as it's called, may cover as much as 10 percent of the Amazon basin, or 55 million hectares (136 million acres). Many of these fertile soils are still farmed today.[14] Contemporary use of biochar is still experimental and limited in scale, although it is expanding rapidly.

THE BIOCHAR CONTROVERSY

Biochar has been at the center of a controversy reminiscent of the food-versus-biofuels debate, in which people argued about whether it was legitimate to use land to grow biofuel crops that could be used to grow food. With biochar, there is a similar concern about land grabs. These *carbon grabs* are defined as "large scale appropriations of land and resources for global climate change mitigation benefits and profits from carbon markets." In other words, there is a risk that large agribusinesses and investors will buy out small farmers to produce dedicated biomass crops for biochar production. Indeed, such industrial biochar farms have been proposed as a "silver bullet" for carbon sequestration, and there is concern that biochar, along with biofuels, will compete for land that is needed for food production.[15]

Fortunately there is an alternative. Biochar can be integrated into small-scale agroecological farming systems. For example, crop residues, farm wastes, and unwanted invasive plant species can be charred and utilized on-farm. In this scenario, rather than a silver bullet, biochar is one component of an integrated strategy.[16]

The IPCC rates biochar application as having high global mitigation potential, being moderately easy to adopt by farmers, and still at the research-and-development stage.[17] A meta-analysis of multiple studies found an average crop yield increase of roughly 15 percent in biochar-amended soils. Amending manure and fertilizers with biochar may reduce emissions of methane and nitrous oxides. Biochar in soils is likely to persist for 100 to 1,000 years.[18] It is particularly useful in tropical and sandy soils where nutrients are easily leached away because it provides binding sites for nutrients. The UN Convention to Combat Desertification is promoting the use of biochar to help restore degraded land.[19]

Indigenous Land Management

Despite a widespread belief that indigenous hunting-and-gathering people have, and had, a passive impact on the land, in fact they have a long tradition of actively managing it. Indigenous people have used controlled fires, the planting and transplanting of useful plants, and the active management of game populations to cultivate productive and low-maintenance landscapes to meet their needs.

According to Bill Gammage in *The Biggest Estate on Earth: How Aborigines Made Australia*, for millennia most of Australia was burned every one to five years; even remote ridges would be burned several times during a person's lifetime.[20] This use of managed burning (and the absence of predators) gave people control over where and when new grass would emerge, which allowed for a precise knowledge of where grazing animals, such as kangaroos, could be found.[21] In California indigenous people used fire and horticultural practices on "wild" plants to produce a diverse and productive mosaic of habitats for perhaps 12,000 years,

FIGURE 9.5. Bur oak savanna was once a widespread anthropogenic ecosystem under the management of indigenous people. This photograph is from a savanna in Wisconsin that is managed with fire to restore it to its pre-colonization state.

until forced off their land.[22] In fact, indigenous oak ecosystem management in the United States is classified as an agroforestry practice by the US Department of Agriculture in *Indigenous Uses, Management, and Restoration of Oaks in the Far Western United States*.[23] M. Kat Anderson documents such indigenous systems in pre-colonial United States in her book *Tending the Wild: Native American Knowledge and the Management of California's Natural Resources.*

Scientists have also estimated that 11.8 percent of the non-flooded Amazon forest was anthropogenic, essentially varying degrees of multistrata agroforest featuring the 69 species of domesticated native Amazonian trees.[24] In Central Amazonia alone, this comes to 21 million hectares (52 million acres). Many scholars today speculate that the entire Amazon forest may have been managed in this fashion—and much of it still is today.[25]

These management systems transcend the distinction Western cultures draw between nature and agriculture and provide a model of sustainable land management that can meet people's needs for thousands of years while maintaining high levels of soil carbon—a critical example of the underacknowledged importance of Indigenous Knowledge (IK) to humanity.

Little data is available on the carbon sequestration value of indigenous management practices. However, we do know that Australia's soils have suffered carbon losses of between 20 and 70 percent since colonization. The Australian *State of the Environment 2011* report asserted that the positive impact of Aboriginal management on soils was "profound."[26] The American prairies lost about 50 percent of their soil carbon after conversion from indigenous management to agriculture.[27] As an adaptation practice, indigenous management has much to recommend it, especially since Australian indigenous management systems, for example, produced more reliably than farming. Even with lower yields, having some predictable food can be better than gambling on rain-requiring annual cereals, given Australia's highly unpredictable rainfall.[28]

The Australian government's Department of the Environment has acknowledged the carbon sequestration value of indigenous fire management and, as part of its Carbon Farming Initiative, pays Aboriginal communities for fire management. This fire management also reduces the destructive impact of uncontrolled fires; the fire-related methane and nitrous oxide emissions are more than offset by vigorous, healthy postfire plant regrowth. This program has provided revenue and recognition of the ecosystem services provided by Aboriginal land managers.[29] Data on yields is likewise lacking, although broad-scale indigenous management is an extensive practice that likely has lower crop yield than industrial agriculture—but also fewer inputs and less required labor.

Productive Restoration

Many techniques have been developed to restore healthy ecosystems in degraded, deforested, and otherwise damaged habitats. Productive restoration is a still largely experimental subset of these practices that intentionally incorporates useful organisms such as edible native fruit and nut trees. Basically, the idea is that when replanting an area to native vegetation, why not emphasize the edible and useful members of the local flora, fungi, and fauna?

FIGURE 9.6. Michelle Gabrieloff-Parish of the Woodbine Ecology Center in Colorado using productive restoration to heal a riparian area following fire suppression by managing for useful plants and planting useful native species.

One ambitious proposal for productive restoration aims to restore the "buffalo commons" in the US central prairie region. This vast region covers 162 million hectares (400 million acres) and today is largely annual cropland.[30] What once was a wild population of 30 to 60 million bison had declined to a few thousand by 1880. Today there are perhaps 150,000 American bison.[31] The buffalo commons proposal would restore between 4 and 8 million hectares (10 and 20 million acres) of dry prairie—which is not well suited to annual crops—to native prairie, including wild buffalo.[32] This may be the largest proposed productive restoration effort and very much worth implementing.

Although there are productive restoration efforts around the world, there is little data available on the extent of these systems. Carbon sequestration would likely be similar to restoration of the baseline habitat (a savanna emphasizing native nut trees, for instance, would likely have carbon impact similar to that of ordinary savanna restoration). In the case of buffalo commons, restored prairies could potentially regain 30 to 40 tons of carbon per hectare over the course of decades.[33] Yields in these systems, even those designed to provide a yield, would probably be much lower than yields of the same land dedicated to agriculture, although with much less labor and fewer inputs.

CHAPTER TEN

Introduction to Species

The practices I describe in chapters 6 through 9 provide broad patterns of land use. The chapters in parts 3 and 4 describe species, and groups of species, of perennial crops for use in carbon farming practices. In this chapter I cover some issues that cut across crop categories and introduce the typology of cultivation status and climate used in this book.

Any and all perennial crops can be appropriate for carbon farming strategies. Part 3 (chapters 11 through 17) focuses on perennial staple crops that provide protein, fats, and carbohydrates. This is a category of crops that has received little systematic attention. Part 4 (chapters 18 through 24) profiles perennial industrial crops that can be used for materials, chemicals, and energy. In both cases I profile standout individual species and genera in the text; numerous additional species are listed in table form only. Appendix A lists more than 600 useful perennial species for all the world's climates. In addition to staple and industrial crops, these tables also include multipurpose agroforestry support species, including nitrogen fixers, contour hedgerow species, and livestock fodder plants.

As nice as it would be to include them, I've chosen not to cover perennial fruits, vegetables, pasture species, or other crop categories. This is in part due to limited space but, more important, because there are already comprehensive resources on fruits, perennial vegetables, and pasture species. To my knowledge, until now there has never been a comprehensive global inventory of perennial staple crops, nor a comprehensive global inventory of perennial industrial crops. The final reason I've chosen to focus on perennial staple and industrial crops is that most of the world's cropland is devoted to

annual staple and industrial crops, so the prospects for perennializing this area will have greatest impact.

A realistic assessment of the plants I've included in parts 3 and 4 and their potential means acknowledging that there are no intrinsically "bad plants" and no "superplants" waiting to save us. Any of these plants could be managed poorly—grown in industrial monocultures, displacing people and causing human suffering and ecological damage, for example. Clearing healthy forest to plant any of them would likely result in significant net emissions (an effect that would be greatly exacerbated on peat soils). In other words, although these plants have incredible potential, they can't be divorced from sound management practices, which is why I so strongly recommend planting them in the polycultures and other integrated systems that I cover in chapters 6 through 9. Likewise, no individual

FIGURE 10.1. North Korean farmers display the results of their participatory tree species selection process. Chestnut has the highest score. Photograph courtesy of the World Agroforestry Centre.

plant will "save" humanity or any given region. The most important element of addressing climate change is building a powerful movement of people. Crops are tools we can use to help achieve these goals, but their impact ultimately depends on the social, political, and economic context in which they are embedded.

Cultivation Status Categories

In the chapters that follow, I've classified the plants in this book into nine cultivation status categories in order to distinguish among those that are ready and available now for cultivation, those that are in the midst of research and breeding efforts and may soon be ready for cultivation, and those that are still a glimmer in the eye of hopeful researchers and plant breeders. It's important to draw these distinctions because there are only a few short decades to retreat from the climate tipping point; a plant's readiness for prime time should be a major factor in project development.

The nine cultivation status categories are as follows:

1. **Global perennial staple and industrial crops (global).** These crops are already grown or traded around the world. The annual value of each is more than $1 billion US. Examples include coconuts, almonds, and bananas.

2. **Minor global perennial staple and industrial crops (minor global).** These crops are already grown or traded around the world, but on a smaller scale than the global perennial staple and industrial crops. The annual value of a minor global crop is under $1 billion US. Examples include shea, carob, Brazil nuts, and fibers such as ramie and sisal.

3. **Regional perennial staple and industrial crops (regional).** These crops have been domesticated and cultivated regionally but have not been adopted elsewhere and are typically not traded globally. Examples in this large category include perennial cottons and many nuts and staple fruits.

4. **Historic perennial staple and industrial crops (historic).** These crops were once cultivated but have been abandoned. The reasons for abandonment may include colonization, genocide, market pressures, arrival of superior crops from elsewhere, and so forth.

5. **Wild perennial staple and industrial plants (wild).** Some wild plants have strong historic or contemporary use. Although they are not cultivated crops, they may be wild-managed. I have only included wild plants that are relied on as staple foods or industrial feedstocks. Examples include mongongo nuts and North American piñon pines.

6. **New perennial staple and industrial crops (new).** Most new crops were important wild plants until recently, although some are the result of hybridization. They have been developed in the last few decades. What they have in common is that they are currently cultivated by farmers. Examples include baobab, argan, and buffalo gourd.

7. **Perennial staple and industrial crops under development (under development).** Plant breeders are actively working to domesticate these plants for cultivation, but they are not yet commercially available as crops. Examples include most of the perennial cereal grains.

8. **Experimental perennial staple and industrial crops (experimental).** Plant breeders are testing these plants to see if they could be domesticated for cultivation, but they are still in an experimental phase. Examples include milkweed and leafy spurge.

9. **Hypothetical perennial staple and industrial crops (hypothetical).** These are perennial plants that could potentially be developed for cultivation. Some,

TABLE 10.1. Global Perennial Staple and Industrial Crops (Global)

Category	Crops
Basic starch	Banana, plantain
Balanced carbohydrates	Chestnut
Protein oil	Walnut, pistachio, hazelnut, cashew, almond
Edible oil	Oil palm, olive, avocado, coconut, cacao
Sugar	Sugarcane, date
Industrial oil	Oil palm, castor oil
Hydrocarbons	Rubber

Source: FAO Statistical Division online 2012 data.

such as cycads (for industrial starch), are simply neglected; others, such as buckwheat and soybeans, are annual crops that could potentially be perennialized by crossing with relatives.

In part 3 I've included virtually all perennial staple crops that are reported in the literature. The exceptions are a few cases in which there are many cultivated species within a single genus. In these cases, I've included a representative sample of the most important cultivated species from diverse geographic origins and climate tolerances. The perennial industrial crops that I cover in

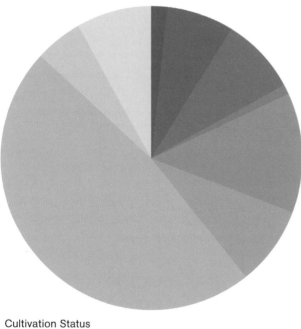

Cultivation Status

■ Experimental crop   ■ Hypothetical crop   ▨ Regional crop
■ Global crop   ■ Minor global crop   ▨ Under-development crop
■ Historic crop   ▨ New crop   ▨ Wild staple

FIGURE 10.2. This is a breakdown of 671 perennial plants and crops profiled in this book by cultivation status category. It's worth noting that the majority are already under cultivation as regional or minor global crops. Along with new crops, these regional and minor global crops are of particular importance to carbon farming because they are already available for cultivation. (All told, several *thousand* perennial species are cultivated worldwide, mostly as regional fruit, nut, vegetable, and industrial crops.)[1] Unlike domesticating new crops, we don't need to wait decades to use these species to perennialize agriculture.

part 4 are somewhat less comprehensive because there is a greater abundance of candidate species. I've selected those industrial crops and crop candidates to ensure that all cultivated species are listed, but promising wild species are included only when strong cultivated alternatives are lacking for their climate. (This is not to undervalue the amazing resources that these "orphan crops" can provide; they are worthy of attention.)

Climate Categories

There are several international climate classification systems. Unfortunately, the available data on most of the crops covered in this book is not nearly specific enough to apply the Köppen climate classification system or the Holdridge Life Zone system to classify the crops I cover by climate. Given this constraint, my goal is to simply provide a set of species so that farmer groups, governments, scientists, or NGOs that would like to explore possibilities for their region have a place to start, although this approach doesn't account for differences such as the length of a rainy season or frost-free periods. My first set of climate categories looks at broad climate types.

Tropical. Fully frost-free with temperatures always above 1°C (34°F).
Highland tropical. Above 1,500 meters (4,921 feet) in elevation and more subtropical, temperate, or even boreal (depending on elevation) than tropical.
Subtropical. Tropical with a colder season that includes minor frost and low temperatures from 0 to −6°C (32 to 21°F).
Warm temperate. Low temperatures from −7 to −17°C (19 to 1°F). (Often these climates have high summer temperatures, but for our purposes this also includes cool temperate climates that stay cool in summer, such as southern Britain.)
Mediterranean. Warm temperate or subtropical with winter rains and a summer dry season.
Cold temperate. Low temperatures from −18 to −34°C (0 to −29°F).
Boreal. Bitterly cold in the winter with low temperatures from −35 to −50°C (−31 to −58°F). (This category also

SALT-TOLERANT AND HALOPHYTE CROPS

Halophytes are extremely salt-tolerant plants. They tend to come from two distinct contexts: coastal areas and dry, salty deserts. Many other plant species are tolerant of salt, though not to the extreme that halophytes are. Irrigated agriculture in dryland regions has salted roughly half of current and former agricultural land, often resulting in its abandonment.[2] There has been a lot of investigation into salt-tolerant plants as part of a search for suitable crops in these regions. Should sea-level rise continue as projected, salt-tolerant coastal species will be of increasing utility as well. Perennial halophytes not only make salty soils more productive but can actually reduce salt levels in soils.[3]

TABLE 10.2. Perennial Salt-Tolerant and Halophyte Crops

Function	Species[a]
Balanced carbohydrate	Ceratonia siliqua, Distichlis palmeri, Inocarpus fagifer, Panicum turgidum, Porteresia coarctata, Prosopis chilensis, Prosopis glandulosa, Prosopis juliflora, Prosopis pallida, Prosopis tamarugo, Sporobolus fimbriatus
Protein	Acacia victoriae, Parkinsonia aculeata
Protein and oil	Adansonia digitata, Balanites aegyptiaca, Chrysobalanus icaco, Pistacia vera, Salvadora persica, Santalum acuminatum, Terminalia catappa
Oil	Cocos nucifera, Elaeis oleifera, Lepidium spp., Mauritia flexuosa, Moringa oleifera, Moringa peregrina, Olea europaea
Sugar	Cocos nucifera, Phoenix canariensis, Phoenix dactylifera
Biomass	Acacia tortilis, Albizia lebbeck, Arundo donax, Casuarina equisetifolia, Cocos nucifera, Elaeagnus angustifolia, Eucalyptus camaldulensis, Eucalyptus globulus, Euphorbia tirucalli, Leucaena leucocephala, Mauritia flexuosa, Melia azedarach, Phragmites australis, Pongamia pinnata, Populus euphratica, Prosopis chilensis, Prosopis glandulosa, Prosopis juliflora, Prosopis pallida, Prosopis tamarugo, Sesbania grandiflora, Typha spp.
Industrial oil	Melia azedarach, Pongamia pinnata, Ricinus communis, Simmondsia chinensis
Hydrocarbons	Calotropis spp., Euphorbia tirucalli, Pistacia terebinthus
Fibers	Adansonia digitata, Calotropis spp., Hibiscus tilliaceus, Raphia vinifera, Typha spp., Yucca filamentosa

Function	Species[a]
Other industrial uses	Acacia seyal (gum), Albizia lebbeck (pesticide), Eucalyptus camaldulensis (tannins), Eucalyptus globulus (medicinal), Euphorbia antisyphilitica (wax), Melia azedarach (pesticide), Simmondsia chinensis (wax)
Livestock fodder	Acacia leucophloea, Acacia nilotica, Acacia saligna, Acacia seyal, Acacia tortilis, Albizia lebbeck, Arundo donax, Atriplex spp., Balanites aegyptiaca, Ceratonia siliqua, Elaeagnus angustifolia, Leucaena leucocephala, Moringa oleifera, Populus euphratica, Prosopis chilensis, Prosopis glandulosa, Prosopis juliflora, Prosopis pallida, Prosopis tamarugo, Sesbania grandiflora, Tamarindus indica
Windbreak	Acacia nilotica, Acacia tortilis, Acacia victoriae, Atriplex spp., Casuarina equisetifolia, Leucaena leucocephala, Populus euphratica, Ricinus communis, Sesbania grandiflora
Living fence	Chrysobalanus icaco, Euphorbia tirucalli, Hibiscus tilliaceus, Leucaena leucocephala, Moringa oleifera, Sesbania grandiflora, Yucca filamentosa
Nitrogen fixation	Acacia leucophloea, Acacia nilotica, Acacia saligna, Acacia seyal, Acacia tortilis, Acacia victoriae, Albizia lebbeck, Casuarina equisetifolia, Elaeagnus angustifolia, Inocarpus fagifer, Leucaena leucocephala, Pongamia pinnata, Prosopis chilensis, Prosopis glandulosa, Prosopis juliflora, Prosopis pallida, Prosopis tamarugo, Sesbania grandiflora
Crop shade	Acacia saligna, Albizia lebbeck, Leucaena leucocephala, Olea europaea, Ricinus communis, Sesbania grandiflora

Note: See parts 3 and 4 for descriptions of these species. There are many more for which data is unavailable. Halophyte species are also identified as such in appendix A.
[a] eHALOPH Database online; Dov, "Introduction of drought and salt tolerant plants for landscaping."

includes some alpine regions even if they are outside of boreal latitudes.)

Precipitation is also a variable.

Arid. An average of 0 to 250 millimeters (0 to 9.8 inches) of rainfall annually.

Semi-arid. An average of 250 to 1,000 millimeters (9.8 to 39.4 inches) of rainfall annually.

Humid. An average of 1,000 millimeters (39.4 inches) or more of rainfall annually.

Breeding Perennial Crops

Our planet is home to approximately 250,000 species of plants. Approximately 20,000 are edible for humans.[4] Approximately 6,000 are cultivated for food, fodder, materials, chemicals, and energy.[5] Fully domesticated crops tend to show reduced toxicity, better flavor, greatly increased yields, and easier harvest. They also usually become more dependent on irrigation, pest control, and weeding. Of the 6,000 cultivated species, few are fully domesticated—perhaps 100 food plants and 30 industrial crops.[6] That means we're relying heavily on only 0.5 percent of the plants we could potentially be using.[7]

Plant domestication is a process of people—whether consciously or unconsciously—selecting for useful traits. In return, the plants become increasingly dependent on us for their survival and dispersal. For example, an individual tree might produce larger nuts with thinner shells, higher and more consistent yields, and more concurrent ripening relative to other individual trees.[8] These traits might make it less suited to survival in the wild, but if it exhibits these traits on a farm it may be protected from pests, may have competing plants weeded from its neighborhood, and may even be irrigated and fertilized. In this way domestication is a two-way street, a mutually beneficial arrangement between humanity and our crops. When it goes far enough, both species become dependent on the other for survival.[9]

Climate change mitigation has a tight time line. We have only a few decades to retreat from the tipping point

to 350ppm or below.[10] The question of plant breeding and how long it takes to domesticate a new perennial crop is important in this context. During World War II the United States found itself cut off from imported rubber supplies. A search for a domestic rubber source identified a promising candidate (a goldenrod). Not surprisingly, it proved impossible to start with a wild, completely undomesticated plant and domesticate it to the point of serving as a commercial rubber source in just four years.[11] Mass-planting a crop that is not yet ready for release, as happened with the industrial oil crop jatropha, which was planted for biodiesel production, can lead to disappointing yields and economic disaster.[12]

It looks like there are three waves of perennial crops coming to our climate rescue. First are those that are already in production, such as our global, minor global, and regional crops. Most could benefit from additional breeding work to improve yields and mechanical harvest ability, but there are already good varieties in production. This wave of perennial crops is ready for widespread planting now. It is for this reason that this book places such a strong emphasis on regional and minor global crops.

The second waves may take 5 to 10 years. This includes crops such as perennial rice that researchers and plant breeders have been working on for decades, the results of which are nearly ready to release. It also includes some crops that require some more basic research or market development before they're ready for broad-scale adoption by farmers, or they may perhaps require innovations in harvesting or processing equipment. Breeders estimate that developing and releasing a new variety of an established crop can take between 8 and 20 years.[13]

The third wave may take decades. Perennial maize could be ready in 10 to 40 years if funding were available.[14] Although this might seems like too far in the future to be of much use from a carbon farming perspective, all climate projections I've seen indicate that we'll need to continue sequestering surplus carbon through this century and beyond, especially since the current rate of fossil fuel emissions is likely to continue or increase. In other words, this third wave doesn't offer a short-term solution, but that doesn't mean we should abandon our efforts just because we won't see the fruits of those efforts for decades. We may still need them at that point.

International law does not currently have a mechanism to protect the intellectual rights of indigenous people and poor farmers who make contributions to plant breeding.[15] Plant Breeders Rights (PBR) are rights granted to the breeder of a new variety that give the breeder control over the propagation and harvest material of a new variety for 20 years (and 25 years for trees and vines), but these rights are typically granted by national offices to large seed companies, creating a tension not only between the well-funded patent efforts by large corporations and the rights of individual smaller farmers, but internationally as well.[16] In fact, corporations have patented numerous species and varieties that were developed by poor farmers. Perhaps the most famous such biopiracy effort was a Texas company's outrageous attempt in the 1990s to patent basmati rice, a set of varieties developed by farmers in India over thousands of years.[17] A fair trade group called PhytoTrade Africa is helping farmers develop markets for agroforestry products without losing intellectual property rights.[18] For example, they helped a women's marula oil producers cooperative in Namibia develop a shared patent for their product that protects their rights and secures them a market.[19]

There are ongoing efforts to develop an intellectual property model suited to small farmers, rather than a model that only serves the interests of large corporations, but to date no agreed-upon model has emerged. A modified form of PBRs would provide a 25-year patent to farmer groups. NGOs may be able to assist farmer-breeders by creating a registry of varieties that includes the GPS location of the original tree, the history of the variety, and a tree's "genetic fingerprint."[20]

Three Standout Breeding Efforts

Some perennial crops are not yet competitive with annuals, particularly in cold climates. Around the world, plant breeders are hard at work improving yields, harvestability, and other key aspects of up-and-coming perennial crops. These scientists, farmers, and backyard

FIGURE 10.3. A Cameroonian nursery offering elite fruit tree varieties selected by the participatory, farmer-led breeding work described in the text. Photograph courtesy of the World Agroforestry Centre.

breeders are unsung heroes of carbon farming. Here we profile three outstanding breeding efforts.

PROPAGATING THE "TREES OF LIFE"

Roger Leakey is a global expert on agroecology, tree domestication, and sustainable development who spent five years as the director of research at the International Centre for Research in Agroforestry (today the World Agroforestry Centre) and was a coordinating lead author in the International Assessment of Agricultural Science and Technology for Development (IAASTD). His book *Living with the Trees of Life: Towards the Transformation of Tropical Agriculture* was published in 2012 and offers a strategy for a transition to tree-based agriculture based on his decades of experience.

Although breeding is typically performed in government, university, or corporate labs, Leakey argues that it makes more sense to base domestication work on farms and in communities. He envisions crop-improvement work as a "participatory, self-help process in the community, done by the community," with the assistance of scientists.[21] Farmers know that without a market, new crops will be limited to household use only, so Leakey argues that breeding should be "farmer-driven and market-led."[22] At the World Agroforestry Centre, he and his colleagues developed a simple, reliable vegetative propagation technique that requires no electricity for misting. It is a marvelous example of appropriate technology.[23] This allows the rapid multiplication of exact clones of the best individual trees found in the wild or on farms.

Leakey describes "Cinderella trees," which offer important foods but are ignored by formal science.[24] These species need a "fairy godmother" to assist farmers in domesticating them for a "woody plant revolution."[25] Because wild plants have high genetic diversity, the best individuals ("plus-trees") can be selected for propagation. Leakey has found that local communities always know exactly where the best individual trees are.[26] Almost everywhere this farmer-led domestication process took place, farmers prioritized nuts and nutritious tree fruits for their breeding efforts, although they could have picked any wild plant species they wanted (timber, vegetables, medicinals, and so on).[27]

Once a plus-tree is successfully cloned in a nursery, it can be mass-propagated and widely planted. In Leakey's system each village domesticates its own local varieties, so that even if many villages undertake this work, a wild crop's genetic diversity is not lost.[28] Typically, these plus-trees are planted as components of diverse farms, not in large-scale monocultures. Crossing these elite plus-trees with one another can "[give] rise to tree cultivars that are as distinct as dog breeds . . ."[29] This kind of tree improvement is not widely practiced because it is slow—trees can take years to start producing and showing the positive or negative traits needed to select the following generation.[30] But selecting superior individuals of a species and vegetatively propagating them is an ancient practice that continues to this day around the world, often by dedicated individuals and clubs lacking funds and government or institutional support. As more farmers learn the techniques, they can begin to domesticate other species. Participatory cultivar development is also much less expensive compared with a Green Revolution–like approach. Leakey would like to see governments, NGOs, and businesses play a positive role in participatory plant domestication in the future.[31]

Leakey's work offers the prospect of "waves" of new perennial crops. The first wave would contain the elite plus-trees selected from the wild. They become available quickly. The second wave would come from crossing these elite individuals to one another. Finally, these crops could be fully domesticated in a third wave in which they reach their full potential.

PERENNIALIZING GRAINS

Perennial grains have been a dream of the sustainable agriculture movement for some time. They promise to marry the carbon and agroecological benefits of perennials to the annual staples we know and love. The task of perennializing grains is a slow and laborious process, but thanks to visionaries such as Wes Jackson and his colleagues at the Land Institute in Salina, Kansas, we are relatively far along in the process with some crops because they began working on it decades ago with the goal of developing perennial cereals, pulses, and oilseeds. Likewise the Yunnan Academy of Agricultural Sciences in China has made incredible progress in efforts to perennialize rice. A big advantage of developing perennial grains is that, unlike cloned tree crops, grains are usually planted from seed and each plant is a genetic individual. This kind of breeding takes longer, but it means a given crop will have greater genetic diversity, resulting in relatively less pest and disease pressure than would be typical in an orchard or plantation of cloned crops.

There are a few different approaches to perennializing grains. One approach is to begin with promising wild

FIGURE 10.4. Land Institute staff emasculating sunflowers (removing the male flower parts) as part of perennial sunflower breeding efforts. Photograph courtesy of perennialgrains.org.

perennial grains and select for desired traits for many years until they become domesticated. The Land Institute has used this technique to domesticate a perennial wheatgrass they call Kernza, which they project will yield the same as annual wheat by 2033.[32]

Another approach, referred to as wide crosses, involves crossing annual crops and their wild perennial relatives. So how long will it be before we have new field-ready perennial grains using the knowledge and techniques we have today? There are a number of factors that make this a difficult prediction. Some wide crosses, such as wheat, are very difficult, while others, such as sorghum, go rather smoothly. In fact, wide crosses of sorghum have been so successful that perennial sorghum looks poised to be the next perennial grain to arrive on the world scene. Some plants bear seed the first year, while others take multiple years to bear, slowing the process significantly. Available funding is also a key variable.

How long does it take to develop a new perennial grain yielding at least 1.2 tons per hectare (the Land Institute's current target, based on yields of quinoa and other marketable specialty grains)? How long to yield as well as its annual competition? David Van Tassel of the Land Institute was courteous enough to give me the short answer ("It depends") and the long answer: "Remember that the yield of major annual grains also 'depends.' So, for example, achieving parity with grains in African subsistence agriculture could be much easier than achieving parity with corn or soybeans in Illinois. Rice and maize in particular have very high yield potential so that phenomenal yields can be obtained under high-input, optimal conditions. Average yield everywhere else can be quite different."[33] Van Tassel's own words on where perennial grains are and are not likely to be able to compete:

> *It will take much longer to develop perennial grains that can compete with annual grains in both extremely productive environments and extremely unproductive environments. This is because very few grains (annual or perennial) have the yield potential necessary to take full advantage of extremely favorable conditions. [These "prime" areas represent about a third of world cropland.[34]] Massive investment in breeding would be required*

for any other crop to catch up to these few annual superstars which themselves continue to receive the lion's share of agronomic investment.

Annual crops may also be difficult to displace in very drought-prone environments where soils are commonly left bare of vegetation for long periods of time (e.g., for a full year) until soil moisture has been recharged enough to allow a short-lifecycle grain to produce a single crop. Alternatively, crops may be sown each season with the expectation that they will fail if the rains do not come. It may be very difficult to breed perennials to remain dormant for a full year or to green up only when the rains come.

In general, perennials adapted to environments with regular drought-stress are likely to be slower to domesticate because their adaptations (long roots, succulent tissues) are likely to require early investment that delays the transition to sexual reproduction. This implies longer breeding cycles and therefore slower progress. Moreover, finding and maintaining an optimum plant density for soil water management is likely to be more difficult than in less water-limiting environments.[35]

Van Tassel further clarifies that these estimates are "assuming we're talking about perennials that can set seed the first or second year (not trees that take five years to begin flowering), and assuming that we are talking about parity with major crops outside of the super-special breadbasket regions (i.e., maybe wheat but not maize in the upper Midwest, maize or sunflower but not rice in Thailand; any annual grain in Africa, any unirrigated grain in Kansas)." He notes that annual crop yields are a moving target since ongoing annual crop breeding is well funded and ongoing.[36]

With all these caveats in mind, Van Tassel ventured his best estimate. Starting today, with current knowledge, he estimates that to bring promising undomesticated wild grain yields to 1.2 t/ha would take between 8 and 12 years, and 15 to 25 years to bring them to yield parity with grains outside of the special breadbasket regions. For wide crosses of annual grains with perennial relatives, he estimates 5 to 20 years and 10 to 40 years respectively.[37] Van Tassel notes that the

TABLE 10.3. Estimated Number of Years Before New Perennial Grain Crops Will Be Field-Ready

Method	Years to Achieve 1.2 t/ha if Started Today[a]	Years to Achieve Competitive Yields with Major Grains if Started Today[a]	Years to Achieve Parity with Highest Grain Yields
Domestication of wild perennial grains	8–12	15–25	50–75
Wide-crossing of annual grains with perennial relatives	5–20	10–40	50–75

[a] "If started today" because enough has been learned in the development of the partially domesticated perennial grains we have today that, if we started today with what we now know, this is how long it would take. We do in fact have some crops quite a bit further along than this already.

ranges are greater for wide crosses because they are more unpredictable than simply domesticating wild perennial grains.[38] See chapter 11 for an overview of perennial grain development efforts and chapter 13 for a crop-by-crop update on the status of perennial cereal grain domestication and perennialization. Note that many perennial grains have already been worked on for many years, so we are not starting from scratch with crops, including rice, rye, wheat, and sorghum.

Breeding work alone is not sufficient to bring perennial grains to life. How will weeds, pests, and fertility be managed in perennial grain fields? Development of management and agronomy practices needs to go hand in hand with breeding to ensure sustained yields.[39]

HYBRID SWARM BREEDING

Back in the 1980s breeder Phil Rutter of Badgersett Research Corporation in Minnesota decided to dedicate his life to developing woody perennial staple crops to fight climate change. He calls his approach "woody agriculture," and you can read more about this production model in chapter 8. Here I focus on Badgersett's remarkable breeding practices.

A hybrid swarm is a phenomenon found in nature when two or more species in a genus have overlapping ranges, and cross with each other and their hybrid progeny repeatedly. This creates an extremely diverse population that can form the basis of a new species.[40] Rutter's vision was to create hybrid swarms of staple tree crops to develop new food sources, a process he calls "accelerated guided evolution."

Badgersett began with interspecific hybrids of various nut crops, meaning each was the result of a successful cross between two related species. Their chestnut and hickory-pecan swarms each contain genes from five species, and their hazels from three.

Rutter says that interspecific hybrids are "promiscuous," meaning they cross more easily than pure species, leading to exponential increases in diversity in the hybrid swarm.[41]

In each generation Badgersett selects for two desirable traits. This involves huge amounts of record keeping and culling undesirable plants. The first round of hazels selected for blight resistance and cold hardiness; the second selected for yield and annual cropping (producing good yields every year instead of every other year). And so on. Today they have planted out the fifth generation—bear in mind that it takes five years or more until hazelnuts produce nuts.[42]

Rutter and his colleagues envision their neohybrid plants becoming totally new crops. To Badgersett, no temperate nut crops are properly domesticated yet, but it's possible. Already they have chestnuts with 7, 9, and even 12 nuts per burr instead of the normal 3 or 4.[43] And they are seeing interesting traits come up in their hazels. For example, they are finding shells with as many as four nuts inside, a step on the way to their goal of hazelnut clusters the size of ears of maize. Ultimately, they want to see hazels come as far from where they are now as maize has come from its wild ancestor, which had only a few kernels per head.[44] These "hazelnuts on the cob" could transform the temperate food system. This transformation is Badgersett's goal.

All of Badgersett's crops are seed-grown (although the organization is beginning to select some individuals for clonal propagation). Rutter says, "Clonal crops are a death trap" due to monoculture vulnerability to disease. Genetically diverse plantings are also more resilient to climate change. Finally, seed-planted trees are much less expensive than their clonal counterparts, making it possible for large-scale farmers to convert to woody agriculture.[45] Rutter also points out that, unlike

Figure 10.5. Neohybrid chestnut with seven nuts per burr instead of the usual three or four. Photograph courtesy of the Badgersett Research Corporation.

perennial grains, nut trees are already edible and cultivated to begin with. Their higher market prices can offset their lower yields.

Genetically Modified Organisms

Genetically modified organisms (GMOs) have been promoted as an important strategy for feeding the world through agricultural adaptation to climate change. They are reported to increase yields and reduce losses to pests, diseases, and weeds; they may also someday provide drought-proof crops.[46] Many movements have been extremely critical of GMOs, however. Concerns include loss of crop diversity, genetic contamination of non-GMO crops and wild crop relatives, and the potential of food allergies and other deleterious effects on human health. GMO transnationals are quick to brand such talk as "anti-science," although some claims are backed up by multiple studies.[47] It is apparent to me from discussions in my classes and workshops that there is sometimes an anti-science bias among GMO opponents that clouds their understanding of the issue, but it is more frequently manifested as confusion between GMO technology and traditional plant breeding. People with this perspective are often primarily concerned about the food they eat and may promote labeling GMO food as the solution. Labeling is one element of an anti-GMO strategy but to me does not sufficiently target the real issue: control and power over seeds themselves, the most basic means of production.

In recent decades there has been tremendous consolidation in the seed industry, such that agrichemicals and seed are now sold by the same corporations. Due to virtual monopolies on seed, these corporations make it difficult to obtain non-GMO seed of staple crops. Moreover, they have sued hundreds of farmers for saving seed and are lobbying to clamp down further on farmer seed saving and seed exchanges.[48] Fundamentally, leaving the world's seed supply to a handful of companies seems the opposite of food security. A dispersed, decentralized network of breeders, small companies, seed banks, seed exchange networks, and individual seed savers would be far more resilient, especially in the context of an unpredictable, changing climate. And it would be far more democratic as well.

GMO crops have also failed to bring promised improvements. Drought-resistant GMO maize, for example, yields no better than conventionally bred maize.[49] The Union of Concerned Scientists reports that after 20 years of GMO research and untold billions of dollars, GMOs have had little impact on crop yields, whereas traditional breeding and agroecological approaches account for 86 percent of gains in the United States.[50]

Many scientists have proposed that while GMOs have been used for undesirable purposes, the technology is neutral and could be used for agroecological purposes. Such proposals include greater water-use efficiency, increased cooperation within and between crop species and their microbial partners, improved nitrogen fixation, and enhancing the natural pest control abilities of crops.[51] Several of the reviewers of this book noted that genetic engineering could potentially convert the oils in oil crops to high-omega-3 fish oils, or remove bad flavors or toxins from otherwise promising crops. A public domain patenting process would be necessary in this scenario to avoid the consolidation that plagues the industry today. Although I admit to being intrigued, I have zero confidence in the industry as it is today. I also prefer techniques that can be decentralized and do not rely on incredibly expensive technology. GMOs don't fit the bill.

Invasive Plants

Whenever people discuss new crops, the issue of aggressive naturalization frequently comes up: "Will it become an invasive species?" Already, anthropogenic dispersal of species, both intentional and accidental, has spread species of organisms around the world at a rate unprecedented in recent ecological time. Agriculture and agroforestry have been part of the problem, from kudzu in the eastern United States to *Leucaena* everywhere in the tropics. Certainly, some anthropogenic introductions of "invasive" species have caused extinctions and have had negative impacts on ecosystem function, mostly on islands and in lakes.[52]

One of the first steps in any carbon farming strategy or project development should be to assess the potential of the native flora. It seems unlikely that any part of the world lacks for nitrogen-fixing plants, windbreaks, or biomass producers, yet non-native species are often used and sometimes introduced specifically for these and other functions that could be served by native plants. Appendix A gives each species' region of origin (sometimes broadly), which can be used to narrow down an initial list of potential candidate species. Virtually every populated region of the world is inhabited by a multitude of fascinating useful plants, and they should not be overlooked in a rush to bring in exotic-sounding "miracle plants" from far away. As we'll see in chapter 13, researcher Richard Felger searched the world over to find crop candidates for salty soil before determining that nipa, a species in his own bioregional backyard, was the best candidate.[53]

That said, I am unaware of any region of the world today that meets its food needs with exclusively native crops. Our global palette of food crops has been pretty thoroughly homogenized. Imagine Italian food without the tomato (Mesoamerica), Mexican food without rice (Southeast Asia), Thai food without cilantro (Mediterranean). A perennial food system should not be held to a higher standard in this regard than the existing annual-based agricultural model.

Domestication also usually reduces the competitiveness of crops in the wild. Van Tassel notes, "Domestication is really the opposite of the evolutionary forces selecting

plants to become weedier."[54] He reports that the Land Institute's crosses of the weedy johnsongrass with its annual relative sorghum have resulted in much less aggressive plants. Domesticated plants have lost many of their natural defenses as well, making them less able to compete in the absence of farmers and gardeners to care for them.

Of the roughly 675 species featured in this book, 63 are listed as invasive by the Global Invasive Species Database (GISD), which was developed as part of an initiative to combat invasive species, and 9 are featured in the GISD's "Top 100" invasive species list.[55] But all of these species are native to *some* part of the world—often huge swaths of it. At the very least, they should be put to use in regions of the world to which they are native and are already naturalized.

Unlike the overwhelming consensus among scientists that climate change is real and anthropogenic, not all ecologists agree that invasive organisms are a threat to be eradicated. In 2011 the journal *Nature* published a comment from 19 ecologists called "Don't Judge Species on Their Origins" stating that the preoccupation with native versus alien species has become counterproductive. Instead they argue that species should be judged on beneficial or harmful impact, regardless of their origin.[56] These scientists propose that we "organize priorities around whether species are producing benefits or harm to biodiversity, human health, ecological services and economics."[57] Carbon-sequestering food and industrial plants are critical to human health, economies, and ecosystem services, and it's time to think about the potential benefits an alien species might offer alongside its potential for invasion. Will it help to reduce emissions, sequester carbon, reduce the amount of tilled cropland? These are the questions we should be asking rather than "Where does it come from?" and "Does it belong here?"

The planet is headed for devastating loss of biodiversity. Coral reefs and Arctic ecosystems are already on their way out.[58] At 3°C (5.4°F) warming, 10 to 50 percent of our ecosystem types will no longer exist anywhere on the planet. Efforts to eradicate invasive species from those ecosystems will become moot at that point. If invasive species, or potentially invasive species, could keep that from happening, we should at least consider planting mesquites and edible air potatoes outside of their native range.[59]

In an ironic development, citizen-biologists are beginning a process of "assisted migration" by moving species to new regions where—due to climate change—they are now adapted in the hope of averting their extinction.[60] Scientists actually anticipate that this activity will become increasingly important as climate change continues.[61] In fact, many scientists are delving into the emerging field of novel ecosystems. According to Richard Hobbs, Eric Higgs, and Carol Hall in *Novel Ecosystems: Intervening in the New Ecological World Order*, "Novel ecosystems are composed of non-historical species configurations that arise due to anthropogenic environmental change, land conversion, species invasions or a combination of the three. They result as a consequence of human activity but do not depend on human intervention for their maintenance."[62] Although people are attached to the idea of a "pure" or "original" ecosystem, today novel ecosystems represent the majority of our natural ecosystems—28 to 36 percent of all ice-free land.[63]

Novel ecosystems are not actually a new phenomenon. In fact new, mixed-species ecosystems have always arisen in response to climate changes. The difference now is the anthropogenic cause. Most ecosystems we know today arose as novel ecosystems after the last ice age and are not, as we sometimes think, timeless and unchanging.[64] How do novel ecosystems compare with historic ecosystems? Even in Hawaii, where 100 flowering plants have gone extinct and 1,000 new plants have escaped cultivation (or otherwise arrived) and joined the flora since human arrival, ecosystem functions continue.[65] Novel ecosystems can provide similar ecosystem services, and sometimes better.[66] In fact, a number of scientists think that novel ecosystems are an important part of climate change adaptation because they can often respond better to change than historic assemblies.[67]

Mark Davis, author of *Invasion Biology*, a comprehensive review of the science of invasion biology, writes: "Nature is continually changing, but it does not have a goal. Nor is there only one nature that can or should exist at a particular place and point in time . . . As scientists . . . [we need to] work toward remedying harm where it truly exists, without becoming compulsive and parochial in our perspectives and behavior, without mistaking change for harm."[68] He goes so far as to propose the abolition of invasion biology as a science due to its flaws and biases.[69]

Science seems to be heading, albeit slowly, in the direction of appreciation of "invasive" plants and the novel ecosystems they help to form. Regulation has not kept up. There are many legal barriers to the international shipping of plants. Some are in place to quarantine diseases and pests of crops—an important function. Others target plants on invasive lists. Many individuals and institutions interested in perennial crops, including myself, have found the paperwork and restrictions to be a major barrier to the development of carbon farming.

PART THREE

Perennial Staple Crops

For farmers, getting over the "establishment" hump is a major barrier. Annual crops provide a yield the same year they are planted. Waiting three to five years or more for a yield is difficult for any farmer, but especially for farmers who subsist on what they grow. Some perennial staples don't begin to produce for 20 years or more! Lack of long-term land tenure is also an obstacle. Who wants to plant and care for trees or other perennials on rented or otherwise insecure land? We are also entrenched in a global industrial food system that stacks—via subsidies and structural adjustment loans—in favor of large-scale annual crop production. Like annual staple crops such as maize and wheat, many perennial staple crops require specialized equipment or infrastructure for harvesting, processing, and storage. A global conversion to perennial staple crops would require us to develop and manufacture new or modified low-tech and high-tech equipment.

We also need a lot more research. Even for crops where the selection and breeding work is relatively complete, do we know yet what species and varieties are best suited to a given region? What agronomic practices work best? How can perennial staples be integrated with nitrogen-fixing species, livestock, and annual crops? This kind of research should be funded and prioritized as part of global and regional carbon sequestration efforts.

Finally, there is a psychological factor to adopting new foods. People like to eat what we're used to and won't necessarily eat something new just because it is good for us, grows easily, and helps fight climate change. Fortunately, many perennial staple crops taste excellent. This obstacle can also be partially overcome by replacing annual livestock feed crops with perennial ones. Hogs and cattle are a lot less picky about their foods than people are!

CHAPTER TWELVE

Basic Starch Crops

Basic starch crops provide carbohydrates with only 0 to 5 percent protein and oil. Global annuals in this category include tubers like cassava, taro, yam, and sweet potato. Bananas and plantains are global perennials in this class. Production of basic starch crops accounts for about 3 percent of global cropland.[1] Perennial basic starches include fruits, aerial tubers, and edible starchy trunks. In terms of yield these perennials compare very favorably with their annual "competition."

Starchy Fruits

Bananas, plantains, breadfruit, and their relatives are important carbohydrate crops. Although bananas and plantains are probably already grown virtually everywhere they can be, there is a lot of opportunity to extend production of breadfruit and other *Artocarpus* species elsewhere.

BREADFRUIT (*ARTOCARPUS ALTILIS*)

Breadfruit was domesticated more than 3,000 years ago, probably in the New Guinea lowlands. It was carried through the Pacific as a "canoe crop" and thrived on most islands. It prefers humid tropical lowlands with summer rains and a dry season of fewer than three months. It can produce well with as little as 1,000mm (39 inches) of annual rainfall if the rain is well distributed throughout the year but is limited to humid tropics with high temperatures all year.[2] (Miami is much too cold for it!) Breadfruit is a minor global crop.

There are hundreds of varieties, and year-round production is possible by selecting the right ones.[3]

FIGURE 12.1. Breadfruit is a tremendous starch producer and one of the world's finest perennial staple crops. Photograph by Jim Wiseman, © Jim Wiseman.

Unfortunately very few varieties—between only two and five—have made it out of the Pacific region! This has prevented people from fully appreciating this incredible fruit and its potential. Breadfruit is an outstanding perennial staple crop. It is widely grown in the Pacific

Harvest and processing. Bananas and plantains are typically picked when they are underripe. They are usually handpicked and carefully handled to prevent bruising.

Carbon farming applications. Bananas and plantains are among the most common plants in tropical homegardens around the world.[18] They can tolerate up to 50 percent shade, a trait that suits them well to polycultures, and can themselves be used to shade crops such as coffee.[19] They are commonly grown beneath and between breadfruit, macadamia, avocado, coconut, and chachafruto. Intercrops and understory species include papaya, pigeon pea, coffee, sweet potato, pineapples, kava, and ginger. Bananas are also used in alley cropping systems.

Bananas and plantains are heavy feeders and drinkers. Some production systems irrigate bananas with gray water or otherwise fertilize and irrigate them from the waste stream. The leaves, stalks, and fruits are all used as fodder for livestock. Bananas are commonly grown in polycultures and are well suited to agroforestry systems at the home and farm scale.[20]

Other issues. Bananas are one of the most widely traded food commodities in the world. For almost a century they have been ubiquitous and very inexpensive. This availability, particularly in the cold-climate countries where they do not grow, has come at tremendous social and ecological cost. Rain forests have been cleared for chemical-intensive monocultures. The social cost has been even worse. To keep labor costs down and profits high, banana companies such as United Fruit (today known as Chiquita) supported dictatorships throughout Latin America. In 1953 United Fruit persuaded the CIA to help overthrow the democratically elected government of Guatemala in response to its efforts to protect workers and farmers—part of the chain of events that led to a devastating 36-year civil war.[21] If we factored in the social and ecological costs of raising and shipping bananas around the world, they would be a terribly expensive fruit.

Importantly, however, bananas and plantains show us the productivity that perennial staple crops are capable of. They also provide a sobering reminder of the dangers of depending on a single cultivar and that any plant can be misused to cause great harm to people and the planet.

Crop development. Another drawback of industrial banana production has been intense problems with pests and diseases. In the early days of banana commodity production, essentially all production was a variety called 'Gros Michel'. When 'Gros Michel' was wiped out by Panama disease (*Fusarium* spp.), it was replaced by 'Cavendish'. Today massive plantings of 'Cavendish' are being wiped out by a new strain of Panama disease. Many species of insects, fungi, bacteria, viruses, and nematodes attack bananas. Monoculture plantings are at most risk, but even bananas in integrated polycultures are vulnerable. Breeding work has been focused for decades on developing varieties resistant to pests and diseases. Unfortunately, breeding bananas is extremely difficult in part because commercial varieties produce no viable seed; many are sterile as well. Unlike most crops there is little role for amateur backyard breeders, because advanced techniques are needed to achieve results.[22]

Air Potato

Air potato (*Dioscorea bulbifera*) and its relatives are a class by themselves. No other cultivated crop I'm aware of produces storable tubers aboveground—a no-till root crop. Rather than producing tubers belowground like typical root crops, air potatoes bear large yields of tubers on the aboveground vines, as if they were fruits.

This minor global crop is native from Australia through Asia, India, the Middle East, and into Africa. It appears to have been independently domesticated in Africa and Asia. African varieties tend to have an irregular shape and firm texture; they are somewhat bitter. Asian varieties tend to have a rounder shape, softer texture, and, to some tastes, better flavor.[23]

Dioscorea bulbifera is tolerant of a wide range of climates, from lowland tropics to 2,500m (8,200 feet) elevation, and well into subtropical (handling −6°C/20°F or lower). It prefers humid conditions with

FIGURE 12.3. Edible air potato at Las Cañadas in Mexico. This species is apparently the only edible aerial tuber grown on a global scale.

a long rainy season. Other *Dioscorea* species with edible bulbils include the tropical *D. alata*, *D. hirtiflora*, and *D. semperflorens*, and the temperate *D. polystachya*.[24] They are worthy of investigation, but those I've grown and observed have yields far lower than *D. bulbifera*.

Uses. These tubers or bulbils can be as heavy as a kilogram (2.2 pounds). The bulbils are 80 to 95 percent starch dry weight, and some varieties are as high as 10 percent protein. This protein, however, is poor quality, lacking important amino acids.[25]

Yield. Yields range from 1 to 19 t/ha fresh, corresponding to 0.3 to 5.1 t/ha dry weight.[26]

Harvest and processing. Air potato can bear in as little as three months after planting. The tubers continue to ripen for an additional four to five months, dropping from the vines when they are mature. They may be the only root crop that does not need to be washed free of dirt. The tubers will keep in storage for months. It is a great crop for homegardens and is popular in many regions of the world, from Ethiopia to India to Southeast Asia. Air potato is rarely grown commercially. This may be in part because the tubers mature over a period of many months and commercial farmers often want simultaneous maturation for ease of harvest and transport to market.

Carbon farming applications. Air potato must be trellised or planted on living trees that can handle very vigorous vines. The trellises are often kept low to make it easier to harvest unripe (but still edible) tubers. Ripe tubers eventually fall to the ground. The vines die back in the cold or dry season.

Other issues. Wild forms of air potato are typically toxic to people and perhaps to wildlife. These wild forms have become rampant in the tropics and subtropics and are featured on lists of the world's worst invasive species. They can climb 20m (66 feet) or higher into trees and smother forests in a lush green blanket of growth. The edible forms are much less aggressive and tend not to survive in the wild.[27] Ricardo Romero has been cloning edible air potatoes at Las Cañadas in Veracruz, Mexico, and the result has proven productive, delicious, and well behaved when trellised on established nitrogen-fixing trees. Because air potato's native range stretches from Australia to Africa, it can make a tremendous contribution to carbon sequestration and food security even if its production spreads no farther.

Crop development. In *Tropical Yams and Their Potential*, tropical crop specialist Franklin Martin wrote of air potato:

The potential of D. bulbifera *is undoubtedly in the home garden, where its health and vigor recommend it . . . The fact that an edible crop is borne over 4 or 5 months makes this yam particularly desirable. No other edible species [of yam] is equal in this respect . . . The best Asian varieties of* D. bulbifera *merit collection, trial, and widespread distribution. Because of their potentially high yields, they should be magnificent producers of edible flesh and starch. Their commercial potential has not been well tested . . . The necessary research does not seem to be in progress anywhere. Thus their potential is still undiscovered.*[28]

Decades after Martin's writing, little has still been done. One bright spot is a recent paper from Ethiopia analyzing yields and associated characteristics of more than 47 edible accessions (collected types, may be of the same or different varieties) of air potato collected from farms in the southern part of the country.[29] This indicates that air potato is an important starch provider in some regions.

Starchy Trunks

Starchy trunks as staple crops probably sounds strange to people from cold climates, but there are many tree-like plants around the world that store up starch for years, then put it to use in dramatic flowering events. People in Africa, Oceania, and Mesoamerica independently figured out that this starch is at its most abundant when the plants are just about to flower. Species in the palm, banana, and agave families were domesticated for this starch production—and all remain important foods today. Interestingly, all of these species are tree-like monocots—giants of the plant group that gives us grasses, palms, gingers, and bananas.

The starch typically has to be separated from fibers before eating. Some starchy-trunked plants, particularly palms and some cycads, are destructively harvested for their starch, but if you cut down a plant to harvest the starch and it dies, you've ended its carbon-sequestering potential along with its life. The species profiled here are all colony forming—allowing for the harvest of some but not all trunks—or resprout after harvest, meaning they are candidates for non-destructive harvest.

Sago Palm (*Metroxylon sagu* and spp.)

This phenomenal group of palms—of which *Metroxylon sagu* is the most domesticated—composes a perennial staple crop that has already far exceeded the starch production potential of most annual crops. Sago is extremely important as a basic source of carbohydrates to more than one million people in New Guinea and the Western Pacific, where it is an important regional crop.[30] It could be much more widely grown in humid tropical lowlands around the world as a source of starch for people, livestock, and industrial uses. Sago is one of the many crops domesticated in New Guinea, along with bananas, breadfruit, winged beans, taro, sugarcane, and pandanus nuts. It prefers humid lowlands and freshwater swamps (though not flooding year-round) and can even be found at the freshwater end of mangroves.[31] Staple crops that are tolerant of flooding and salt water should be of particular interest, given projections for sea-level rise.

Uses. Sagos are multistemmed palms whose trunks are composed of carbohydrates and fibers. When an individual trunk matures, it can be harvested without damaging the rest of the trunks or their future

TABLE 12.3. Trunk Starch Crops

Latin Name	Common Name	Origin	Climate	Status
Agave americana	Maguey	Mesoamerica	Warm temperate to tropical, arid to semi-arid	Regional
Agave cantala	Cantala	Mesoamerica	Tropical, subtropical	Regional
Agave murpheyi	Hohokam agave	Southwest North America	Warm temperate to subtropical highlands, arid to semi-arid	Historic
Agave parryi	Parry's agave	Mexico and southwest North America	Boreal to warm temperate, arid to semi-arid	Historic
Agave salmiana	Maguey	Americas	Warm temperate to tropical, arid to semi-arid	Regional
Agave tequilana	Tequila agave	Americas	Subtropical to tropical highlands, arid to semi-arid	Regional
Ensete ventricosum	Enset	Africa	Tropical highlands	Regional
Eugeissona utilis	Nanga palm	Asia	Humid tropical lowlands	Regional
Metroxylon amicarum		New Guinea, Oceania	Humid tropical lowlands	Regional
Metroxylon paulcoxii, M. salomonense, M. vitiense, M. warburgii	Minor sago palms	Oceania	Humid tropical lowlands	Regional
Metroxylon sagu	Sago palm	New Guinea	Humid tropical lowlands	Minor global

FIGURE 12.4. Micronesian sago (*Metroxylon amicarum*), a multi-stemmed palm with edible starchy trunks. Photograph courtesy of Forest and Kim Starr.

productivity. Processing—which can range from Stone Age to modern techniques—separates the edible starch from the fibers. Sago fronds are the finest thatching material for building. The thatch lasts 10 years instead of the usual 1 to 4 for most palms. Some sagos also produce hard starch-like seeds that could be used as a bioplastic feedstock.[32]

Yield. Once established, which takes between five and eight years, a well-managed sago stand can produce 25 tons or more of edible starch per hectare for many years to come.[33] Without fertilization and management, yields are much lower, around 5 to 6 tons per hectare.[34]

Harvest and processing. Using traditional hand labor, it takes about 150 to 160 hours per year to produce enough starch to feed one adult for a year, just a month of 40-hour weeks.[35]

Carbon farming applications. Older plantings are typically allowed to phase out their production and are interplanted with fruit trees as they age. Sago starch is used for pig feed and as part of commercial livestock feed blends in Southeast Asia. In New Guinea edible palm grubs (*Rhynchophorus* spp.) feed on the trunks. The grubs' high protein and fat helps complement the sago starch in the human diet. Sago can protect coastlines from erosion and storm damage and serve as windbreaks. The spiny palms can also be used as a living fence.[36]

Other issues. Other cultivated sago species include *M. amicarum*, *M. paulcoxii*, *M. salomonense*, and *M. warburgii*—all of which are less productive than *M. sagu*. Although many other palms have starchy trunks, most of them are single-trunked, and thus are killed by harvest.

ENSET (*ENSETE VENTRICOSUM*)

Enset is a banana relative that may date back as far as 10,000 years. Although it is little known outside of Ethiopia, it remains an important regional perennial staple crop in the Ethiopian humid tropical highlands, where today 10 million people rely on it.[37] Though Ethiopia has been struck by terrible famines, enset regions have not experienced hunger, in part because the plants can survive long droughts, though they don't actively grow without water. They can be harvested at any time, so the trunks (and tubers) provide "living starch storage" for times of need. In some regions of Ethiopia, grains or tubers are the primary food crops, but enset is grown as a fallback crop for times of need.[38] Growing a banana relative for edible roots and trunks may seem strange, but scientists think this may have been the original reason that bananas and plantains were domesticated as well, before their larger, seedless fruits developed.[39]

Uses. Enset is "a vegetable taller than a house" with enormous (1m by 3m/3 by 10 feet) starch-filled trunks and huge (1m by 1m) starchy tubers.[40] This

FIGURE 12.5. Enset is a banana relative with edible starchy trunks. Photograph courtesy of Forest and Kim Starr.

starch is processed into many products, including fermented *kocho* and *injera* bread.

Yield. Yields are around 5 tons of dry starch per hectare. A single plant can feed a family of five for a month.[41]

Harvest and processing. When the trunk is harvested, up to several hundred shoots will sprout from the base, a wonderful form of non-destructive harvest. Enset is not without drawbacks, however. The starch must be separated from the fibers, which is currently done via painstaking manual labor. The fibers themselves are also a useful product, but there is great need for appropriate technologies to reduce the tedious labor of enset processing.[42]

Carbon farming applications. Families manage enset patches with plants of different ages, in integrated systems with other crops and cattle. Cow manure is particularly important as a source of fertility for enset plants.

Crop development. Some ornamental cultivars of enset are widely grown. These are not as edible as enset, and die after flowering.[43] It is not clear whether edible enset clones have made it to other tropical highland areas outside of Ethiopia. Enset, along with chachafruto and a few other key species, should be widely grown throughout the world's humid tropical highlands. While in other regions, *kocho* and *injera* may take a while to catch on, enset could be a great way to replace annual grains in the diets of livestock.

AGAVE HEARTS (*AGAVE* SPP.)

Agaves are large, spiny desert succulents. A number of species have been grown or domesticated for the edible starchy hearts, which can be sizable, although not all agave species can be used this way. Some are bitter, while others are too small. The primary species cultivated for alcohol in Mexico today are from tropical or subtropical arid and semiarid deserts, including *A. americana, A. cantala, A. salmiana,* and *A. tequilana.*[44] Some grow at high tropical elevations. At least two other species, *A. parryi* and *A. murpheyi,* were historically cultivated in what is today the southwestern United States.[45] These species are more cold-tolerant, with some *A. parryi* clones tolerating temperatures as low as −40°C (−40°F). I suspect many years typically pass before they flower

FIGURE 12.6. Mezcal production with agave hearts, probably *A. americana*. Photograph courtesy of Gérard Janot/Creative Commons 3.0.

in such cold regions. Edible agaves are a regional crop, while the fiber agaves are minor global crops.

Uses. Today agave hearts or "piñas"—so called because they resemble pineapples when trimmed of their leaves—are the raw material for distilled alcohol products such as tequila and mezcal. Historically, they were used as staple energy food for people in extremely dry regions of Mesoamerica and southwestern North America. Agave hearts were roasted in outdoor pits for several days to convert some of the starches to sugars.[46]

Some agaves (mostly *A. americana* and *A. tequilana*) are also used to make a sweetener called agave nectar. Textile fiber is commercially extracted as a by-product, although some agaves are grown primarily for their fiber. Agaves are also grown as spiny living fences. *Agave americana* is on the global invasive species lists because it has successfully naturalized in dry regions around the world.[47]

Yield. In commercial production systems, agaves are grown on rotation, with new plants being planted every year. This way, although each plant only flowers once in a 5- to 10-year cycle, others are always yielding. This seems like a long time to wait, but in the extremely dry and sometimes high-elevation or cold areas where they survive, it is still quite welcome.

Harvest and processing. The plants grow five years or more before swelling in preparation for flowering. At this time, or when the flower stalk is still only a meter (3 feet) or less in height—they can grow to 7 meters (23 feet) or more!—the spiny leaves are removed, exposing the heart. After cutting, many or most species produce offsets, or small suckers, at the base. This is the next generation, thanks to the possibility of a non-destructive harvest.

Carbon farming applications. Agaves are often used in living fences. Their spiny leaves make them effective once established. They are also grown as living terrace edges in Mexican *metepantli* systems.

CHAPTER THIRTEEN

Balanced Carbohydrate Crops

Balanced carbohydrate crops have high levels of starch or sugar. Both protein and oil are under 15 percent, but at least one is over 5 percent. Global annual crops in this class include rice, maize, wheat, potatoes, barley, millet, oats, rye, and other cereal grains. The only global perennial in this class is the chestnut.[1] This is a very important class of foods; they are probably the most important human food and account for 675 million hectares (1.7 billion acres), 45 percent of all cropland. Perennial balanced carbohydrate crops include nuts, beans, cereals, fruits, and woody legume pods. Many of these perennials yield extremely well, with many equal to or better than their annual counterparts.

Nuts

Botanically speaking, nuts are hard, dry fruits with seeds inside (usually just one), which do not open when they become ripe. Not all nut crops described in this book meet that definition. Some, like jakfruit, are encased in soft, edible flesh, while others, like chestnuts, have more than one seed per fruit. Nuts may be starchy, oily, high in protein, or a combination of the three. In this chapter we'll look at those with balanced starch.

Balanced starch nuts include some of my very favorites. I love the starchy sweetness of fresh-roasted chestnuts. One of the best foods I've ever eaten was a fresh-roasted bunya bunya nut. These grow in enormous cones on huge, beautiful trees. Imagine the size and texture of a chestnut and the aromatic, savory flavor of a pine nut. In their homelands many of these species are important foods today. In the past many—from acorns to chestnuts to the various breadnuts—were the staff of life.

CHESTNUT (*CASTANEA* SPP.)

It is easy for us cold-climate dwellers to feel discouraged when we look at the marvelous diversity of perennial crops that is available to tropical growers. When someone asks me what tree staples are best for cold, humid climates, the chestnut is often first on my list. Chestnuts have served as important starchy staple crops for humans for thousands of years. Writers have described "chestnut cultures" in Asia, the Mediterranean, and eastern North America.[2] Between the various species, there are chestnuts that range from the cold end of cold temperate (–17°C/1.4°F) through the subtropics. Chestnut is the only perennial balanced carbohydrate crop of global importance, with 536,000 hectares (1.3 million acres) of chestnuts producing 2 million tons of nuts, valued at $4.7 billion US in 2012.[3]

There are many cultivated species. The primary ones of importance are the Chinese (*C. mollissima*) and Japanese (*C. crenata*), European or sweet (*C. sativa*), and American (*C. dentata*). Chinese and Japanese chestnuts have been staples in Asia for millennia. Chinese chestnut is probably the most common parent of hybrid chestnuts. European chestnuts are widely grown through Europe and Australia.

American chestnut was once the most common tree in the eastern United States, until it was

TABLE 13.1. Balanced Carbohydrate Crop Yields Compared

Latin Name	Common Name	Climate	Avg. Global Yield	Yield t/ha	Product
Prosopis spp.	Tropical mesquites	Arid to semi-arid tropics		9.0–49.5[a]	Pods
*Zea mays	Corn, maize	Worldwide	4.4	0.9–18.0[b]	Grain
Bactris gasipaes	Peach palm	Humid tropics		8.8–13.2[c]	Fruit
*Oryza sativa	Rice	Humid worldwide	3.8	2.6–13.1[d]	Grain
*Solanum tuberosum	Potato	Humid worldwide	5.3	5.5–8.3[b]	Tubers
Ceratonia siliqua	Carob	Mediterranean	1.7	1.7–8.3[b]	Pods
*Triticum aestivum	Wheat	Cold temperate, Mediterranean	2.7	2.6–7.9[b]	Grain
Quercus ilex	Encina	Mediterranean, warm temperate, semi-arid to humid		3.1–7.7[e]	Acorns
Brosimum alicastrum	Mayan breadnut	Humid tropics		6.6–7.5[f]	Nuts
Oryza sativa	Perennial rice 'PR23'	Humid tropics and subtropics		3.4–6.3[g]	Grain
Gleditsia triacanthos	Honey locust	Boreal to subtropical, lowlands and highlands, semi-arid to humid		6.0[h]	Pods
Inocarpus fagifer	Tahitian chestnut	Humid tropics		0.7–4.3[i]	Nuts
Artocarpus camansi	Breadnut	Humid tropical lowlands		3.9[j]	Nuts
Prosopis spp.	Cold-tolerant mesquites	Cold arid		1.8–3.6[k]	Pods
*Sorghum bicolor	Sorghum	Arid worldwide		2.7[b]	Grain
Cordeauxia edulis	Yeheb	Arid tropics		1.4–2.2[b]	Beans, extrapolated
Castanea spp.	Chestnut	Cold temperate and Mediterranean	1.3	1.0–1.7[l]	Nuts
Distichlis palmeri	Nipa	Tropics and subtropics		1.0[m]	Grain
Phleum pratense	Timothy	Temperate		0.4–0.5[n]	Grain

Note: Here we compare perennial versus annual balanced carbohydrate sources. Annual comparison crops are marked with an asterisk (*). All yields calculated based on dry weight of edible portion. Crops are arranged in descending order of yield.

[a] Pasiecznik, *The* Prosopis juliflora–Prosopis pallida *Complex*, 92–93.
[b] Plant Resources of Tropical Africa online database.
[c] Clement, "The potential use of the pejibaye palm in agroforestry systems," 203–04.
[d] Plant Resources of Tropical Africa online database; Uphoff, "Agroecological alternatives," 6.
[e] See the Encina section on page 154 for details.
[f] Janick and Paull, *The Encyclopedia of Fruits and Nuts*, 492; Leung and Flores, *Food Composition Table for Use in Latin America*.

[g] Zhang, et al., "The progression of perennial rice breeding and genetics research in China," 37.
[h] Seibert et al., "Fuel and chemical co-production from tree crops," 54.
[i] Elevitch, *Traditional Trees of Pacific Islands*, 421.
[j] Janick and Paull, *The Encyclopedia of Fruits and Nuts*, 479–80.
[k] Felger and Logan, "Mesquite," 4.
[l] Wilkinson, *Nut Grower's Guide*, 123.
[m] Pearlstein et al. "Nipa (*Distichlis palmeri*)," 60.
[n] Sands, Pilgeram, and Morris, "Development and marketing of perennial grains with benefits for human health and nutrition," 211

devastated by disease. (The blight probably arrived with introduced Chinese chestnut trees in the late 1800s.) Today, researchers are nearly ready to release Chinese-American hybrids that are close to the original American chestnut; we may soon see this magnificent tree return to some of its former glory. In fact, there are many hybrid chestnut varieties; chestnut is one of the species that Badgersett is developing as a neohybrid crop, drawing on the genetic resources of five chestnuts from three continents.

Uses. The nuts are sweet and starchy, about 1 to 3 percent oil and 5 to 12 percent protein.[4] They can be eaten raw but are typically roasted or boiled. The nuts are borne in spiny burrs, typically with three to four nuts per burr. Breeders Phil Rutter of Badgersett Research Corporation and Luther Burbank each independently increased this to as high as 12.[5] Unfortunately the Burbank nuts were very small, and the Badgersett 12-nut tree died after a heavy yield (literally producing itself to death).[6]

Yield. Unlike many cold-climate nuts, chestnuts yield consistently every year. Fresh weight yields up to 5 tons per hectare, with a global average yield of 3.7 t/ha.[7] However, after accounting for the weight of the shells, and removing the substantial water content, chestnut yields are a far more modest 1.0 to 1.7 t/ha, with a global average of 1.3.

Harvest and processing. Harvesting can be fully mechanized, with machines sweeping the fallen nuts into windrows, then vacuuming them up in flexible tubes. Care must be taken to avoid damaging the nuts, however, so most growers hand-harvest. Cleaning and some of the processing can also be mechanized to some extent.[8] There is a Korean acorn-peeling technology that might be used to improve chestnut processing.

Chestnuts tend to have a short shelf life, which can be extended with refrigeration and other methods. Traditionally the nuts were candied or ground into flour (or fed to hogs and converted into hams) to preserve the harvest. Badgersett has developed a chestnut "polenta," which offers the exciting potential of eating and selling chestnuts year-round.

Carbon farming applications. Chestnuts are intercropped with annual crops in China.[9] In North Korea they are grown on contour and intercropped with annuals.[10] They have a long history of use in coppice and silvopasture systems in Europe. In the Badgersett vision of woody agriculture, chestnut is the species that replaces corn. Whether in coppiced rows or orchard-style, this is a crop with a critical role to play in temperate humid climates in any perennial carbon farming scenario.

RAMON OR MAYAN BREADNUT (*BROSIMUM ALICASTRUM*)

The Mayan breadnut is the primary mulberry family nut crop of the Americas. It is known as ramon, ojoche, ujuxte, and many other names throughout its wide range, which runs from southern Mexico through Brazil with native populations in the Caribbean as well. Many researchers believe that ramon was an important staple of the pre-Columbian Maya people, a perennial counterpoint to annual maize.[11] It certainly grows abundantly around Mayan ruins today. Ramon is grown in Mayan homegardens, is a regional crop in Mexico, and is in the midst of a renaissance, thanks to the work of the Maya Nut Foundation.[12] It grows in semi-arid to very humid tropical lowlands and tolerates a long dry season. It can also handle coastal salt spray. Several other *Brosimum* species produce edible nuts.

Uses. Unlike the other nuts in the family, there is only one nut inside each small, sweet ramon fruit. The edible nuts are about 2 centimeters (0.8 inch) across. They are boiled, roasted, or ground and used for tortillas. Mayan breadnuts are about 11 percent protein and 2 percent fat.[13] Roasted and ground seeds are used as a coffee substitute. Ramon produces a latex that provides some hydrocarbons and is used as an edible milk.

FIGURE 13.1. The author with a 100-year-old chestnut tree in California. This chestnut tree has seven varieties grafted on to it. Photograph courtesy of Brock Dolman.

FIGURE 13.2. Mayan breadnuts in Guatemala. Photography courtesy of Marikler Girón Toensmeier.

TABLE 13.2. Balanced Carbohydrate Nuts

Latin Name	Common Name	Origin	Climate	Status
Araucaria angustifolia	Pinheiro	South America	Subtropical humid lowlands and highlands	Regional
Araucaria araucana	Monkey puzzle	Chile and Argentina	Warm temperate humid	Minor global
Araucaria bidwillii	Bunya bunya	Australia	Tropics and subtropics, Mediterranean, humid	Minor global
Artocarpus camansi	Breadnut	New Guinea, Pacific	Humid tropical lowlands	Minor global
Artocarpus heterophyllus	Jakfruit	India	Tropical and subtropical lowlands, humid	Global
Artocarpus integer	Champedak	Southeast Asia, New Guinea	Tropical humid lowlands	Regional
Artocarpus mariannensis	Dugdug	Pacific	Tropical humid lowlands	Regional
Barringtonia edulis	Vutu kana	Fiji	Humid tropical lowlands	Regional
Barringtonia novae-hiberniae	Cutnut	New Guinea, West Pacific	Humid tropical lowlands	Regional
Barringtonia procera	Cutnut	New Guinea, Southwest Pacific	Humid tropical lowlands	Regional
Brosimum alicastrum	Ramon, Mayan breadnut	Mesoamerica	Tropical lowlands, humid to semi-arid	Regional, historic staple
Brosimum utile	Milk tree, cow tree	Americas	Tropical lowlands, humid	Regional
Castanea spp.	Hybrid and neohybrid chestnuts	Hybrid	Cold to warm temperate humid	Global
Castanea crenata	Japanese chestnut	Asia	Cold to warm temperate humid	Regional
Castanea dentata	American chestnut	North America	Cold to warm temperate humid	Historic staple (disease-resistant under development)
Castanea mollissima	Chinese chestnut	Asia	Cold to warm temperate	Minor global
Castanea sativa	European chestnut	Europe	Warm temperate to Mediterranean	Global
Castanopsis acuminatissima		Asia, New Guinea	Tropical highlands, humid	Regional
Castanopsis javanica		Java	Humid tropical lowlands	Regional
Castanopsis inermis	Berangan	Malacca, Sumatra	Humid tropical lowlands	Regional
Castanopsis tribuloides		Southeast Asia	Tropical highlands, humid	Regional
Cordeauxia edulis	Yeheb	Somalia	Arid tropical lowlands	Wild staple, minor new crop
Euryale ferox	Gorgon fruit	Asia	Perennial in tropics; aquatic	Regional
Ginkgo biloba[a]	Ginkgo	Asia	Cold temperate to subtropical, humid	Regional
Gnetum gnemon	Jointfir, bago	Asia	Humid tropical lowlands	Minor global
Inocarpus fagifer	Tahitian chestnut	Oceania	Tropical humid lowlands	Minor
Nelumbo lutea	American lotus	North America	Cold temperate to tropical, aquatic	Historic
Nelumbo nucifera	Chinese water lotus	Asia, Australia, Middle East	Warm temperate to tropical; aquatic	Minor global
Quercus spp.	Hybrid oaks	Hybrid	Boreal to subtropical, arid to humid	Under development
Quercus hybrids	Burgambel oak	Natural hybrid in western North America	Boreal to warm temperate, semi-arid	Under development

Latin Name	Common Name	Origin	Climate	Status
Quercus acutissima	Sawtooth oak	East Asia	Warm temperate to subtropical, humid	Regional
Quercus brantii	Barro	Turkey, Iran	Warm temperate, semi-arid	Regional
Quercus emoryi	Emory oak	Southwest North America, northern Mexico	Warm temperate to subtropical, arid to semi-arid	Wild staple
Quercus ilex	Encina	Mediterranean	Mediterranean, warm temperate, subtropical, semi-arid	Regional
Quercus insignis	Chicalaba	Mesoamerica	Tropical highlands, humid	Wild
Quercus kelloggii	California black oak	Western North America	Cold temperate to subtropical, Mediterranean, semi-arid	Wild staple

ª Toxic in large amounts, and even small amounts raw, according to some sources. Nonetheless widely consumed in Asia, according to Duke's *Handbook of Nuts*.

Yields. Ramon is a high-yielding nut crop, with yields of 6.6 to 7.5 tons per hectare dry weight from unselected plants.[14] Improved forms could yield considerably more.

Carbon farming applications. Ramon leaves are also an important livestock fodder, often used in the dry season. The fallen fruits and nuts are important feed for cattle and pigs as well.

ASIAN BREADNUTS (*ARTOCARPUS* SPP.)

There are important mulberry family nuts from tropical Asia (*Artocarpus*), Africa (*Treculia*), and the Americas (*Brosimum*). There is even a temperate breadnut awaiting development—the Osage orange (see chapter 18 for details). As a group, they represent some of the most productive and potentially important perennial staple crop producers in the world. The most diverse and developed of these are the members of the genus *Artocarpus*. These species hail from the region between India and the Western Pacific. All of them are limited to tropical humid lowlands. They include:

Breadnut (*A. camansi*). Among the highest-yielding nut crops in the world at 11 tons of nuts per hectare. The nuts are 13 to 20 percent protein and 6 to 12 percent oil, which is typical for *Artocarpus* nuts.[15] Breadnut fruit flesh is not particularly edible by humans.

Jakfruit (*A. heterophyllus*). Widely grown in tropical homegardens in Asia, Africa, and Latin America.[16] It is native from India through Malaysia. The mature fruits are enormous, from 5 to 30kg (11 to 66 pounds) each, with the occasional 50kg (110-pound) specimen. Immature fruits are eaten like breadfruit. Once mature, the fruit flesh is no longer a staple but becomes a wonderful dessert. The nuts of mature fruits are cooked and eaten, tasting to me a lot like large peanuts.

Dugdug (*A. mariannensis*). Native to the Marianna Islands and Palau. Dugdug nuts and underripe fruits are staples in Micronesia.

Champedak (*A. integer*). Native to Southeast Asia and New Guinea. It has similar uses and yields to jakfruit.

Marang (*A. odoratissimus*). Another jakfruit-like crop from Borneo with sweet dessert fruit and edible nuts when ripe. Underripe fruits are cooked like breadfruit.

Uses. Cooked nuts of all these species are used similarly. They are good-sized (2 to 4 centimeters/0.8 to 1.6 inches) and have an appealing flavor and texture. In all of the species listed here, the smaller young green fruits are cooked as starchy vegetables, while the nuts from mature fruits are used as staples. Some also provide sweet fruits when ripe. The young leaves of many are edible.

FIGURE 13.3. Asian breadnut at Maya Mountain Research Farm in Belize. Photograph courtesy of Christopher Nesbitt.

FIGURE 13.4. Encina oaks in recently tilled *dehesa* in Iberia. Photograph courtesy of © AGROOF scop/www.agroof.net.

Yields. Asian breadnuts produce as much as 11 tons of nuts per hectare each year, but these are 10 percent shell and 61 percent water. The edible dry weight yields are a more modest 3.9 t/ha.

Harvest and processing. Harvesting the very large fruits of this genus can be challenging, especially in the case of the enormous jakfruit. And efficiently removing the nuts from the large fruits represents a processing challenge. All Moraceae nuts should be cooked before eating; usually they are boiled or roasted. They can be ground into flour and used in many recipes.

Carbon farming applications. Jakfruit is common in homegardens around the world. Breadnuts and other *Artocarpus* species are grown in homegardens in their home regions. Breadnuts are interplanted with many tropical crops and used as a shade overstory species. The fruit and nuts are used as livestock fodder.[17]

ENCINA (*QUERCUS ILEX* VAR. *BALLOTA*)

Encina is an edible-acorn oak that has fed humans and livestock, especially hogs but also sheep and cattle, for millennia. At some point "sweet" selections were made from the bitter ancestor, and the process of domestication began. Farmers have identified and grafted superior selections.[18] Today the acorns are largely free

of bitterness and are produced consistently in large amounts every year, at least among good varieties in the domesticated subspecies *ballota*.

Encina is the primary acorn-producing species in the *dehesa* silvopasture system in Spain and Portugal. Although it was once much more widespread, *dehesa* systems still occupy 5.5 million hectares (13.6 million acres) in Spain and Portugal.[19] However, almost no new oaks have been planted in *dehesa* for a century, so the system is considered endangered.[20] This ancient regional crop is primarily limited to Mediterranean climates. It can handle cold down to −17°C/1°F.

Uses. Acorns of *Q. ilex* ssp. *ballota* are virtually free of bitterness and can be eaten out of hand like almonds. They can also be eaten as nuts, ground for flour, and pressed for oil. Today their primary use is in feeding livestock, particularly pigs. The processed meat from such hogs is called *jamón ibérico* and is one of the world's most highly prized meats. Encina trees are also used for firewood, construction timber, charcoal, and tannins.[21]

Yields. In *Tree Crops*, J. R. Smith reported that tracts of encina yield consistently, although individual trees bear in alternate years. He noted that individual trees yielded 100kg to 400kg (220 to 882 pounds) per tree.[22] Some superior forms of encina consistently yield 300kg (661 pounds) per tree per year.[23] Because the spacing of *Q. ilex* in Mediterranean *dehesa* silvopasture systems is between 20 and 50 trees/ha, we can extrapolate that a hectare of grafted elite *Q. ilex* trees in dehesa would yield an extremely impressive 6 to 15 t/ha annually fresh weight.[24] After accounting for shells and water, this converts to 3.1 to 7.7 t/ha, still very impressive.

Harvest and processing. Today most encina acorns are consumed by hogs when they fall to the ground. Encina acorns do not require leaching but are otherwise used like other acorns for human and livestock consumption.

Carbon farming applications. Encina is an anchor component of *dehesa* silvopasture systems. It is also used as a windbreak.[25]

OTHER OAKS (*QUERCUS* SPP.)

Encina is not the only edible oak. How do other oaks compare as potential perennial staple crops? Certainly oaks were a critical staple for humanity for millennia. Indigenous people throughout North America wild-managed oaks with fire, pruning, and other horticultural techniques to maintain productive savannas for acorns, wildlife, and medicinal and edible herbs with mushrooms in the understory.[26] (M. Kat Anderson's *Tending the Wild* is required reading on the subject.) Although acorn-eating traditions are still alive in California and the Mediterranean, North and South Korea are the only countries where acorns are still an important starch for the majority of the human population. As of 1991, Koreans consumed 1 to 2 million tons of acorns annually.[27]

Why has humanity largely shifted away from the acorn as food? I'm aware of three pretty good reasons. First and most important, most acorns contain bitter tannins that must be removed by leaching or other processing. Second, most oaks do not yield well every year; in cold climates a good harvest from any individual tree may come only every three to five years. Finally, even in a good year, most oak yields are not fantastic. Aside from the remarkable encina, oaks have not yet been developed into a "superoak" that supplies non-bitter acorns and yields both heavily and annually. And encina

FIGURE 13.5. These giant acorns of *Q. insignis* in Mexico demonstrate the potential of domesticated oaks.

is largely limited to Mediterranean climates. Historically, the abundance of oaks seems to have rendered them important enough that they overcome many of these drawbacks, although most cultures have abandoned acorns as food.

However, within the great diversity of oak species there is one for almost every non-tropical region, from cold deserts to cloud forest to humid temperate climates. Descriptions of a few of the less bitter and cultivated species follow. They offer raw material for breeding and hope in the long term for potential perennial staples for many challenging climates.

Sawtooth oak (*Q. acutissima*). Native to Japan, China, and Korea. Along with several other species of oak, it is wild-collected and cultivated in Korea for acorn starch extraction. The starch is used to make a popular tofu-like jelly called *dotorimuk*. Sawtooth oak is productive and bears more regularly than most oaks but is high in bitter tannins. This is not a barrier in Korea and China, where commercial processing facilities remove the tannins and produce acorn flour.[28]

Barro (*Q. brantii*). A cold- and drought-tolerant oak native to Turkey and Iran.[29] It is cultivated for its acorns in northern Kurdistan.

California black oak (*Q. kelloggii*). One of the most important species managed by indigenous people in western North America. Black oak savannas were burned regularly to manage diseases and pests, reduce competition, and cycle nutrients. The acorn was the most important staple in California. Today, native management continues for this and other species in some parts of California, and acorn mush is still consumed regularly by many native people.[30]

Emory oak (*Q. emoryi*). Native to the arid southwestern United States and northern Mexico and historically very important to native people of the region. I've eaten the small acorns of Emory oak and found them relatively lacking in bitter tannins, as were the acorns of the Emory oak hybrid wavy-leaf oak (*Quercus × undulata*).

Chicalaba (*Q. insignis*). Native to Mesoamerican cloud forests. Although not cultivated to my knowledge, it produces what may be the world's largest acorns—5 centimeters (2 inches) in diameter. I've processed and eaten them on several occasions; once the tannins are extracted, chicalaba is good eating. Ricardo Romero of Las Cañadas reports that chicalaba seems to bear relatively regularly each year as well. These traits suggest that *Q. insignis* should be part of efforts to breed large-fruited, annual-bearing oaks. Las Cañadas has been grafting elite local forms for their nursery.

Hybrid oaks. Produced naturally anywhere that oak species overlap in the wild. Breeders have also actively made many crosses. Among the most promising is the burgambel oak, a hybrid of two incredibly tough species, Gambel oak (*Q. gambelii*) and bur oak (*Q. macrocarpa*). Burgambel acorns are low in tannins and relatively quick to bear, between three and six years.[31]

Uses. Acorns are high in starch with 2 to 9 percent protein and 1 to 31 percent oil.[32] Bitter tannins are present in most species, which must be leached out with copious amounts of water. The tannins themselves are an important industrial product, used in tanning and dyeing. Leached acorn can be cooked in a variety of ways or pressed for an edible oil.[34]

Yield. In one case, yields of acorns in a good year were reported at 3 to 6 t/ha fresh weight, although this is very far out of line of most reported yields.[35] Typically even "heavy" acorn yields are low, and these are followed by several years of even lower yields.

Harvest and processing. In Korea and China, commercial processing facilities remove the tannins and produce

KEN ASMUS: OAK BREEDER

Ken Asmus runs the Oikos Tree Crops nursery in the north-central United States. He grows and breeds many useful perennial crops but has a special focus on oaks, only because, he says, there was really no one else working on it. Ken has collected seedlings and grafted clones of superior wild strains of native oaks and low-tannin and high-yielding oaks from around the world. His specialty is hybrid oaks. Ken does a lot of crosses himself as well as exchanging hybrid acorns with others. This year Oikos is offering 17 different hybrid oaks, and Ken is trialing many more in their testing plots.

Ken's vision is of sweet acorn orchards for humans and livestock, although he notes that most of the people working on acorn improvement in the United States are deer hunters looking to improve wildlife habitat. Ken reports that oak yields are often limited by pollination, and plantings with multiple species and hybrids are the best way to increase pollination (and therefore yields). He says that the next step would be to plant out grafted trees of the best hybrids and begin crossing them with one another to develop second-, third-, and fourth-generation crosses.[33] With many neglected crops like acorns, it is dedicated individuals like Ken and small nurseries like Oikos that are safeguarding diversity and moving crop development forward despite no institutional or government support.

acorn flour for use in *dotorimik* and other dishes.[36] Elsewhere people grind or crack acorns and leach the tannins with water in a labor-intensive process.

Carbon farming applications. Indigenous oak management systems in California are classed as agroforestry systems.[37] Sawtooth oak is intercropped with annual crops in northern China.[38] Oaks are potential candidates for woody agriculture.

Crop development. Given the ubiquity of oaks and the historic importance of acorns as food, some breeding efforts have been undertaken in recent decades. Ken Asmus of Oikos Tree Crops and Miguel Marquez are two leaders in this effort. Given the ease with which oaks hybridize even in the wild, the potential to develop neohybrid oaks with low tannins and high annual yields is very real, although utterly unfunded and probably still decades away. Oak grafting is not reliable, and this is one obstacle to overcome. Anyone looking to make a contribution to carbon farming in non-tropical climates would do well to explore this area.

TAHITIAN CHESTNUT (*INOCARPUS FAGIFER*)

Tahitian chestnut is a nut tree native to Melanesia, Micronesia, and Polynesia. It was more widely cultivated historically than it is today, although it remains a regional crop for local markets. *Inocarpus* grows in humid lowland tropics with a minimal dry season and is tolerant of shade.[39]

Uses. Tahitian chestnuts are starchy with 5 to 6 percent protein.[40] They are typically cooked before consumption.[41]

Yields. Fresh weight yields are outstanding at 4 to 30 tons per acre, with 60 percent of that being edible kernel, resulting in a remarkable 3 to 18 tons of kernels.[42] However, at 76 percent water, this converts to a more modest 0.7 to 4.3 t/ha.

Carbon farming applications. Tahitian chestnut makes a good living fence and works well for erosion control and coastal protection. This species is tolerant of 20 to 80 percent shade. The trees can be used to shade other crops such as *Gnetum gnemon* and cacao

FIGURE 13.6. A Tahitian chestnut at Fruit and Spice Park in Florida. This staple species grows in shade and fixes nitrogen.

and can also be used as a trellis for vines such as betel. Tahitian chestnut coppices strongly and has potential as a woody agriculture crop.[43]

Harvest and processing. Processing technology is currently lacking, so the nuts must be cut out of the shells by hand. Mechanized methods would require the identification or breeding of cultivars with uniform nut size.[44]

Crop development. This high-yielding, nitrogen-fixing, shade-tolerant tree crop should be a high priority for carbon-sequestering agriculture efforts.

YEHEB (*CORDEAUXIA EDULIS*)

Yeheb is a nitrogen-fixing legume shrub of the Danakil desert in Somalia and adjacent Ethiopia. It is a very important wild-collected staple for nomadic herders in the area, and today is cultivated on a small scale in the region and beyond. Some nuts are exported.[45] This is a species for arid and semi-arid tropical lowlands. In the wild, overgrazing and overharvesting has had such an impact that yeheb has been declared rare.[46]

Uses. The nuts are roughly 12 percent protein and 12 percent fat. They are eaten raw or cooked like chestnuts.[47]

Yields. Yields are 5 to 8 kg per plant, which at 320 plants per hectare would produce 1.6 to 2.5 tons per hectare—but may be zero in drought years. In dry weight terms this is equivalent to 1.4 to 2.2 t/ha/yr. In Somalia, where there are two rainy seasons, yeheb gives two corresponding harvests per year.[48]

Harvest and processing. Currently yeheb is hand-harvested and processed.

Carbon farming applications. Yeheb is planted in fodder banks and is an important wild fodder as well.[49] Because it coppices and has durable "nuts," it could be investigated as a crop for woody agriculture.

Crop development. Efforts to cultivate yeheb and to protect existing wild stands have been initiated, although political turmoil in its native Somalia is an obstacle to the preservation and development of this important dryland crop. *Cordeauxia edulis* could be a promising potential source of food and dry-season fodder for other hot, arid regions. However, the limited availability of viable seed and the shortage of knowledge on the crop, especially its propagation, agronomic practices, and potential for selection and breeding, are significant constraints.[50]

Perennial Cereals and Pseudocereals

Perennial cereals are the (mostly hypothetical) edible seeds of perennial grasses. Here we'll also profile some pseudocereals, which is the term for cereal-like edible seeds that do not come from grasses. Some examples of annual pseudocereals include quinoa, buckwheat, chia, and amaranth. Virtually all fall within the balanced carbohydrate crop category, although a few of the grass seeds have surprisingly high protein.[51] None of these crops is yet ready to step up as a major contributor to human and livestock carbohydrate needs. However, because cereals make the most significant contribution to human sustenance, it is important to visit this group. A major advantage of perennial cereals, once they are ready, will be ease of mechanical harvesting.

RICE (*ORYZA SATIVA*)

The perennial grain that is closest to competing with annual grains for yield is rice. (In fact, under some conditions, some annual rice plants will "rattoon," or resprout, for several years. Annual rice has several perennial relatives, including an African perennial rice and a strain of the wild ancestor of annual rice. In the 1990s the International Rice Research Institute in the Philippines initiated perennial rice breeding efforts. In 2007 the Yunnan Academy of Agricultural Sciences in Kunming, China, picked up their work and carried it forward.[52]

Yunnan Academy breeder Fengyi Hu has developed the 'PR23' line of rice that has produced grain for three years in a row, with yields that are competitive with annual rice. (See table 13.4.) Over the three-year trial only 3.5 to 5.4 percent of plants died.[53] A recent Land

TABLE 13.3. Current Status of Perennial Cereals and Pseudocereals

Competitive yields in trials. These perennial grains produce yields competitive with annual cereals although they are not yet in commercial production.	Rice (*Oryza* spp.), nipa (*Distichlis palmeri*), sorghum (*Sorghum* spp.)
Minor and regional crops. These perennial grains are actually in production, although yields are low or unreported.	Wangunu (*Eragrostis eriopoda*), Indian ricegrass (*Oryzopsis hymenoides*), timothy (*Phleum pratense*), rye (*Secale* spp.), fountaingrass (*Pennisetum setaceum*), highlands pitpit (*Setaria palmifolia*), sand dropseed (*Sporobolus fimbriatus*), weeping rice grass (*Erharta stipoides*)
Historic crops. These perennial grains were cultivated in the past but largely or completely abandoned.	Bull Mitchell grass (*Astrebla squarrosa*), good King Henry (*Chenopodium bonus-henricus*), native millet (*Panicum decompositum*), arroz bravo (*Rhynchoryza subulata*)
Under development. There is active breeding work on these species though they are not presently in commercial production.	Kernza (*Elymus hispidus*), wheat (*Triticum* spp.), maize (*Zea* spp.), barley (*Hordeum* spp.)

Note: Some of these crops are annual staples that breeders are working to perennialize by crossing them with perennial relatives or ancestors. Others are traditional regional perennial grain crops. (Some have been abandoned and are now historic crops only.) Some wild perennial cereals are being domesticated, as is at least one pasture grass. A few are important wild staple carbohydrates.

FIGURE 13.7. Perennial rice grains at the Yunnan Academy of Agricultural Sciences, China. Photograph courtesy of David Van Tassel.

FIGURE 13.8. Breeder Fengyi Hu with perennial rice in China. Rice is poised to be the first major perennial grain. Photograph courtesy of David Van Tassel.

Institute publication notes that these plants are now going strong in their fourth year and continuing to yield well.[54] In 2015 perennial rice trials took place on more than 600 hectares (1,483 acres) of farms in China and Laos.[55]

'PR23' is a paddy rice variety suited to flooded rice production (a major contributor to methane emissions). Researchers also hope to develop perennial versions of the upland rice—grown on slopes throughout Asia—that might have a more powerful climate impact, as upland rice is methane-free. The incredible progress of perennializing rice is exciting and promising news and could have tremendous impact in tropical and subtropical areas of the world. Although not a heavy yielder, one of the Yunnan crosses has lived for 10 years.[56]

NIPA (*DISTICHLIS PALMERI*)

Nipa is a perennial salt-tolerant grass of the Sonoran desert deltas. The flavor of its grain is purportedly excellent. Once a staple of the Cocopa people, wild populations of nipa have been greatly reduced due to dams and other watershed disruptions. Wild patches of nipa have been estimated to yield 1.25 tons per hectare, making this one of the most productive perennial grains on the planet.[57] As a C_4 grass it is particularly efficient at photosynthesis. (C_4 plants use a metabolic pathway that allows for more efficient photosynthesis under hotter and drier conditions.)

Dr. Richard Felger, a researcher with the University of Arizona herbarium and the Sky Island Alliance, has been researching the potential of nipa as a salt-tolerant perennial grain for decades. Although other researchers attempted to commercialize the crop before it was ready, Felger is confident that it will become a major world crop, comparable to short-grain rice in grain size and flavor.[58] But nipa is still undomesticated and poses

TABLE 13.4. Yields of Perennial Rice 'PR23' Over Three Years in Two Locations

Location	Yield Year One t/ha	Yield Year Two t/ha	Yield Year Three t/ha	Plant Death Rate
Jinghong, Yunnan, China (500m/1,640 feet elevation)	5.6	3.9	4.0	3.5%
Simao, Yunnan, China (1,300m/4,265 feet elevation)	7.3	Very low due to cold	6.7	5.4%

Source: Zhang et al., "The progression of perennial rice breeding and genetics research in China," 37.
Note: Compare these with the global average of 4.4 t/ha for annual rice.

TABLE 13.5. Perennial Cereals and Pseudocereals

Latin Name or Genus	Common Name	Origin	Climate	Status / Cultivation / Wild Utilization / Notes
Astrebla squarrosa	Bull Mitchell grass	Australia	Semi-arid tropics and subtropics	Historic crop[a]
Avena hybrids	Perennial oats	Hybrid	Boreal to warm temperate, semi-arid to humid	Annual *A. sativa* crosses with *A. macrostachya*
Chenopodium hybrids	Perennial quinoa	Hybrid	Temperate to highland tropics, semi-arid to humid	Under development
Chenopodium bonus-henricus	Good King Henry	Europe	Boreal to warm temperate	Historic pseudocereal crop, minor vegetable today
Coix lacryma-jobi	Job's tears	Asia	Tropical, subtropical	Annual forms cultivated as cereal, could be crossed with perennial forms
Distichlis palmeri	Nipa	Mesoamerica	Tropical and subtropical, humid lowlands. halophyte	Historic staple, under development as new grain crop
Elymus hispidus	Kernza	Asia	Boreal to warm temperate, semi-arid	Under development as new cereal
Eragrostis eriopoda	Wangunu	Australia	Tropical to subtropical, arid to semi-arid	Historic wild staple, today a minor crop
Erharta stipoides	Weeping rice grass	Australia	Warm temperate, Mediterranean, subtropical, semi-arid	New crop
Fagopyrum hybrids	Perennial buckwheat	Eurasia	Boreal to warm temperate, arid to humid	Hypothetical, annual *F. esculentum* could potentially be crossed with *F. cymosum*
Glyceria fluitans	Manna grass	Eurasia, Africa	Boreal to Mediterranean	Historic grain crop
Hordeum hybrids	Perennial barley	Hybrid	Boreal to warm temperate, semi-arid to humid	Under development, annual *H. vulgare* crosses with *H. jubatum*
Hyptis suaveolens	Chan	Central America	Tropical lowlands, semi-arid to humid	Wild-collected perennial chia
Leymus arenarius	Beach wildrye	Europe	Boreal to warm temperate, humid	Wild-collected as food, grown for beach stabilization
Oryza hybrids	Perennial rice	Hybrid	Temperate to tropical, humid	Under development as new cereal, annual *O. sativa* crosses with its perennial ancestors, *O. longistaminata* and *O. rufipogon*
Oryza longistaminata	African perennial rice	Africa	Humid tropics	Wild-harvested
Oryzopsis hymenoides	Indian ricegrass	North America	Boreal to warm temperate, arid to humid	Former wild staple, under development, some commercial cultivation
Panicum decompositum	Native millet	Australia	Arid to semi-arid, tropics to warm temperate, Mediterranean	Historic crop[a]
Panicum turgidum	Afezu	Africa through South Asia	Arid to semi-arid, tropics	Historic wild staple in Sahara, cultivated for dune stabilization, potential cross with annual millet (*P. miliaceum*)
Pennisetum spp.	Perennial pearl millet	Hybrid	Tropics and subtropics, semi-arid to humid	Annual *P. glaucum* crossed with *P. purpureum*, grown as fodder crop
Pennisetum setaceum	Fountain grass	North Africa, Middle East	Tropics and subtropics, semi-arid to humid	Minor grain crop
Porteresia coarctata	Uri	South Asia	Humid tropical lowlands, halophyte	Wild-harvested grain, grown for erosion control and coastal protection
Rhynchoryza subulata	Arroz bravo	South America	Tropical humid lowlands	Historic grain crop
Salvia spp.	Perennial chia	Hybrid	Temperate to tropical, arid to humid	Annual chia *S. hispanica* potentially crossable with perennial *Salvia* spp.

Latin Name or Genus	Common Name	Origin	Climate	Status Cultivation Wild Utilization Notes
Secale hybrids	Perennial cereal rye	Hybrid	Boreal to warm temperate, semi-arid to humid	Under development, annual *S. cereale* crosses with *S. montanum*, some cultivars available
Setaria palmifolia	Highlands pitpit	New Guinea to Philippines	Tropical highlands	Minor crop as vegetable and grain
Sorghum hybrids	Perennial sorghum	Hybrid	Tropics to temperate, semi-arid to humid	Under development, annual *S. bicolor* crosses with *S. halepense*, *S. propinquum*
Sporobolus fimbriatus	Sand dropseed	South Africa	Tropical to warm temperate, lowlands and highlands, semi-arid to humid	Minor crop as grain, fodder, and erosion control
Stipagrostis pungens	Drinn	Africa	Arid to semi-arid tropics	Historic wild staple of Sahara
Triticum hybrids	Perennial wheat	Hybrid	Temperate and tropical highlands, semi-arid to humid	Under development, annual *T. aestivum* crosses with *Agropyron*, *Elymus*, *Leymus*, and *Thinopyrum* spp.
Urochloa mosambicensis	Sabi grass	Africa	Tropics to subtropics, semi-arid to humid lowlands	As pasture grass, important wild cereal
Zea hybrids	Perennial corn	Hybrid	Temperate to tropics, semi-arid to humid	Hypothetical, under development; annual *Z. mays* crosses with *Z. diploperennis*, *Tripsacum dactyloides*, *T. floridanum*

Sources: Hanelt, *Mansfeld's Encyclopedia of Agricultural and Horticultural Crops*; National Research Council, *Lost Crops of Africa*, vol. 1: *Grains*; FAO, *Perennial Crops for Food Security.*

Note: This table includes annual cereals that are being or could be perennialized, perennial grains that have been cultivated or wild-harvested, and some pseudocereals such as quinoa that might theoretically be developed. All of these for which yield data exists except nipa yield substantially less than 1 t/ha as of 2013.

[a] Gammage, *The Biggest Estate on Earth*, 293–95.

several challenges. Roughly half of plants are seedless males. It can also take several years after planting until full yields are achieved.[59]

Nipa tolerates salty conditions, including irrigation with salt water, and is adapted to flooding. In the wild, nipa salt marshes are inundated twice daily by tidal seawater, although it does not require salt or waterlogging. As unsustainable irrigation practices and sea-level rise result in increasing salinization of coastal plain farmlands, nipa could become prominent in regions like the Colorado, Ganges, Indus, Murray, and Nile deltas. It is adapted to tropical and subtropical conditions. Although there had been hopes for crossing it with the cold-hardy saltgrass *D. spicata*, those efforts have not been successful so far.[60]

This perennial grain, already a wild staple for ages, has great promise for salty tropical areas and beyond. Nipa development should be a high priority to agencies and individuals concerned with food security, salinization, and climate change. It may also offer an opportunity to develop productive revegetation of barrier islands, to provide protection to coastal areas from extreme weather events.

SORGHUM (*SORGHUM* SPP.)

Sorghum is already weakly perennial in the tropics. It "rattoons," or resprouts, for several years in ideal conditions.[61] Perennial sorghum breeding at the Land Institute has focused on crosses with the perennial weed johnsongrass (*S. halepense*), while other breeders are using *S. propinquum*.[62] As with maize, there are challenges in overwintering tender rhizomes, which would not be an issue in the tropics.[63]

Perennial sorghum is further along than most of the other perennial versions of major grains but is not yet ready for prime time. Perennial sorghum could be bred not just for grain but also for sweet syrup, which was once made from the stalks of annual analogs across the American Midwest. Sorghum is very versatile in terms of the climates to which it is suited, but it is particularly

FIGURE 13.9. Perennial sorghum breeding work at the Land Institute. Photograph courtesy of Scott Seirer.

FIGURE 13.10. Perennial wheat breeding trials at the Land Institute. Photograph courtesy of Scott Seirer.

appropriate to dry regions, where it can outperform maize. Some promising hybrids are being trialed. Some are fully perennial.[64] In some cases, rhizomes of *S. halepense* have nitrogen-fixing endophytic bacteria.[65] This would certainly be a desirable trait to pass into a perennial grain sorghum.

WHEAT (*TRITICUM* SPP.)

Perennial wheat breeding efforts began in the Soviet Union almost 100 years ago, but they've only recently begun to bear results with improved techniques. Many institutions are working on perennial wheat breeding, including the Land Institute, the Future Farm Industries Cooperative Research Center in Australia, Washington State University, Texas A&M, and University of Manitoba.[66] International coordination and seed exchange is underway between these and other researchers.[67]

Perennial wheats are typically crossed with wheatgrass species (*Elymus*). One wheat hybrid yielded 5.8 t/ha (versus 9.0 t/ha for the annual wheat control crop) in the first year but was only weakly perennial.[68] Some wheats have yielded for three years, and some have lived for four.[69]

Perennial wheat is expected to yield somewhat less than annual wheat, but it also requires less labor. In economic projections, this hypothetical crop appears economically competitive with annual wheat from the standpoint of costs per hectare.[70] An Australian economic study has shown that perennial wheats could be economically viable if they yielded just 40 percent as well as annual wheat (providing good fodder for several years after for grazing sheep).[71]

Until recently the Land Institute has had no success in perennial wheat survival in Kansas (nor have I in my micro-trials in Massachusetts). In Australia, some varieties have lived for multiple years, but the yields are low and decline each year, although recently some perennial wheats produced grain for their fourth year there.[72] After almost 100 years of work, and with hundreds of breeding lines, perennial wheat is still not sufficiently perennial.[73] The work continues to perennialize the world's number one crop.

MAIZE (*ZEA* SPP.)

Maize (also known as corn in the United States and Canada) is one of the most important staple crops on the planet. Perennial maize could slow or reverse the degradation of sloping lands around the world that are inappropriately used to grow annual maize. Scientists and backyard breeders have been working toward this goal for many years and have made some limited progress. Diploid perennial teosinte (*Z. diploperennis*) is a wild relative that is crossable with annual corn.[74] Scientists have recently discovered several other wild corn relatives. Maize can also be crossed with more distantly related hardy perennials, including Eastern gammagrass (*Tripsacum dactyloides*) and the related dwarf Fakahatchee grass (*T. floridanum*).

The Land Institute has made substantial progress toward developing perennial maize. One challenge is that the perennial rhizomes that overwinter the plants are not cold-hardy, so their breeding is focused on deeper rhizomes that survive below the frost line.[75] This doesn't matter in the tropics, where hundreds of millions of people rely on maize as a staple.

One interesting new trait emerging in perennial maize breeding is the "ear forest"—the production of multiple ears at the base of the plant. These are produced until the plant is killed by drought, cold, or heat. The ear forest trait might be of use in livestock-harvested perennial grain systems.[76] Researchers anticipate between 10 and 40 more years before perennial maize is ready and posit it as a good public-sector project.[77] Recently the US Department of Agriculture has begun to show interest in perennial maize breeding.[78]

Rye (*Secale* spp.)

Rye is another perennial cereal that is relatively close to commercial viability. Annual rye has been crossed with a wild rye (*S. montanum*), and several varieties have been developed, including 'Permontra' and 'ACE-1'. An Australian variety called 'Black Mountain Family 10' has been yielding well.[79] Most perennial rye hybrids are forage varieties.[80]

In a recent study perennial ryes yielded 73 percent as well as their annual counterparts in years one and two. Not enough plants came back for a third year to make further measurement possible.[81] Annual rye itself yields quite a bit lower than annual wheat. Nonetheless, we are probably closer to seeing perennial rye in production on real farms than most other global cereals.

Additional Perennial Grains Worth Exploring

Indian ricegrass (*Oryzopsis hymenoides*). This perennial North American native was a major staple to indigenous peoples of the West. Discovery of a non-shattering clone allows it to be grown today on a commercial scale in Montana, producing a specialty gluten-free flour marketed as Montina. High prices make up for low yields (0.1 t/ha), and about 1,200 hectares (2,965 acres) are in production.[82] Little breeding work has been done for this remarkably drought- and cold-tolerant perennial grain.

Kernza (*Elymus hispidus*). The Land Institute has been working for several decades to domesticate this perennial wild grain. They have had relatively rapid success, and intermediate wheatgrass is currently undergoing a 12-hectare (30-acre) field trial. The research fields are burned annually to control weeds, and apparently the crop can also be grazed to provide a non-seed yield. Production is still low, though researchers aim to see it reach 1 ton per acre. *Thinopyrum* species are also used as the perennial parent in attempts to develop perennial wheat.[83] Nomenclature of this species is tricky. The Latin name was recently changed from *Thinopyrum intermedium* to *Elymus hispidus*, and the wild form is

Figure 13.11. Kernza, a new perennial grain, at the Land Institute. Photograph courtesy of David Van Tassel.

often called intermediate wheatgrass, while the new, domesticated crop has been named Kernza by the Land Institute.

Kernza was chosen by the Rodale Institute in the 1980s as a perennial cereal candidate, and selected for two generations after emerging as winner from almost 100 perennial cereals that were tested. The task (and seeds) were eventually passed on to the Land Institute.[84] The University of Manitoba joined the Kernza-breeding effort in 2011.[85]

Kernza is an example of domesticating a promising wild grain as opposed to perennializing an annual staple.[86] It is also used as the perennial parent in efforts to perennialize wheat. Researchers estimate that Kernza could be yielding as well as wheat within 20 years. However, seed size will still be smaller; it might take 130 years to reach equivalent seed size unless wheat genes are bred in.[87] Gluten quality is also not as good as with wheat.[88]

Job's tears, adlai grass (*Coix lacryma-jobi*). Wild Job's tears is a perennial from South and Southeast Asia. The seeds of the wild forms (var. *stenocarpa* and var. *monilifer*) have thick, hard shells that are often used as beads. These forms are grown around the world as ornamentals and have naturalized widely. Although they have edible seeds, the shells make these forms impractical for use as human food (I imagine pigs would be quite happy crunching on them). Farmers in India domesticated an annual or mostly annual form with thin, soft shells (var. *ma-yuen*) between 3,000 and 4,000 years ago, which by 2,000 years ago was being grown in China. Before the arrival of maize, Job's tears was an important grain in subtropical Asian highlands. Annual Job's tears yields a respectable 2 to 3.5 tons per hectare and tolerates acid, poor, and waterlogged soils. Crossing annual varieties with perennial strains could result in a new perennial grain for the tropics and subtropics, including highland areas.[89]

Markouba grass, or afezu (*Panicum turgidum*). A wild perennial grass ranging from the heart of the Sahara desert, where it's an important staple, through Pakistan. It grows in areas with as little as 25mm (1 inch) of precipitation. Fallen stems root into the soil, spreading the plants vegetatively. Efforts at domesticating this species could serve the dual functions of feeding people in very arid tropical areas and reversing desertification. Breeders should attempt to cross afezu with the annual staple proso millet (*P. miliaceum*), a fellow member of the genus *Panicum*.

Pearl millet (*Pennisetum glaucum*). An important world grain, thriving in arid environments and poor soils. It has been successfully crossed with the vigorous perennial napier grass, *P. purpureum*, to create forage grasses, but not, to my knowledge, with the goal of developing a perennial grain.[90] Starting with these successful hybrids could give a good head start to perennial millet development. A perennial millet could be significant in African and Indian drylands, where millions of people rely on millet for survival.

Wangunu (*Eragrostis eriopoda*). An Australian wild edible that has been an important staple to indigenous people there for millennia. It is reported to have been the most important native grass seed, in part because of the ease of processing seeds, high yields, and holding on the plant for months.[91] This species is now cultivated on a small scale by "bush tucker" (wild edibles) enthusiasts in Australia.[92] This is one of several Australian perennial grains that were cultivated by Aboriginal people before colonization.[93]

Timothy (*Phleum pratense*). This widely grown pasture grass has recently been taken into cultivation as a commercial perennial grain crop. The seeds are very small, and yields are low at 0.4 to 0.5 t/ha. Protein is 17 percent. As it is being marketed as gluten-free flour, the processing and high price make up for these deficiencies.[94]

Fruits

Peach palm seems to be the only "fruit" in the balanced carbohydrate category, although botanically speaking the edible woody legume pods are also fruits. The fruit of baobab also almost makes the cut at 5 percent protein dry weight. Acaí is close at 4.3 percent dry weight.[95]

PEACH PALM (*BACTRIS GASIPAES*)

Peach palm, also known as pejibaye, pejivalle, or chontaduro, was once a major staple in the Americas and has tremendous worldwide potential. *The Encyclopedia of Fruits and Nuts* states, "Throughout western Amazonia and extending up to Costa Rica, the peach palm appears to have been as important as maize and cassava in much of this region. The date palm is the Old World's dry tropical domesticate with similar importance in subsistence."[96] Peach palm is also sometimes grown in large monocultures not for its fruit but for the hearts of palm, for which the trunks are cut and peeled. Fortunately peach palm is multistemmed and well adapted to this type of management.

Peach palm gives us some insight into why many cultures have abandoned tree crops for short-lived annuals: vulnerability to war. During the conquest of what is today Costa Rica by the Spanish, 30,000 to 50,000 peach palms were cut to subjugate the indigenous population of the Sixaola River valley.[97] Tree crops are vulnerable in war and, once cut, require years of labor before bearing again. It may be easier in such circumstances to switch to annual crops, which will provide food in a matter of months instead of years.

FIGURE 13.12. Peach palm fruiting in the overstory of a multistrata agroforestry system at Maya Mountain Research Farm in Belize. Photograph courtesy of Christopher Nesbit.

Uses. The cooked fruits are an important food, high in beta-carotene and starch, much like orange-fleshed sweet potatoes or winter squash. They are 3 to 15 percent protein and 5 to 9 percent oil. (Some varieties are as high as 61 percent oil and are grown exclusively for that purpose.)[98] Their primary drawback is that they are highly perishable with a short shelf life.

Yield. Yields are outstanding, up to 30 tons per hectare of fresh fruit, or 13.2 t/ha dry weight. This is higher than potatoes or rice. Breeders project that this could be increased to 50 t/ha fresh (22.0 dry).[99] Oil yields from elite clones are estimated at 2 to 3 t/ha, among the highest oil yields in the world.[100] Peach palm takes 3 to 5 years to bear and yields for 50 to 75 years.[101]

Harvest and processing. Harvesting is a challenge because the palms are tall and the trunks are extremely spiny. To preserve the harvest, the fruits are dried and ground into flour, fermented, and otherwise stored for later consumption.

Carbon farming applications. In its native range, peach palm is grown in diverse homegardens. In multistrata systems, it is grown with cacao and black pepper vines as shade and trellis respectively. The palms are also planted as windbreaks.[102]

Crop development. Peach palm should be much more widely grown in the humid tropics. It is only grown experimentally outside of the Americas, although it could be significant in Africa, Asia, and the Pacific. This beta-carotene-rich staple crop could make a big impact in areas with vitamin A deficiency. Breeding work has resulted in some dwarf, spineless forms with superior ease of harvest. There are many other *Bactris* species grown for fruit and oil.

Edible Woody Pods

Some legume trees produce edible woody pods. In some species, yields are remarkably high. These pods evolved to attract large mammals to act as dispersal agents; they are particularly common in savanna ecosystems. In general, woody legume pods consist of sugar rather than starch, along with a decent amount of protein. The beans are often much higher in protein than the rest of the pod but are too hard for humans to chew. The way around this problem is to grind the whole pods, including the seeds, to make flour or meal.

There are also legume trees with inedible pods and edible beans inside. These are profiled in other chapters. Some produce pods with a sweet pulp inside (like the famous ice cream beans, *Inga* spp.). I consider these to be fruits, so I've chosen not to include them in this book, although I would certainly plant them.

MESQUITE (*PROSOPIS* SPP.)

Mesquites are nitrogen-fixing trees that grow in deserts around the world. Their high-yielding and nutritious pods have been a staple crop for indigenous peoples for millennia and have outstanding potential as a future food for humanity and its livestock. The better-quality mesquites are very aggressive and are reviled by many where they have become invasive and even where they are native.

A 2000 study on the influence of mesquite on soil nitrogen and carbon found that wild mesquites sequestered an additional 5 to 73 t/ha of lifetime soil carbon at 30cm (1-foot) depth compared with adjacent soils—a fairly impressive amount for desert conditions.[103]

Some mesquites are truly tropical, including the elite edibles *P. pallida* and *P. juliflora*. Many more are subtropical, and some can handle warm temperate climates.[104] In the United States the honey mesquite (*P. glandulosa*) grows in areas with temperatures as low as −20°C (−4°F).[105] Some mesquites grow at very high elevations—*P. chilensis* at 2,900 meters (9,514 feet) and *P. ferox* at a remarkable 3,700 meters (12,139 feet), both in the Andes. These are trees of arid environments. Many fruit reliably with less than 250 millimeters (9.8 inches) of annual rainfall. In the Chilean Atacama desert, *P. tamarugo* grows with a remarkable 2 millimeters (0.08 inch) of rain a year, although it catches coastal fog and some subsurface runoff from the Andes. Terrible soils are not a problem for mesquite; some species can handle alkaline pH as high as 11 and very salty soils. In fact,

TABLE 13.6. Woody Pods with Balanced Carbohydrates

Latin Name	Common Name	Origin	Climate	Cultivation
Albizia saman	Monkeypod	Tropical Americas	Tropical lowlands, semi-arid to humid	Yes for fodder
Bauhinia thonningii	Camel's foot	Africa	Tropical lowlands, semi-arid	As agroforestry tree, pods as snack or survival food
Ceratonia siliqua	Carob	North Africa, Mediterranean	Mediterranean, semi-arid subtropics	Yes for pods
Faidherbia albida	Apple-ring acacia	Africa, Middle East	Tropical arid to humid lowlands, highlands	Common as agroforestry, pods minor food
Gleditsia triacanthos	Honey locust	Eastern North America	Boreal to subtropics, Mediterranean, tropical highlands, semi-arid to humid	Widespread ornamental, experimental for pods
Prosopis alba	Algarrobo blanco	Argentina	Tropical lowlands, arid to semi-arid	Yes for pods
Prosopis chilensis	Algarrobo de Chile	South America	Warm temperate to tropical, highlands, arid to semi-arid, halophyte	Yes for pods
Prosopis glandulosa	Honeypod mesquite	North America, Mesoamerica	Cold temperate (−20°C/−4°F) to tropics, including highlands, arid to semi-arid, halophyte	Yes for pods
Prosopis juliflora	Mesquite	Central and South America	Tropical lowlands, arid to semi-arid, halophyte	Yes for pods
Prosopis pallida	Mesquite	Central and South America	Tropical lowlands, arid to semi-arid, halophyte	Yes for pods

some mesquites can be irrigated with water as salty as seawater and still thrive.[106]

Mesquite occupied 30 million hectares (74 million acres) of the southwestern United States in 1990.[107] There are approximately 44 species in the genus *Prosopis*. The best edible mesquites are from North and South America. Though there are native species in India, the Middle East, and Africa, unfortunately their pods are not of good quality. The best mesquites are in the subgroup Algarobia. This group includes the South American *P. pallida*, considered the very best all-around mesquite, as well as *P. juliflora* and the temperate-climate *P. glandulosa*.[108]

Uses. The fresh pods can be eaten raw and are quite good eating, but humans do not digest the small, protein-rich beans inside very well. Ground mesquite pod

FIGURE 13.13. Mesquite pods, shown here in Arizona, are one of the most productive perennial carbohydrate crops. Photograph courtesy of Brad Lancaster.

flour is more digestible and ranges from 7 to 17 percent protein, 1 to 6 percent fat, with lots of sugar and fiber.[109] I enjoy eating the pods raw, and I use mesquite flour in stews and baked goods, where I find it sweet and nutty. A sugary syrup called algarrobina, which is similar in taste and texture to dark molasses, is made from mesquite pods in South America.[110] Some mesquites have blue or purple pods that are high in anthocyanins, like blueberries.[111]

Mesquites are an extremely important source of firewood, timber, and charcoal. Sometimes they are the only tree that will grow in arid regions with poor soils.[112] Food and fuel are not the only important products. The trees exude a gum of almost as high a quality for food processing as gum arabic, and the bark is used for tannins.[113] The flowers are an important food for bees and are essential to the production of desert honey in many regions.[114]

Yield. Mesquites have been called an "unfailing crop" because they produce well every year, although occasionally they yield less well during wet years.[115] In some zones mesquites produce almost year-round, while in others there are two or three harvests.

Reported pod yields vary. Mesquite experts Richard Felger and Neil Logan write that it is reasonable to expect yields of 9 t/ha for tropical species and 1.8 to 3.6 t/ha for temperate species, dry weight.[116] One study of 40-year-old trees measured a pod yield of 14.8 tons per hectare (dry weight). Other studies of older trees have shown yields as high as a staggering 49.5 tons per hectare dry weight.[117] If these remarkable higher figures could be consistently achieved, it would make mesquite one of the top-yielding staple crops on the planet, although experts such as Felger don't think this is likely.

Harvest and processing. Mechanical harvest is probably possible by adapting nut-harvesting machinery. Although it is enjoyable to eat mesquite pods whole, they are typically dried and ground to make flour. Techniques for doing this by hand have been around for millennia. In recent years large facilities have been established, particularly in South America. There remains a need for intermediate-scale processing technology.[118]

Carbon farming applications. Thousands of hectares of mesquite have been planted in Chile for silvopasture. In parts of India and Africa where the South American species have naturalized aggressively, farmers and land managers are thinning them to a silvopasture spacing for livestock fodder production. They are intercropped with other forage species like saltbush (*Atriplex* spp.) and prickly pear (*Opuntia* spp.) in various regions.[119] The use of the pods as fodder is traditional and widespread.[120] However, livestock that are fed only mesquite for extended periods can become ill or even occasionally die.[121]

Today mesquites are grown in short-rotation coppice systems for firewood in India and elsewhere.[122] They are grown for shade, windbreak, and erosion control and as a spiny living fence.[123] Alley cropping with mesquite in dry regions is probably not advisable due to competition with crops for water. In India, however, the native *P. cineraria* is the most common species intercropped with annual crops in traditional parkland agroforestry.[124]

Other issues. Because they grow in terrible soils, thrive in arid conditions, fix nitrogen, and bear seeds that are dispersed by livestock who enjoy their pods, mesquites can spread rapidly in both new and familiar environments. They are considered a major weed of pastures and are classed as noxious or invasive species in a number of regions. Ironically, it is the mesquite species that have the most potential as human food that are also the most problematic in this regard (they are also the thorniest mesquites, including *P. juliflora* and *P. pallida*).

Although it wasn't the best selections of South American mesquites that were brought to Africa and India and naturalized, they are beloved by poor farmers there who rely on them for their daily fuel. They tend to be disliked by wealthier people, who often cook with gas and may not appreciate the importance of firewood and charcoal.[125] Ranchers in the United States are learning that planned rotational grazing keeps new mesquite seedlings from establishing, even if there are lots of seeds in the manure because cattle are eating the pods: With proper management the cattle trample and consume the small seedlings.[126]

Crop development. There is a lot of variation between and within mesquite species. Many species also hybridize freely, creating new possibilities. Superior varieties can be propagated with cuttings and grafting.[127] There are plantations of superior strains in Peru for seed production.[128] Mesquite seems like an ideal candidate for Badgersett-style neohybrid breeding. Such an ideal candidate would grow and bear quickly, with high yields of sweet pods. It would coppice readily, be spineless, and be minimally invasive, perhaps having a tendency toward sterile seeds. We need such varieties for each of the climates to which mesquite is adapted.

Mesquite should be promoted strongly where it is native and managed and utilized where it has naturalized. Techniques have been worked out to convert overgrown, invasive mesquite patches into managed agroforestry systems.[129] Dr. Richard Felger, a champion of nipa salt grass domestication, has also spent decades investigating the potential of this important and high-yielding desert food crop. He feels that within the near future it will become a major world crop alongside such staples as rice and wheat, especially given its tolerance of salt as well as drought.[130]

A 2001 monograph on mesquite that was funded by the UK Department of International Development states, "*Prosopis* has been the species of the politically disenfranchised, who do not have the influence to request research and development funding for genetic improvement, processing technology and marketing. *Prosopis* has the . . . qualities and environmental attributes to be a species of worldwide commercial importance . . . [We must] ensure that such benefits reach the poor people of arid lands who truly deserve them."[131]

CAROB (*CERATONIA SILIQUA*)

Carob is a traditional tree crop from the Mediterranean, North Africa, and the Middle East. The pods are primarily fed to livestock but are also consumed by humans. Although it grows in arid regions, 350mm (14 inches) per year of rainfall is the minimum for fruit to set.[132]

Figure 13.14. Carob has been grown as food for people and feed for livestock for millennia. Photograph courtesy of Riu Chixoy/ Creative Commons 3.0.

Carob is a minor global crop, with 162,000 tons worth $60 million US grown on 81,000 hectares (200,000 acres) in 2012.[133]

Uses. Carob has been grown for centuries for its edible pods. These are used as livestock fodder, the source of a sweet syrup, and whole or ground as human food. When I was a child, carob bars were promoted as a substitute for chocolate; I was not convinced. However, as an adult I have come to appreciate both the raw, whole pods and the flour as fine foods on their own merits.

The whole ground pods are 5 percent protein and about 1.5 percent fat. As with honey locust and mesquite, the seeds are higher in protein than the rest of the pod. The seeds are uniform in size and are said to have been the original sample weight for 1 carat.[134] One and a half kilos (3.3 pounds) of pods yields 1 kilo (2.2 pounds) of sweet syrup.[135] Locust bean gum, a minor ingredient in many processed foods and industrial products, is made from carob seeds.[136]

Yield. Carob prefers semi-arid subtropics and Mediterranean climates. It can be quite productive, from 1.7 to 8.3 tons of pods per hectare dry weight.[137] Usually yields are much lower because carob is rarely a priority crop; the global average dry weight yield is 1.7

t/ha.[138] One reason for lower yields is that carob is often grown on the very worst, driest, stoniest slopes.

Carbon farming applications. Carob is used as a windbreak. Unfortunately it is not nodulated and presumably does not fix much if any nitrogen. Carobs are traditionally interplanted with olives, grapes, almonds, and annual crops in the Mediterranean.[139]

HONEY LOCUST (*GLEDITSIA TRIACANTHOS*)

Honey locust is a tree native to North America that has become a widespread ornamental and invasive species outside of its original habitat. It is adaptable, ranging from the warmer end of boreal climates (−40°C/−40°F), through cold and warm temperate climates, and also succeeding in Mediterranean regions and tropical highlands.[140] Honey locust grows in semi-arid and humid climates. It is a non-nodulating legume and as such probably does not fix much, or any, nitrogen. Honey locust pods have been viewed with great promise as a perennial staple crop for humans and livestock for a century, including by J. Russell Smith, but have been sorely neglected.[141] Archaeologists believe that honey locust may have been an important staple from 8,000 to 1,000 BC in what is today the eastern United States.

Uses. Honey locust produces edible woody pods, similar but inferior to mesquite and carob. These are a traditional food and sweetener in the tree's native range. The pods contain up to 15.4 percent protein and 42.3 percent sugar.[142] The pods have been used to produce sugar.[143] Some pods are thick and sugary like carob, while others are thin and papery. Where I live in the northeastern United States, I rarely find a honey locust with sugary pods (goats ate the grafted portion of the improved varieties I purchased). This isn't because sugary varieties cannot tolerate cold; several sweet varieties have been shown to tolerate at least −29°C (−20°F).[144] It is simply because I live outside the honey locust's native region and only ornamental, spineless, paper-podded types have been brought here.

Honey locust seeds have been the subject of controversy. Early researchers reported that ruminants could not digest the seeds unless the pods were ground first, reducing their feed value considerably. Harvesting, grinding, then feeding the pods to animals proved uneconomical. It was later determined that ruminants can digest almost all of the seeds within a given pod, however, so allowing the pods to fall directly into the pasture is a viable strategy. Seed coat hardness also varies, with some forms too hard for good digestion by sheep.[145] *Gleditsia triacanthos* coppices well and is a high-quality resprouting firewood.[146]

Yields. Pod yields are variable, and honey locust trees often bear them in alternate years, a dilemma that can be overcome by planting different complementary varieties, so that half of the trees yield pods in a given year.[147] Grafted trees can begin bearing in as little as three years, although only about 30 percent of grafting attempts succeed. How much could an orchard or silvopasture planting of soft-seeded, sweet-podded, high-yielding honey locust trees produce? One elite 25-year-old tree was recorded as bearing 113kg (250 pounds) of dry matter in a season.[148]

About 96 trees can be planted per hectare. It is easy to jump from this to assume that a hectare of grafted trees would yield 17.25 t/yr, but no one has published results of such an experiment. In one experiment, a fodder orchard of nine-year-old trees yielded 4.6 tons of pods dry weight annually.[149] In another, nine-year-old grafted trees bore 6 t/ha/yr dry weight, although they bore in alternate years.[150] Assuming not all trees are as productive, researchers speculate a yield of 1.5 tons per hectare (dry) from improved selections. Further breeding could presumably increase this. Were this to happen, honey locust could be the

FIGURE 13.15. Honey locust tree producing well in dry, cold Colorado.

first tree domesticated for livestock fodder pods. The existence of a number of other *Gleditsia* species around the world holds great promise as well, although not all have woody pods.

Carbon farming applications. Some wild forms are intensely spiny and were used as living fences in the United States before the introduction of barbed wire. Honey locust is used in silvopasture as well as in fodder banks. Though not nodulated, it is believed to fix some nitrogen.[151]

CHAPTER FOURTEEN

Protein Crops

Protein crops consist of 16 percent or more protein, but 15 percent or less fat. Global annual protein crops include beans, chickpeas, cowpeas, peas, pigeon peas, lentils, and favas. There are currently no global perennial protein crops. Together these crops are grown on more than 61 million hectares (15.1 million acres), or about 4 percent of world cropland.[1] Perennial candidates include beans, nuts, tree milks, and leaf protein concentrate. These perennials compete well, with one undomesticated species, the

TABLE 14.1. Protein Crop Yields Compared

Latin Name	Common Name	Climate	Avg. Global Yield[a]	Yield t/ha	Product
Treculia africana	African breadnut	Humid tropical lowlands		19.8–47.4[b]	Nuts, extrapolated
*Cajanus cajan	Pigeon pea (annual types)	Tropics	0.7	0.9–6.8[c]	Beans
*Vigna unguiculata	Cowpea	Global annual	0.5	0.9–6.2[d]	Beans
Erythrina edulis	Chachafruto	Highland tropics		1.9–5.1[e]	Beans
*Phaseolus vulgaris	Common bean (pole)	Global annual		0.9–4.8[d]	Beans
Phaseolus lunatus	Lima bean (vine types)	Tropics		2.7–4.5[d]	Beans
Canavalia gladiata	Sword bean[f]	Semi-arid to humid tropics		0.9–4.5[d]	Beans
Phaseolus coccineus	Runner bean (vine types)	Tropical highlands		2.6–4.4[d]	Beans
Lablab purpureus	Lablab bean	Tropical, semi-arid to humid		1.4–1.9[d]	Beans
Cajanus cajan	Pigeon pea (perennial types)	Semi-arid to humid tropics and subtropics		1.8[g]	Beans
Sphenostylis stenocarpa	African yambean	Humid tropical lowlands		1.8[d]	Beans
*Phaseolus vulgaris	Common bean (bush type)	Cold temperate, tropical	0.7	0.9–1.8[d]	Beans
Medicago sativa	Alfalfa	Temperate to Mediterranean, semi-arid to humid		1.2[h]	Leaf protein concentrate
Acacia spp.	Edible acacias	Arid tropics		1.0[i]	Beans
Urtica dioica	Stinging nettle	Humid, boreal to warm temperate		0.6[h]	Leaf protein concentrate

Note: Here we compare perennial versus annual protein sources. Annual comparison crops are marked with an asterisk (*). All yields calculated based on dry weight of the edible portion. Crops are arranged in descending order of yield.

[a] For 2012. FAOStat, accessed April 8, 2014.
[b] Janick and Paull, *The Encyclopedia of Fruits and Nuts*, 509.
[c] PROTA4u.org, accessed November 3, 2014; El Bassam, *Handbook of Bioenergy Crops*, 276.
[d] Plant Resources of Tropical Africa online.
[e] Duarte, *Guia para el cultivo y aprovechamiento del Chachafruto o Balú: Erythrina edulis*, 14; Florez, "*Erythrina edulis*, an Andean giant bean for human consumption," 26–27.

[f] *Canavalia* beans are moderately toxic and must be processed properly to be consumed safely. See species profile for details.
[g] Daniel and Ong, "Perennial pigeonpea."
[h] Pirie, *Leaf Protein and Its By-Products in Human and Animal Nutrition*, 47.
[i] Extrapolated; see profile for details.

TABLE 14.2. High-Protein Perennial Beans

Latin Name	Common Name	Origin	Climate	Cultivation
Acacia colei	Cole's wattle	Australia, Asia	Tropical lowlands, semi-arid	New crop, historic wild staple
Acacia cowleana		Australia	Tropics, arid to semi-arid	New crop, historic wild staple
Acacia holosericea		Australia	Semi-arid to humid tropics	New crop, historic wild staple
Acacia murrayana		Australia	Subtropics to warm temperate, arid to semi-arid	New crop, historic wild staple
Acacia victoriae	Elegant wattle	Australia	Warm temperate to tropical, arid to semi-arid	New crop, historic wild staple
Albizia lucidior	Tapria siris	South and Southeast Asia	Humid tropical lowlands	Regional
Amphicarpaea bracteata	Ground bean	Eastern North America	Cold to warm temperate, humid to semi-arid	Historic crop
Amphicarpaea bracteata ssp. edgeworthii	Saekong	East Asia	Cold to warm temperate, humid	Regional
Apios americana	Groundnut	North America	Cold temperate, warm temperate, subtropical	Historic (beans) minor global (tubers)
Bauhinia petersiana	Kalahari white bauhinia	Africa	Tropics and subtropics, semi-arid	Wild staple
Cajanus cajan	Perennial pigeon pea	India	Subtropics to tropics, lowlands to highlands, semi-arid to humid	Global (annual), regional (perennial)
Canavalia gladiata	Sword bean	Africa, Asia	Semi-arid to humid tropics, lowlands and highlands	Minor global
Caragana arborescens	Siberian pea shrub	Asia	Boreal to cold temperate, semi-arid to humid	Undomesticated
Cercidium microphyllum	Paloverde	Southwest North America, Mesoamerica	Subtropical, arid	Wild staple
Desmanthus illinoensis	Illinois bundleflower	North America	Cold to warm temperate, semi-arid to humid	Under development
Erythrina edulis	Chachafruto	Andes	Tropical highlands, semi-arid to humid	Regional
Guibourtia coleosperma	African rosewood	Africa, South Asia	Tropical and subtropical lowlands, semi-arid	Wild staple
Lablab purpureus	Lablab bean	Africa	Semi-arid to humid tropics, lowlands and highlands	Minor global
Lupinus polyphyllus	Washington lupine	Western North America	Boreal to warm temperate, semi-arid to humid	New crop
Parkinsonia aculeata	Jerusalem thorn	Tropical Americas	Tropics to subtropics, semi-arid	Wild staple
Phaseolus coccineus	Runner bean	Mesoamerica	Tropical highlands, warm temperate, humid	Minor global
Phaseolus coccineus ssp. polyanthus	Cache bean	Mesoamerica	Tropical highlands, semi-arid	Minor global
Phaseolus polystachios	Thicket bean	North America, Mesoamerica	Cold temperate to tropical, humid	Historic crop
Phaseolus lunatus	Lima bean	Tropical Americas	Tropical lowlands to highlands, semi-arid to humid	Minor global
Phaseolus ritensis	Perennial tepary bean	Mesoamerica	Arid to semi-arid subtropics and tropics	Regional
Sphenostylis stenocarpa	African yambean	Africa	Tropical lowlands and highlands, humid	Regional

Sources: Bryan, *Leguminous Trees with Edible Beans;* Plant Resources of Tropical Africa online; Hanelt, *Mansfeld's Encyclopedia of Agricultural and Horticultural Crops;* House and Harwood, *Australian Dry-Zone Acacias for Human Food.*

African breadnut, offering spectacular yields if other challenges to growing it can be overcome.

Beans

Legumes are virtually the only plants that produce a dry, storable staple that is high in protein but low in oil. (One exception is African breadnut, profiled on page 179.) A fair number of legume trees, vines, and perennial herbs produce edible beans. A few are important crops at least regionally, but this is not the strongest category of perennial staple crops for most climates. Some are toxic to humans and animals, while many others are simply not yet domesticated. Table 14.2 looks at some of the best perennial protein staple crops. Note that some beans with other nutrient profiles are covered in other chapters, such as Tahitian chestnut profiled in chapter 13 and winged beans and others profiled in chapter 15.

CHACHAFRUTO (*ERYTHRINA EDULIS*)

Chachafruto, the "tree bean of the Andes," is a unique and productive perennial protein source. While most other species in the genus *Erythrina* have small, poisonous beans, chachafruto produces enormous pods full of large, soft edible beans. This is a regional crop for humid tropical highlands.

Uses. The edible beans are 18 to 23 percent protein.[2] They are cooked and used like a larger version of common bean. They are 84 percent water.[3]

Yield. Chachafruto trees begin to bear at the end of the second year after planting. By the 10th year, chachafruto yields 170 kilograms (375 pounds) of pods per tree, half of which is edible beans. At 8m by 8m (26 by 26 feet) spacing, this gives a yield of 13 tons of cleaned bean seed per hectare, a dry weight of 27 kilos (60 pounds).[4] By the 20th year, the trees can yield 240kg to 420kg (529 to 926 pounds) of pods per tree;[5] by extrapolation, at the same spacing that would yield 18 to 32 tons per hectare of cleaned bean seed, a dry weight of 2.8 to 5.1 tons per hectare. These yields are in line with top-producing annual beans.

Harvest and processing. There are many methods for preparing and preserving chachafruto beans. The length of season is remarkably long, with a harvestable crop for most of the year. The authors of a 1998 paper on chachafruto noted, "This guarantees a constant harvest for anyone who owns more than

TABLE 14.3. Prospects for Perennializing Annual Bean and Pulse Crops

Annual Crop (Latin Name)	Annual Crop (Common Name)	Perennial Relatives	Perennialization Status
Cajanus cajan	Pigeon pea		Perennial varieties exist; see Perennial Pigeon Pea entry
Cicer arietinum	Chickpea	*C. microphyllum* is perennial, edible, and tolerant of cold and arid climates	Not actively under development
Glycine max	Soybean	*G. tomentella*	Crosses have been made, but not for perennialization
Lens culinaris	Lentil		Not actively under development
Lupinus albus; L. luteus; L. mutabilis	White lupine, yellow lupine, tarwi	Many perennials such as *L. perennis* and "tree" types, toxic seed caution	Perennial *L. polyphyllus* is under domestication
Phaseolus vulgaris	Common bean, kidney bean	*P. coccineus; P. lunatus; P. polyanthus; P. polystachios*; all are edible and crossable	Not actively under development
Pisum sativum	Pea		Not actively under development
Vicia faba	Fava bean	There are many perennial *Vicia* species such as the toxic-seeded hairy vetch (*V. villosa*)	Not actively under development
Vigna angularis; V. mungo; V. radiata; V. umbellata; V. unguiculata	Adzuki bean, black gram, mung bean, rice bean, cowpea	*V. luteola* *V. vexillata*	Not actively under development

Note: As with cereal grains, there are opportunities to develop perennial analogs of annual beans, although the research is far more advanced with cereals. Still, there is a lot of opportunity. Here are some current and potential lines of work.

five trees."[6] Although this is undesirable for large-scale industrial production, it is ideal for small-scale intensive growers and subsistence farmers.

Carbon farming applications. Chachafruto is a nitrogen-fixing legume and is easily grown from seed and live stakes (firewood-diameter cuttings). It is used as a living fence, fodder bank, fodder pod, and crop shade tree. It is intercropped with understory species such as bananas, coffee, aroid tuber crops, tree tomatoes, and blackberries. Perennial peanut (*Arachis pintoi*) is a popular ground cover in these systems.[7] Because chachafruto resprouts readily, it could be suited to woody agriculture.

Crop development. Sadly for most of us, chachafruto production is confined to the humid highland tropics, from 1,000 to 3,000 meters (3,281 to 9,843 feet). Multiple efforts to cultivate it at Las Cañadas at 1,500 meters (4,900 feet) have thus far met with failure. Chachafruto has received little breeding attention

and is difficult to even acquire for testing outside of its native Andes. This species should be in every backyard in the highland tropics. Widespread testing, breeding, and distribution of the Andean tree bean should be a top priority for carbon farming.

PERENNIAL PIGEON PEA (*CAJANUS CAJAN*)

Annual varieties of pigeon pea are a global crop grown on 5.3 million hectares (13 million acres) globally and yielding 4.2 million tons valued at $1 billion US.[8] This crop originated in India and spread to Africa 4,000 years ago or more. Pigeon pea is especially adapted to the semi-arid tropics and subtropics but also produces well in humid regions.

Most breeding work has focused on the annual varieties, which yield well the first year, then persist for a few years without much yield if not plowed under. In the 1980s the International Crops Research Institute

FIGURE 14.1. Chachafruto is a remarkable tree bean. Photograph courtesy of Calliope Jen/Creative Commons 3.0.

FIGURE 14.2. Pigeon pea is a global staple with perennial varieties. Photograph courtesy of Forest and Kim Starr/Creative Commons 3.0.

for the Semi-Arid Tropics (ICRISAT) began work on identifying perennial varieties that would yield well for multiple years. These perennial pigeon peas have been grown in homegardens for millennia—and remain common in India and elsewhere—but are largely ignored by the research community.[9] Perennial pigeon pea seed is difficult to obtain commercially outside of India.

Uses. The dry seeds are 21 percent protein and 1.5 percent fat.[10] The fresh green seeds are also eaten as a vegetable.

Yield. Yields of annual pigeon pea range from 0.9 to 6.8 t/ha dry weight, though the average global yield is a far more modest 0.7 t/ha/yr.[11] Perennial pigeon pea yields 1.8 t/ha dry weight for three to four years; the plants are often coppiced annually to rejuvenate them. Early-flowering varieties have the best success as coppiced crops.[12]

Carbon farming applications. Pigeon peas are among the most widely used agroforestry species. They serve as nitrogen fixers, crop shade, windbreaks, alley crop species, living trellises, fodder bank, and hosts for lac insects (the source of natural shellac).[13] Perennial types would be great candidates for woody agriculture with machine harvest, perhaps on an annual coppicing cycle. Pigeon peas are ubiquitous in tropical homegardens around the world.[14]

Crop development. Perennial pigeon pea is one of the few perennial protein crops grown on a significant global scale. Although the plants are not as long-lived as we might like, they have an important role to play. Wider distribution and promotion of the perennial varieties in regions where pigeon pea already grows represent two of the "low-hanging fruits" of carbon farming implementation. Genetically modified pigeon peas have been developed for disease resistance.

PERENNIAL VINING BEANS (*PHASEOLUS*, *LALAB PURPUREUS*, *SPHENOSTYLIS*, *CANAVALIA*)

There are several genera of cultivated beans with perennial forms. Vining, climbing, or pole, beans are poorly suited to mechanical harvest, which may partially explain the underutilization of these crops. However, these perennial beans offer the advantage of familiar cultivation, harvest, processing, and cooking. Annual *Phaseolus* and lablab species are bush beans while the vining varieties are perennial (in the right climate). It is interesting that even in the case of the annual common bean, climbing varieties yield as much as 3 tons more per hectare than bush types.[15] Common bean (*P. vulgaris*) is a global crop, grown on 29.2 million (72.2 million acres) hectares. This cropland produces 23.5 million tons worth $15.5 billion US.[16]

Perennial "true" beans (*Phaseolus coccineus, P. lunatus*, and others). Runner beans (*P. coccineus*) and lima beans (*P. lunatus*) are minor global crops, although secondary in importance to the annual common bean. The vining varieties can live for many years, typically dying back to the ground and resprouting in the cold or dry season. Runner beans are suited to tropical highlands and cool temperate climates. I have seen plants 35 years old that are still bearing well at the Occidental Arts and Ecology Center in California. Lima beans are more drought-tolerant than runner beans with somewhat shorter life spans. A long-lived African perennial lima variety called '7 Year' is distributed by the Educational Concerns for Hunger

FIGURE 14.3. A 35-year-old runner bean planting at Occidental Arts and Ecology Center in California.

Organization (ECHO) in Florida. Yields of perennial lima beans and runner beans range from 2.4 to 4.5 dry weight tons per hectare, although they may decline after the first year.[17] The wild *P. ritensis*, a species for arid and semi-arid tropical highlands, is also very promising.

Thicket beans (*P. polystachios*). Historically cultivated by Native Americans, thicket beans are reliably perennial in my Massachusetts garden (−23°C/−9.4°F). That said, thicket beans need more work before they will be ready for prime time.

Lablab beans (*Lablab purpureus*). Lablab beans are native to Africa but became important long ago as a vegetable and dry bean in India and Southeast Asia. The green pods are a popular vegetable in much of the tropics. The dry beans are 20 to 28 percent protein and 1 to 2 percent fat.[18] Lablabs grow in tropical lowlands and highlands, in semi-arid and humid regions, and produce 1.5 to 2t/ha.[19] Although they are often grown as annuals, many cultivated forms are short-lived perennials. The ancestral wild forms are true perennials; some backcrossing may be necessary to develop lablabs as a long-lived perennial bean crop.[20] Some of the fodder and green manure forms are toxic to humans.

African yambean (*Sphenostylis stenocarpa*). African yambean is primarily grown as a tuber crop, but if the tubers are left unharvested, it can be perennial. The dry beans are considered very tasty and are 20 to 25 percent protein and 1 percent fat. Yields are 2 t/ha.[21] In some regions, such as Nigeria, the beans are the primary crop. Nigerian varieties would be a good place to begin efforts to breed even more productive perennial bean strains. Like winged bean, African yambean is an extremely efficient nitrogen fixer, fixing from 120kg to 150kg (265 to 331 pounds) per hectare each year. It grows from sea level to 2,300 meters (7,546 feet) or higher in humid tropics.[22]

Sword bean (*Canavalia gladiata*). Sword bean is a minor crop from the Old World tropics. It is strongly perennial, growing up to 10 meters (32 feet). Dry-weight yields are fantastic at 0.9 to 4.5 tons per hectare, making it one of the highest-yielding of all perennial beans.[23] Sword bean is also tolerant of some of the most nutrient-depleted, highly leached tropical soils. The dry beans are 22 to 29 percent protein and 2 percent fat.[24] Why isn't everyone growing sword bean? The seeds contain a lectin called concanavalin A that is a moderately hazardous cell toxin.[25] Traditionally, the seeds are boiled for three hours in several changes of water, and the seed coats are removed after boiling. Researchers have determined that concanavalin A is completely destroyed if the beans are broken in pieces and cooked for one hour, or for 15 minutes in a pressure cooker.[26] Nonetheless, breeding non-toxic sword beans should be a priority. The young pods and green seeds are also eaten as vegetables. Sword bean grows with as little as 700 millimeters (28 inches) of rainfall in tropical lowlands and highlands.

Harvest and processing. For all of these beans, annual bush varieties yield less but are easier to harvest mechanically. The mechanical harvest issue may be the primary reason there has been little effort to domesticate perennial beans; trellising and handpicking beans on a large scale is tedious, labor-intensive, and therefore expensive.

Carbon farming applications. All of the vining bean species profiled here fix nitrogen. Lablabs are important as fodder and green manure plants, often grown in the understories of coconut, oil palm, and rubber plantations.

As complete speculation, I wonder if these perennial vining beans could be grown on woody plants that vigorously resprout after coppicing. If woody trellis plants could be found that grow early and fast enough to keep ahead of the vine growth, perhaps the whole field could be machine-harvested with a heavy combine or biomass harvester and the dry beans separated out from a biomass harvest for mulch, fuel, or the many other products made from biomass. Perhaps a high-protein woody leaf crop might serve as trellises and the leaves could be harvested and separated by a modified combine.

Wattleseed Acacias (*Acacia* spp.)

For the world's driest climates and poorest soils, wattleseed acacia trees and shrubs offer a high-protein seed crop. Wattleseed acacias are native to Australian drylands, known for terrible soils and unpredictable, limited rainfall. Some will even bear with less than 250 millimeters (9.8 inches) of rainfall, the point at which semi-arid becomes arid. Some are fully tropical, while others tolerate warm temperate winters.[27]

Uses. Australian Aboriginal people have relied on wattleseed acacias as a wild-managed staple for thousands of years. The beans store well. The dozens of edible species range from 16 to 25 percent protein and 4 to 16 percent fat.[28] In the 1980s researchers became interested in trialing wattleseed acacias in dry regions of Africa as human food. Aboriginal experts collected seed and shared information with researchers on utilization of wattle seeds. Australia's Commonwealth Scientific and Industrial Research Organization (CSIRO) sent many species and accessions to dry countries such as Niger. (Trials have also been carried out in Karachi, Pakistan.) Many perished, but some species produced well, notably *A. colei*. A key lesson was that although yields might begin to slow by the third year, a heavy pruning or coppicing rejuvenated plants and restarted production—while providing important firewood for cooking.[29]

Yield. Wattleseed acacias often bear the year they are planted. They bear fairly well for desert crops in poor soils, up to 1.0 t/ha (at 2 kg/tree, two harvests/year, and 6m/20-foot spacing).[30] Other reports indicate a substantially lower yield of 50 to 100kg/ha.[31] Few other crops can do much better under the same arid conditions.

Figure 14.4. Wattleseed acacias intercropped with sorghum in Niger. Photograph courtesy of Peter Cunningham.

Harvest and processing. The dry wattleseed beans are roasted and ground. In some species, the young green seeds are eaten.

Carbon farming applications. Wattleseeds are excellent windbreaks. They would be worth investigating as woody agriculture species for machine harvest. Wattleseed acacias are an integral component of the Farmer-Managed Agroforestry Farming System.

Crop development. Wattleseeds have received little breeding attention beyond trialing different wild species and accessions from different regions of Australia. Tropical species include *A. colei* (widely considered the best species in arid Africa), *A. cowleana*, and *A. tumida*. Warm temperate and subtropical species include *A. murrayana* and *A. victoriae*.

SIBERIAN PEA SHRUB (*CARAGANA ARBORESCENS*)

This shrub from Siberia and other semiarid parts of northeastern Asia is remarkably cold-hardy, tolerating temperatures below −40°C (−40°F), the brutally cold temperature where Celsius and Fahrenheit overlap. It is particularly common in the Canadian prairies, where hundreds of kilometers of pea shrub windbreaks have been planted.

Uses. Canadian farmers use the beans as survival food, boiling them in several changes of water in lean years to remove the bitterness. That doesn't quite meet my definition of edible. However, Facciola's *Cornucopia II: A Source Book of Edible Plants* reports the dry beans contain up to 36 percent protein, very similar to soybeans.[32] Somewhere in the wilds of Siberia or in a lonely windbreak in Saskatchewan, the perfect edible pea shrub may already exist—a perennial pigeon pea for cold and dry regions.

Yield. Siberian pea shrub produces fairly high yields of small beans.

Harvest and processing. All the Siberian pea shrubs I've seen "shatter" when ripe, meaning the pods burst open and expel the seeds. This makes harvest a challenge and needs to be bred out.

Carbon farming applications. *C. arborescens* is used as a windbreak, nitrogen fixer, livestock fodder,

FIGURE 14.5. A Siberian pea shrub shows promise as a potential new perennial bean crop for cold climates.

and erosion control. The shrubs might serve well in woody agriculture.

Crop development. This crop offers the potential for a long-lived woody protein crop for the world's boreal forests and grasslands, semi-arid mountains, and other inhospitable climates. Screening existing populations for edibility and breeding lines of truly edible pea shrubs should be a high priority for researchers in cold and arid climates, including backyard and amateur breeders. Breeding techniques were worked out in the 1960s in Canada for shelterbelt purposes, which gives us a head start.[33]

Nuts

I am only aware of one nut species that fits in the protein category; most are oil, protein-oil, or balanced carbohydrates. That nut is the remarkable and promising African breadnut.

AFRICAN BREADNUT (*TRECULIA AFRICANA*)

African breadnut has been largely neglected by researchers and is scarcely grown outside of tropical Africa. Nonetheless this relative of ramon and Asian breadnut may well be the world's most productive source of edible protein. This is a regional crop, growing in sun and shade in the humid tropics.[34] Fast-tracking of research and development of this crop, which may be able to produce roughly three to seven times more protein per hectare than the best-yielding annual bean crops, should be a high priority.

Uses. The fruit flesh is inedible (although it can be fed to livestock), but each fruit is filled with up to 10kg (22 pounds) of peanut-sized nuts. These tasty nuts are 12 to 23 percent protein and 6 to 15 percent fat and have been an important part of indigenous diet for centuries. Today they are becoming neglected in favor of annual peanuts, although there are also efforts to commercialize them for export.[35] According to the *Encyclopedia of Fruits and Nuts*, a softening agent (a salt called trona) needs to be added to the cooking water to prepare the hard seeds. According to *Congo Native Fruits: Twenty-Five of the Best*, they are fried in oil with no special preparation.[36]

Yield. Yields are extremely high, with an estimated 250 to 600 kilograms (550 to 1,300 pounds) of nuts per tree (fresh weight).[37] If we were to extrapolate this to 100 trees per hectare at 10m by 10m (33 by 33 feet) spacing (as recommended in *Congo Native Fruits*[38]), that gives 25 to 60 tons of nuts per hectare, or 19.8 to 47.4 t/ha/yr of dry, deshelled nuts. This would make African breadnut one of the world's most productive staple crops of any kind, though published data on orchard production is lacking. African breadnut trees in West Africa begin to bear eight years after planting.[39]

Harvest and processing. So why are we not all eating African breadnuts daily? One reason is that there is a fair amount of processing involved. First the nuts must be separated from the inedible flesh of the enormous fruit. The seeds are roasted, and the seed

FIGURE 14.6. Removing the edible seeds from the high-yielding African breadnut fruit in Malawi. Photograph courtesy of Günter Baumann, http://www.africanplants.senckenberg.de.

coat must then be individually peeled from each small nut. All of this work is done by hand. Recently researchers developed a small device for separating the seed hulls, but there is still a need for machinery to remove the seeds from the flesh.[40] Another limitation: Once the nuts are shelled they do not last long in the hot, humid tropics.[41]

Carbon farming applications. African breadnut is used as a living fence.[42]

Other issues. Because the fruits are up to 0.5 meter (20 inches) in diameter, they should not be planted near houses or pathways. The falling fruit can be fatal.

Tree Milks

Among perennial staple crops, there are many strange and unusual foods. Starch-filled tree trunks, aerial tubers, edible woody pods, and other marvels abound. Perhaps none is so strange as the trees that produce edible milk.

These trees are tapped in a similar manner to maple syrup, turpentine, or rubber. The milky latex that seeps out is drunk like cow's milk. Milk trees may be the most neglected and underutilized of all the food categories in this book. There is very little research on nutrition or yield. I'm not aware of any comprehensive list of species. Many trees produce latex, and some of those are used for chewing gum. These non-toxic latexes might be a good place to start testing, first for edibility (in the lab) and then for palatability (with cautious but curious human subjects).

COW TREES (*BROSIMUM* SPP.)

These milk-producing relatives of the Maya breadnut are actually rather widely used in their native Central and South America, and even cultivated on a small scale in South and Southeast Asia.[43] Sometime in the last few thousand years, as indigenous people were learning to tap trees for rubber, resin, and medicine, someone was sampling the edible qualities of the latex. Many *Brosimum* species have passed the test, including *B. alicastrum*, *B. parinarioides*, and *B. utile*.

Uses. *Brosimum* milks are drunk like cow's milk but can also be used for cheese, ice cream, and other products. They are also used as medicine and a source of rubber. (See chapter 13 for information on the edible nuts of *B. alicastrum*.) The milk of *B. parinarioides* is 5 to 7 percent protein, almost double that of cow's milk or soy milk, and is also higher in calcium than both cow's and soy milk. Some trees are sweet, while others have bitter latex.[44]

Yield. There is very little yield data, but one source reported 15 to 20 liters (4 to 5 gallons) from one tree.[45] (It's not clear what period of time this covers, nor is it clear, more generally, how frequently these trees can be tapped. More data is needed

FIGURE 14.7. Cow trees are tapped just like these rubber trees, but they produce drinkable, high-protein milk. Photograph courtesy of International Center for Tropical Agriculture (CIAT)/Creative Commons 2.0.

before the relative protein-producing capacity of milk trees is understood.)

Carbon farming applications. Cattle enjoy the leaves and branch tips, and the fodder provided by large groves is considered equivalent to the best pasture. The nuts and fruits are edible by humans and are also often used as pig feed.

Crop development. Although there is currently very little data to work with, this is an area that merits further investigation. Vegan carbon farming enthusiasts of the tropics, take note!

TABLE 14.4. Trees with Edible Milk

Latin Name	Common Name	Origin	Climate	Cultivation
Brosimum alicastrum	Ramon, Mayan breadnut	Mesoamerica	Tropical lowlands, semi-arid to humid	As nut crop and fodder
Brosimum parinarioides	Leite de amapá	South America	Tropical humid lowlands	Wild-harvested
Brosimum utile	Milk tree, cow tree	Americas	Tropical lowlands, semi-arid to humid	Minor crop in South and Southeast Asia
Manilkara huberi	Cow tree	South America	Tropical humid lowlands	As rubber crop

Sources: Hanelt, *Mansfeld's Encyclopedia of Agricultural and Horticultural Crops*; Galuppo, "Documantacao do uso y valoriza de olea de piquia (*Caryocar villosum*) e do leite do amapa-doce (*Brosimum parinarioides*) para a comunidade de Piquituba."
Note: In all likelihood there are many other species.

Leaf Protein Concentrate

Many edible leaves contain a respectable amount of protein, although you could never get all the protein you need by eating these crops because they contain too much fiber. You'd need a huge belly like an elephant or giant ground sloth to process it all. Two hundred years ago researchers developed a technique to extract the protein from edible leaves and create a tofu-like product called leaf protein concentrate (LPC), which is 50 percent protein and concentrates the vitamins and minerals from the leaves.[46] LPC is fully digestible by people and gets all that bothersome fiber out of the way so we can use leaves—including perennial leaves—as a true protein staple.[47] In France alfalfa has been reported as yielding 1.2 t/ha of dried LPC, while nettles have been reported as yielding 0.6 t/ha of dried LPC.[48]

Making a leaf protein concentrate doesn't work for all crops. Some don't form good curds, while in others it concentrates toxins. Some just taste bad. Good species for making LPC include alfalfa, lablab bean leaf, and moringa. Few crops, particularly perennials, have been tested for LPC suitability.

I have Crohn's disease and frequently require a low-fiber diet. This can make it difficult to get necessary vitamins from vegetables. I have made LPC a handful of

FIGURE 14.8. Fresh alfalfa leaf protein concentrate, a leaf cheese or tofu.

LEAF PROTEIN FROM INDUSTRIAL BIOMASS CROPS

The 2008 article "Protein Feeds Coproduction in Biomass Conversion to Fuels and Chemicals" lays out the potential of a massive upscaling of LPC production. The authors propose that it would be simple to extract LPC during the processing of perennial biomass crops like giant grasses. This would not negatively impact the biomass fuel and feedstock uses of the raw material. They propose that, in a scenario with greatly increased biofuel production, this LPC could replace up to 20 percent of soybean meal in livestock diets (24 million tons/yr). They note that perennial biomass LPC yields of protein/ha can equal or beat those of soybeans.[49] The IPCC is enthusiastic about the potential of this strategy.[50]

---

times. For home use, I just juice the leaves of appropriate species such as tree collards and bush clover, then boil them for a minute or two. The curds rise to the surface, and I strain them off just as I would for cheese or tofu making.

LPC is not delicious on its own, although the flavor varies widely among species. It is usually mixed with other foods and can be dried and stored indefinitely. It may be that its primary use is as a substitute for annual soy in the diet of non-ruminant livestock such as chickens and pigs, on which it has been trialed successfully.[51] The by-products of alfalfa LPC processing are whey and finely chopped fiber, which are of equivalent and sometimes superior feed value to the original alfalfa and are important fodder products.[52]

In France a company called France Lucerne is making leaf protein concentrate at the factory scale. An organization called Leaf for Life has been instrumental in the popularization and spread of LPC. As of 2011 about 40,000 people consume dried LPC daily in hunger-stricken communities. LPC has been particularly useful for people living with HIV and AIDS.[53] One of the main limiting factors to LPC in rural poor areas

TABLE 14.5. Perennial Leaf Protein Concentrate Sources

Latin Name	Common Name	Origin	Climate
Boehmeria nivea	Ramie	Asia	Warm temperate to tropical, humid
Brassica carinata	Ethiopian kale	Africa	Tropical lowlands, highlands, semi-arid to humid
Brassica oleracea	Perennial kales and collards	Mediterranean	Warm temperate, Mediterranean, tropical highlands, humid
Clitoria ternatea	Butterfly pea	Asia	Humid tropics
Cnidoscolus aconitifolius	Chaya	Mesoamerica	Arid to humid tropics, highlands and lowlands
Lablab purpureus	Lablab bean	Africa	Semi-arid to humid tropics, lowlands and highlands
Lespedeza bicolor	Bush clover	E. Asia	Cold to warm temperate, semi-arid to humid
Medicago sativa	Alfalfa	Central Asia	Boreal to warm temperate and tropical highlands, humid to semi-arid
Moringa oleifera	Moringa	India	Tropics and subtropics, lowlands, semi-arid to humid
Panicum maximum	Guinea grass	Africa, Middle East	Humid tropical lowlands and highlands
Pennisetum purpureum	Napier grass	Africa	Warm temperate, subtropical, tropical semi-arid to humid
Phalaris arundinacea	Reed canary grass	Europe, Asia, North Africa, North America	Cold and warm temperate, semi-arid to humid
Phaseolus lunatus	Lima bean	Tropical Americas	Tropical lowlands to highlands, semi-arid to humid
Populus spp.	Poplar	Northern temperate	Boreal to warm temperate, humid to semi-arid
Psophocarpus tetragonolobus	Winged bean	New Guinea to India	Tropical lowlands and highlands, humid
Pueraria montana var. *lobata*	Kudzu	Asia	Cold to warm temperate humid
Saccharum officinarum	Sugarcane	New Guinea	Humid tropical lowlands
Sesbania sesban	Sesbania	Africa	Tropical lowlands, semi-arid to humid
Trifolium repens	White clover	Eurasia	Boreal to warm temperate, tropical highlands, semi-arid to humid
Urtica dioica	Stinging nettle	Eurasia	Boreal to subtropics, humid

Sources: Telek, *Leaf Protein Concentrates*, 52–111; Pirie, *Leaf Protein and Its By-Products in Human and Animal Nutrition*, 38–39.
Note: This table shows species that can be used to make leaf protein concentrate. Many species are not suited to LPC production due to toxicity, foamy or slimy consistency, or producing tiny protein granules that won't aggregate. These include some species of *Agave, Eucalyptus, Indigofera, Leucaena, Morus, Musa,* and *Salix.*[a] Due to my own experiments I can add *Lespedeza bicolor* to the list.
[a] Pirie, *Leaf Protein and Its By-Products in Human and Animal Nutrition*, 38–39.

is access to appropriate technology for juicing leaves quickly and efficiently.[54]

Of course, leaf crops would be an incredibly important part of the human diet even if they were 0 percent protein because they are full of fiber, vitamins, minerals, and antioxidants. Because I decided to cover only staple crops, and not fruits and vegetables in this book, I'll refer the reader to The World Vegetable Center (AVDRC) and books such as *Lost Crops of Africa*, vol. 2: *Vegetables*; *Edible Leaves of the Tropics*; the *Leaf for Life Handbook*; and my own *Perennial Vegetables*.

MORINGA (*MORINGA OLEIFERA*)

Of the staple crops profiled in this book, *Moringa oleifera* is among the few that have been getting

substantial attention in the last few years. This is for good reason. Moringa trees are astonishingly nutritious and grow in difficult conditions. It is a major vegetable in India (where it hails from), as well as in the Philippines. It is increasingly taking its rightful place as a crop around the world. I recently saw a sign outside a Miami gas station advertising the miraculous superfood moringa available inside. Although I'm always skeptical of hyped trends in food and supplements, moringa is truly a world-class resource. The Educational Concerns for Hunger Organization (ECHO) has done an outstanding job promoting moringa utilization and has shipped free seeds of moringa and other tropical crops to development projects around the world for decades.

Figure 14.9. A coppice block of *Moringa oleifera* for leaf production, with *M. stenopetala* (cabbage tree) in the background at the ECHO Global Farm in Florida.

Uses. Moringa leaves are 30 percent protein on a dry-weight basis and full of minerals and vitamins.[55] Dried and powdered leaves are known as "mothers' best friend" because they are eaten by nursing mothers in dry regions of Africa.[56] Moringa produces pods like long green beans. When they are pencil-thin they can be cooked and eaten, although when they are fatter the vegetable portion must be scraped off the tough pod like an artichoke. Edible horseradish-flavored seeds produce a valuable oil. This oil is similar in composition to olive oil and is used in cooking and as a fuel.[57] Moringa has many other uses. The seeds (and the seedcake residue left after pressing oil) are used to purify drinking water. They are both a flocculant (removing solids) and an antimicrobial.[58]

Yield. The yield of fresh leaves per hectare is 10 to 50 tons. Pod yields can be 31 tons per hectare. When mature, the pods fill with small edible seeds. These can be pressed for oil at a quantity of 250 liters (66 gallons) per hectare.[59]

Harvest and processing. Like most woody leaf crops, moringa plants are typically coppiced for ease of harvest when they are being managed for leaf production. Moringa leaves are very well suited for use as leaf protein concentrate.[60]

Carbon farming applications. Moringa is a great fodder bank species for livestock, including aquaculture fish. The trees serve as a living fence, alley crop, and contour hedgerow. Their light shade is welcome in agroforestry systems, and they are used as living trellises for yams and other vine crops.[61] As an oil crop they would be suited to woody agriculture. There are improved varieties of moringa selected for pod or leaf production. Improved oil and water-purifying seed varieties would also be of use. These improved varieties are easily propagated by cuttings and live stakes.

Other issues. Moringa is already grown as an ornamental in much of the world's tropics. In fact, in some regions it has begun to naturalize.

Crop development. There are a number of other species of moringa. The African moringa (*M. steno-petala*) was domesticated in the Ethiopian highlands and can grow at altitudes of 2,000 meters (6,500 feet) or higher. Its leaves are larger and easier to harvest. Ben tree (*M. peregrina*) grows from North Africa to the Middle East. This species is tolerant of extremely arid conditions. It is the source of ben oil. Ben tree is rare in the wild, and adopting it as a new crop may be a good strategy for conservation.

CHAYA (*CNIDOSCOLUS ACONITIFOLIUS*)

Chaya, or Mayan spinach, is a shrub of the tropics and subtropics. It tolerates arid to humid conditions, in lowlands and highlands. It was domesticated by Maya people in ancient times, who discovered and propagated non-stinging forms with a variety of leaf shapes.

Uses. Although the raw leaves contain cyanide, cooking for 10 minutes deactivates it, and the cooked leaves are very good. The fresh leaves are 4 to 8 percent protein.[62] Chaya is quite safe for use in leaf protein concentrate.[63]

Yields. Chaya has proven itself to be a reliable producer in most tropical climates. It is easily propagated by cuttings and is commonly grown as an ornamental

FIGURE 14.10. Chaya, or Mayan spinach, is high in protein and makes good leaf protein concentrate.

in the tropics for its beautiful flowers. Because it is already widely distributed, it is mostly a matter of education to teach people to safely consume it by cooking thoroughly.

Harvest and processing. Some varieties of chaya sting and must be handled carefully. Chaya is managed as a coppiced plant for best leaf production.

Carbon farming applications. Chaya is used as a living fence. This adaptable and resilient species should be grown much more widely.

CHAPTER FIFTEEN

Protein-Oil Crops

Protein-oil crops have more than 16 percent in both protein and oil. Global annuals in this group are soybeans, sunflowers, peanuts, and egusi (melonseeds). Global perennials include cashews, almonds, walnuts, hazelnuts, and pistachios. Crops in this class are grown on 139 million (344 million acres) hectares, of which 104 million hectares (257 million acres) is soybeans. This is about 9 percent of world cropland.[1] Perennials in this class include nuts, beans, seeds, and fruits. In many cases their yields are highly competitive with annuals. Many of these crops are also pressed for oil. The leftover presscake from the oil extraction is typically high in protein and fed to livestock.

Nuts

Our planet abounds with nuts rich in oil and protein. This group is one of the strongest categories of perennial staple crops and includes a number of global crops such as almond, walnut, pistachio, cashew, and hazel. Several emerging crops such as yellowhorn, mongongo, and inche also show great promise. Many of these species yield very well. Among them, we have candidate species for just about every climate of the world. Processing and allergy challenges apply here, as they do with all nuts.

ALMOND (*PRUNUS DULCIS*)

Almonds were cultivated in Central Asia as early as 4,000 BC.[2] Today they are a global crop, with 1.6 million hectares (3.9 million acres) producing 1.9 million tons, worth $7.9 billion US.[3] About 80 percent of that is grown in the United States in California, in industrial monocultures. During the 2015 drought, California almonds were criticized for heavy water use.[4] Almonds are adapted to semi-arid, warm temperate climates and require some chilling hours.[5]

Uses. The nuts are 16 to 21 percent protein and 50 to 54 percent fat. They are consumed raw and roasted and in desserts and baked goods. Bitter varieties are grown for their edible oil. Almond milk has been commercialized in recent decades.[6]

Yield. Under the highly intensive California production system, yields can be high as 6.7 t/ha of clean kernels (6.4 dry weight).[7] Global average yield is 0.5 t/ha of dry clean kernels.[8]

Carbon farming applications. Almonds could be considered for woody agriculture.

FIGURE 15.1. An almond orchard in Spain. Almonds are a top-yielding protein-oil producer. Photograph courtesy of Plácid Pérez Blu/Creative Commons 2.5.

TABLE 15.1. Protein-Oil Crop Yields Compared

Latin Name	Common Name	Climate	Avg. Global Yield[a]	Yield t/ha	Product
Prunus dulcis	Almond	Temperate, Mediterranean	0.5	6.4[b]	Nuts
Pinus pinea	Italian stone pine	Mediterranean		6.2[c]	Nuts
Caryodendron orinocense	Inche	Lowland humid tropics		2.3–5.8[d]	Nut kernels
*Helianthus annuus	Sunflower	Global annual	1.4	1.0–5.7[e]	Seeds
*Glycine max	Soybean	Cold temperate, tropical	2.1	0.9–4.6[f]	Beans
Psophocarpus tetragonolobus	Winged bean	Humid tropical lowlands and highlands		1.1–4.6[g]	Beans
Dacryodes edulis	Safou, butterfruit	Humid to semi-arid tropical lowlands		4.5[e]	Fruit
Citrullus colocynthis	Colocynth, perennial egusi	Arid tropics and subtropics		4.5[h]	Seeds
Bertholletia excelsa	Brazil nut	Lowland humid tropics		4.4[i]	Nuts
*Arachis hypogaea	Peanut	Cold temperate, tropical	1.0	0.7–3.3[e]	Peanuts
Telfairia pedata	Oyster nut	Humid tropical lowlands and highlands		1.0–2.4[e]	Clean seeds
Cucurbita foetidissima	Buffalo gourd	Arid cold to tropical		1.7–2.0[j]	Seeds
Juglans regia	Persian walnut	Mediterranean, warm temperate semi-arid to humid	0.8	0.2–1.8[k]	Nuts
Corylus avellana	Hazelnut	Cold to warm temperate, Mediterranean, semi-arid to humid	0.6	1.0–1.4[l]	Nuts
Pistacia vera	Pistachio	Mediterranean	0.7	1.0[m]	Nuts
Xanthoceras sorbifolium	Yellowhorn	Cold to warm temperate, semi-arid to humid		1.0[n]	Nuts
Anacardium occidentale	Cashew	Tropical lowland, semi-arid to humid	0.7	0.8–1.0[o]	Nuts
Parkia biglobosa	African locust bean	Tropics, semi-arid		0.3–0.8[p]	Beans
Blighia sapida	Akee	Humid tropical lowlands		0.5–0.7, extrapolated[e]	Fruit
Corylus spp.	Neohybrid hazel	Boreal to warm temperate, humid to semi-arid		0.7[q]	Nuts
Schinziophyton rautenenii	Mongongo	Tropics and subtropics, lowlands, semi-arid		0.1–0.7[e]	Nuts
Pinus koraiensis	Korean pine	Boreal to warm temperate, humid		0.3[r]	Nuts
Pinus sibirica	Siberian pine	Boreal to warm temperate, semi-arid to humid		0.2[c]	Nuts

Note: Here we compare perennial versus annual protein-oil sources. Annual comparison crops are marked with an asterisk (*). All yields calculated based on dry weight of the edible portion. Crops are arranged in descending order of yield.

[a] FAO Statistics Division online.
[b] University of California, *Sample Costs to Establish an Almond Orchard and Produce Almonds,* 5.
[c] Crawford, "Nut pines."
[d] Duke, *Handbook of Nuts,* 78.
[e] Plant Resources of Tropical Africa online database.
[f] Plant Resources of Southeast Asia online database.
[g] National Research Council, *The Winged Bean,* 16; French, *Food Plants of Papua New Guinea,* 54.
[h] Duke, *Handbook of Energy Crops.*
[i] Janick and Paull, *The Encyclopedia of Fruits and Nuts,* 450.
[j] Duke, *Handbook of Energy Crops,* under profile for *Citrullus colocynthus;* El Bassam, *Handbook of Bioenergy Crops,* 137.
[k] Duke, *Handbook of Nuts,* 194–97.
[l] Wilkinson, *Nut Grower's Guide,* 129.
[m] Ibid, 185.
[n] World Agroforestry Center, *Participatory Agroforestry Development in DPR Korea,* 141.
[o] Duke, *Handbook of Nuts,* 21.
[p] Plant Resources of Tropical Africa online; National Research Council, *Lost Crops of Africa,* vol. 2: *Vegetables,* 214.
[q] Rutter, Weigrefe, and Rutter-Daywater, *Growing Hybrid Hazelnuts,* 77.
[r] World Agroforestry Center, *Participatory Agroforestry Development in DPR Korea,* 21.

Other issues. In more humid regions, almonds are vulnerable to diseases. There are some cultivars selected for adaptation to cold and humid climates. For example, 'Hall's Hardy' almond does well here in humid cold temperate Massachusetts. The main problem in extending the range of the almond is that the flowers emerge early and are easily killed by frost. Another challenge in almond production is that they are pollinated by insects. In commercial orchards, large numbers of honeybee hives are brought in at the appropriate time to perform this essential service.[9]

Related species. Almonds have a number of relatives that are raised as minor or regional crops, including apricots (*Prunus armeniaca*) with edible seeds. Some, such as the Siberian apricot (*P. sibirica*) and cherry prinsepia (*Prinsepia utilis*), are cultivated in boreal climates for their oil.[10]

WALNUT (*JUGLANS* SPP.)

The Central Asian ("Persian") walnut (*Juglans regia*) has been cultivated since ancient times. Typical Persian walnuts grow in semi-arid to humid warm temperate and Mediterranean climates. Some selections from the eastern part of the wild range in the Carpathian Mountains of Eastern Europe are substantially more cold-hardy and may grow in the cold temperate zones. These Carpathian walnuts are less fully domesticated than their warmer-climate counterparts.[11] Other cultivated walnut species are discussed below. Walnuts are a global crop, with 995,000 hectares (2.5 million acres) in production. Global yield in 2012 was 3.4 million tons (fresh weight and in-shell), valued at $14 billion US.[12]

Uses. Persian walnuts are 13 to 18 percent protein and 63 to 67 percent fat, which seems on par with other walnut species that have been analyzed for this.[13] The nuts are a major global commodity and are also used to produce an edible oil. Many walnut species are also important timber crops, although they are, by necessity, destructively harvested.

Yield. Walnuts are among the highest-yielding temperate nut crops.[14] Yields can be as high as 6.5 to 7.5 t/ha in fertile valleys, which converts to 1.8 t/ha of dry, deshelled nut kernels.[15] Average global yield is 3.4 t/ha in-shell, or 0.8 dry clean kernels.[16]

Harvest and processing. The modern walnut industry was built around varieties that bear nuts at the tips of the branches. These tip-bearing walnuts are mechanically harvested with tree shakers and nut sweepers but must be hand-pruned. In recent decades lateral-bearing varieties have been developed. Instead of bearing at the tips of the branches, these forms bear on the smaller side branches. Lateral-bearing walnuts bear in four to five years (tip-bearing require seven years or more), and yields

FIGURE 15.2. A Central Asian "Persian" walnut, a global protein-oil crop. Photograph courtesy of Friedrich Böringer/Creative Commons 2.5.

are higher. The trees are smaller and are grown in mechanically trimmed hedgerows, minimizing pruning labor. Because more trees are grown per hectare, the establishment costs of these lateral-bearing varieties are somewhat higher.[17]

Carbon farming applications. In north and northeastern China, both *J. regia* and *J. mandschurica* are intercropped with annual crops.[18] In North Korea, *J. ailanthifolia* is intercropped with annuals on marginal land, and *J. regia* is grown on contour with annual crops.[19] Some walnuts coppice strongly, so they might be developed as a woody agriculture species.

Additional species descriptions. Many other walnuts are cultivated around the world. To my knowledge most have received little breeding work, although some hybrids have been developed. Walnuts could be a great candidate for neohybrid crop development. We need lateral-bearing walnuts for boreal through tropical climates; this could be done using the many cultivated *Juglans* species.

Heartnut (*J. ailanthifolia*) is native to Japan and China, where it is a crop of regional importance. The

Perennial Staple Crops

TABLE 15.2. High Protein-Oil Nuts

Latin Name	Common Name	Origin	Climate	Cultivation
Adansonia digitata	Baobab	Africa	Tropical lowlands, arid to semi-arid	Wild staple, new crop
Anacardium occidentale	Cashew	Brazil	Tropical lowlands, semi-arid to humid	Global
Balanites aegyptiaca	Balanites	Africa, Asia	Tropical to subtropical, arid to semi-arid	Regional
Bertholletia excelsa	Brazil nut	South America	Humid tropical lowlands	Global
Buchanania cochinchinensis	Almondette	South and Southeast Asia	Tropics, semi-arid to humid	Regional
Carya hybrids	Hybrid and neohybrid hickories	Hybrid origin	Variable	Regional
Carya cathayensis	Chinese hickory	China	Cold to warm temperate, humid	Regional
Carya illinoinensis	Pecan	North America, Mesoamerica	Cold temperate to subtropical, highlands, humid	Minor global
Carya laciniosa	Shellbark hickory	East North America	Cold to warm temperate, humid	Regional
Carya ovata	Shagbark hickory	North America, Mesoamerica	Cold temperate to subtropical, tropical highlands, humid	Regional
Caryodendron orinocense	Inche	Northwest South America	Tropical lowlands, humid	Experimental, very promising
Chrysobalanus icaco	Cocoplum	Americas	Tropical lowlands, humid	Regional
Corylus hybrids and neohybrids	Hybrid and neohybrid hazel	Hybrid	Boreal through warm temperate, Mediterranean, semi-arid to humid	Minor global
Corylus avellana	Hazelnut	Europe	Cold to warm temperate, Mediterranean, humid	Global
Corylus colurna	Turkish tree hazel	Central Eurasia	Cold to warm temperate, Mediterranean, semi-arid to humid	Regional
Corylus heterophylla	Siberian hazel	Northeast Asia	Boreal to warm temperate	Regional
Corylus maxima	Giant filbert	Central Eurasia	Cold to warm temperate, Mediterranean, semi-arid to humid	Regional
Corylus sieboldiana	Manchurian hazel	Northeast Asia	Boreal to warm temperate, humid	Regional
Jatropha curcas	Edible jatropha	Tropical Americas	Tropical lowlands, semi-arid to humid	Regional as food, minor global as industrial
Juglans ailanthifolia var. *cordiformis*	Heartnut	East Asia	Cold to warm temperate, humid	Regional
Juglans cinerea	Butternut	Eastern North America	Boreal to warm temperate, humid	Minor global
Juglans neotropica	Andean walnut	South America	Tropical highlands, subtropical lowlands, humid	Regional
Juglans regia	Persian walnut	Central Asia	Warm temperate, Mediterranean, semi-arid to humid	Major
Juglans nigra	Black walnut	Eastern North America	Cold to warm temperate, humid	Regional
Juglans olanchana	Nogal	Tropical Americas	Tropical lowlands, humid	Regional
Lecythis corrugata	Guacharaco	Amazonia	Tropical lowlands, humid	Regional
Lecythis minor	Coco de mono	Northern South America	Tropical lowlands, humid	Regional
Lecythis pisonis	Castanha e sapucaia	Amazonia	Tropical lowlands and highlands, humid	Minor
Lecythis zabucajo	Monkey pot	Northern South America	Tropical lowlands, humid	Minor
Maclura pomifera	Osage orange	North America	Cold to warm temperate, semi-arid to humid	Wild
Pachira glabra	Pachira nut	Central America	Tropical lowlands, semi-arid to humid	Experimental
Pinus cembroides	Mexican piñon pine	Mesoamerica	Warm temperate to subtropics, lowlands, highlands, semi-arid	Regional

Latin Name	Common Name	Origin	Climate	Cultivation
Pinus edulis	Colorado piñon pine	Western North America, Mesoamerica	Cold temperate to subtropical highlands, arid to semi-arid	Wild-managed
Pinus koraiensis	Korean nut pine	Northeast Asia	Boreal to warm temperate, humid	Global
Pinus monophylla	Singleleaf piñon	Western North America	Cold temperate to subtropical highlands, arid to semi-arid	Wild-managed
Pinus pinea	Italian stone pine	Mediterranean	Warm temperate to subtropics, Mediterranean, semi-arid	Global
Pinus sibirica	Siberian stone pine	Northeast Asia	Boreal to warm temperate, semi-arid to humid	Wild-harvested
Pistacia vera	Pistacio	Central Asia	Warm temperate to subtropical, Mediterranean, semi-arid	Global
Plukenetia conophora	Conophor	Africa	Tropical lowlands, humid	Regional
Plukenetia volubilis	Sacha inchi	South America	Tropical lowlands, humid	Minor global
Prunus armeniaca	Edible-pit apricot	Central Eurasia	Cold to warm temperate, semi-arid to humid	Minor global
Prunus dulcis	Almond	Mediterranean	Cold temperate to subtropics, Mediterranean, semi-arid to humid	Global
Ricinodendron heudelotii	Djansang	Africa	Tropical lowlands, humid	New crop, wild staple
Salvadora persica	Salvadora	North Africa through Pakistan	Tropics, arid to semi-arid	New crop
Santalum acuminatum	Quandong	Australia	Tropics and subtropics, arid to semi-arid	New crop
Schinziophyton rautanenii	Mongongo nut	Africa, South Asia	Tropics and subtropics, lowlands, semi-arid	Wild staple, new crop
Sclerocarya birrea	Marula	Africa	Tropical and subtropical lowlands, semi-arid to humid	Regional
Terminalia catappa	Tropical almond	Pacific	Tropical lowlands, semi-arid to humid	Minor global
Terminalia kaernbacchii	Okari nut	New Guinea	Tropical lowlands, humid	Regional
Torreya grandis	Chinese kaya	China	Warm temperate to subtropics, humid	Regional
Torreya nucifera	Japanese kaya	Japan	Cold to warm temperate, humid	Regional
Xanthoceras sorbifolium	Yellowhorn	China	Cold to warm temperate, semi-arid to humid	Minor global

Sources: Hanelt, *Mansfeld's Encyclopedia of Agricultural and Horticultural Crops*; Duke, *Handbook of Nuts*; Janick and Paull, *The Encyclopedia of Fruits and Nuts*; French, *Food Plants of Papua New Guinea*; National Research Council, *Lost Crops of Africa*, vol. 2: *Vegetables*; National Research Council, *Lost Crops of the Incas*, Elevitch, *Traditional Trees of Pacific Islands*; Northern Nut Growers Association; Plant Resources of Tropical Africa; Wilkinson, *Nut Grower's Guide*; Gymnosperm Database online.

variety *cordiformis* has heart-shaped nuts that are easily removed from the shell. Heartnut is considered the best walnut for cold temperate climates.[20]

Butternut (*J. cinerea*) is native to eastern North America. It is hardy into the boreal zone and used in breeding work to impart hardiness. It is experimentally cultivated in Estonia and the former Soviet Union as a new cold-climate nut crop.[21]

Andean walnut (*J. neotropica*) is native to high-elevation South America. It is cultivated for the nuts as well as the timber. Andean walnut grows in tropical highlands well over 2,500 meters (8,200 feet).[22] I have seen plants said to be *J. neotropica* fruiting well in tropical Homestead, Florida, although they might have actually been *J. olanchana*.

Black walnut (*J. nigra*) is native to eastern North America. It is grown on a small scale for the flavorful nuts, although timber is the primary crop. Black walnut is widely grown as an alley crop species in North America. It produces a botanical herbicide that suppresses the growth of some neighboring species.[23]

Nogal (*J. olanchana*) is a Central American species cultivated for timber and edible nuts. Nogal is suited to humid tropical lowlands.[24]

HAZEL, FILBERT (*CORYLUS* SPP.)

Hazelnuts are among the best perennial protein and oil staple crops for temperate and boreal climates. The primary cultivated species is the European hazel (*C. avellana*), but many other species are cultivated regionally in Eurasia and North America. Hybrid and neohybrid hazels are of increasing importance as well. Hazelnuts are a global crop, with 600,000 hectares (1.5 million acres) producing 914,000 tons at a value of $2.6 billion US.[25]

Uses. The nuts of *C. avellana* are 16 to 20 percent protein and 54 to 64 percent oil.[26] Hazelnuts are used fresh, in pastries and desserts, and for a high-quality edible oil.

Yield. Yields are high, reaching 2.5 to 3.5 t/ha of fresh, in-shell nuts.[27] This is equivalent to 1.0 to 1.4 t/ha of dry, deshelled nuts. The more cold-hardy neohybrid hazels are currently yielding 1.6 t/ha (0.7 t/ha of dry, clean kernels).[28] Unlike many temperate-climate nuts, hazels yield consistently every year.

Harvest and processing. Although it is a multistemmed shrub, when grown commercially *C. avellana* is typically pruned to a tree-like form. This allows a mechanized harvest, similar to that for walnuts and chestnuts.[29] Badgersett Research Corporation's woody agriculture model instead grows neohybrid hazels in close hedgerows, which are coppiced every 10 years. These dense rows are straddled by blueberry-harvesting machines for a more efficient harvest and healthier plants.[30]

Carbon farming applications. Siberian hazel is intercropped with annual crops in southern China.[31] It is the basis of temperate woody agriculture.

Species descriptions. Many species of hazel are cultivated around the world. (This is not a comprehensive list.)

FIGURE 15.3. Neohybrid hazelnuts. Hazels are one of the most cold-tolerant perennial staples.

Hazelnut (*C. avellana*) is the primary cultivated species. It is native to Europe and suited to temperate and Mediterranean climates of semi-arid and humid conditions. This is a multistemmed species.

Giant filbert (*C. maxima*) is a regional crop from central Eurasia. A shrub crop like *C. avellana*, giant filbert has similar climate requirements.[32]

Siberian hazel (*C. heterophylla*) and **Manchurian hazel** (*C. sieboldiana*) are both regional crops in eastern Asia. Both shrub species are hardy into the frigid boreal climate zone.[33]

Turkish tree hazel (*C. colurna*) is the main cultivated tree-type hazel. It is a regional crop in central Eurasia. Good varieties such as those that are cultivated in Turkey bear annually, with nuts the same size as *C. avellana*.[34]

Neohybrid hazels (*C. hybrids*). Badgersett Research Corporation is taking things to another level with their neohybrid swarm breeding program. These are distinct from hybrid hazels, which are *C. avellana* with a few American or Asian genes for disease resistance.

PISTACHIO (*PISTACIA VERA*)

Pistachios are an ancient nut crop. They were cultivated as early as 4,000 years ago in their native West and Central Asian high deserts.[35] In fact, archaeologists have found remains of pistachios, almonds, acorns, and

Figure 15.4. A pistachio orchard in Iran. Photograph courtesy of A. Davey, Creative Commons 2.0.

other nuts with nutcracking stones in Israel that were 780,000 years old![36]

Pistachios require a semi-arid warm temperate or subtropical climate with long, dry summers and 600 to 1,000 "chilling hours." This measurement indicates the number of hours below 7°C (45°F) necessary for plants to set fruit.[37] With that said, I have also seen trees in dry but fully tropical regions of Guatemala where I was told they produce well. This is a global crop. Pistachios are grown on 494,000 hectares (1.2 million acres), with a global yield of 1 million tons in-shell, worth $7.9 billion US.[38]

Uses. Pistachio nuts are 20 percent protein and 40 to 56 percent oil.[39] They are mostly eaten fresh as a snack or used in desserts and baked goods. They are also pressed for an edible oil. Tannins are extracted from the shells.[40]

Yields. Yields can be as high as 3 t/ha fresh in-shell, or 1.0 dry, deshelled nuts There is a strong tendency to bear heavily on alternate years.[41] Global average yield is 2 t/ha or 0.7 t/ha dry, cleaned kernels.[42]

Harvest and processing. Pistachio harvest is fully mechanized in countries such as the United States and Australia.[43] My analysis of California staple crops found pistachios had the worst labor efficiency of the species reviewed.

Carbon farming applications. Pistachios are inter-cropped with annual crops in northwest China.[44] Pistachios coppice and could be used in woody agriculture.

Nut Pines (*Pinus* spp.)

The Italian stone pine (*Pinus pinea*) has been a Mediterranean staple for millennia. Around the world nut pines are adapted to some of the planet's most difficult conditions, including cold, arid deserts and even extremely cold boreal and alpine regions with temperatures as low as −51°C/−60°F (although evergreen plantations in high boreal regions actually cause more warming from albedo than they mitigate from carbon sequestration). Though they are often slow to bear, nut pines are long-lived; some species live as long as 900 years.[45] Most are low yielding, but *P. pinea* has impressive protein-oil yields.

Uses. Many species of pines from Europe, Asia, and the Americas produce edible nuts. Their protein content ranges from 14 to 34 percent and oil from 48 to 65 percent depending on species and variety.[46]

Yield. Nut pines tend to yield well only every two to three years (and more like three to seven for the North American species). Italian stone pine has yielded as much as 6.2 t/ha of dry, deshelled nuts—extremely impressive, though not a consistent yield.[47] Recorded yields of other species are generally quite low.

Carbon farming applications. *Pinus koraiensis* is traditionally irregularly intercropped with annuals in North Korea.[48]

Crop development. To my knowledge there has been little work on domestication or cultivation techniques to improve the yields and frequency of bearing of nut pines besides *P. pinea*. Some basic work could improve food security and carbon sequestration in some extremely difficult conditions.

Species descriptions. This is only a selection of the many cultivated and wild-managed species.

Italian stone pine (*P. pinea*) is adapted to Mediterranean and semi-arid warm temperate climates. It is by far the highest-yielding pine nut at 6.2 t/ha of clean, dry nuts, though it takes a few years off after a heavy yield. This species is widely cultivated in the Mediterranean and North Africa.[49]

Korean pine (*P. koraiensis*) is an important food in eastern Asia. It is suited to boreal to warm temperate humid climates. It is a biennial bearer, with yields of 0.3 t/ha (clean, dry kernels) in the good years. Improved varieties are cultivated in Korea in hillside agroforestry systems.[50]

Piñon pines (*P. cembroides*, *P. edulis*, and *P. monophylla*) are native to the cold, dry high-altitude deserts of North America and Mexico. Good yields may come as infrequently as every four to seven years. These species grow in truly arid environments with rainfall under 250 millimeters (9.8 inches) and temperatures as low as −28°C (−18.4°F).[51]

Siberian stone pine, Siberian cedar (*P. sibirica*), is an important wild staple in northern Asia. It is also cultivated, with nuts sold internationally. Heavier yields are every two to three years and come in around 0.2 t/ha of dry, clean nuts. This species handles cold as low as −51°C (−60°F) and rainfall as low as 250mm, making it a strong candidate for the toughest perennial staple crop on the planet.[52]

YELLOWHORN (*XANTHOCERAS SORBIFOLIUM*)

Yellowhorn is atypical of cold-climate nuts in that it is a rather small tree and tolerant of quite dry conditions.

FIGURE 15.5. Wild piñon pine in Colorado.

FIGURE 15.6. Yellowhorn, a promising nut crop for cold, dry regions, at Denver Botanic Gardens in Colorado.

It is promoted in the colder parts of the United States as a xeriscape (low-water-use) ornamental, with nice plants at the semi-arid, high-elevation Denver Botanic Garden. Until recently yellowhorn has been a fairly minor nut crop in its native China. Recently production has expanded beyond China as a nut crop and a biodiesel feedstock.[53]

Uses. The hazel-sized nuts are up to 60 percent oil and 30 percent protein.[54]

Yield. Yields can be as high as 1.0 t/ha of clean, dry kernels. Oil yields are 0.3 to 0.4 t/ha—as high as olives! Improved varieties have been selected in North Korea.[55]

Carbon farming applications. The trees are somewhat short-lived at 25 years but bear in as little as 3 years. They are used in alley cropping systems in North Korea on a small scale.[56] Yellowhorn is intercropped with annual crops in north and northwest China.[57]

CASHEW (*ANACARDIUM OCCIDENTALE*)

The cashew is a global crop but due to multiple limitations is not the most promising species for carbon farming at this time. Cashew is adapted to humid and semi-arid tropics and likes a dry season of six months or so. It is native to Brazil.[58] Cashews are produced on 5.3 million hectares (13.1 million acres), yielding 4.1 million tons of nuts in-shell, worth $2.3 billion US.[59]

Uses. The nuts themselves are tasty and 17 percent protein, 43 percent fat.[60] Cashew fruits ("apples") are edible as well. Cashew nutshell oil (CNSO) has many industrial uses but causes terrible skin rashes on contact as well as gastroenteritis and jaw paralysis.[61] *Mansfeld's Encyclopedia of Agricultural and Horticultural Crops* lists "varnishes, paints, adhesives, gums, impregnations, brake linings, floor coverings, and other technical products" among the uses of CNSO.[62]

FIGURE 15.7. Cashews are delicious and widely grown, though yields are poor and processing is challenging. Photograph courtesy of V. R. Vinayaraj/Creative Commons 3.0.

Yield. The trees are precocious, bearing in as little as 18 months. The yields are not terribly good, however—between 0.8 and 1.0 t/ha of dry, shelled nuts.[63] Global average yield is 0.7 t/ha dry-weight, clean kernels.[64] It is surprising that a global crop performs so poorly in terms of yields.

Harvest and processing. Cashew processing is challenging. The nuts and shells are attached outside of an edible "fruit." It is the nutshell that causes problems, because it is full of toxic cashew nutshell oil. No one has yet developed a mechanical cashew sheller. Each nut must first be carefully roasted (allowing the toxic oil to slowly drip out), then cracked by hand. Australian cashew producers ship their harvest outside the country for processing because it is too labor-intensive.[65]

Carbon farming applications. Cashews are common in homegarden systems throughout the tropics.[66] Cashews coppice well and might be suited to woody agriculture.

Crop development. To me, for the cashew to move from a specialty food to a staple it must overcome several barriers. We need high-yielding varieties (there are many named clones already). Selection and breeding should also investigate the possibility of non-toxic types. There is also a major need for processing technology for the farm, village, and semi-industrial scale. This technology should speed up the process and reduce or eliminate worker contact with toxins.

BRAZIL NUT
(*BERTHOLLETIA EXCELSA*)

These members of the Lecythidaceae family hail from the humid tropical lowlands of South America. Brazil nut trees can live to be 1,000 years old and can reach 45 meters (148 feet) tall.[67] In the Peruvian Amazon alone, Brazil nuts are collected from 970,000 hectares (2.4 million acres) of wild forest.[68] Many species in the related genus *Lecythis* are also promising nut crops.

Uses. Brazil nuts are edible and also pressed for an edible oil. They are 14 to 17 percent protein and 64 to 70 percent fat.[69]

FIGURE 15.8. Brazil nuts are globally traded and still mostly wild-harvested. Photograph courtesy of Lior Golgher/Creative Commons 3.0.

Yields. Yields can be very high, with 7.5 t/ha recorded.[70] This converts to 4.4 t/ha of dry, shelled nuts.

Harvest and processing. When ripe the huge, woody fruit opens its "lid," revealing large, delicious nuts. As the trees can be rain forest giants, the fruit may fall quite a distance and can be hazardous to harvesters below. Although they are not cultivated on a large scale, wild-harvested Brazil nuts are a major item of world commerce.

Carbon farming applications. Brazil nuts require pollination by orchid bees. This limitation has resulted in the failure of efforts to establish monoculture plantations, in which the bees do not thrive.[71] Some agroforestry systems apparently simulate natural forest enough to allow successful Brazil nut fruit-set. These are mostly multistrata systems incorporating such tree crops as rubber and peach palm, with understories including cacao, cupuacu, banana, cassava, rice, and black pepper.[72] The flowers of the fruit trees help increase populations of orchid bees.[73]

MONGONGO
(*SCHINZIPHYTON RAUTANENII*)

Mongongo grows in the semi-arid tropical and subtropical savannas and deserts of Central and southern

FIGURE 15.9. A mongongo tree in South Africa. Photograph courtesy of Sean Spender.

Africa. It is a critical wild staple to this day for the San people, who eat 100 to 300 nuts per day for most of the year in some areas.[74] One San man, when asked why he did not farm, famously replied, "Why should we plant, when there are so many mongongo nuts in the world?"[75]

Uses. Mongongo nuts have been important wild staple foods for millennia. The fruit that surrounds the nuts is sweet and edible. The seeds also yield an edible oil. Mongongo nuts have 26 to 29 percent protein and 57 to 58 percent fat.[76]

Yield. Wild plants in Namibia were recorded to yield 0.1 to 0.7 t/ha of dry, clean kernels.[77]

Carbon farming applications. Mongongo is used as a living fence.[78]

Crop development. These wild African nuts are the subject of domestication efforts by the World

Agroforestry Center.[79] Mongongo is under trial in Israel and Australia and is being commercially produced in Namibia and Cameroon.[80]

INCHE (*CARYODENDRON ORINOCENSE*)

Inche might be thought of as a New World counterpart of the mongongo, as it is also a nut of the Euphorbiaceae family. Like its African relative, inche is attracting interest as a new nut and oil crop. Inche is native to humid tropical lowlands of Colombia, Ecuador, and Venezuela.

Uses. Inche nuts are 50 to 54 percent fat and 20 percent protein.[81]

Yield. Yields have been estimated at 6 to 15 t/ha of fruits, of which 40 percent is kernel, resulting in a very impressive 2.3 to 5.8 t/ha of dry, deshelled kernels.[82] Like many nuts, inche also takes a year or more off after a heavy yield.[83] Oil yields are estimated at 0.8 to 3.0 t/ha, among the world's best.[84] Much remains to be learned, however, before inche can step onto the world stage.

Beans

Some beans, like soybeans, are high in oil as well as protein. This is the case for perennial beans as well. The best of these is the winged bean, but many others await development.

WINGED BEAN (*PSOPHOCARPUS TETRAGONOLOBUS*)

This crop was domesticated in New Guinea, like so many useful perennial crops. Winged beans are suited to the humid tropics from sea level to 2,000 meters (6,500 feet). Winged beans are grown much like other perennial vining beans, as a pole bean with a trellis. Winged beans are common in tropical homegardens around the world.[85]

Uses. Almost all parts of the plant are edible, including leaves, flowers, green pods, green seeds, dry beans, and tubers.[86] It is the dry beans that concern us here. These

TABLE 15.3. High Protein-Oil Perennial Beans

Genus	Common Name	Origin	Climate	Cultivation
Adenanthera pavonina	Condori wood	South Asia through Australia	Tropical humid lowlands	Wild
Bauhinia petersiana	Kalahari white bauhinia	Africa	Tropics and subtropics, semi-arid	Wild staple
Caesalpinia digyna	Teri-pod	Asia	Warm temperate to tropical, semi-arid to humid, lowlands and highlands	Wild
Entada phaseoloides	Gogo bean	Africa and Asia	Humid tropical lowlands	Regional, perhaps not all forms edible
Glycine hybrids	Perennial soybean	Hypothetical	Unknown	Hypothetical
Guibourtia coleosperma	African rosewood	Africa, South Asia	Tropical and subtropical lowlands, semi-arid	Wild staple
Parkia biglobosa	African locust bean	Africa	Tropical lowlands, semi-arid	Regional
Pentaclethra macrophylla	African oil bean	West Africa	Humid tropical lowlands	Regional
Psophocarpus tetragonolobus	Winged bean	New Guinea to India	Tropical lowlands and highlands, humid	Regional, new global crop
Tylosema esculentum	Marama bean	East and South Africa	Subtropics, arid to semi-arid	Wild staple, new crop

FIGURE 15.10. Winged beans at the ECHO Global Farm in Florida.

are 30 to 42 percent protein and 15 to 20 percent oil, making them comparable to soybeans and peanuts.[87]

Yield. Dry bean yields are very impressive at 1.1 to 4.6 t/ha.[88]

Harvest and processing. The need for trellising makes mechanical harvest of winged beans difficult.

Carbon farming applications. Winged beans fix nitrogen.

AFRICAN LOCUST BEAN (*PARKIA BIGLOBOSA*)

African locust bean is an important tree in the parkland agroforestry system of African semi-arid savanna lowland regions. A major benefit is that the beans ripen during the dry season when there is little else to eat. It is also a reliable producer, regardless of drought or other challenges.[89]

Uses. This legume tree produces pods filled with sweet pulp. The seeds are usually not eaten in huge quantity. The fermented product, known as dawadawa (and by other names as well) has a strong smell, like a stinky cheese. It is used in soups as a seasoning, and is eaten daily in some regions. Dawadawa is mixed with moringa leaves to make a highly nutritious sauce. The fermented beans are 35 percent protein and 29 percent fat.[90]

Yield. Yields of the seeds are 0.3 to 0.8 t/ha (dry, clean seed) for this undomesticated crop and reportedly higher in experimental plantations.[91]

Harvest and processing. The beans are toxic but made edible through a traditional process of fermentation, which also preserves them in the tropical heat.

Carbon farming applications. *Parkia* is a major component of African parkland agroforestry systems.[92] It is a potential candidate for woody agriculture. Locust bean does not form nitrogen-fixing nodules, although some *Parkia* species do.[93]

Crop development. Breeding priorities could focus on reduced toxicity and improved yield. There is a need for appropriate technology for seed and pulp processing. This tree is already critically important to millions of people in dry regions; with some development work, it could be an even more important carbon farming resource than it is today.

FIGURE 15.11. Women preparing African locust beans for market in Burkina Faso. Photograph courtesy of the World Agroforestry Centre.

AFRICAN OIL BEAN (*PENTACLETHRA MACROPHYLLA*)

African oil bean is a large tree of West African lowland forests and savannas. This is one of the better candidates for tree beans to domesticate for the humid tropics. At present it is a regional crop with little domestication work done to date.

Uses. The large beans are 17 to 22 percent protein and 35 to 52 percent oil.[94] The beans are boiled, roasted, fermented, and pressed for cooking oil.[95]

Yield. Yields are unreported.

Harvest and processing. The authors of *Congo Native Fruits: Twenty-Five of the Best* report that the beans are scarified (nicked) and soaked before cooking to release bitter tannins.[96] A challenge is harvesting the beans from trees up to 35 meters (115 feet) tall; skilled harvesters reportedly charge high rates for this dangerous task.[97]

Carbon farming applications. In West Africa, oil bean is an important component of indigenous agroforestry systems, often planted at the edge of fields for the food and fertility benefits of this nitrogen-fixing legume. Some individual African oil bean trees drop their leaves in the rainy season like apple-ring acacia, making this bean crop a potential overstory candidate for evergreen agriculture.[98] Oil bean fixes nitrogen.[99]

PERENNIALIZING THE SOYBEAN

Annual soybeans (*Glycine max*) are a top global crop; perennializing them would be a major carbon farming coup. Their perennial relatives are distantly related, and sophisticated techniques are required to produce viable offspring. With that said, some researchers at the University of Illinois have had success in crossing annual soybeans (*G. max*) with an Australian perennial (*G. tomentella*). Their goal is not a perennial soybean, but adding disease resistance to annual soybeans. Thanks to their efforts, however, we have some techniques for working toward this intriguing possibility.[100]

Crop development. African oil bean has been recommended for domestication. Improving methods to safely harvest the beans is a priority.[101] For example, breeders might select dwarf forms, or others that drop their pods when ripe.

Seeds

This category includes many promising species under development for cold climates, including some for extremely dry and cold conditions. These include buffalo gourds and perennial sunflowers. Others are already regional crops in the dry tropics, like perennial egusi. Of course the standout species is from humid tropical regions, the oyster nut, which has very high yields of protein- and oil-rich seeds.

STAPLE-SEEDED PERENNIAL CUCURBITS (*ACANTHOSICYOS, CITRULLUS, CUCURBITA,* AND *TELFAIRIA* SPP.)

Probably everyone on the planet is familiar with cucurbits, which include squashes, cucumbers, and melons. Fewer people know how important the high-protein, high-oil seeds are in many traditional diets. Among this group of life-sustaining cucurbit seeds are some interesting perennials that, while somewhat obscure, can compete with some of the world's finest crops, often in extremely dry conditions and poor soils. Annual egusi (melonseeds) are a global crop, grown on 906,000 hectares (2.2 million acres). Global yield is 788,000 tons, worth $1.1 billion US. Average global yield of annual egusi is 0.9 t/ha fresh weight.[102]

OYSTER NUT (*TELFAIRIA PEDATA*)

Oyster nut is an East African perennial cucurbitaceous climbing vine. It is a regional crop for humid tropical lowlands and highlands. This cucurbit seed, little studied or known outside its native range, is a high-yielding protein-oil crop. Oyster nut is native to Central and West Africa and prefers humid lowlands and highlands, though it yields less at higher elevations.[103]

Uses. Oyster nut vines produce large fruits up to 15kg (33 pounds) full of edible nuts. Oyster nuts taste excellent and can be stored up to eight years. The seeds are pressed for oil.[104] The seeds are 27 percent protein and 66 percent fat.[105]

Yield. Seed yield is high, 3 to 7 tons per hectare, or 1.0 to 2.4 t/ha of dry, deshelled nuts. Yields are lower at high elevations.[106]

Carbon farming applications. Oyster nut is grown in sophisticated agroforestry systems in East Africa alongside bananas and coffee.[107]

Other issues. Oyster nuts can be very aggressive outside of their native range.[108]

PERENNIAL EGUSI (*CITRULLUS COLOCYNTHIS*)

Melon and squash seeds are important staple foods in many parts of the world. Perennial egusi, or colocynth, is an herbaceous vine that resembles its annual cousin the watermelon. It is tolerant of extremely dry conditions, even growing throughout the Sahara desert. Its native range runs from North Africa through West Asia to India and Pakistan. Perennial egusi grows in arid and semi-arid tropical and subtropical lowlands. In some regions it is cultivated and in others wild-harvested. Improved varieties of this ancient crop have been selected.[109] It is grown on 361,000 hectares (892,000 acres) in Nigeria alone.[110]

Uses. The seeds of perennial egusi are 47 to 53 percent oil and 23 to 28 percent protein. The seeds are eaten and pressed for oil. Currently there is some interest in it as a biodiesel feedstock.[111]

Yield. Yields as high as 6.7 t/ha (5.4 dry, clean seed) are reported.[112]

Carbon farming applications. Perennial egusi is intercropped with annual staple crops such as maize, cassava, and yam.[113] It is also used for sand dune stabilization.[114]

Crop development. Typically the fruit flesh is bitter, but some of the West Asian subspecies *insipidus* lack bitterness and may be the result of crosses with annual watermelon.[115] Perennial egusi has been crossed successfully with annual watermelons for the purpose of conferring disease resistance to the popular annual fruit crop. This would imply the

TABLE 15.4. High Protein-Oil Seeds

Latin Name	Common Name	Origin	Climate	Cultivation
Abelmoschus esculentus	Perennial okra	Africa	Tropical lowlands, semi-arid to humid	Regional as perennial, experimental as protein-oil seed
Acanthosicyos horridus	Nara melon	Africa	Tropics and subtropics, arid	Wild staple
Citrullus colocynthis	Perennial egusi	Africa through South Asia	Subtropics and tropics, arid to semi-arid	Regional
Crescentia alata	Jícaro	Mesoamerica	Semi-arid tropics	New crop
Crescentia cujete	Calabash tree	Mesoamerica	Semi-arid tropics	New crop
Cucurbita digitata	Coyote gourd	Southwestern North America, Mesoamerica	Subtropical, arid to semi-arid	New crop
Cucurbita ficifolia	Figleaf gourd	Mesoamerica	Tropical highlands, humid	Regional
Cucurbita foetidissima	Buffalo gourd	Western North America	Cold temperate to subtropical, arid to semi-arid	New crop
Helianthus hybrids	Perennial sunflower	Hybrid	Boreal to warm temperate, semi-arid to humid	Perennials under development
Helianthus cusickii	Cusick's sunflower	North America	Cold to warm temperate, arid to semi-arid	Under development
Helianthus maximiliani	Maximilian sunflower	North America	Boreal to warm temperate, semi-arid to humid	Under development
Helianthus pauciflorus	Stiff sunflower	North America	Boreal to warm temperate, semi-arid to humid	Under development
Helianthus tuberosus	Sunchoke	North America	Boreal to subtropical, tropical highlands; semi-arid to humid	Under development as seed crop
Silphium integrifolium	Prairie rosinweed	North America	Boreal to warm temperate, semi-arid to humid	Under development
Telfairia pedata	Oyster nut	Africa	Humid tropical lowlands and highlands	Regional

FIGURE 15.12. Perennial egusi, a staple seed crop of tropical and subtropical deserts. Photograph courtesy of Philippe Birnbaum, http://www.africanplants.senckenberg.de.

potential of developing a true perennial watermelon, which, while not a staple crop, would be a lovely addition to the world's crops. However, the flesh of *C. colocynthis* can be very toxic so breeders should proceed with caution.

BUFFALO GOURD (*CUCURBITA FOETIDISSIMA*)

Here is a perennial staple for that most difficult climate—the cold drylands. Buffalo gourd is native to western North America and Mexico. It is tolerant of both intense cold (–29°C/–20°F) and heat. I have overwintered one for many years in my snowy Massachusetts garden (–23°C/–9°F). It is also suited to arid tropical highlands. The related Calabacilla or coyote gourd (*C. digitata*) is similar but grows in warmer deserts and arid lowlands in the southwestern United States and northern Mexico. The plants can live 40 years.[116]

FIGURE 15.13. Buffalo gourd, a protein-oil crop for dry, cold regions. Photograph courtesy of Thure Johnson/Creative Commons 2.0.

FIGURE 15.14. Nara melon in the Namib desert. Photograph courtesy of Sean Spender.

Uses. Though the fruit is hard and bitter, the seeds are edible, much like cultivated pumpkin seeds. Protein content is 30 to 65 percent and oil 34 percent.[117] Brad Lancaster of Desert Harvesters reports that the seeds are soaked in water before eating to remove bitterness.[118]

Yield. Seed yields can reach 2.5 to 3 t/ha, or 1.7 to 2.0 dry, clean seed.[119] Buffalo gourd yields are competitive with sunflowers and peanuts when all are grown on barren land.[120]

Harvest and processing. Buffalo gourd can be harvested mechanically.[121]

Crop development. Buffalo gourd domestication and cultivation efforts have been under way for several decades. There are two separate lines under development—an annual type for root starch and a perennial line for seed oil.[122]

NARA (*ACANTHOSICYOS HORRIDUS*)

Nara is a spiny cucurbit shrub that grows on sand dunes in the intensely dry Namib desert of Namibia. Its distribution is currently limited to coastal Namibia where some subsurface water is available. In this region it receives 100 millimeters (4 inches) of rainfall with

additional ocean fog. The seeds are traded as far as South Africa.[123] The plants can spread widely, sometimes covering 1,500 square meters (0.4 acre). Nara plants can live over 100 years.[124]

Uses. The fruits are the size of small melons and have been eaten for at least 8,000 years. They have been compared favorably with macadamias and are also used as the source of an edible oil. These nuts are known as botterpitte or "butter pips" for their rich, buttery flavor.[125] The seeds are 30 percent protein and 57 percent oil. The fruit pulp is also eaten fresh and dried. Before the arrival of maize from the Americas nara seeds were an important staple.[126]

Yield. Yield information per hectare on nara is lacking. A single bush can produce 2 to 100 kg (4 to 220 pounds) of fresh seed.[127]

Harvest and processing. The fruit is dried into a leather, and the edible seeds are dried and stored.[128]

Crop development. Attempts to cultivate nara outside of its native region have mostly failed thus far. There are some other similar habitats, such as the Chilean Atacama desert, where it might be tried.[129] It might be crossed with the much more widely adapted Gemsbok cucumber (*A. naudinianus*).[130]

PERENNIAL SUNFLOWER (*HELIANTHUS* SPP.)

The annual sunflower was domesticated in the central United States thousands of years ago and has been the most successful of the suite of crops known as the Eastern Agricultural Complex. Annual sunflowers (*Helianthus annuus*) are a top world oilseed crop. Sunflowers are suited to cold and semi-arid regions.[131] There are many perennial sunflowers that also have protein- and oil-rich seeds, and work is under way to develop perennial edible sunflowers with competitive yields.[132] Annual sunflowers are global crops, with 24.8 million hectares (61.3 million acres) producing 37.4 million tons of seed, valued at $20 billion US.[133]

FIGURE 15.15. Land Institute breeder David Van Tassel with perennial sunflowers. Photograph courtesy of Scott Seirer.

Uses. The seeds are a popular snack and a historic staple food. Annual sunflower seeds are 23 to 24 percent protein and 47 to 50 percent oil.[134]

Yield. Perennial seed sunflowers are still under development. Annual sunflowers can yield as high as 5.7 t/ha of dry, clean seed.[135]

Carbon farming applications. The Land Institute envisions this crop as part of perennial polyculture systems.[136]

Crop development. The Land Institute began work in the 1980s to develop perennial oilseed sunflowers. There is clearly a need for perennial oil crops for cold, even boreal climates. Sunflower is an excellent candidate. Today the Land Institute has been joined by university efforts in Wisconsin, Minnesota, and Manitoba.[137]

David Van Tassel and Sheila Cox of the Land Institute have been following several strategies to develop perennial grain sunflowers. These include developing "amphiploid" hybrids with the full chromosome set of both species; moving perennial genes into annual sunflower; moving annual crop sunflower genes into perennial species; and domesticating wild perennial sunflowers. Domestication of the wild species may be the most successful approach, since many of the hybrids demonstrate low fertility. One challenge is that some of the wild species have been aggressively spreading tubers. Although this is hard to manage in the fields, it does offer the possibility of planting fields of vegetatively propagated elite varieties using mechanized potato planters.[138]

Several perennial sunflowers are being worked with.[139] Maximilian sunflower (*H. maximiliani*), Jerusalem artichoke (*H. tuberosus*), and stiff sunflower (*H. pauciflorus*) are prairie natives that can handle boreal through warm temperate semi-arid conditions. Cusick's sunflower (*H. cusickii*) is a species of the arid western mountains of North America. It may be relatively less tolerant of boreal conditions, but it grows in terrible soils. The related prairie rosinweed (*Silphium integrifolium*) is also under development at the Land Institute.[140] One especially impressive individual rosinweed plant in their trials yielded the equivalent of 1.1 t/ha of seed.[141]

PERENNIAL OKRA (*ABELMOSCHUS ESCULENTUS*)

Okra is a global annual pod vegetable of African origin. Okra seeds, if allowed to ripen past the vegetable stage, are high in oil and protein. Perennial forms up to 5 meters (16 feet) high can be found in the African tropics and could be the basis of a multipurpose perennial

Figure 15.16. Okra seeds are a potential new oil-protein crop. Photograph courtesy of Planteur/Creative Commons 3.0.

soybean analog. They are coppiced annually.[142] A hypothetical staple-seeded perennial okra's range would be tropical and perhaps subtropical lowlands, semi-arid to humid. Unfortunately the seeds may contain gossypol (a toxin found in cotton seeds). This issue needs to be resolved before adoption of annual or perennial okra seeds as staple crops or a technique to remove it must be developed, as has been developed for cottonseed oil.[143] If not, it could be used as an industrial oil.

Uses. Okra seeds contain 18 to 27 percent protein.[144] Oil content is about 20 percent.[145] The oil is of high quality, similar to olive oil.[146] A team led by the late plant genius Dr. Frank Martin made a delicious tofu from okra seeds in the 1970s. This tofu was 43 percent oil and 53 percent protein.[147] Okra has many industrial uses. It is a good fiber crop, for example, and reportedly makes a good combustible fuel. Its industrial uses parallel those of its relative kenaf, an important feedstock for paper, fiberboard, and other fiber and biomass products.[148]

Yield. Seed yields of annual okras range from 0.5 to 1t/ ha fresh weight.[149] Annual okra seeds produced 0.6 t/ha of oil and 0.6 t/ha of protein in a trial in Puerto Rico.[150] Pod and seed yields of perennial types are not published.

Crop development. Okra has remarkable untapped potential as a staple food and industrial crop. Research and some selection for non-toxic seed

forms may be necessary. To add cold hardiness, perennial seed okras might be crossed with hardier *Abelmoschus* and *Hibiscus* species.

Fruits

Few fruits are high in both oil and protein. I am aware of only two: butterfruit and akee.

SAFOU, BUTTERFRUIT (*DACRYODES EDULIS*)

Safou is a popular fruit in West and Central Africa, but few people outside the region have ever heard of it. It is cultivated in "most or possibly all West and Central African villages." Safou grows in humid and semi-arid climates with rainfall as low as 600 millimeters (24 inches).[151] It prefers to grow in the shade.[152]

Uses. Safou has edible fruit, described as sour and slippery, more a vegetable than a fruit. Roger Leakey describes the flesh as a cross between olive and avocado.[153] An edible oil is pressed from the seed; the skin and flesh can be a variety of colors. The dry weight is 30 to 60 percent oil and 14 to 26 percent protein.[154]

Yield. Yields of the fresh fruit can be as high as 10 t/ ha.[155] This translates to about 4.5 t/ha of dry, deseeded flesh. Male and female flowers are on separate trees. Not all trees bear annually.[156]

Harvest and processing. The fruits must be cooked for a few minutes to soften before eating. The pulp and seeds are also pressed for oil.

Carbon farming applications. Safou is very common in homegardens and other indigenous agroforestry systems, including as a shade tree for coffee and cacao. It is itself a shade crop as well.[157]

Crop development. Fruit size and flesh-to-seed ratio vary widely. Indigenous farmers have selected for fruit size and depth of flesh, with a 67 percent improvement.[158] Until recently there was little or no research attention on this important African fruit crop. In the late 1990s the African Safou Network was formed to research, breed, and promote the

FIGURE 15.17. A Cameroonian child with a buttery safou fruit. Photograph courtesy of the World Agroforestry Centre.

crop. Today there remains much to learn about this important tree staple with global promise.[159]

AKEE (*BLIGHIA SAPIDA*)

Akee is popular in West Africa and the Caribbean. In fact, it is the national fruit of Jamaica. In West Africa, where it is native, it is one of several similar species. It was brought to the Americas during the slave trade and became quite popular in some countries. It has naturalized in the Americas to a certain extent.[160] In Jamaica, annual trade in the fruit was valued at $400 million US in 2005. Jamaicans treat the fruit, high in protein and fat, something like eggs, often cooking it

with salt cod.[161] The Spanish name *ceniza* translates as "vegetable brains."

Akee, however, has a terrible flaw. The unripe fruit is fatally poisonous, particularly to children. Unripe akee arils (flesh) are considered a highly hazardous cell poison and their sale is banned in the United States.[162] Typically, by selecting perfectly ripe fruits and avoiding consumption of the seeds and skin, there is no hazard. However, economic botanist Julia Morton reports that during Jamaican winters, some arils contain small, undeveloped seeds, which can be overlooked. (Economic botanists specialize in the study of useful plants.) In this case, even if the aril is ripe, akee can cause illness or death.[163]

This staple fruit remains popular in West Africa, where it is native. For example, in a recent survey in Benin—where it is a popular tree in parkland agroforestry systems—akee was scored as a high priority for domestication. The same survey looked at uses of and attitudes about akee. No one mentioned its toxicity as a concern. Perhaps it has been used and valued there for so long that there is little possibility for mistake.[164] If not for its toxicity, I would recommend akee to everyone in the humid tropical lowlands as a must-have plant. Efforts to promote akee growing in new areas would require extensive education efforts.

Uses. The arils (the fleshy fruit part) are 9 to 24 percent protein and 19 to 45 percent fat, an almost unheard-of level of protein for a fruit, and twice the fat of an avocado.[165] In fact these levels are closer to a bacon cheeseburger than to most fruits. Akee is also quite the multipurpose tree. The fruit capsule (shell) is full of saponins, which are used to make an important commodity soap in Benin. This is not surprising since akee is related to soapnuts. Additionally, an inedible oil can be pressed from the seeds and used for fuel and other industrial purposes. The

TABLE 15.5. High Protein-Oil Fruits

Latin Name	Common Name	Origin	Climate	Cultivation
Blighia sapida	Akee	Africa	Tropical lowlands, semi-arid to humid	Minor global crop
Dacryodes edulis	Safou	Africa	Tropics and subtropics, semi-arid to humid	Regional crop

ripe fruit and leaves are used as fodder. Medicine and pest repellents are also produced from akee.[166]

Yield. Little yield information is available. Trees in Florida yield 45kg to 68kg (100 to 150 pounds) of fruit annually, with about a quarter of that being edible flesh.[167] If we were to extrapolate that at a conservative 10-meter (33-foot) spacing (100 trees/ha), it would yield 1.2 to 1.7 tons of edible flesh per hectare each year (0.5 to 0.7 dry weight)—a fairly respectable amount for a tree that has received little formal breeding attention.

Harvest and processing. As mentioned, special care must be taken to ensure proper ripeness to avoid toxic, even fatal, reactions.

Carbon farming applications. Akee is grown in African parkland systems.[168] I've seen it in homegardens on several Caribbean islands.

Crop development. I am not aware of any efforts to select or breed less toxic akee trees. I'd recommend efforts to look for differences in toxicity in akee and its wild relatives, with the long-term goal of breeding

FIGURE 15.18. Akee fruits are high in fat and protein but are deadly if eaten before they are ripe. This akee tree is found at the Fruit and Spice Park in Florida.

non-toxic or at least non-fatal forms. At a minimum, varieties that do not produce undeveloped seeds in ripe arils would be a good place to start. This would be a major benefit to millions of people in the areas where akee is grown—and beyond.

CHAPTER SIXTEEN

Edible Oil Crops

Edible oil crops are those that have 0 to 15 percent protein and more than 16 percent oil. Some of these crops are grown for oil alone, while others have significant protein. (When searching for oil crops, don't neglect the species in chapter 15; many of those crops are also pressed for edible oils.) These edible oil crops include those that are eaten fresh and those that are pressed for oil. Some are used in both ways. Three of the top five most widely grown perennial crops in the world are oil crops: oil palm, coconut, and olive.[1] Oil crops are grown on 281 million hectares (694 million acres) of world farmland, representing 26 percent of staple crop farmland and 18 percent of all cropland.[2] Calculating precise numbers is difficult,

as many oil crops serve a dual purpose. For example, maize, soybeans, and peanuts are all used for oil and for other purposes.

Global annual oil crops include rapeseed (including canola), sesame, linseed, safflower, and mustard. Much of the world's sunflower, peanut, and soybean crops are grown for oil as well. Global perennials include oil palm, coconut, olive, and avocado. Perennial oil crops include nuts, fruits, and pressed oils (from fruits and seeds that are not themselves eaten). Many perennial oil crops yield far better than the best annuals. Today 41 percent of world vegetable oil is already from perennial crops, making this likely the most perennialized of any staple food category.[3]

TABLE 16.1. Edible Oil Crop Yields Compared

Latin Name	Common Name	Climate	Avg. Global Yield[a]	Yield t/ha	Product
Elaeis guineensis	African oil palm	Humid tropical lowlands		13.5–34.2[b]	Fruit bunches
Persea americana	Avocado	Tropical and subtropical lowlands and highlands, Mediterranean, humid	1.7	0.9–6.0[c]	Fruit
Olea europaea	Olive	Mediterranean	0.6	0.4–3.6[d]	Fruit
Cocos nucifera	Coconut palm	Humid tropics	2.6	1.1–3.2[e]	Copra
Elaeagnus rhamnoides	Sea buckthorn	Boreal to warm temperate, semi-arid to humid		2.0–2.8[f]	Fruit
Macadamia ternifolia	Macadamia	Humid tropics and subtropics		2.2[g]	Nuts
Carya illinoinensis	Pecan	Cold temperate to subtropical, tropical highlands, humid		1.5[h]	Nuts
Caryocar villosum	Piquia	Humid tropical lowlands		0.9[i]	Edible portion

Note: These fruits and nuts are eaten directly but can also be pressed for oil. All yields calculated based on dry weight of the edible portion. Crops are arranged in descending order of yield.

[a] FAO Statistics Division online.
[b] Duke, *Handbook of Nuts*, 150.
[c] Plant Resources of Southeast Asia online database.
[d] Plant Resources of Tropical Africa online database.
[e] Ohler, *Modern Coconut Management*, 6.
[f] Janick and Paull, *The Encyclopedia of Fruits and Nuts*, 342.
[g] Wilkinson, *Nut Grower's Guide*, 145; Elevitch, *Specialty Crops for Pacific Islands*, 308.
[h] Wilkinson, *Nut Grower's Guide*, 171.
[i] Shanley et al., *Fruit Trees and Useful Plants in Amazonian Life*, 118.

TABLE 16.2. Edible Oil Yields Compared

Latin Name	Common Name	Climate	Oil yield t/ha
Elaeis guineensis	African oil palm	Humid tropics	5–8.7[a]
Persea americana	Avocado	Tropical and subtropical lowlands and highlands, Mediterranean, humid	0.2–6.6[b]
Cocos nucifera	Coconut	Humid tropics	2–4[c]
Dacryodes edulis	Safou, butterfruit	Humid to semi-arid tropical lowlands	2–4[d]
Bactris gasipaes	Peach palm	Humid tropical lowlands	2–3[e]
Caryodendron orinocense	Inche	Humid tropical lowlands	0.8–3 extrapolated[f]
Sapium sebiferum	Chinese tallow tree	Warm temperate to subtropical, semi-arid to humid	2.8 of edible tallow, also 2.6 of inedible oil from seed[g]
*Arachis hypogaea	Peanut	Worldwide	1.7[h]
*Carthamus tinctorius	Safflower	Annual	1.6[i]
Macadamia integrifolia	Macadamia	Tropics, subtropics, tropical highlands, humid	1.5[j]
*Helianthus annuus	Sunflower	Worldwide	0.2–1.5[i]
*Brassica napus	Rapeseed	Cold temperate	0.5–1[i]
Elaeagnus rhamnoides	Sea buckthorn	Boreal to warm temperate, semi-arid to humid	0.6–0.8 extrapolated[k]
Cynara cardunculus	Cardoon	Mediterranean, warm temperate, semi-arid to humid	0.5–0.6 extrapolated[l]
Theobroma cacao	Cacao	Humid tropical lowlands	0.4–0.6[m]
Attalea speciosa	Babassu palm	Humid tropics	0.5[n]
*Papaver somniferum	Poppy	Cold temperate, Mediterranean	0.5[i]
*Zea mays	Maize	Annual	0.02–0.5[o]
Olea europaea	Olive	Mediterranean	0.3–0.4[h]
Xanthoceras sorbifolium	Yellowhorn	Cold to warm temperate, semi-arid to humid	0.3–0.4[p]

Note: Here we compare perennial versus annual oil sources by examining the reported yields of oil. Annual comparison crops are marked with an asterisk (*). Crops are arranged in descending order of yield. This table includes reported oil yields from protein-oil crops in chapter 15 and the partially edible Chinese tallow tree (*Sapium sebiferum*) from chapter 21.

[a] Miccolis et al., "Oil palm and agroforestry systems."
[b] Human, "Oil as a byproduct of the avocado," 5.
[c] Ohler, *Modern Coconut Management*, 6.
[d] El Bassam, *Handbook of Bioenergy Crops*, 311.
[e] Clement, "The potential use of the pejibaye palm in agroforestry systems," 204.
[f] Duke, *Handbook of Nuts*, 78.
[g] Ibid, 264.
[h] Plant Resources of Tropical Africa online.

[i] Duke, *Handbook of Energy Crops*.
[j] Wilkinson, *Nut Grower's Guide*, 145; Elevitch, *Specialty Crops for Pacific Islands*, 308; Duke, *Handbook of Nuts*, 209.
[k] Janick and Paull, *The Encyclopedia of Fruits and Nuts*, 342.
[l] Foti et al., "Possible alternative utilization of *Cynara* spp.," 219–28.
[m] Janick and Paull, *The Encyclopedia of Fruits and Nuts*, 885.
[n] El Bassam, *Handbook of Bioenergy Crops*, 113.
[o] National Corn Growers Association, *2012 World of Corn Statistics Book*.
[p] World Agroforestry Centre, *Participatory Agroforestry Development in DPR Korea*, 141.

Oily Nuts

These species are eaten as nuts, but many (or even most) are also pressed for oil. (Seeds that are only pressed for oil are featured under Pressed Oils, on page 216.) Macadamia, pecan, and coconut are examples. Not all oils are interchangeable. You wouldn't want to replace olive oil with coconut oil or cocoa butter in a Mediterranean salad, for example. Nonetheless there is a great diversity of edible oils, a diversity that most of us barely sample.

COCONUT (*COCOS NUCIFERA*)

Coconuts turn up on any short list of the world's most useful plants. They originated in Southeast Asia and Melanesia and were spread around the Old World coastlines (and perhaps to the Pacific coast of South America) prehistorically by people and ocean currents. Today they are an important global crop in humid and semi-arid lowland tropics everywhere. Coconuts are highly tolerant of salt and wind.[4] This global crop is grown on 12.1 million hectares (29.9 million acres).

TABLE 16.3. Nuts High in Oil

Latin Name	Common Name	Origin	Climate	Cultivation
Acrocomia aculeata	Macauba palm	Tropical Americas	Humid tropical lowlands	Minor
Afraegle paniculata	Nigerian powder-flask fruit	West Africa	Humid tropical lowlands	Regional
Allanblackia floribunda	Allanblackia	West Africa	Humid tropical lowlands	New crop
Attalea maripa	Maripa palm	Northern South America	Humid tropical lowlands	Regional
Attalea speciosa	Babassu palm	Northern South America	Tropical humid lowlands	Regional
Borassus flabellifer	Palmyra palm	Asia	Tropics and subtropics, semi-arid to humid	Regional
Canarium indicum	Ngala	New Guinea, West Pacific	Tropical humid lowlands	Minor global
Canarium luzonicum	Elmi canary tree	Philippines	Tropical humid lowlands	Regional
Canarium ovatum	Pili nut	Philippines	Tropical humid lowlands	Minor global
Canarium schweinfurtii	African elemi	Tropical Africa	Tropical humid lowlands	Regional
Carya hybrids	Hybrid and neohybrid hickories	Hybrid origin	Variable	Regional
Carya cathayensis	Chinese hickory	China	Cold to warm temperate, humid	Regional
Carya illinoinensis	Pecan	Eastern North America, Mesoamerica	Cold temperate to subtropical, highlands, humid	Minor global
Carya laciniosa	Shellbark hickory	Eastern North America	Cold to warm temperate, humid	Regional
Carya ovata	Shagbark hickory	North America, Mesoamerica	Cold temperate to subtropical, tropical highlands, humid	Regional
Caryocar brasiliense	Pequi	Brazil, Bolivia, Paraguay	Semi-arid tropical lowlands	Regional crop
Caryocar nuciferum	Souari-nut	Northern South America	Tropical humid lowlands	Regional
Caryocar villosum	Piquia	Brazil, French Guiana	Tropical humid lowlands	Regional
Cocos nucifera	Coconut palm	Asia	Tropical lowlands, semi-arid to humid	Global
Elaeis guineensis	African oil palm	West Africa	Tropical humid lowlands	Global
Gevuina avellana	Chilean hazelnut	Chile	Warm temperate, Mediterranean	Regional
Irvingia gabonensis	Dika nut	West Africa	Tropical humid lowlands	Regional
Jubaea chilensis	Chilean wine palm	Chile	Warm temperate to tropical, Mediterranean, semi-arid	Regional
Macadamia hildebrandii	Celebes nut	Sulawesi	Humid lowland tropics	Regional
Macadamia integrifolia	Macadamia	Eastern Australia	Semi-arid to humid tropical lowlands	Global
Macadamia ternifolia	Macadamia nut	Eastern Australia	Subtropical, tropical highlands, semi-arid to humid	Minor global
Madhuca longifolia	Butter tree	India	Subtropics, semi-arid	Regional
Mesua ferrea	Surli nut	South and Southeast Asia	Ttropical humid lowlands	Regional
Pandanus julianettii	Karuka	New Guinea	Tropical highlands, humid	Regional
Parajubaea cocoides	Quito palm	Andes	Tropical highlands, semi-arid to humid	Regional
Parinari curatellifolia	Mabo cork-tree	Central Africa	Semi-arid tropics	Regional
Raphia farinifera	Madagascar raffia palm	Africa, South Asia, Madagascar	Humid tropics, subtropics	Fiber crop with edible oil
Raphia hookeri	Ivory Coast raphia palm	West Africa	Tropical humid lowlands	Fiber crop with edible oil
Syagrus coronata	Ouricury palm	Brazil	Humid tropics and subtropics	Regional
Vitellaria paradoxa	Shea butter tree	Africa	Tropical lowlands, semi-arid to humid	Minor global

PRODUCTIVE PALMS

The palm family (Arecaceae) is an absolute stand-out in the world of oil crops. Many palms have edible fruits and nuts high in oil. Others have inedible fruits or nuts but are pressed for edible oils. The world's best oil crop is a palm (African oil palm), as is the ubiquitous and multipurpose coconut. These are just the tip of the iceberg.

The diversity and abundance of edible oily palms is so great that it is not possible to comprehensively list even just the cultivated species here. These oil crop genera include *Acrocomia*, *Attalea*, *Astrocaryum*, *Bactris*, *Oenocarpus*, and *Syagrus*.[5] Many more uncultivated, productive palms await cultivation and domestication.

FIGURE 16.1. Coconuts, a global multipurpose oil crop, in Florida. Photograph courtesy of Marikler Girón Toensmeier.

Total world production is estimated at 60 million tons, worth $9.7 billion US.[6]

Uses. Coconut fruits are useful in many stages. Young coconuts provide a sterile source of water. Coconut water is a health food darling today but is also an important source of drinking water after disasters. It has even been used as an emergency source of intravenous fluids.[7] This is followed by the "jelly" stage when the young fruit is sweet and tender. Finally the coconuts ripen and the fruit dries into copra, an edible nutmeat. Copra is 41 to 75 percent fat and 4.5 percent protein.[8] Coconut oil is pressed from copra. Copra is also made into coconut milk and cream, as well as shredded coconut and many other products. The flower stalks can be tapped for sugar and toddy.[9]

Coconut husks are the world's most important perennial source of fiber. The husks are a valuable biomass resource, used for mulch, potting soil, fuel, and activated carbon filters. The leaves are an excellent thatch.[10]

Yield. A hectare of coconuts produces 2 to 6 tons of copra, or 1.1 to 3.2 t/ha dry weight. Oil yields are reported of up to 4 t/ha.[11] Average global yield is 2.9 t/ha.[12] When tapped for sugar, coconuts provide 12 to 23 t/ha/yr of it.[13] Taller varieties bear in five years, while dwarf types bear more but smaller nuts in three years. Coconuts can live to be 100 years old.[14]

Harvest and processing. Most varieties fall to the ground when they're ripe and are picked up by hand. Today 95 percent of coconut production occurs on small farms.[15] I speculate that mechanical harvesters are used on larger plantations.

Carbon farming applications. This crop is common in agroforestry systems wherever it can be grown. Coconuts are frequently included in homegardens around the world.[16] They are used to provide shade to crops like cacao and coffee. Groves that are more than 40 years of age start to cast less shade and are then used as overstory for crops such as coppiced firewood legumes.[17] Coconuts are good windbreaks and are used for coastal protection. They are used at times as a living trellis although reportedly not a perfect one. Copra is used to feed livestock, and coconut–cattle silvopasture is widespread in the Pacific and elsewhere.[18]

The percentage of intercropped coconut production is 85 percent in Tonga, 70 percent in Papua New Guinea, and 43 percent in the Philippines.[19] Annual crops are intercropped, with a mere 8 percent yield reduction in maize under coconuts in one study, although impacts were stronger on other grains, legumes, and sweet potatoes.[20] Other crops, such as ramie and chili peppers, increase yields when grown under coconuts. Coconut yields themselves can increase in polycultures; the rich leaf litter of cacao boosts coconut production by 95 percent![21] Coconut is grown in multistory plantation systems and commonly managed as silvopasture with cattle beneath.[22] *Modern Coconut Management* summarizes a passage on intercropping by noting, "Most of the annuals, biennials, and perennials tried in coconut-based multiple cropping systems are compatible with coconuts. Not only were most of them found beneficial to coconut productivity because of consequential site enrichment, but also their productivity was often comparable to that when grown in the open."[23]

Cattle are used to mow the understory of coconut plantings. In traditional systems the cattle are something of a minor bonus, but in larger plantations they are an important secondary product or even the main one.[24] Coconut yields often improve with grazed understories, partly because dropped nuts are easier to find in close-grazed undergrowth. These coconut–cattle silvopastures are widespread, especially in Asia.[25]

Related species. The related palms *Parajubaea cocoides* and *Jubaea chilensis* have similar nuts, although they are much smaller. These species are suited to Mediterranean and tropical highland climates, extending the range of coconut-like fruits to an even wider portion of the globe.

PECANS AND HICKORIES (*CARYA* SPP.)

Pecans and hickories are in the genus *Carya* and are closely related to walnuts. There are many species native to the Americas, as well as some in eastern Asia. Generally these species are adapted to cold and warm temperate humid climates, but some *Carya* species, including pecan and shagbark hickory, extend to the subtropical highlands of central Mexico.[26] The pecan is a global crop, with the other species serving as regional crops in North America and Asia. Historically, *Carya*

FIGURE 16.2. Part of a 10,000-tree pecan orchard, Texas. Photograph courtesy of US Department of Agriculture, Natural Resources Conservation Service.

species were cultivated in what is now the United States prior to colonization.[27]

Uses. Pecan nuts are 73 to 75 percent fat and 9 percent protein.[28] Other species are nutritionally similar. They are eaten fresh, in confections, and used to produce milks and oils.

Yield. Mature, grafted pecans can yield 3 t/ha of nuts, of which 1.5 t/ha is the dry, edible portion.[29] Most varieties bear on alternate years. Newly developed pecan varieties can yield in as little as three years from planting.[30] Other hickories can be expected to be somewhat inferior in yield, alternate bearing, and precocity.

Harvest and processing. Commercial pecans are harvested with tree shakers and sweepers or other orchard floor harvesters.[31]

Carbon farming applications. Pecans are widely used in alley cropping in the southern United States.[32] Typically annual crops are grown between the pecan rows until the pecans begin to shade them out. Pecans are intercropped with annual crops in northern China.[33] *Carya* hybrids show good promise as woody agriculture species in initial trials at the Badgersett Research Corporation test farm.[34]

Crop development. Although pecans have had a fair amount of attention, the other species of *Carya* are in need of breeding work.

Species descriptions.

Shellbark hickory (*C. laciniosa*) is native to eastern North America, where it is a rather minor regional crop. The nuts are among the largest of all hickories.[35]

Chinese hickory (*C. cathayensis*) is native to temperate eastern Asia, where it is a regional crop.[36]

Pecan (*C. illinoinensis*) is native from the central Mexican highlands to the midwestern United States. This global crop is by far the most commercially important *Carya* species, and though the nuts are delicious they're not the best tasting of this genus.[37]

Shagbark hickory (*C. ovata*) is native from the central Mexican highlands to southern Canada. It is a minor regional crop.[38]

Hickory hybrids have been developed, often with goals of large nuts, thin shells, and excellent flavor. Badgersett Research Corporation is developing neohybrid hickories.

MACADAMIA (*MACADAMIA* SPP.)

Macadamias have become an important global crop over the last century. The two primary species are smooth-shell (*M. integrifolia*) and rough-shell (*M. tetraphylla*). Both hail from eastern Australia; rough-shell tolerates somewhat cooler weather. Both are tropical and damaged by frost, and both require humid but not terribly wet weather.[39] Macadamias will yield in tropical highlands as long as they are frost-free. A rain forest species for truly wet lowlands is *M. hildebrandii*, a regional crop from Sulawesi.[40] World production is around 93 billion tons of wet, in-shell nuts.[41]

FIGURE 16.3. Macadamias bearing well at Las Cañadas in Mexico.

Uses. Like many other nuts macadamias are eaten fresh, roasted, and in baked goods. They are 7 to 8 percent protein and 71 to 76 percent fat.[42]

Yield. Yields are high, up to 6t/ha in-shell for elite varieties.[43] The kernels are 38 percent of the in-shell weight, meaning kernel yield is up to 2.2 t/ha dry weight.[44] Because 70 percent of the weight of nuts can be extracted as oil, this permits us to extrapolate an oil yield of 1.5 t/ha.[45]

Harvest and processing. Macadamias can be harvested mechanically. Typically machinery is used to pick up nuts off the orchard floor, but tree shakers are challenging to implement successfully: Unripe nuts will drop, and macadamias ripen over a long period of time.[46]

Carbon farming applications. Macadamias cast dense evergreen shade, somewhat limiting their intercropping options. Young trees are intercropped with shorter-lived sun-loving crops for the first 7 to 10 years. Mature trees are used as shade for coffee. Studies in Hawaii have found that forage peanut (*Arachis galbrata*) is a good understory legume for macadamia.[47]

SHEA (*VITELLARIA PARADOXA*)

Over an area encompassing 1 million square kilometers (247 million acres) of African savanna, shea is the primary, daily source of fat for millions of people.[48]

FIGURE 16.4. Shea fruits ripening on the tree. Photograph courtesy of Karen Hahn-Hadjali, http://www.africanplants.senckenberg.de.

Shea butter, made from the oily kernels, is an important item of international commerce. Nonetheless, shea needs much breeding and processing equipment development to reach its potential. Shea grows in semi-arid to humid tropical savanna, with as little as 400 millimeters (16 inches) of rainfall per year.[49] It would be an interesting crop candidate for the Brazilian *cerrado* savanna and similar climates. Shea nuts are a minor global crop. FAO estimates they are grown on over half a million hectares (1.2 million acres), with global production of 728 thousand tons worth $70 million US.[50]

Uses. Shea "nuts" are encased in a sweet edible fruit. The kernels inside are 31 to 62 percent oil and 7 to 9 percent protein. They can be eaten fresh or roasted. The flowers are also edible.[51] Oil is extracted from the kernels. This "shea butter" is used in many of the ways butter and lard are in temperate countries. It does not fully melt even in tropical heat. Shea butter is an international commodity used in cosmetics and some foods. The oil is also used locally for waterproofing in traditional natural building.[52]

Yield. Shea is slow to bear, taking 12 to 15 years for first fruiting and 30 to 50 to come into full production. Yields are irregular, often only one out of three years for any given tree. For these reasons shea is rarely planted but instead is allowed to remain when new areas are cleared.[53] Average yield according to the FAO is 1.4 t/ha fresh weight.[54]

Harvest and processing. Shea butter processing is mostly done by hand. Although it is an important source of rural income and export revenue, processing of shea butter is an incredibly laborious process, taking eight hours per liter of shea butter produced.[55] There is a great need for development of appropriate, village-scale technology for shea butter production. In some more developed areas, shea is processed with industrial oil presses. Most shea nuts are exported without being processed for foreign oil extraction, resulting in lost opportunity to add value and retain more income.[56]

Carbon farming applications. Shea is an important anchor of the parkland agroforestry system. Shea nut parkland covers an incredible 22.9 million hectares (56.6 million acres) in semi-arid West Africa.[57]

Crop development. Grafting and other vegetative propagation techniques could shorten the time to bearing but require some specialized techniques. There has been a lot of experimentation on shea propagation lately with some promising results.[58] Given that it grows in some of the poorest communities on earth, its predominance in one of the world's largest agroforestry systems, and its impressive qualities for food and industry, investment in development of improved shea varieties and processing equipment should be a high carbon farming priority.

Oily Fruits

These trees and shrubs produce high-oil fruits that are eaten fresh or cooked, avocado probably being the most frequently consumed. Oil can also be pressed from them. There are a number of promising underutilized species in this category. Tropical South America is particularly abundant in oily fruit species. (I wonder what extinct megafauna were once their dispersal partners.) Fifty years ago few people outside of Mesoamerica had ever eaten an avocado, and now this oily fruit is commonplace. Perhaps soon we will enjoy more of these energy-rich fruits.

AFRICAN OIL PALM (*ELAEIS GUINEENSIS*)

African oil palm grows well in tropical humid lowlands. It is native to West and Central Africa. It is a major global crop, producing $45 billion US in 2012 from 17.2 million hectares (42.5 million acres) in production. This area produced 249 million tons of fruit, plus 50 million tons of oil, and an additional 14.6 million tons of palm kernels.[59]

Oil palm yields are so high that as of 2010 it produced 38 percent of the world's vegetable oil on only 5 percent of globe's oil-producing cropland.[60] This shows oil palm's potential for mitigation through intensification—growing more food on the farmland we have, to avoid clearing of additional forest. Unfortunately, this is not the way in which this crop is being planted.

TABLE 16.4. Fruits High in Oil

Latin Name	Common Name	Origin	Climate	Cultivation
Acrocomia aculeata	Macauba palm	Tropical Americas	Humid tropical lowlands	Regional
Attalea maripa	Maripa palm	Northern South America	Humid tropical lowlands	Regional
Borassus aethiopum	African palmyra palm	Tropical Africa	Tropical lowlands, semi-arid to humid	Regional
Canarium schweinfurtii	African elemi	Tropical Africa	Humid tropical lowlands	Regional
Caryocar brasiliense	Pequi	Brazil, Bolivia, Paraguay	Semi-arid tropical lowlands	Regional
Caryocar nuciferum	Souari-nut	South America	Humid tropical lowlands	Regional
Caryocar villosum	Piquia	Brazil, French Guiana	Humid tropical lowlands	Regional
Elaeagnus rhamnoides	Sea buckthorn, seaberry	Eurasia	Boreal to warm temperate, semi-arid to humid	Minor global
Elaeis guineensis	African oil palm	West Africa	Humid tropical lowlands	Global
Gustavia superba	Membrillo	Tropical Americas	Humid tropical lowlands	Regional
Iryanthera laevis	Cumala	South America	Humid tropical lowlands	Regional
Mauritia flexuosa	Buriti palm	Tropical South America	Humid tropical lowlands	Regional
Oenocarpus bataua	Pataua palm	South America	Humid tropical lowlands	New crop
Olea europaea	Olive	Mediterranean	Mediterranean, warm temperate to tropics, semi-arid	Global
Pandanus conoideus	Marita	New Guinea, Moluccas	Humid tropical lowlands, highlands	Regional
Persea americana	Avocado	Mesoamerica	Tropical lowlands to highlands, humid	Global

Sources: Hanelt, *Mansfeld's Encyclopedia of Agricultural and Horticultural Crops*; Janick and Paull, *The Encyclopedia of Fruits and Nuts*

FIGURE 16.5. African oil palm fruit can be roasted and eaten, as well as pressed for oil. Photograph courtesy of Luisa Reynolds.

African oil palm is a great example of how even an extraordinary plant can be terribly misused. The drive for biofuel and cheap oil for processed and fried foods is driving an ecological and social catastrophe. Tens of thousands of hectares of rain forest in Indonesia and elsewhere have been cleared for oil palm monocultures, representing a great loss of stored carbon. The peat soils on which many of these forests grew quickly oxidize their rich stores of carbon. These huge emissions in the name of an alternative to fossil fuels amount to a sad irony. Indigenous people and other small farmers have been displaced to make way for plantations.[61]

None of this is the plant's fault, and it is not deserving of anger. Instead a great respect for its productivity, tempered with full considerations of carbon life-cycle analysis and social and ecological impacts, is what's needed.

Uses. The unrefined oil is red due to its extremely high levels of beta-carotene. The oil is used for cooking. Traditional uses include as an illuminant, soap, and candles. Modern uses include refined vegetable oil and margarine, chemical feedstock, and especially biodiesel. Different kinds of oil are pressed from the fruit and the kernel.[62] The fruits are 23 to 30 percent oil, and the kernels 3 to 5 percent.[63] Roasted African oil palm fruits are eaten in Africa. The edible coconut-flavored nut kernels are popular with children.

It would be interesting to see new breeding lines developed that emphasize the fruit and nut potential of this crop, rather than just using it as a source of pressed oil.[64] The fruits and presscake can be fed to livestock. Male flowers can be tapped for sugar.[65]

Yield. Yields of fruit bunches range from 15 to 38 t/ha or more, 13.5 to 34.2 dry weight.[66] This translates to oil yields of 3 to 5 t/ha/yr or more.[67] In Brazilian polycultures oil palm yields even better, at 6.4 to 8.7 t/ha/yr versus 5 t/ha/yr in adjacent monocultures.[68]

Carbon farming applications. Oil palm polycultures can be highly successful, not only increasing palm yields but also adding products such as timber, fruits, cacao, passion fruit, and black pepper. Nitrogen-fixing trees and legume ground covers are also effective companions.[69]

Other issues. In oil palm we have a brutal but clear case study of how not to implement carbon farming practices. Even an amazing perennial crop can be terribly destructive when greed is the driver and human and environmental costs are ignored. This crop could and should be grown on lands that were degraded, with healthy forests left in place. Intercropping and other techniques can turn it into a benign presence. In chapter 18 we will review the facts: There isn't enough land to meet our energy needs with biofuels; we need to look elsewhere for clean energy. With that said, oil palm could be a fine provider of feedstocks for bioplastics and lubricants on a much-reduced footprint, in addition to providing healthy unrefined high-carotene edible oil.

Related species. The similar American oil palm (*E. oleifera*) is a promising South American species.

AVOCADO (*PERSEA AMERICANA*)

By the time of colonization, cultivation of the avocado had already spread from its southern Mesoamerican center of origin to a region spanning the Rio Grande to the north and northern South America to the south. Since then it has become a global crop and a splendid example of how perennial staple crops can be integrated into the human diet.

There are three "races" of avocado. Guatemalan avocados are suited to tropical and subtropical highlands.

Mexican types are even more cold-tolerant. West Indian avocados thrive in humid tropical lowlands. Many hybrids are cultivated. Among them, there are varieties for tropical and subtropical highlands and lowlands, and even for warmer Mediterranean zones.[70] Avocados are a global crop. On 486,000 hectares (1.2 million acres), 4.3 million tons of fruit are grown, valued at $3.8 billion US.[71]

Uses. The fruit of the avocado is creamy and soft. Like most oily fruits, it is eaten more as a vegetable. The flesh is 5 to 23 percent fat and 1 to 4 percent protein.[72] In Guatemala where the tree is native, construction workers bring avocados and fresh tortillas for lunch. Avocado flesh can also be consumed as a fruit, for example in three-layer smoothies with avocado, papaya, and mango. Avocado flesh is pressed for oil as well. In fact one-third of Brazilian avocados are used for edible and cosmetic oil production.[73]

Yield. Here is a perennial staple crop that has come into its own. Yields are high, ranging from 5 to 32 t/ha fresh, or 0.9 to 6.0 dried edible portion (no skins or pits). Global average yield is 1.7 t/ha dry flesh.[74] Oil extraction is about 90 percent, giving an oil yield range of 0.2 to 6.6 tons/ha/yr.[75]

Carbon farming applications. The avocado is a standard tropical homegarden crop around the world.[76] In Puerto Rico they are so abundant that they are grown over pig runs as a fodder tree.

OLIVE (*OLEA EUROPAEA*)

Olive cultivation began as early as 4,000 BC. Then, as now, olive is the global standard for quality edible oils.[77] Olives (*Olea europaea*) were domesticated in the Mediterranean and thrive in similar climates in Chile, Australia, South Africa, and California. The subspecies *O. europaea* ssp. *cuspidata* ranges throughout tropical Africa and east as far as China.[78] Some varieties handle warm temperate winters of –12°C (10°F), while others fruit in humid subtropical conditions.[79] The olive is one of the world's top perennial crops, ranking number 22 in terms of hectares devoted to crops, with more than 10 million hectares (25 million acres) in olive cultivation worldwide. This global crop produces 16.5 million tons of fruit annually, valued at $17.8 billion US.[80]

Uses. Olive fruits are 20 to 30 percent oil and 1.3 to 1.5 percent protein. Most olives are pressed for oil, with about 8 percent used as table olives. Raw olives

FIGURE 16.6. Avocado, the world's favorite oily fruit for fresh eating, in Guatemala.

FIGURE 16.7. An olive orchard in California. Olives have been cultivated as an oil crop for 4,000 years.

are inedible, but through soaking in brine, alkali, or other solutions they become edible and delicious.[81]

Yield. Yields range from 1 to 10 t/ha (0.4 to 3.6 dry flesh). About one-fifth of the weight of the fruit becomes oil, with a global average of 350 to 400kg/ha of oil produced.[82]

Carbon farming applications. In Greece 650,000 hectares (1.6 million acres) of olives are intercropped with annuals and grapes.[83] Olives are a strong candidate for woody agriculture. In fact, olives are already produced on a 10-year coppice rotation for ease of harvest in some areas.[84]

SEA BUCKTHORN, SEABERRY (*ELAEAGNUS RHAMNOIDES*)

If, like me, you live in a cold climate, you might feel discouraged by the diversity and abundance of perennial staple crops for tropical climates. Seaberry is part of the answer to our prayers. This tough, spiny, suckering shrub is native to much of Eurasia. It handles cold as low as −40°C (−40°F) and grows with as little annual rainfall as 300 millimeters (12 inches). Infertile, rocky, and even salty soils can accommodate this nitrogen-fixing species. It is an important fruit crop in Siberia and Saskachewan.[85]

Uses. Sea buckthorn fruits are small and tart. They are high in vitamins and antioxidants. The fruits and seeds are oily, with about 6 percent of the combined fruit and seed weight being oil (fresh weight). This oil is of a high quality for eating. Currently it is a very expensive specialty product in the United States.[86] Sea buckthorn was recently moved from the genus *Hippophae* to *Elaeagnus* by the International Council of Botanical Nomenclature.

Yields. Yields per hectare reach 10 to 14 tons, or 2.0 to 2.8 tons dry weight.[87] That could theoretically produce 0.6 to 0.8 t/ha of oil, a highly competitive yield especially for cold climates.

Carbon farming applications. Sea buckthorn is intercropped with annual crops in northern and western China.[88] About 10,000 hectares (25,000 acres) of sea buckthorn alley cropping is practiced in North Korea.[89] Sea buckthorn has additional roles in agroforestry as a fodder bank, windbreak, and nitrogen fixer. It is a high-quality coppiced firewood species. It has strong potential as a woody agriculture species.

Sea buckthorn is also a critically important species for combating desertification and stabilizing sand dunes in north and northwest China. One county in China had planted 66,700 hectares (164,800 acres) by 1999, making it "the largest artificial seabuckthorn forest in the world."[90] China plans to revegetate more than 6 million hectares (14.8 million acres) with seabuckthorn by 2020.[91] This type of regeneration of degraded land is a key aspect of carbon sequestration. While monocultures of any kind are undesirable, in some dry, cold areas of northern China, this is the only woody plant that will grow. As conditions improve, other species can be added.

Crop development. To my knowledge, there has been no effort to breed sea buckthorns for oil yield; efforts have been focused on fruit yield and overall harvestability. Many cultivars have been developed in the former Soviet Union and Eastern Europe. The search for high-oil clones could start with assessing this rich diversity. Ease of mechanical harvest has been a priority in breeding and management practices.

FIGURE 16.8. A seaberry shrub in Scotland, one of few oily fruits for cold climates. Photograph courtesy of Mark Longair/Creative Commons 2.0.

PIQUIA (*CARYOCAR VILLOSUM*)

Caryocar species hail from tropical South America, from humid and semi-arid climates. Although a lot of domestication work needs to be done, they are wild staples with great potential. The primary species of importance appears to be piquia (*C. villosum*), which is native to Amazonian Brazil and French Guiana. The trees can grow to an impressive 40 to 50 meters (130 to 165 feet). Piquia is an important wild staple and minor regional crop in Amazonia.[92] Although it is considered a new crop, indigenous people may have planted it for centuries or more. Today there is an emphasis on planting and protecting piquia in natural forests.[93] Some small-scale plantings also exist.[94]

Uses. Both the fruits and nuts are edible. The fruits are the size of grapefruit.[95] An average fruit is 65 percent inedible rind, 30 percent edible pulp, and 5 percent edible seed.[96] The edible pulp is 64 to 73 percent oil, three times higher than an avocado.[97] This savory pulp must be cooked before eating. The flesh also contains 3 percent protein, which is respectable for a fruit.[98] The nut is an important food and is 45 to 61 percent oil. They are said to be among the most delicious nuts in the tropics.[99] Oil is pressed from the fruit and seeds.[100]

Yield. Although they do not yield annually, they can bear 6 t/ha of fruit.[101] At 35 percent edible and 50 percent water, that comes to 0.9 t/ha dry edible portion.

Harvest and processing. The fruits, which are cooked before eating, are savory rather than sweet. The flesh must be consumed with caution because the nut inside is protected with sharp spines.[102]

Related species. Pequi (*C. brasiliense*) is used similarly but is native to the semi-arid *cerrado* savanna. The souari (*C. nuciferum*) is reported to have the largest nuts in the genus. Cagui (*C. amygdaliferum*) has almond-flavored fruits and nuts.[103]

FIGURE 16.9. The savory grapefruit-sized oily fruits and nuts of piquia. Photograph public domain.

Pressed Oils

Pressed oil plants are those whose seeds or fruits are not eaten but are pressed for edible oil. The category includes some promising new oil crops like argan and cardoon, as well as some temperate and even boreal species. Although its primary use is for chocolate, cacao also produces an edible oil that is used for cocoa butter and other food and industrial uses, such as cosmetics. Some of these may also have edible fruits, nuts, or seeds (although I have not turned up such a reference). Note also that all or virtually all of the nuts and fruits profiled above can be pressed for oil, as can protein-oil crops. Even some balanced carbohydrate crops like rice and corn are sometimes pressed for oil.

CACAO (*THEOBROMA CACAO*)

Most of us think of chocolate when we think of cacao products. The cacao bean is also the source of an edible oil, cocoa butter. So important was cacao (mostly as a beverage) in the pre-Columbian Americas that the beans were used as currency.[104] Today cacao is a global

TABLE 16.5. Perennial Pressed Edible Oils

Latin Name	Common Name	Origin	Climate	Cultivation
Adenanthera pavonina	Condori wood	South Asia through Australia	Tropical humid lowlands	Undomesticated
Allanblackia parviflora	Vegetable tallow tree	West Africa	Tropical humid lowlands	New crop
Allanblackia stuhlmannii	Mkanye	East Africa	Tropical humid lowlands	New crop
Argania spinosa	Argan	Morocco	Tropics and subtropics, arid to semi-arid	New crop
Astrocaryum vulgare	Tucuma palm	Northeastern South America	Tropical humid lowlands	Regional
Attalea funifera	Piassava palm	Tropical Americas	Tropical humid lowlands	Regional
Bactris gasipaes	Peach palm	Tropical Americas	Tropical humid lowlands	Regional
Brassica carinata	Ethiopian kale	Africa	Tropical lowlands, highlands, semi-arid to humid	Minor global
Cynara cardunculus	Cardoon	Mediterranean	Warm temperate, Mediterranean, semi-arid	Minor global
Elaeis oleifera	American oil palm	Tropical Americas	Tropical humid lowlands	Regional
Gossypium spp.	Perennial cottons	Tropics worldwide	Tropics, arid to humid	Regional
Leopoldinia piassaba	Piassava palm	Tropical South America	Tropical humid lowlands	Regional
Lepidium spp.	Perennial lepidium	N. temperate	Boreal to warm temperate, semi-arid to humid	Under development
Linum hybrids	Perennial flax	Hypothetical	Boreal to warm temperate, semi-arid to humid	Under development
Moringa oleifera	Moringa	India	Tropics and subtropics, lowlands, semi-arid to humid	Minor global
Moringa peregrina	Ben tree	Middle East, North Africa	Tropics, arid to semi-arid	Historic
Moringa stenopetala	African moringa	Africa	Tropics and subtropics, lowlands and highlands, semi-arid to humid	Regional
Oenocarpus distichus	Bacaba palm	Tropical South America	Tropical lowlands, semi-arid to humid	Regional
Prinsepia utilis	Cherry prinsepia	Himalayas	Cold to warm temperate, semi-arid to humid	Regional
Prunus armeniaca var. *mandshurica*	Manchurian apricot	East Asia	Boreal, cold temperate, humid	Regional
Prunus sibirica	Siberian apricot	East Asia	Boreal, cold temperate, humid	Regional
Sapium sebiferum	Chinese tallow tree	China	Warm temperate, Mediterranean, semi-arid to humid	Minor global (fruit oil edible, seed oil industrial)
Sarcocornia fruticosa	Oilseed glasswort	Old World	Mediterranean	Regional
Schleichera oleosa	Macassar oil tree	South and Southeast Asia	Semi-arid to humid tropical lowlands	Regional
Tetradium daniellii	Evodia	Korea	Cold to warm temperate, humid	Regional
Theobroma cacao	Cacao	Tropical Americas	Tropical humid lowlands	Global
Theobroma grandiflorum	Cupuacu	Tropical Americas	Tropical humid lowlands	Regional

Sources: Hanelt, *Mansfeld's Encyclopedia of Agricultural and Horticultural Crops*; Janick and Paull, *The Encyclopedia of Fruits and Nuts*

crop grown on 9.9 million hectares (24.5 million acres). Five million tons of cacao beans are traded annually at a value of $9 billion US.

Uses. Chocolate is, of course, a splendid use of *Theobroma cacao*, but it is not the only one. (The oil pressed from the seeds, known as cocoa butter, is used in chocolate making.) It also has industrial uses, in soaps, cosmetics, and pharmaceuticals.[105]

Yield. A hectare of cacao can yield 0.8 to 1 ton of cocoa beans.[106] These are 45 to 60 percent fat, giving the potential to yield 0.4 to 0.6 t/ha of oil.[107]

FIGURE 16.10. Cacao is a shade-loving crop well suited to the understory of multistrata agroforestry systems. Photograph courtesy of World Agroforestry Centre.

Carbon farming applications. Cacao has the great advantage of being a shade crop. This makes it a splendid intercropping species. In pre-Columbian tropical America, cacao agroforests were widely grown.[108] Today multistrata cacao agroforests are grown on an estimated 7.8 million hectares (19.3 million acres) globally.[109]

In the Pacific, establishing cacao trees are often intercropped with bananas or the nitrogen-fixing proverbial "mother of cacao" (*Gliricidia sepium*). Once the trees are mature, they are grown in the shade of nitrogen-fixing trees like *Inga* and *Albizia* species, in homegardens, or in multistrata plantation polycultures with overstories of coconut, rubber, or fruit trees.[110] In Brazil they are grown beneath peach palm trees that themselves serve as living trellises for black pepper.[111]

EVODIA (*TETRADIUM DANIELLII*)

Evodia is an important oil crop in Korea and China. Outside of Asia, it is mostly grown by beekeepers because the flowers are attractive to honeybees and other pollinators. It is one of the few edible oil tree crops for cold climates and likewise one of the few dedicated oil crops (that is, not otherwise used for food as a nut or seed) for temperate climates. It is hardy to −34°C (−29°F) and through warm temperate regions as well.[112]

Uses. Evodia seeds are not eaten but contain 34 percent edible oil. The oil is used for cooking, soap, cosmetics, and industrial applications.[113]

Yields. Yields are 300 to 600kg/ha, indicating the need for breeding work. Improved varieties have been selected and distributed in North Korea.[114]

FIGURE 16.11. Korean evodia, a tree oilseed for cold climates. Photograph courtesy of Jerzy Opiola/Creative Commons 3.0.

Carbon farming applications. Evodia is grown in agroforestry systems in North Korea in both contour strips and irregular intercropping systems.[119]

Crop development. I can imagine evodia playing a significant role in the humid temperate lands of the future. These favorite trees of honeybees might someday fill in for annual soybeans, sunflowers, peanuts, and canola.

ARGAN (*ARGANIA SPINOSA*)

These native Moroccan trees have enjoyed a boost in popularity lately. Argan trees grow in inhospitable arid and semi-arid tropical deserts. In recent years, argan has been trialed in arid regions outside its native range as a potential new crop.[120] Promising crops for truly arid lands are relatively rare. Argan is one of the finest candidates currently being evaluated.

Uses. A high-quality edible oil is pressed from the inedible fruits and nuts. The fruit and nuts are 4 to 5 percent oil.[121]

Yield. Most argan is wild-harvested today. Wild trees yield around 8kg (18 pounds) per tree of fruit, while

PERENNIAL OILSEED BRASSICAS

Rapeseed or canola (*Brassica napus*) is one of the top oilseeds in the world with 34 million hectares (84 million acres) in production.[115] It would be useful to have perennial analogs—and there are several options, including some under development. Canola itself is an annual, although there are some perennial populations of *B. napus* like 'Western Front' kale.[116] Ethiopian kale (*B. carinata*) has annual forms grown as oilseeds, and perennial forms grown as leaf crops.[117] Perhaps a perennial oilseed type might be developed. Perennial *Lepidium* and *Lesquerella* oil crops have also been proposed.[118] I would add *Crambe maritima*, *Diplotaxis* spp., and *Bunias orientalis* as candidates worth trying. All of these set a lot of seed as perennials in my home garden.

FIGURE 16.12. Argan, an arid-climate oil tree under development. Photograph courtesy of Creative Commons 3.0.

elite selections made in Israel by Dr. Elaine Solowey yield 50kg (110 pounds) per tree. The Israeli test orchard is laid out at 100 trees per hectare, which could potentially yield 5 t/ha.[122] This could produce 0.2 t/ha of oil—not great but not bad for arid conditions, either. Additional crop improvement would be beneficial.

Carbon farming applications. Given its strong coppicing ability, argan might be a good woody agriculture species.

CARDOON (*CYNARA CARDUNCULUS*)

Cardoon is the wild ancestor of globe artichokes, a popular perennial vegetable.

Uses. Cardoon itself is sometimes raised as a vegetable for the edible leaf midribs. Recently it has been adopted as a source of edible oil and a biodiesel and biomass product.[123]

Yields. Seed yields have been recorded as high as 2.6 t/ha, impressive for a species that was domesticated as a stem vegetable rather than a grain crop. Oil content has been reported at about 25 percent.[124] This would yield about 0.5 to 0.6 t/ha of oil.

FIGURE 16.13. Cardoon, a minor perennial vegetable now under development as an oilseed.

Other issues. Cardoon is considered highly invasive in the South American prairies of Argentina.[125] It is a robust perennial relative of thistles and as such should be treated with respect and caution.

CHAPTER SEVENTEEN

Sugar Crops

It has been challenging to determine exactly what section of this book sugar crops belong in. Sugar is not exactly a staple crop as I see it, although dates have been a staple for millennia due to their sugar content. Certainly sugar accounts for many of the calories consumed by humanity today, although not the healthiest source of calories. Sugar is also used as an industrial feedstock, although mostly for ethanol, a low-priority product in my scheme since its use today is mostly limited to biofuel. Ethanol does have other uses, including production of synthetic rubber.[1]

Sugar crops are grown on over 32 million hectares (79 million acres).[2] This is about 2 percent of cropland. The only global annual dedicated sugar crop is sugar beets, though much of the maize harvest is converted to corn syrup. Perennials include sugarcane (a short-lived perennial) and dates. Perennial sugar crops compare favorably to annuals.

Sugar crops come in several categories. Some are saps—tapped in the case of palms, agaves, and maples, and extracted from biomass in the case of sugarcane. Many are fruits or pods, including dates and carob. Sugary legume pods are covered in chapter 13 since they also have significant protein. Dates are profiled here due to their historic significance. Any sweet fruit could potentially be listed here, although I have chosen not to profile them for simplicity's sake since there are hundreds or thousands of such species. In the United States, natural products often list white grape juice or apple juice as sweeteners, for example. Honey, a livestock product from several genera of bees, is also an important sweetener, as is corn syrup, made from the annual grain maize.

SUGARCANE (*SACCHARUM OFFICINARUM*)

More tons of sugarcane are grown than any other crop in the world. In 2012 a whopping 1.8 billion tons of

ALTERNATIVE SWEETENERS

Much of the world's sugar production goes to produce sweetness for populations that do not need extra calories. Most of the sugar consumed in the United States is not needed as staple calories, but instead contributes to diabetes and other chronic diseases. There are some perennial crops with sweet flavors that do not come from sugars.

Stevia (*Stevia rebaudiana*) is a tropical and subtropical perennial herb. Its non-caloric compounds are 30 times sweeter than sugar in the dried leaves, and 300 times stronger as an extract. At 2.6 t/ha of dried leaf, stevia produces the equivalent sweetness of 78 t/ha of sugar—more than 10 times the average global sugar yield of sugarcane and almost 3 times higher than the best-recorded sugar yields of any kind.[3] Perennial crops such as the cucurbit *Siraitia grosvenorii* and the "miracle fruit" *Synsepalum dulcificum* are similar, powerful non-caloric sweeteners.

These crops could permit production of sweetness for non-caloric purposes on greatly reduced land (one-third to one-tenth the land in the case of stevia), permitting the former sugar cropland to be used for food, forage, feedstocks, or reforestation.

TABLE 17.1 SUGAR CROP YIELDS COMPARED

Latin Name	Common Name	Climate	Sugar Yield in t/ha
Saccharum officinarum	Sugarcane	Humid tropical and subtropical lowlands	7.0[a] average, 28 theoretical maximum[b]
*Beta vulgaris	Sugar beet	Annual	Global average 9, maximum 26[c]
Arenga pinnata	Sugar palm	Humid tropical lowlands	25[d]
Cocos nucifera	Coconut	Humid tropical lowlands	12–23[e]
Nypa fruticans	Mangrove palm	Humid tropical lowlands	15–20[f]
Borassus flabellifer	Palmyra palm	Tropics, subtropics, lowlands, semi-arid to humid	19[g]
*Zea mays	Maize	Annual	0.6–11.8 syrup[h]
Phoenix dactylifera	Date palm	Tropics and subtropics, lowlands, semi-arid	6.7 t/ha dry fruit and 4.5 tons sugar[i]
Ceratonia siliqua	Carob	Mediterranean, arid to semi-arid subtropics	1.3–6.6 t/ha syrup[j]
Acer saccharum	Coppiced sugar maple	Boreal to cold temperate, humid	4–6 syrup[k]
Acer saccharum	Unselected sugar maple	Boreal to cold temperate, humid	0.4–0.6 syrup[k]

Note: Here we compare perennial versus annual sugar sources. Annual comparison crops are marked with an asterisk (*). Crops are arranged in descending order of yield.

[a] FAO Statistics Division online.
[b] Duke, *Handbook of Energy Crops*.
[c] Re, "The effective communication of agricultural R&D output in the UK beet sugar industry"; *Farmers Guardian*, "Attention to detail essential in meeting beet yield targets," 2010.
[d] Flach and Rumawas, *Plant Resources of Southeast Asia*, vol. 9, 58.
[e] Ohler, *Modern Coconut Management*, 6.
[f] Flach and Rumawas, *Plant Resources of Southeast Asia*, vol. 9, 136.
[g] Ibid., 62.
[h] National Corn Growers Association, *2012 World of Corn Statistics Book*.
[i] Morton, *Fruits of Warm Climates*, 9.
[j] Solowey, *Growing Bread on Trees*, 193; Janick and Paull, *The Encyclopedia of Fruits and Nuts*, 389.
[k] Nickerson, "A jolt for the science behind harvesting maple sap."

sugarcane biomass was harvested from 26 million hectares (64 million acres) of cropland.[4] It is the world's 11th most widely grown crop and the most widely planted perennial crop (although a short-lived one).[5] This giant grass grows up to 6 meters (19 feet) in height and thrives in tropical and subtropical humid conditions. This global crop was domesticated in New Guinea about 8,000 years ago.[6]

Sugarcane has some unique biological properties that make it so productive. It has C_4 metabolism, allowing for efficient photosynthesis in tropical heat. Some sugarcane also fixes nitrogen, not through root nodules but thanks to a bacterial endophyte called *Gluconacetobacter diazotrophicus* that lives between the cell walls inside the stem.[7] Sugarcane is the only perennial crop of which I am aware that has the double advantage of C_4 and nitrogen fixation. (A few annual sorghums do, too.)

"Noble" cane types are optimized for sugar production. Energy cane types have three times the fiber for use as biofuel.[8] Many energy canes are the result of crosses with biomass grasses like *Miscanthus*,[9] raising the interesting possibility of cold-tolerant sugarcane.

Sugarcane production is so widespread because it feeds two addictions: sugar and liquid fuel. It has had, and still has, an enormous environmental impact due to conversion of wild land to cane. Sugarcane production has probably caused more biodiversity loss than any other crop, as entire islands and coastal areas were cleared for it. Today a dozen countries use 25 percent or more of their cropland for sugarcane production.[10] Expansion in Brazil to produce agrofuel is projected to deforest 10 million hectares (25 million acres) of the unique *cerrado* savanna unless policies are changed to promote conversion of degraded farm- and pastureland.[11] Social conditions in sugarcane production can be terrible. Cane was a major crop of slave plantations, and even today workers on some large farms in Brazil are enslaved.[12]

Mitigating climate change means reducing consumption. Here is a case where reduced consumption of a food we don't really need and a fuel that can't hope to replace our fossil fuel use could shrink the footprint of a major world land use. Think what else might be done with those 26 million hectares. Sugarcane is a uniquely productive crop that will always have a role to play in agriculture, but a smaller role is both possible and desirable.

Uses. The primary use of sugarcane is extraction of sugar for food and ethanol. Other products include

FIGURE 17.1. A sugarcane plantation in Hawaii. Note the burning fields in preparation for harvest, not a climate-friendly practice. Photograph courtesy of Derek van Vliet/Creative Commons 2.0.

molasses, rum, and wax. The residue left after sugar extraction is called bagasse, and it is among the world's top biomass feedstocks. Much of it is burned to fuel sugar processing. Bagasse is also used in paper and cardboard production, plastics, plywoods, and composite materials, or it can be returned to the fields as mulch.[13] Fresh sugarcane is a good fodder for livestock including cattle, often fed in the dry season when it is still green.[14] Too much sugarcane can be fatal to horses.[15] Energy cane types are utilized as biofuel but could be a productive biomass feedstock for material and chemical applications instead of or as well.

Yield. Global sugarcane biomass yields averaged 70.2 t/ha in 2012.[16] Sugarcane is about 10 percent sugar.[17] Sugar yields typically range from 4.1 to 7.8 t/ha.[18] In theory, the maximum sugar yield is 280 tons of biomass and 28 t/ha of sugar.[19]

Harvest and processing. Sugarcane is first cut 14 to 18 months after planting. Thereafter it is often cut every 12 to 14 months for a total of four to five cuttings.[20] Plantings can last as long as eight years, but three years is more typical.[21] Most sugarcane is harvested by hand. Often fields are burned before manual harvesting to remove snakes, remove sharp-edged leaves, and improve access for laborers. Burning removes a lot of organic matter and reduces yields by about 5 percent. It has an undesirable carbon impact as well.[22] Mechanical harvest as practiced in wealthy countries like Australia is unburned but uproots about 10 percent of the plants, requiring tilling and replanting the field sooner.[23]

Carbon farming applications. Sugarcane need not be grown in large monocultures. It is common in homegardens around the world.[24] It is also used as a fodder bank.

Crop development. Sugarcane and energy cane breeding is active and ongoing. GMO sugarcane is under way.[25]

WHAT ABOUT MAPLE SYRUP?

I am a huge fan of maple syrup, a delicious forest product from the frozen northland. Maple trees (mostly sugar maple, *Acer saccharum*) are tapped when their sap begins to rise as they come out of winter dormancy. However, in terms of yield, maple syrup pales in comparison with tropical sugar crops. One hectare of maple trees produces 0.4 to 0.6 t/ha of syrup—compared with sugarcane and sugar palms, which yield 4 to 28 t/ha of sugar.[26] Of course, maple syrup is wonderful on pancakes and profitable for farmers, and there's nothing wrong with that.

Several developments may improve syrup yields. Improved varieties of sugar maple have been selected.[27] Increasing sugar content in this fashion is important because huge amounts of energy are needed to evaporate the water out of the sap, which in unselected sugar maples means 39 out of every 40 liters must be boiled away. High-sugar forms of the related silver maple (*A. saccharinium*) have also been selected, with higher sugar content than sugar maple.[28] Silver maple is also more adaptable to warmer and drier climates—good news in a changing climate since the more delicate sugar maple is retreating northward.[29]

Researchers at the University of Vermont have also developed a promising new technique that uses a vacuum pump to suck sap from recently coppiced young trees instead of tapping mature trees. These trees can be coppiced multiple times with sustained yields. Yields could be an order of magnitude higher at 4 to 6 tons/ha.[30]

This impressive jump is still less than the sugar yield from annual sugar beets, the main source of local sugar in cold climates. Sugar beets produce a global average of 9 t/ha of sugar and up to 26 t/ha in ideal conditions.[31] Coppiced maple trees will sequester far less carbon than maple forest and may require considerably more management. Nor will they have the great beauty of a stand of mature sugar maples in the fall.

DATE PALM (*PHOENIX DACTYLIFERA*)

Dates have been cultivated for at least 5,000 years and were originally domesticated in southern Iraq. They have been the essential staple for the survival of people in West Asian deserts for millennia.[32] Dates fruit for 40 to 50 years and sometimes for 150.[33] They need a long, hot, dry period for the fruit to ripen but can handle brief exposure to temperatures as low as −7°C (19°F).[34] Tropical, subtropical, and Mediterranean climates are appropriate.

Dates are a global crop with 1.1 million hectares (2.7 million acres) in production. Global yield is 7.5 million tons of fresh dates, worth $6.2 billion US.

Uses. Dates are 72 to 88 percent sugar, dry weight. They contain almost no starch.[35] Moisture ranges from 31 to 78 percent.[36] Research is under way to develop an ethanol industry based on culled dates of inferior quality.[37] The seeds can be pressed for oil or fed to livestock.[38]

Yield. Dates can yield 11 to 17 t/ha or more fresh weight, or up to 6.7 t/ha dry weight.[39] Roughly 67 percent of

FIGURE 17.2. Date palm oasis agroforestry system in Morocco. Photograph courtesy of Martin and Kathy Dady/Creative Commons 2.0.

TABLE 17.2. Perennial Sugar Crops

Latin Name	Common Name	Origin	Climate	Cultivation
Acer spp.	Maples	Northern temperate	Boreal to subtropical, humid	Regional
Agave americana	Maguey	Mesoamerica	Warm temperate to tropical, arid to semi-arid	Regional
Agave tequilana	Tequila agave	Americas	Subtropical to tropical highlands, arid to semi-arid	Regional
Arenga pinnata	Sugar palm	Asia	Tropics, lowlands, humid	Minor global
Borassus aethiopum	African palmyra palm	Tropical Africa	Tropical lowlands, semi-arid to humid	Regional
Borassus flabellifer	Palmyra palm	Asia	Tropics, subtropics, lowlands, semi-arid to humid	Regional
Caryota urens	Fishtail palm	Asia	Tropical, lowlands and highlands, humid	Regional
Cassia fistula	Golden shower tree	South and Southeast Asia	Humid tropical lowlands	Regional
Cassia grandis	Carao	Tropical Americas	Tropical	Regional
Ceratonia siliqua	Carob	North Africa, Mediterranean	Mediterranean, arid to semi-arid subtropics	Minor global
Cocos nucifera	Coconut palm	Asia	Tropical lowlands, semi-arid to humid	Global
Corypha utan	Buri palm	Asia	Tropical humid	Regional
Gleditsia triacanthos	Honey locust	Eastern North America	Boreal to subtropics, Mediterranean, tropical highlands, semi-arid to humid	Wild staple
Nypa fruticans	Mangrove palm	Asia	Humid tropical lowlands	Regional
Oenocarpus distichus	Bacaba palm	Tropical South America	Tropical lowlands, semi-arid to humid	Regional
Parajubaea cocoides	Coquillo palm	South America	Tropical highlands	Regional
Phoenix canariensis	Canary Island date palm	Canary Islands	Mediterranean	Regional
Phoenix dactylifera	Date palm	North Africa, western Asia, India	Tropics, lowlands, semi-arid	Global
Prosopis alba	Algarrobo blanco	Argentina	Tropical lowlands, arid to semi-arid	Regional
Prosopis chilensis	Algarrobo de Chile	South America	Warm temperate to tropical, highlands, arid to semi-arid, halophyte	Regional
Prosopis glandulosa	Honeypod mesquite	North America, Mesoamerica	Cold temperate (−20°C/−4°F) to tropics, including highlands, arid to semi-arid, halophyte	Regional
Prosopis juliflora	Mesquite	Central and South America	Tropical lowlands, arid to semi-arid, halophyte	Minor global
Prosopis pallida	Mesquite	Central and South America	Tropical lowlands, arid to semi-arid, halophyte	Regional
Raphia farinifera	Madagascar raffia palm	Africa, South Asia, Madagascar	Humid tropics, subtropics	Regional
Raphia hookeri	Ivory Coast raphia palm	West Africa	Humid tropical lowlands	Regional
Saccharum hybrids	Energy cane	Hybrid of Asian species	Tropical and subtropical, humid	New crop
Saccharum officinarum	Sugarcane	New Guinea	Humid tropical and subtropical lowlands	Global
Sarcocornia fruticosa	Oilseed glasswort	Old World	Mediterranean	Regional
Sorghum hybrids	Perennial sorghum	Hypothetical hybrid	Temperate to tropical, semi-arid to humid	Under development as grain crop

Note: Some palms are destructively tapped for sugar, killing the palm. These species are not listed here.

the fresh yield is dry and edible. A hectare of dates can thus produce roughly 4.5 tons of sugar.

Carbon farming applications. Date palms are the keystone species of ancient desert oasis agroforestry systems. These systems spread from Africa through West and South Asia in the traditional range of the date palm. Dates are the canopy in all, providing shade in the desert sun. Understory species vary by region and climate. In some regions they include apricot, fig, mulberry, tamarind, carob, pomegranate, banana, quince, grape, guava, citrus, olive, and peach. Annual crops such as cereal grains are also grown in the understory.[40]

SUGAR PALM (*ARENGA PINNATA*)

Many palms are cultivated for sugar in the tropics and subtropics. Sugar palm is the highest yielding and probably the most widely cultivated of these. Sugar palm is native to tropical East and Southeast Asia, where it is an important sugar and fiber source.[41]

Uses. The sweet sap is tapped from the flower stalks. It is used for fresh juice, fermented toddy, vinegar, and "jaggery" palm sugar. The juice is 5 to 21 percent sugar. The unripe fruits are eaten after significant processing. Sugar palm is also an important fiber crop.[42]

Yields. Sugar palm yields up to 25 t/ha of sugar annually once producing. This figure could increase significantly with domestication.[43]

Harvest and processing. Sugar palm begins to flower between 5 and 15 years after planting, depending on altitude and other factors (lower elevation is faster). Normally they die two years later, when fruit production is completed. Male flower stalks are cut to capture the sweet sap. The flowerstalks are beaten

FIGURE 17.3. Sugar palm harvest. Photograph courtesy of Kars Alfrink/Creative Commons 3.0.

and swung about before cutting to cause internal breakage and ensure good flow. With good technique, the flowering period can be prolonged for up to 10 or occasionally 15 years.[44] In any given hectare of 200 palms, an average of 34 will be tapped in any given year.[45]

Carbon farming applications. Sugar palm is grown in multistrata agroforestry systems in Southeast Asia with fruit trees (such as durian), tree vegetables (such as *Parkia speciosa*), and timber trees (such as *Litsea* spp.)[46] Intercropping with coppiced woody legumes can provide fertility and firewood for evaporating the sap.[47]

Crop development. Sugar palm is still not properly domesticated.[48] Currently only half of tapped palms produce sap well, and breeding could increase this to 85 percent.[49]

PART FOUR

Perennial Industrial Crops

CHAPTER EIGHTEEN

Industrial Crops: Materials, Chemicals, and Energy

Since the dawn of agriculture, humans have raised crops for more than food. One of the earliest places domestication arose was the coast of Chile, where the first crops to be developed were not food plants, but cotton for weaving fishing nets.[1] Today economic botanists like myself call these non-food plants industrial crops. The name can be confusing because most of these crops—firewood, for example—are actually still used on a home scale rather than as raw material for enormous factories or power plants, as the name suggests.

Industrial crops provide resources in three main categories: materials, chemicals, and energy.[2] Traditional materials include lumber and thatch, paper and cardboard, and textiles. Bamboo alone provides as many as 1,500 uses.[3] Modern materials science has added plastics, rubber, and "composite" materials that combine multiple raw ingredients (plywood, for example).

Although growing crops for chemical production sounds like a recent idea, some ancient chemicals from plants include paints, soaps, glues, varnishes and lacquers, lubricants, dyes, medicines, and solvents. Twentieth-century chemists learned to synthesize thousands of chemicals from petroleum and natural gas. Similarly, plants can provide the raw ingredients ("platform chemicals") for manufacturing stabilizers, dispersants, flocculants, and an unthinkable diversity of products.

In terms of energy, firewood is the most ancient industrial use of plants. Charcoal, candlenuts, and illuminating oils also go back thousands of years.

Today plant-based fuels, though controversial, are part of daily conversation and debate about climate change strategy.

Much of the conversation about slowing climate change has focused on reducing the amount of petroleum that we use for fuel. However, about an eighth of petroleum is used to synthesize materials and chemicals. A full 10 percent of petroleum is used as feedstock to synthesize chemicals, with another 10 percent used to power the process. (A feedstock is a raw material used for industrial processing. For example, cotton fibers are the feedstock for clothing and textiles.[4]) In 2000 the European Union used 16.6 million tons of petroleum for chemicals production.[5]

Plastics are synthesized from petroleum and natural gas. In 2012, 190 million barrels of oil and natural gas were used as feedstocks for plastic manufacturing in the United States, with another million barrels used to fuel the process. This adds up to 2.7 percent of US petroleum and natural gas use.[6] Another estimate reports that 4 to 7 percent of world petroleum is used as feedstock for plastic manufacturing. Lubricants for vehicles and industry account for another 1 percent.[7] Thousands of products, from plastics to household cleaners, are made from fossil carbon.

Before the petroleum age, many of these same products were made from plants. As recently as the 1950s, a third of European cropland was devoted to industrial crops.[8] Petroleum itself is of course made from plants—massive quantities of ancient plants, pressurized deep underground for millions of years.

OSAGE ORANGE: A PROMISING PERENNIAL INDUSTRIAL FEEDSTOCK FOR COLD CLIMATES

Osage orange (*Maclura pomifera*) is a cold-hardy member of the mulberry family and, like its tropical relatives breadfruit and jakfruit, it produces a large green fruit.[9] That is where the resemblance ends, however; Osage orange is inedible and possibly somewhat toxic. Osage orange fruits have promising potential as multipurpose industrial feedstocks providing oil, hydrocarbons, and sugar.

Native only to a small area of Texas and adjacent states, Osage orange is widely adapted to warm and cold temperate climates, from semi-arid to quite humid. This tree certainly has its drawbacks. The inedible fruits are regarded as a nuisance. Their stickiness occasionally causes choking death in cattle that try to eat them (although some horses enjoy them). The vicious thorns can be problematic in many contexts. Finally, Osage orange has naturalized aggressively wherever it has been planted.

Uses. Osage orange fruits have many potential industrial uses. Their sugar content is 15 percent dry weight.[10] The dried fruits also contain protein (16 percent), fiber (12 percent), oil (18 percent), and resin (21 percent).[11] The edible seeds make up 11 percent of the fruit and are 33.9 percent protein and 32.8 percent fat.[12] The seed oil is also being investigated as a biodiesel feedstock.[13] The fruits contain proteolytic enzymes, which are used in cheesemaking and other applications.[14] The fruit resin is composed of hydrocarbon triterpenes, a potentially interesting petroleum replacement.[15] Osage orange coppices strongly and is one of the best firewoods in the United States. It is reputed to be the finest wood for archery bows.

Yields. I grew up with Osage oranges in my front yard and can attest to its high yields. Female trees bear annually, and some bear heavily. Yields as high as 450kg (992 pounds) per tree in undomesticated trees have been reported.[16] Extrapolated to 120 trees per hectare (with 20 being non-fruiting males for pollination),

this gives 45 tons of fresh fruit per hectare. Keep in mind, however, that the fruits are 80 percent water, giving us 9 tons per hectare dry weight.[17] Ethanol yields of 1,073 liters/ha have been extrapolated.[18] Oil yields have been extrapolated at 1,800 liters/ha (about 1.5 t/ha).[19] With resin content at 21 percent we might expect up to 1.9 t/ha of hydrocarbons. This neglected, inedible fruit might, with propagation of elite varieties, produce twice the oil of rapeseed, a bit over half the ethanol of corn, and twice the resin of turpentine pines—simultaneously! Of course there is an enormous difference between extrapolation and reality, but clearly this species deserves further investigation.

Harvest and processing. The large fruits might be well suited to mechanized collection from the ground. Trees can live to 150 years or more, and seedling trees start bearing at 10 years of age.[20] As with most fruits, we could expect grafted trees to bear earlier.

Carbon farming applications. Vast hedgerows of Osage orange were planted throughout the eastern and central United States as thorny living fences before the development of barbed wire. It is an alternative host for silkworms.

Crop development. To my knowledge, no one has ever selected Osage orange varieties for superior fruit production. Given the large feral populations, great candidates surely are already available. The plants are dioecious, with male and female flowers on separate plants, although some female plants apparently set seedless fruit without a male present. Production of female-only clones may offer a strategy for cultivation without the potential for naturalizing outside its current range.

Those of us in cold climates can also dream of a wide cross between Osage orange and jakfruit or breadfruit, aiming for a cold-hardy edible starchy fruit. It has been crossed with the related edible *Cudrania tricuspidata*,[21] although the fruits from that hybrid were not edible when I tried them. (In fact, they looked and tasted a lot like regular Osage orange fruit.) The 16 percent protein dry weight of Osage orange fruit, as well as the small but edible seeds, would be a great future contribution to cold-climate perennial staple crops.

One reason industry likes petroleum is that it is possible to produce uniform results, whereas obtaining uniform products from agricultural products can be more difficult. There are efforts under way to improve product-grading systems to provide industry with the uniform quality it requires.[22]

Some industrial feedstocks are mined products, including metals, stone, sand, and minerals. We can't grow those on trees, but they could be extracted, used, and reused more responsibly. As of 2008 only about 5 percent of industrial feedstocks were biobased.[23] Aside from metals, almost all of today's industrial products from clothing to motor oil are refined from fossil fuels (plastics, chemicals), are produced by annuals (cotton), utilize food crops as feedstocks (corn ethanol and plastics), or are from destructively harvested perennials (pine plantations for paper). Those few that are non-destructively harvested from perennials (*Hevea* rubber, African oil palm) are grown in enormous monocultures with devastating social and ecological consequences. All of that needs to change.

The IPCC has stated that using long-lived wood in buildings instead of concrete and steel can significantly reduce emissions.[24] They also write, "It is theoretically possible to increase carbon storage in long-lived agricultural products (e.g., strawboards, wool, leather, bio-plastics) . . ." The IPCC essentially writes off the potential of these non-wood industrial products not because of lack of potential but because of the negligible impact they currently have.[25] That's not a good reason to ignore the potential of biobased feedstocks, especially of perennial biobased feedstocks, a subject the IPCC doesn't touch on at all.

Why not restrict petroleum use to materials and chemicals, avoiding the creation of gasoline and other carbon-emitting fuels? The answer is that crude oil is broken down into fractions, so that for any barrel of oil, a certain percentage is suited for gasoline, another for asphalt, and another for plastic. So if you want to make plastic from crude oil, you end up with a bunch of gasoline as a by-product.[26] With enough extra cost, energy, and effort, it is possible to convert those other fractions to plastic (or any other product), but it is not usually economically or energetically effective to do so.[27] Using

FIGURE 18.1. Osage orange fruit, a potential multipurpose food and feedstock.

fossil fuels as feedstocks for materials and chemicals also fails to sequester any atmospheric carbon, instead putting more fossil carbon into the atmosphere one way or another.

In the course of writing this book, I've received quite an education about these products, how they are made today from fossil carbon, and the alternatives, both traditional and under development, for a biobased economy. I've learned there are many non-destructively harvested perennial industrial crops that can serve as carbon-sequestering renewable feedstocks for the 21st century.

Perennial Industrial Crops for a Sustainable Biobased Economy

Like staple crops, there is a promising but somewhat neglected body of perennial industrial crops that are non-destructively harvested. I've grouped these species

TABLE 18.1. Perennial Industrial Crop Categories and Products

Category	NDHP Crop Types	Materials	Chemicals	Energy
Biomass	Resprouting woody plants and grasses, bamboo, crop residues	Paper, cardboard, agromaterials, biochar, insulation, green building materials, synthetic fibers	Solvents, chemical feedstocks, resins, stabilizers, dispersants, binders and fillers	Firewood, combustibles, biogas, gasification, pyrolysis, ethanol, methanol
Starch	Pods, starchy fruits, nuts and seeds, starchy trunks	Bioplastics, paper, cardboard, packaging materials, plasterboard	Solvents, paints, glues, binders, coaters, stabilizers, coagulants, flocculants, textile finishing agents, chemical feedstocks	Ethanol
Sugar	Pods, saps, extracted sugars, fruits	Biomass products from crop residues	Solvents	Ethanol
Oils	Oilseeds	Bioplastics, biomass products from residues	Glycerin, soaps, lubricants, surfactants, surface coatings, solvents, paints	Biodiesel
Hydrocarbons	Resprouting plants, saps	Bioplastics, rubber, biomass products from crop residues	Full range of modern chemical industry	Gasoline, jet fuel, diesel, butane, propane, biogas, more…
Fibers	Coppiced crops, barks, fibrous fruit and seedpods	Textiles, cordage, paper and cardboard, bioplastics and composites, biomass products from residues	Residues used like biomass	Residues used like biomass
Other products	Variable	Cork, natural containers	Soaps, waxes, resins, biopesticides, essential oils, pharmaceuticals, dyes, tannins, gums, lacquers	Residues used like biomass

into seven broad categories: biomass, starch, sugar, oil, hydrocarbons, fiber, and other. In most cases the primary chemical constituents of these crops is carbon. These carbon compounds can serve as feedstocks for a range of activities and industries that keep civilization running.

This chapter presents a global inventory of non-destructively harvested perennial crops that produce materials, chemicals, and energy. They can serve as part of the foundation for a post-fossil-fuel civilization, both by sequestering carbon and by providing raw materials for meeting human needs.

AN UNSUSTAINABLE AND UNJUST BIOBASED ECONOMY

You hear a lot these days about switching to a biobased economy as a climate change mitigation strategy. I want to see us transition to a biobased economy—it's the only feasible scenario to address climate change in a meaningful way. But I want a democratic, fair one, based on non-destructively harvested perennial feedstocks. That's *not* what most people mean when they talk about a biobased economy. They are thinking about sugarcane

and GMO corn grown in vast monocultures, processed in energy-intensive and wildly expensive biorefineries, with the resulting products shipped around the world as global commodities.[28]

Industrial crops seem to lend themselves to atrocities even more than food crops. This is nothing inherent in the crops themselves; many were used for centuries or millennia in relatively benign ways. Rather, it is the intense greed of capitalism that drove the industrial revolution and still drives our economy today that is the cause of the colonization, slavery, and genocide that industrial crops have been intimately associated with.

In the profiles that follow, you'll read some of the terrible stories that are associated with some of these crops. During the period of Belgian colonization of the Congo region, for example, millions of people died to fuel the harvest of rubber and resins.[29] Much of the United States' wealth is built on a foundation of cotton that was grown and harvested by enslaved people. Even today thousands of people are held in slave-like conditions in Brazil to serve as labor for the sugarcane ethanol industry.[30]

Modern "land grabs" represent the newest version of the colonialism that has devastated people and the environment for centuries to the benefit of wealthy countries and their elites. Since the world food price crisis of 2008, foreign investors have bought up millions of hectares in the developing world, mostly in Africa and Southeast Asia. In some cases, like Colombia and Brazil, much of the land grabs have been initiated by in-country elites.[31] Land grabs account for an estimated 30 million hectares (74 million acres) worldwide.[32]

Indonesia is one hot spot for land grabs, with much of the land being bought up for oil palm plantations for agrofuel. Large companies buy up and clear land that has been managed by indigenous people for centuries—land that provides their food, homes, and the basis of their culture. In 2011 alone, 5,000 human rights violations were reported from land grabs in Indonesia, including forced evictions, torture, and deaths.[33]

Peasant groups such as Via Campesina are leading the fight against land grabs. Such activists have distinguished between industrial-scale *agrofuels* and small-scale, traditional *biofuels*.[34] The movement for carbon-sequestering agriculture needs to build partnerships with and follow the leadership of these movements. I believe that as we think about transforming global civilization to a post-fossil-carbon model, we'll need perennial industrial crops. We must learn from the lessons of the past (and the present) to ensure that they are used to enable a free and ecological future for all.

FOOD VERSUS FUEL

Biofuel is the subject of a major debate. My position on biofuels came the day I learned that there simply isn't enough land to grow both food and fuel. My desire to eat places me firmly on "team food" in the food-versus-fuel debate. For example, a 2008 Oxfam study reports that converting the entire global crop of carbohydrates and sugar would only replace 40 percent of fossil fuel energy use, and all the world's oilseeds would only replace 10 percent of the diesel we use.[35] Agrofuels simply can't replace the amount of fossil carbon energy we use.

As of 2009, world fossil energy use was 388 exojoules (EJ). Researchers have projected the amount of bioenergy that could be produced without sacrificing food production. Realistic estimates range from 40 to 85 EJ by 2050—meaning perhaps 10 to 20 percent of current energy use. This land use would also compete with biochar, industrial feedstocks, habitat restoration, and other land uses.[36]

Biofuel advocates propose raising their feedstocks on marginal land, such as abandoned farmland and wastelands. While such lands do exist, in much of the world "marginal" land is full of people—often indigenous people, farming, herding, gathering, and making their subsistence. Agrofuel plantations displace these people, including many of the world's poorest. To make matters worse, crops like jatropha that were touted as succeeding in marginal lands have proven to grow there, but to yield poorly.[37]

Agrofuel crops don't just compete for land. In many regions, they compete for water. For example, jatropha requires water (and fertilizer) to produce much on marginal lands in India. As climate change intensifies water scarcity in much of the world's hungriest regions, this becomes a greater and greater concern.[38]

Currently a lot of biofuel is made directly from food plants. The argument is that these plants, particularly maize, are highly domesticated and yield very well, making for an efficient use of space. But the impact on people has been terrible. The ethanol boom raised the price of maize significantly. The amount of maize it takes to fill an SUV tank could feed a person for a year.[39] I've seen with my own eyes how this resulted in smaller tortillas at the same price for my wife's family in Guatemala.

This is in part because we are working to replace fossil fuels, truly remarkable substances that represent the concentrated energy of millions of years of photosynthesis. Bill McKibben reports, "One barrel of oil yields as much energy as twenty-five thousand hours of human manual labor—more than a decade of human labor per barrel."[40]

Finally, many have questioned the carbon efficiency of biofuels. When their full life span is accounted for from production to processing to use, some are about as bad as fossil fuels, especially when raw materials or fuels themselves are transported great distances.[41] Some crops and practices are much better than

FIGURE 18.2. A truckload of oil palm fruit from an industrial plantation in Guatemala. Photograph courtesy of Louisa Reynolds.

TABLE 18.2. Bioenergy Classes and Perennial Feedstocks

Technique	Feedstocks	Energy Product
Direct combustion	Coppiced woody plants, bamboo, giant grasses, crop residues	Heat, steam, electricity
Densification	Crop residues (nutshells, straw, etc.)	Solid fuels (briquettes and pallets) for combustion
Gasification	Coppiced woody plants, bamboo, giant grasses, crop residues	Gas
Pyrolysis	Coppiced woody plants, bamboo, giant grasses, crop residues	Synthetic oil (biocrude), charcoal
Anaerobic digestion	Manure, humanure, crop wastes, hydrocarbon crop biomass	Methane
Ethanol	Sugar crops, starch crops, wood waste, pulp sludge, straw	Ethanol
Methanol	Coppiced woody plants, bamboo, crop residues, wastes	Methanol
Biodiesel	Oils, animal fats	Biodiesel
Hydrocarbon fuels	Hydrocarbon crop biomass, extracted hydrocarbons, tapped resins and latexes	Synthetic oil (biocrude), butane, methane, propane, gasoline, diesel, heating oil, jet fuel, more

Source: Adapted from Chen et al. "Bioenergy industry status and prospects" Singh, ed., in *Industrial Crops and Uses*, Table 2.3, p. 30.

others. Biodigestion, in particular, is a carbon-efficient practice.[42]

Although biofuels can never meet civilization's current energy needs, there is a modest role for them in any green scenario. Many non-destructively harvested products can be used as feedstocks for bioenergy production. The primary classes are coppiced woody plants, bamboo, giant grasses, sugar and starch crops, oils, hydrocarbon plants, and crop residues.

Let's not imagine that food versus fuel is always an either–or situation. Scientists are developing techniques to remove edible leaf protein concentrate (LPC, see chapter 14) from biomass feedstocks during processing. Based on their calculations for global increase in perennial biomass production, these researchers project a by-product of 24 million tons of LPC annually. By producing both biomass feedstock and protein on the same land, they calculate very impressive reductions in cropland requirements: 16 million less hectares in soybeans for feed, and 8 million less hectares for biofuels.[43] That's a tremendous win–win, and a great example of an innovative multifunctional strategy.

Industrial Sovereignty and a Fair, Sustainable, Biobased Economy

We've already introduced the concept of food sovereignty: Each country or region should produce most of its own food, and food needs to be largely removed from the international commodity market if we're going to mitigate climate change and bring about climate justice. The same logic applies to industry. Drastically reducing emissions means shortening supply lines for all commodities. This means each region should produce most of its own *everything*.

Non-destructively harvested perennial crops could and should provide much of the basis for such industry. Addressing climate change can be an opportunity to address the inequality of infrastructure and technology between Global North and South, by developing regionally based industry, grown, processed, marketed, and used in-country or in-region. This is core to the MAD Challenge of mitigation, adaptation, and development.

My proposal is that we reduce our consumption, acquire most of our energy from wind, water, and solar (WWS), and produce materials and chemicals from non-destructively harvested perennials. I propose we grow those crops on a mosaic of diversified farms, with in-country processing for in-country end users, emphasizing decentralized appropriate technology.

I don't think there are any carbon-responsible scenarios for maintaining current levels of consumption of materials, chemicals, and energy. Scientists suggest that any scenario that would return us to 350ppm would necessarily involve a period of "planned austerity" in the wealthy countries.[44] Environmentalists have been pointing out ridiculous levels of waste for decades. Those of us who are accustomed to privilege need to learn to live with a more reasonable level of resource use. By increasing efficiency alone, the United States could reduce energy use by 23 percent.[45]

Integrated Food and Feedstock Systems

Much of the damage caused by industrial crops stems from growing them in huge monocultures, frequently displacing the people and ecosystems that were there before. This is neither the only nor the most desirable way to grow fuels and feedstocks. We can look to traditional practices like growing castor beans in polycultures and at the edges of fields for a glimpse of another approach. Today researchers call this type of farming an integrated food and fuel system.[46] For purposes of this book, as we are deemphasizing fuel and looking at a wider range of feedstocks, let's call them *integrated food and feedstock systems* (IFFSs).

There are two main types of IFFSs. The first combines food and feedstock crops on the same land using techniques such as intercropping and agroforestry. The second type uses crop residues from food production as the feedstock.[47] I suspect the average smallholder farm in the tropics today does some of both already.

FIGURE 18.3. Recently planted small-scale integrated food and feedstock system at Las Cañadas featuring castor, tephrosia (pesticide), perennial cotton, alder (firewood), banana, taro, cassava, and shade crops including cardamom and naranjilla.

IFFSs are compatible with production on large numbers of small farms, with producers using products directly on the farm, bringing them to a central village processing and distribution area, or participating in a regional cooperative or producers association.[48] Unlike rubber plantations, for example, where no food is produced, there is the opportunity for human needs to be met on a mosaic of farmer-managed small parcels. This is more productive and certainly more socially and ecologically desirable. Like some other industrial processes, industrial crops can be toxic, some to ingest, a few to touch; at least one (petroleum plant) even poisons groundwater. This could represent a major challenge in integrated farming systems. Harvesting and processing streams would need to keep crops separated, for example. However, many of these toxic crops, like castor bean, have been successfully grown in integrated farming systems for centuries or millennia.

There are a number of models of small-scale, decentralized technologies. One that is gaining attention in the biofuel world is called Distributed Biomass Energy Production Systems (DBEPSs). These facilities are small-scale and designed for on-farm or regional-level processing. They are more economical than large centralized facilities, in part because of the reduced need to transport bulky raw biomass materials. The DBEPS model could be applied to the production of materials and chemicals, as well, for regional production of bioplastics, lubricants, paper, and other needs. In fact, some models are actually mobile, traveling from farm to farm to do on-site processing.[49]

Feedstocks, Not Fuels

We know that there isn't remotely enough land to meet current energy needs, even if we reduce our energy needs by half. Fortunately, clean energy sources such as solar, wind, geothermal, hydro, and tidal can get it done. According to a 2009 *Scientific American* article, "A large-scale wind, water and solar energy system can reliably supply the world's needs, significantly benefiting climate, air quality, water quality, ecology and energy security . . . the obstacles are primarily political,

not technical."[50] Scientists recently ran more than 28 billion simulations of different combinations and found that a diverse combination of wind and solar, with diverse energy storage strategies, could meet 90 to 99.9 percent of current energy needs at a lower price than we pay today.[51] While some small amount of plant-based liquid fuels are likely to be needed to operate heavy equipment, a reduced number of airplanes, and some heavy industrial machinery, most of our energy needs can be met with solar, wind, and water.

Many biofuel experts point out that from an ecological, land-use, social, and economic perspective, using waste and crop residues as the primary biofuel feedstock is the most desirable strategy.[52] Organic wastes and residues, if converted to biofuels, could offer 40 EJ or more (about 10 percent of the world's 2009 fossil fuel energy use).[53] With the exception of biogas generation, this means those materials are not available for composting and returning to the soil to preserve fertility. Unwanted "invasive" plant species are also a promising raw material.[54]

In other words, energy production isn't the best use of perennial feedstocks. But you can't make plastic or paper from the wind! So even if almost all of our energy needs can be met without biofuels, we still need material and chemical products, which are currently manufactured primarily from fossil fuels. Since materials and chemicals represent roughly 12 percent of petroleum use, we're in a realistic position regarding the available land. Switching the entire world's current plastic production to biobased feedstocks would take 4 to 5 percent of agricultural land.[55] In a reduced-consumption scenario, replacing materials and chemicals synthesized from fossil fuels with perennial feedstocks is reasonable and feasible.

Integrated Technology Assessment

Any discussion of industrial crops needs to include a discussion of the methods we use to process and use those crops. Although I'm not an engineer or an expert in appropriate technology, I've assembled a set of criteria that can be used to assess the suitability of industrial crop-processing technologies. I propose that industrial crops and processing technologies run on a continuum from most to least suited to small-scale, decentralized, locally owned operation.

First we might assess an industrial crop's requirements in terms of skills, required technologies, and cost. Some practices like methanol production require high technology, while firewood and even biogas can be done on a home scale.

Second, we can look at feedstock flexibility. Many of today's biorefineries are being designed to work a single annual food crop as their only input.[56] These Phase I and Phase II biorefineries encourage monoculture production and a landscape of uniform farms. Much more desirable are facilities and equipment that can accept a diversity of products, which can support diversified farming. Phase III biorefineries, still theoretical, are being designed to accomplish this goal.[57] Some practices, like lignocellulosic ethanol, also require GMO crops or fermenting microorganisms.[58] This removes them from my A-list, not only for environmental concerns, but for the prohibitive cost and intellectual property issues that will keep them out of reach of lower-income communities.

Third, some technologies require very large facilities—again, like biorefineries. Others are more scale-neutral and can be performed at a farm, village, or municipal scale. Scale-neutral technologies with the potential for viable small- and medium-sized operations should be prioritized.

Fourth, we are doing this to mitigate climate change. Practices must be carbon-efficient. This probably means being energy-efficient, based on local materials, and producing a long-lasting product with end uses on short supply chains.

Fifth, we need to think about the toxicity of inputs and by-products. It is critical to minimize these. Some processes also provide useful by-products, like residues from oil extraction that can be fed to livestock, or biomass residues from fiber extractions.

Sixth, we want to look at fairness and equity. The Solidarity Economy model offers some ideas on local and worker ownership. We want to avoid extremes of wealth concentration and excessive foreign ownership. What technologies best lend themselves to decentralized democracy with a fair balance of wealth? Probably not extremely expensive ones.

Finally, we want to look at how mature a technology is. Biodigesters already provide energy for 25 million households around the world while petrofarming is still completely theoretical.[59] Although biochar has seen historic use on a vast scale in the Amazon, modern biochar still has few long-term studies to support its efficacy. Efforts to expand a perennial biobased-economy should emphasize crops and technologies with proven track records.

At the most basic scale, then, we might have firewood, along with home production of fibers, oil, and soapnut soap. A step up would include the remarkable diesel tree, whose sap can be filtered and put straight into a diesel engine. Biodiesel is an intermediate technology, requiring some chemistry background and equipment. Another step up is a bioplastic extruder, which requires some moderately complex (and expensive) machinery. Finally at the very far end of the scale comes the full biorefinery, which can take raw plant materials and "crack" and recombine them just like petroleum refineries. Biorefineries can produce many useful products, but most require an enormous facility, with very high technology and skill, dangerous chemicals, and vast investment. That's not to say that a given region might not decide to construct and maintain one as part of a carbon farming effort, but to me they don't belong as the primary focus of our efforts.

This, then, is the context for the profiles of industrial crops that follow. I don't have the road map laid out from here to a full proposal to replace petroleum and all its uses, but I think these species and practices are the building blocks of a post-petroleum civilization. Note that all the food plants we covered in the last section can be used as industrial feedstocks as well, as can their crop residues (banana peels and stalks, nutshells and husks, et cetera). We'll refer to some of them here in the tables, but I only profile species where industrial feedstock is their primary use.

Very few perennial industrial crops have been properly domesticated. Those that have shown significantly increased yields and the other traits that demonstrate true domestication include cotton and sisal (fiber); *Hevea* rubber and guayule (hydrocarbon); African oil palm, candlenut, and castor (inedible oil); and sugarcane (sugar). Other industrial crops that are at least partially domesticated include willow, poplar, switchgrass, and giant miscanthus (biomass); jojoba and tung (inedible oil); slash pine and its hybrids (hydrocarbon); and ramie (fiber).

CHAPTER NINETEEN

Biomass Crops

Biomass is surely the most ancient of industrial plant uses. Plants provided our earliest ancestors with fire, tools, and materials for crafts and building. Today the number one biobased industrial feedstock in the world is cellulose from wood pulp.[1] Interest in energy crops is driving rapidly expending production of biomass crops.

Unlike most industrial crop categories, biomass has never been threatened or replaced by petroleum-based synthetics. However, destructively harvested pines and annual crop residues like cornstalks are important feedstocks today, leaving room for replacement with non-destructively harvested alternatives such as bamboo, coppiced woody plants, and giant perennial grasses.

Carbohydrates are the most common kind of organic matter and are made from carbon, hydrogen, and oxygen. Just one type, polysaccharides, represents 75 percent of planetary organic matter, living and dead. Carbohydrates include sugar, starch, and cellulose. Cellulose alone makes up 40 percent of all organic matter.[2] This is why carbon sequestration with plants works so well—they are full of carbon compounds! Cellulose and hemicellulose are made of long chains of thousands of sugar molecules. Lignin makes up another 20 percent of all organic matter and represents a tough organic material mostly used for fuel but intriguing to chemists as a potential feedstock for other uses.

More than with any other set of plants profiled in this book, each region should assess its native crop candidates for biomass feedstock uses. Most parts of the world have vigorous grasses or resprouting woody plants. Naturalized species are also fine candidates; in fact many biomass fuel advocates think that "harvested" invasive species make the best feedstock.[3] Crop residues, the leftover non-used parts of other crops, are also a readily available biomass feedstock. This includes nutshells, stalks, prunings, and material remaining after extracting sugar (bagasse), fiber, or hydrocarbons.

Of the tens of thousands of candidate species for this category, here we'll review only a small sample, focused on those that are already cultivated as biomass crops of one kind or another. They fall into three broad categories: bamboos, resprouting woody plants, and giant grasses and grass-like plants.

Bamboos

Bamboos are large woody grasses. There are about 1,450 species worldwide. Most grow in Asia and tropical America, but there are a few native species in temperate North America and in Africa. Bamboos are among the handful of plants labeled the most useful on the planet. In China, it is known as the "friend of the people." More than 1,500 uses have been recorded of this global crop, making bamboo the undisputed champion of multipurpose biomass crops.[4]

Bamboos cover much of the planetary surface. Today roughly 23 million hectares (57 million acres) support bamboo growth, storing 727 million tons of carbon. Many more millions of hectares are suited to bamboo cultivation.[5] Bamboo's main limitation is its need for a relatively humid climate, although a few species tolerate semi-arid conditions.

There are two primary types of bamboos. Clumping, or sympodial, species form clumps and do not spread

BIOMASS CROP USES

PROTEIN

- Leaf protein concentrate (see chapter 14) can be extracted from perennial biomass crops during processing to provide edible protein for humans and livestock as a by-product of producing materials, chemicals, and energy.

MATERIALS

- Paper and cardboard are the most important biomass products today. Although they are typically made from felled trees, paper and cardboard can also be made from perennial grasses, coppiced woody plants, and bamboos.
- Natural building materials represent a major (and ancient) use of biomass. Non-destructively harvested perennials can provide thatch, wattles, insulation, roofing tiles, bamboo poles and plumbing, construction timbers, and other materials. Buildings can represent a long-term storage for the carbon embedded in these biomass products.
- Before the widespread use of cardboard and plastic, woven baskets were the container of choice. Reeds, grasses, and woody coppice crops like willows, rattan, and bamboo have all been used for basketry, mats, furniture, and crafts for millennia.
- More recently engineers have developed composite materials that combine multiple raw materials to make products such as plywood, insulation boards, and more. These materials combine, for example, shredded biomass with glues, plastics, or other binders.
- Biomass products can be used to make rayon and other synthetic fibers.
- Biochar is a somewhat controversial type of charcoal used for carbon sequestration and as a soil amendment. Many biomass feedstocks can be used for biochar creation.

CHEMICALS

- Biomass can be used as a feedstock to develop some of the alternative biobased platform chemicals needed as a basis for the chemicals industry.

ENERGY

- Biomass can be used for most kinds of biofuel: combustion, densification, gasification, pyrolysis, anaerobic digestion, ethanol, and methanol.

aggressively. They are mostly tropical and subtropical, with a few temperate species. Clumping bamboos include the largest bamboos, with some species reaching 36 meters (118 feet) with culms 30 centimeters (12 inches) in diameter. They include the best timber and construction species.

Running, or monopodial, bamboos spread by runners, sometimes aggressively. Many are well adapted to temperate climates, including some hardy enough to overwinter in my Massachusetts garden (–24°C/–12°F) with no cold damage. Although most runners are not utilized for timber, they still have many uses. Running bamboos have a bad reputation in some areas as invasive species, although they spread only vegetatively; most bamboos flower and set seed just once a century or so.

Managed bamboo forests have been found to sequester more carbon than wild bamboo. Here is another example of a type of regenerative farming practice that

FIGURE 19.1. At 1,500 recorded uses, bamboo may be the world's most multifunctional crop. Photo courtesy of Raffi Kojian, Gardenology.org.

TABLE 19.1. Biomass Crop Yields Compared

Latin Name	Common Name	Climate	Yield in t/ha/yr dry weight
Saccharum officinarum	Sugarcane	Humid tropical and subtropical lowlands	25–100[a]
Arundo donax	Giant reed	Cold and warm temperate, subtropical, tropical, semi-arid to humid	10–100[c]
Azadirachta indica	Neem	Arid to humid tropical lowlands	10–100[d]
Echinochloa polystachya	Aleman grass	Warm temperate to tropical, humid	40–100[e]
Rumex patientia × *tianschanicus*	Hybrid sorrel	Boreal to warm temperate, Mediterranean, tropical highlands, semi-arid to humid	14–40[f]
Various	Bamboos (mixed species)	Cold temperate to tropics, humid	2–36[g]
Acacia tortilis	Umbrella thorn acacia	Tropical lowlands, arid to semi-arid	27[h]
Miscanthus spp.	Miscanthus grass	Cold and warm temperate, subtropics, humid	5–25[i]
Eucalyptus gomphocephala	Tuart	Subtropical and Mediterranean, lowlands and highlands, humid	3–22[j]
Leucaena leucocephala	Leucaena	Semi-arid to humid tropical lowlands	15–20[j]
Panicum virgatum	Switchgrass	Boreal, cold and warm temperate, semi-arid to humid	10–20[k]
Populus spp.	Poplar	Boreal to warm temperate, Mediterranean, semi-arid to humid	10–15[l]
Sapium sebiferum	Chinese tallow tree	Warm temperate, tropical highlands, semi-arid to humid	13
Sesbania grandiflora	Vegetable hummingbird	Humid tropical lowlands	10–13[j]
Salix spp.	Willow	Boreal to subtropical, tropical highlands, humid to semi-arid	1–13[m]
Alnus acuminata	Aliso	Tropics, subtropics, highlands, humid	5–8[j]
Lespedeza bicolor	Bush clover	Cold temperate, warm temperate, humid to semi-arid	3–6[n]
Robinia pseudoacacia	Black locust	Cold to warm temperate, tropical highlands, semi-arid to humid	3–6
Prosopis juliflora	Mesquite	Tropical lowlands, arid to semi-arid, halophyte	2–5[j]

Note: These are organized from highest to lowest yield. Drier and colder regions tend to be less productive.

[a] Duke, *Handbook of Energy Crops.*
[b] El Bassam, *Handbook of Bioenergy Crops*, 97.
[c] Czako and Marton, "Subtropical and tropical reeds for biomass," 326–27.
[d] Singh, *Neem*, 79.
[e] El Bassam, *Handbook of Bioenergy Crops*, 96–98.
[f] Ibid., 333.
[g] Hunter and Wu, "Bamboo Biomass," 4–5.
[h] National Research Council, *Firewood Crops*, vol. 2. The figures from *Firewood Crops*, volume 2, are given in cubic meters per year, converted to tons per hectare per year as follows. Determined that high-quality firewoods average about 3,800 pounds per cord, or 1.7 metric tons. A cord is 128 cubic feet, or about 3.6 cubic meters. So 1.7 tons divided by 3.6 cubic meters comes out pretty close to 0.5 ton per cubic meter. Based on Kuhns and Schmidt, "Heating with Wood."

[i] Kim et al., "Developing miscanthus for bioenergy," 303.
[j] National Research Council, *Firewood Crops*, vol. 2.
[k] Vogel et al., "Switchgrass," 356.
[l] Dillen et al., "Poplar," 283.
[m] Aylott et al., "Yield and spatial supply of bioenergy poplar and willow short-rotation coppice in the UK," 358–70.
[n] National Academy of Sciences, *Firewood Crops*, 20.

sequesters more carbon than wild nature.[6] Managed bamboo forests have been found to sequester about the same amount of carbon as fast-growing tropical trees like eucalyptus.[7]

Unfortunately the power of bamboo is sometimes overlooked by carbon schemes. For example, bamboo forestry is not currently included in the REDD+ strategy. The International Network for Bamboo and Rattan (INBAR) is working on this and other policy issues to support increased utilization of this remarkable group of plants.[8] This "friend of the people" belongs at the very center of our strategy to meet human needs while sequestering carbon.

Uses. Construction may be the most important use of bamboo. Recent advances in treatment of bamboo culms means they can last for decades. The big timber bamboos can be used to build almost every element of a building, from central timbers to framing to paneling and flooring to roofing shingles. Bamboo is used in bridges, fencing, cables, and reinforced concrete. Bamboo scaffolding is used in

TABLE 19.2. Elite Bamboo Species

Latin Name	Origin	Climate	Form
Arundinaria spp.	Asia, North America	Cold to warm temperate, humid	Running
Bambusa balcooa, B. bambos, B. blumeana, B. heterostachya, B. nutans, B. oldhamii, B. pervariabilis, B. polymorpha, B. textilis, B. tulda, B. vulgaris	Asia	Tropical to subtropical humid lowlands, some species highlands, *B. textilis* warm temperate, all humid	Clumping
Bambusa chungii	China	Humid tropics and subtropics	Clumping
Cephalostachyum pergracile	Asia	Tropical to subtropical, humid	Clumping
Dendrocalamus asper, D. brandisii, D. giganteus, D. hamiltonii, D. hookeri, D. latiflorus, D. membranaceus, D. strictus	Asia	Humid, tropical to subtropical lowlands (some highlands), *D. asper* semi-arid to humid	Clumping
Gigantochloa albociliata, G. apus, G. atroviolacea, G. balui, G. hasskarliana, G. levis, G. pseudoarundinacea, G. verticillata	Asia	Humid tropical and subtropical lowlands, *G. verticillata* semi-arid to humid	Clumping
Guadua angustifolia	South America	Humid tropics and subtropics	Clumping
Melocanna baccifera	Asia	Humid tropics	Clumping
Neololeba atra	Asia	Tropical to subtropical humid highlands	Clumping
Ochlandra spp.	South Asia	Humid tropics	Clumping
Oxytenanthera spp.	Africa, Asia	Humid to semi-arid, tropics and subtropics	Clumping
Phyllostachys bambusoides, P. edulis, P. glauca	Asia	Humid, cold to warm temperate	Running
Schizostachyum spp.	Asia	Humid tropics to subtropics	Clumping
Thyrsostachys siamensis	Southeast Asia	Humid to semi-arid, tropical lowlands	Clumping

Source: Rao et al., *Priority Species of Bamboo and Rattan*.

the construction of even very tall skyscrapers. In the transportation sector it is used for boats, rafts, and bicycle frames. Bamboo construction is particularly important in disaster response. Bamboo housing is unusually resistant to earthquakes.[9]

Bamboo stilts protect 300,000 homes in Guayaqiil, Ecuador, from flooding and landslides. After an earthquake in Indonesia, 70,000 bamboo shelters were built in nine months at a cost of $100 to $200 US each in "one of the largest and most rapid humanitarian post-disaster shelter responses in recent history."[10] Unlike timber trees, bamboo plants are highly resilient to intense storms, and even if they are destroyed by hurricanes, they can resprout and produce a mature timber crop in three to four years. This is a great example of climate change adaptation!

Bamboo is a better feedstock for paper and cardboard than destructively harvested trees. It can yield six times the pulp per hectare of pine plantations. As of 1977, India used 3 million hectares (7.4 million acres) of bamboo as paper feedstock, and China used 1 million hectares (2.5 million acres). Together this accounted for 40 percent of world bamboo use.[11]

Bamboo basketry, crafts, and furniture are also important products. Properly treated, bamboo can be used as plumbing. Sections of bamboo culms make great nursery pots. Synthetic fibers such as rayon are made from bamboo pulp. For energy, bamboo is a fine source of firewood and charcoal. It can be used as a biochar feedstock as well. And humans relish the young shoots of many species as a vegetable. The foliage can be fed to ruminant livestock and even aquaculture fish as fodder.

Harvest and processing. Both types of bamboo sprout new culms every year. These culms take several years to harden off before they are ready for harvest as construction material. Unmanaged, individual culms die after about 10 years, although the clump or stand persists. In managed bamboo stands, culms are cut out on reaching maturity, stimulating new growth. This is what accounts for the superior carbon sequestration in managed versus wild bamboo forests. The cut culms are treated and can store carbon for decades in their new home.[12]

Carbon farming applications. Clumping bamboos are commonly grown on diversified farms and in

FIGURE 19.2. A bamboo building in Mexico. Photo courtesy of Ricardo Romero.

agroforestry systems. Many bamboos grow well in the shady understory of taller crop trees. Some are used as fodder banks and windbreaks. *Dendrocalamus strictus*, a large clumping bamboo, is intercropped with annuals in southern China.[13]

Many systems are in use in China. Bamboos are often intercropped with annuals during establishment years. Clumping bamboos are intercropped with fruit trees and vegetables. Running bamboos like *Phyllostachys edulis* are intercropped with timber trees. Bamboos are planted around aquaculture ponds for use as fish fodder. Goats and cattle are grazed beneath bamboo in silvopasture systems. And bamboo is used as a shade crop for mushrooms and medicinal plants.[14]

Coppiced Woody Plants

Many regions of the world have long traditions of coppicing woody plants, which involves cutting back

FIGURE 19.3. A coppiced firewood planting of mixed species at ECHO in Florida.

TABLE 19.3. Coppiced Woody Biomass Crops

Latin Name	Common Name	Origin	Climate	Cultivation	Nitrogen Fixer	High-BTU Firewood
Acacia angustissima	Timbre	Mesoamerica, North America	Tropics to subtropics, highlands, semi-arid to humid	Regional	Yes	Yes
Acacia saligna	Port Jackson wattle	Australia	Tropical, lowlands, semi-arid	Minor global	Yes	Yes
Acacia senegal	Gum arabic	Africa, South Asia	Tropics, subtropics, lowlands, highlands, arid to semi-arid	Regional	Yes	Yes
Acacia tortilis	Umbrella thorn acacia	Africa	Tropical lowlands, arid to semi-arid	Regional	Yes	Yes
Acer spp.	Maples	Northern temperate climates	Boreal to subtropical, humid	Regional		Yes
Albizia julibrissin	Mimosa, silk tree	Asia	Cold to warm temperate, Mediterranean, subtropics, tropical highlands, semi-arid to humid	Regional	Yes	Yes
Albizia lebbeck	Lebbek tree	Asia	Tropical, subtropics, lowlands, semi-arid to humid	Minor global	Yes	Yes
Alnus acuminata	Aliso	Tropical Americas	Tropics, subtropics, highlands, humid	Regional	Yes	Yes
Alnus glutinosa	Grey alder	Europe	Boreal to warm temperate, humid	Minor global	Yes	Yes
Azadirachta indica	Neem	South Asia	Arid to humid tropical lowlands	Minor global		Yes
Balanites aegyptiaca	Balanites	Africa, Asia	Tropical to subtropical, arid to semi-arid	Regional		Yes
Bursera simaruba	Gumbo limbo	Americas	Tropical semi-arid to humid lowlands	Regional		Yes
Calamus spp.	Rattan	Asia	Tropical humid lowlands	Regional		
Calliandra calothyrsus	Red calliandra	Mesoamerica	Tropical lowlands to highlands, humid to semi-arid	Minor global	Yes	Yes
Caragana arborescens	Siberian pea shrub	Asia	Boreal to cold temperate, semi-arid to humid	Minor global	Yes	Yes
Castanea spp.	Chestnut	Asia, Europe, North America	Cold to warm temperate	Global		Yes
Colophospermum mopane	Mopane	Africa	Tropical and subtropical lowlands, arid to semi-arid	New crop		Yes
Cornus sericea	Silky dogwood	North America	Boreal to warm temperate, humid to semi-arid	Regional		
Corylus avellana	Hazelnut	Europe	Cold to warm temperate, Mediterranean, humid	Global		Yes
Corymbia citriodora	Lemon-scented gum	Australia	Tropics and subtropics, lowlands and highlands, semi-arid to humid	Minor global		Yes
Elaeagnus angustifolia	Russian olive	Eurasia	Boreal to warm temperate, semi-arid	Minor global	Yes	Yes
Elaeagnus rhamnoides	Sea buckthorn, seaberry	Eurasia	Boreal to warm temperate, semi-arid to humid	Minor global	Yes	Yes
Elaeagnus umbellata	Autumn olive	Asia	Cold to warm temperate, humid	Minor global	Yes	Yes
Eucalyptus brassiana	Cape York red gum	Australia, New Guinea	Tropical lowlands, humid	Minor global		Yes
Eucalyptus camaldulensis	Red River gum	Australia	Tropics and subtropics, lowlands, arid to humid	Global		Yes
Eucalyptus gomphocephala	Tuart	Australia	Subtropical and Mediterranean, lowlands and highlands, humid	Minor global		Yes
Eucalyptus grandis	Flooded gum	Australia	Tropics and subtropics, lowlands and highlands, humid	Global		Yes

Latin Name	Common Name	Origin	Climate	Cultivation	Nitrogen Fixer	High-BTU Firewood
Eucalyptus microtheca	Coolabah	Australia	Tropical to subtropical, lowlands, arid to semi-arid	Global		Yes
Eucalyptus tereticornis	Forest red gum	Australia, New Guinea	Tropics and subtropics, lowlands, semi-arid to humid	Minor global		Yes
Eucalyptus urophylla	Ampupu	Indonesia	Tropical, lowlands and highlands, humid	Minor global		Yes
Gleditsia triacanthos	Honey locust	North America	Boreal to warm temperate, Mediterranean, tropical highlands, semi-arid to humid	Minor global		Yes
Gliricidia sepium	Madre de cacao	Mesoamerica	Semi-arid to humid tropical lowlands	Minor global	Yes	Yes
Hibiscus tilliaceus	Beach hibiscus	Pantropical	Tropics to subtropics, lowlands, humid	Minor global		Yes
Inga vera	Ice cream bean	Caribbean	Humid tropical lowlands	Minor global	Yes	Yes
Lespedeza bicolor	Bush clover	East Asia	Cold temperate, warm temperate, humid to semi-arid	Minor global	Yes	Yes
Leucaena leucocephala	Leucaena	Tropical Americas	Semi-arid to humid tropical lowlands	Minor global	Yes	Yes
Maclura pomifera	Osage orange	North America	Cold to warm temperate, semi-arid to humid	Regional		Yes
Melia azedarach	Chinaberry	Asia	Warm temperate to tropics, lowlands and highlands, semi-arid to humid	Minor global		Yes
Morus alba	White mulberry	Asia	Cold temperate through tropical lowlands and highlands, semi-arid to humid	Minor global		Yes
Parkinsonia aculeata	Jerusalem thorn	Tropical Americas	Tropics to subtropics, semi-arid	Minor global		Yes
Pithecellobium dulce	Manila tamarind	Mesoamerica	Tropics and subtropics, lowlands and highlands, semi-arid to humid	Minor global	Yes	Yes
Pongamia pinnata	Pongam	India, Southeast Asia, Australia	Subtropics, tropics, lowlands, highlands, humid	New crop	Yes	Yes
Populus hybrids	Hybrid poplar	Northern temperate	Boreal to warm temperate, humid to semi-arid	Global		
Populus deltoides	Eastern cottonwood	Eastern North America	Cold to warm temperate, semi-arid to humid	Minor global		
Populus euphratica	Desert poplar	Eurasia, North Africa	Cold temperate to subtropical, semi-arid to humid	Regional		Yes
Populus nigra	Black poplar	Eurasia, North Africa	Boreal to warm temperate, Mediterranean, semi-arid to humid	Minor global		
Populus trichocarpa	Black cottonwood	Western North America	Boreal to warm temperate, semi-arid to humid	Minor global		
Prosopis glandulosa	Honeypod mesquite	North America, Mesoamerica	Warm temperate to subtropical, highlands and lowlands, arid and semi-arid	Minor global	Yes	Yes
Prosopis juliflora	Mesquite	Central and South America	Tropical lowlands, arid to semi-arid, halophyte	Minor global	Yes	Yes
Prosopis pallida	Mesquite	South America	Tropical lowlands, arid to semi-arid, halophyte	Minor global	Yes	Yes
Quercus spp.	Oaks	Hybrid	Boreal to subtropical, arid to humid	Minor global		Yes
Robinia neomexicana	New Mexico locust	Western North America	Cold to warm temperate, tropical highlands, semi-arid to humid	Regional	Yes	Yes

TABLE 19.3. continued

Latin Name	Common Name	Origin	Climate	Cultivation	Nitrogen Fixer	High-BTU Firewood
Robinia pseudoacacia	Black locust	North America	Cold to warm temperate, tropical highlands, semi-arid to humid	Minor global	Yes	Yes
Salix hybrids	Hybrid willows	Northern temperate climates, Africa, South America	Boreal to subtropical, tropical highlands, humid to semi-arid	Minor global		
Salix eriocephala	Heartleaf willow	North America	Boreal to warm temperate, semi-arid to humid	Regional		
Salix miyabeana		East Asia	Boreal to warm temperate, humid	Regional		
Salix purpurea	Purple willow	Europe, North Africa	Boreal to warm temperate, humid	Regional		
Salix viminalis	Basket willow	Eurasia	Boreal to warm temperate, semi-arid to humid	Regional		
Sapium sebiferum	Chinese tallow tree	Asia	Warm temperate, tropical highlands, semi-arid to humid	Minor global		Yes
Sequoia sempervirens	Coast redwood	Western North America	Warm temperate, tropical highlands, humid	Minor global		Yes
Sesbania grandiflora	Vegetable hummingbird	Southeast Asia to Australia	Humid tropical lowlands	Minor global	Yes	Yes
Sesbania sesban		Africa	Tropical lowlands, semi-arid to humid	Minor global	Yes	Yes
Styphnolobium japonicum	Japanese pagoda tree	Asia	Cold to warm temperate, tropical highlands, semi-arid to humid	Minor global	Yes	Yes
Trema orientalis	African elm	Asia	Humid tropical lowland	Regional	Yes	Yes
Ulmus pumila	Siberian elm	Northeast Asia	Boreal to warm temperate, semi-arid to humid	Minor global		Yes

Sources: National Research Council *Firewood Crops*; W. A. Geyer, "Biomass production in the Central Great Plains of the USA under various coppice regimes," *Biomass and Bioenergy* 30 (2006): 778–83; Kuhns, Michael, and Tom Schmidt, "Heating with wood: Species, characteristics and volumes," Utah State University Extension, www.forestry.usu .edu/htm/forest-products/wood-heating, accessed March 25, 2014; Jacke and Toensmeier, *Edible Forest Gardens*, vol. 2.
Note: These species are cultivated for firewood, paper pulp, natural building materials, or other biomass products. All coppice strongly. Doubtless many other species are used similarly. Nitrogen-fixing species (noted in the table) are likely to grow more rapidly and also may have a greater chance of aggressively naturalizing. Many of these species have high-quality firewood (noted).

resprouting species for multiple harvests. This tradition goes back millennia in Europe and western North America, for example. Historically, coppiced wood has many uses, serving in some ways as the analog to bamboo in Europe and temperate North America. Products include natural building material, fencing, firewood, basketry and crafts, and charcoal. Some species can even be used as construction timbers. Today short-rotation coppice is rapidly expanding as a bioenergy feedstock. Coppiced wood products are fine feedstocks for combustion, densification, pyrolysis, ethanol, and methanol, as well as biochar production.[15]

For the best-quality timber, unfortunately noncoppiced, destructively harvested trees still seem to be the best option. Although the first cut of a coppiced tree may yield excellent wood, the quality of subsequent cuts typically suffers from growing so quickly. However, at least one coppiced tree makes construction-grade wood: the coast redwood. Information on species that produce relatively good coppiced timber is not widely available, although many quality lumber species such as oaks and eucalypts coppice strongly.[16] Meanwhile, between redwood and timber bamboos, there are options for the world's humid tropical through humid warm-temperate climates.

Firewood shortage is a major cause of deforestation. For several decades, farmers have been planting high-density coppice plots of quality firewood. Many of the best species are nitrogen-fixing trees that grow quickly.[17] Nitrogen-fixing coppiced firewoods are thus a productive, self-fertilizing, and perennial fuel source. Intensive blocks of these species can produce a tropical family's cooking fuel needs on 0.15 hectare (0.37 acre).[18] Use of rocket stoves and other conservation technologies can reduce the required growing area even further. Cultivation of these or similar species can take enormous pressure off natural forests in areas where people rely on firewood to cook and heat. As with grasses, there are thousands of potential crop candidates; each region should assess its own native, and perhaps naturalized, species for the best selections. Table 19.3 features many of these coppiced, nitrogen-fixing, high-BTU firewood crops.

COPPICED EUCALYPTS

Eucalypts are top global timber and pulpwood crops. Of the more than 500 species, only a select few are cultivated as coppiced paper pulp and firewood crops. Most eucalypts come from Australia, with some from New Guinea and a few from Indonesia. Coppicing eucalypts are adapted to many climates, from tropical and subtropical to Mediterranean, lowland and highlands, and arid to humid. These species include *Corymbia citrioclora, E. brassiana, E. camaldulensis, E. gomphocephala, E. grandis, E. microtheca, E. tereticornis,* and *E.*

FIGURE 19.4. Coppiced eucalyptus in California. Photograph courtesy of Christopher Peck.

urophylla.[19] As a timber crop, only the first cutting is of good quality. The coppiced resprouts are used as firewood and especially as pulp feedstock for papermaking. Most eucalyptus plantations outside of Australia are coppiced for pulpwood.[20]

Eucalyptus plantations have all the social and ecological consequences of any monoculture. Eucalypts also have some additional challenges. They consume a fair amount of water because they grow so quickly. Eucalypts produce biological herbicides (a phenomenon known as allelopathy) that limit agroforestry and intercropping opportunities. Both water consumption and allelopathy are only serious problems in climates with less than 400 millimeters (16 inches) of rainfall a year.[21] They have a reputation as invasive species, although none is listed on the Global Invasive Species Database.[22]

Despite their troublesome reputation and the clear undesirability of monocultures, eucalypts can grow, and grow fast, in conditions where few if any other trees will survive. Development worker Rodney Babe reports on a deforested and goat-infested region in Haiti where *E. camaldulensis* was the only tree that could establish. Once their planting had established, farmers were able to establish pigeon peas and eventually other annual crops between the trees. The farmers are now successfully establishing mangoes and native timber trees, which received little competition from the light shade of the eucalypts. Babe argues that the allelopathic reputation of eucalyptus is overstated.[23]

Uses. Coppiced eucalypts are not good for timber, even though the same species are among the world's finest timbers. Paper pulp and firewood are important global uses of coppiced eucalypts. They could be used as biomass feedstocks for any number of applications. The essential oil extracted from some eucalypts is important as an insecticide, medicinal plant, and oil for cleaning. Some eucalyptus logs make excellent shiitake mushroom–producing substrates, notably *E. camaldulensis.*[24] Hydrocarbons can be harvested from the foliage of *E. globulus.*

Harvest and processing. In Brazil, coppiced eucalypts are first harvested at seven to eight years after seeding. Thereafter they are managed on a five- to six-year rotation.[25]

Carbon farming applications. Eucalypts are used as windbreaks. Many species are important nectar sources for honeybees. The trees are often intercropped with nitrogen-fixing trees to speed their growth. Young plantings of eucalypts are intercropped with annual crops. Some growers report success intercropping fruits and other perennial crops with younger eucalypts, although others report strong allelopathic effects.[26]

LEUCAENA (*LEUCAENA LEUCOCEPHALA* AND RELATED SPECIES AND HYBRIDS)

In the 1970s and early 1980s, leucaena (primarily *L. leucocephala*) was touted as a "miracle tree" due to its multiple agroforestry functions. Leucaena was mass-planted in tropical climates around the world. By 1990 it covered 2 to 5 million hectares (5 to 12 million acres) worldwide. These monoculture plantings turned out to be the perfect buffet for a psyllid insect, with a devastating impact. Leucaena itself also became a problem and was added to the lists of the world's worst invasive species.[27] Today leucaena can be seen along disturbed roadsides in much of the American tropics and beyond. This is an important lesson for carbon farming: There are no miracle trees, and monoculture plantings of anything are not a good idea. Many of the countries where leucaena was introduced already had their own native multipurpose legume trees that were overlooked in the rush. Today there is still a place for leucaena and its relatives—certainly in its native range, and perhaps where it is already rampant.

Leucaena leucocephala, the most commonly planted species, hails from southern Mesoamerica. It thrives in semi-arid to humid tropical lowlands. For highland areas, *L. diversifolia* is a good substitute. In drier areas, *L. esculenta* is well adapted. Its young seeds are eaten as a vegetable called *guaje*.[28]

Uses. As a fodder, leucaena's toxic mimosine content means it is primarily a food for ruminants rather than hogs or horses. Even so, it should be no more than 30

FIGURE 19.5. Bob Hargrave of ECHO standing in front of a coppiced leucaena fodder bank.

percent of ruminant diets and 10 percent of aquaculture fish diets. It is an excellent coppiced firewood.[29]

Yields. Leucanea can yield 15 to 20 t/ha of firewood.[30]

Carbon farming applications. Leucaena is truly a remarkable multipurpose agroforestry tree—a fodder bank crop, alley crop, contour hedgerow crop, windbreak, crop shade, and living trellis for crops such as black pepper and passion fruit. It is among the world's most efficient nitrogen fixers. It has been reported to fix as much as 500kg (1,100 pounds) per hectare of nitrogen, almost double any other I have seen reported.[31]

Other issues. Leucaena is ranked as one of the world's 100 most invasive species by the Global Invasive Species Database.[32]

COAST REDWOOD (*SEQUOIA SEMPERVIRENS*)

Coast redwood is a contender for tallest tree in the world, growing as tall as 100 meters (328 feet). The trees can live 2,000 years or more. Although their ancient range covered much of North America, in recent millennia coast redwoods have been restricted to a narrow, foggy band of the California and Oregon coast. Despite this, they have adapted well to many cool-temperate and Mediterranean climates around the world under cultivation.[33] They can even be grown in highland tropics with sufficiently cool temperatures.[34] Coast redwood forests were decimated by logging due to the immense size and fantastic rot resistance of the wood. Fortunately, the trees resprout vigorously after cutting, a trait relatively rare among conifers.

Uses. The lumber quality of resprouted redwood is not as extraordinary as first-cut redwood timber, but highly competitive with other species nonetheless.[35] It contains more knots and less heartwood.[36]

Harvest and processing. Coppiced redwood shoots grow very quickly, a trait few other quality timber trees possess. This makes coast redwood an excellent candidate for carbon-sequestering coppiced timber production. The second-growth wood is easier to harvest.[37]

WILLOW (*SALIX* SPP.)

Willows are fast-growing woody plants that are found along streambanks and in wetlands throughout the northern temperate region. There are even a few species in South America and Africa. There are tree, shrub, and ground cover willows, although I focus on the shrubs for carbon farming. Europe leads the way in willow production, with thousands of hectares in production. Impressively, many of these willow parcels are simultaneously treating wastewater, turning a waste product into a fertilizer for fuel and feedstock. How does this work? Willows are adapted to wet, even seasonally flooded conditions. They can also take up excess nutrients in the water and use them as fertilizer. They are thus perfectly suited to converting wastewater to useful biomass.[38] In far northern boreal areas with little sunlight much of the year, they are a good option.

Uses. Willows have been wild-managed for millennia, and cultivated for centuries, at least, for their strong, light, flexible wood. Traditional uses, all of which are still active uses, include basketry, fencing, and natural building. Willow bark is medicinal, and it was the basis for the development of aspirin. The leaves are a good fodder for livestock.

FIGURE 19.6. A "fairy ring" of resprouting redwood. Photo courtesy of Edward Z. Yang/Public Domain.

FIGURE 19.7. Short-rotation coppice of willow. These provide biomass and also wastewater purification, converting excess nitrates and phosphates into wood chips for fuel. Villeneuve d'Ascq, France. Photo courtesy of Lamiot/Creative Commons 3.0.

Harvest and processing. For most of these uses, willows are coppiced or pollarded. (Pollarding refers to coppicing higher on the trunk, out of reach of livestock). Willows respond well to coppicing, with some coppiced plants living 500 to 800 years, far beyond their natural life span.[39] Traditionally willows were coppiced manually. In modern short-rotation coppice, they are harvested by large chipping machines.

Carbon farming applications. In recent decades willows have received a lot of attention as a biomass crop. Typically they are grown in the short-rotation coppice system. Unfortunately, when they are grown in this kind of monoculture system, they live only 20 to 30 years instead of the centuries in traditional coppice systems. This increases labor and decreases carbon sequestration value. Willows are also used as a fodder bank species. Some are used as windbreaks and snow fences.[40] Willows are also frequently used today for erosion control and bank stabilization along streams and rivers. They help to treat and filter contaminated water. Willows might also make good contour hedgerows in sufficiently humid environments. They are already often planted on contour as part of erosion control and riverbank stabilization efforts.

Crop development. Breeding efforts for biomass production begin in the 1980s. Willows are easy to breed and easy to hybridize. European willow breeding emphasizes *Salix viminalis* and four or five others, while North American work is based on *S. purpurea* and several other species. Breeding work has increased biomass yields 50 percent in the last few decades.[41]

POPLARS (*POPULUS* SPP.)

Poplars are native to much of the northern temperate region and thrive from boreal through warm temperate climates as well as tropical highlands. Some are adapted to semi-arid sites although most prefer humid conditions. Like willows (*Salix* spp.), their close relatives, poplars are known for fast growth and rapid biomass accumulation. Most coppice well and are easily

FIGURE 19.8. Industrial-scale integrated food and feedstock system in Germany, with annual cereals intercropped with biomass poplar and black locust. Photograph courtesy of AgForward.

propagated from cuttings. Poplar leaves are a fine fodder for livestock. They have been grown as pollarded fodder trees in European pastures for centuries. Although their shade can reduce pasture productivity, this is compensated for by the availability of fresh-cut foliage during dry summers. The fallen foliage in autumn is also a quality fodder.

Uses. Poplars are used for a variety of biomass applications.

Yields. Poplars yield 10 to 15 t/ha of firewood annually.[42]

Carbon farming applications. Poplars are commonly grown as windbreaks. China's Three-North Shelterbelt, intended to hold back the expansion of the Gobi desert, covers 35.6 million hectares (88 million acres)—and 60 percent of its trees are poplars.[43] Fifteen or more species of poplars are intercropped with annuals across northern China.[44] In Italy poplars are pruned to be living trellises for grape growing.

Crop development. Today GMO poplars with modified lignins are in trials around the world. Poplars are grown in short-rotation coppice systems like willows, although they are vulnerable to pests and diseases in such intense monocultures. Nevertheless, coppiced poplars are an important feedstock for biofuel and other biomass products such as plywood and particleboard.

RATTAN (*CALAMUS* SPP.)

Rattans are vine-like, often spiny palms that can climb for 100 meters (328 feet) or more in humid tropical forest understories. The stems have been a valuable raw material for basketry, furniture, mats, clothing, and natural building for millennia. Mature canes resprout after cutting. Today they are a major item of commerce and the second most valuable product, after timber, to come out of Southeast Asia's forests. In fact, overharvesting of wild rattan has become an ecological crisis. Efforts to cultivate and domesticate rattan intensified in the 1980s.[45]

Uses. Rattans are used in craft applications as a high-value material rather than a bulk biomass feedstock. Some species of *Calamus* and the

FIGURE 19.9. Rattan (*Daemonorops jenkinsiana*), a vine-like palm, in Thailand. Photograph courtesy of Rick Burnette.

related *Daemonorops* have edible shoots or fruits. Some *Daemonorops* species exude a resin called "dragon's blood," which has medicinal uses and is also used as violin dye.[46] Scientists have found that rattan is ideal as a basis for "artificial bone" for implantation into (presently) sheep and (perhaps someday) humans.[47]

Carbon farming applications. As a high-value product for the shady understory, rattans have great potential in multilayered tropical agroforestry systems. In Indonesia, rattan is a critical component of traditional swidden systems.[48]

Giant Grasses and Grass-Like Plants

Perennial giant grasses are among the most widely used dedicated biomass crops. Grasses can be an extremely productive source of biomass. Many of the best use the C_4 metabolism, an alternative respiratory approach that allows efficient, fast growth in hot, dry conditions. Historically, giant grasses have been used in natural building (as thatch and more), basketry and mat weaving, and papermaking, as well as for fodder, charcoal, and "firewood." Sedges, rushes, and cattails, which are not grasses but are grass-like, have similar uses. Today much of the interest in giant grasses is as a

TABLE 19.4. Perennial Biomass Grass Crops

Latin Name	Common Name	Origin	Climate	Cultivation
Arundo donax	Giant reed	Europe, Asia	Cold and warm temperate, subtropical, tropical, semi-arid to humid	Biomass crop
Carex meyeriana		East Asia	Boreal to warm temperate, aquatic	Regional crop
Carex morrowii	Iwashiba	East Asia	Boreal to warm temperate, aquatic	Regional crop
Chrysopogon zizanioides	Vetiver	India	Warm temperate to tropical, semi-arid to humid	Minor global crop
Cortaderia selloana	Pampas grass	South America	Warm temperate to tropical, semi-arid to humid	Minor global crop
Cynara cardunculus	Cardoon	Mediterranean, North Africa	Warm temperate, Mediterranean, semi-arid	Minor vegetable; new crop as oilseed and biomass
Cyperus cephalotes		South Asia through Australia	Warm temperate to tropical, aquatic	Regional crop
Cyperus giganteus	Capim de esteira	Mexico to Argentina	Warm temperate to tropics, aquatic	Regional crop
Cyperus papyrus	Papyrus	North Africa	Mediterranean, subtropical, aquatic	Regional crop, historic
Cyperus textilis		Southern Africa	Mediterranean, warm temperate, aquatic	Regional crop
Echinochloa polystachya	Aleman grass	Americas	Warm temperate to tropical, humid	New crop
Eulaliopsis binata	Sabai grass	South to Southeast Asia	Tropics, subtropics, highlands; humid	Regional crop
Gynerium sagittatum	Wild cane	Tropical Americas	Humid tropical lowlands	Regional crop
Juncus effusus	Mat rush	East Asia, North America	Cold to warm temperate, aquatic	Regional crop (Asia), wild-managed
Miscanthus sacchariflorus	Amur silver grass	East Asia	Cold to warm temperate, humid	Minor global crop
Miscanthus sinensis	Eulalia	East Asia	Cold temperate to subtropical, humid	Minor global crop
Miscanthus × giganteus	Giant miscanthus	Hybrid of Asian species	Cold and warm temperate, subtropics, humid	Important new crop
Panicum virgatum	Switchgrass	North America	Boreal, cold and warm temperate, semi-arid to humid	Major crop
Pennisetum purpureum	Napier grass	Africa	Warm temperate, subtropical, tropical, semi-arid to humid	Important fodder grass
Phalaris arundinacea	Reed canary grass	Europe, Asia, North Africa, North America	Cold and warm temperate, semi-arid to humid	New crop
Phragmites australis	Common reed	Europe, Asia, Africa, North America	Cold temperate through tropical, humid	Biomass crop
Saccharum hybrids	Energy cane	Hybrid of Asian species	Tropical and subtropical, humid	New crop
Saccharum officinarum	Sugarcane	New Guinea	Humid tropical and subtropical lowlands	Yes
Schoenoplectiella mucronata		Eurasia, Australia	Warm temperate to subtropical, Mediterranean, aquatic	Regional crop
Schoenoplectus californicus ssp. tatora	Totora	Bolivia, Peru, Chile	Tropical highlands, aquatic	Regional crop
Stipa tenacissima	Esparto grass	Northern Africa, Mediterranean	Semi-arid, warm temperate to subtropical	Minor crop, historic importance
Typha angustifolia	Cattail	Europe, North America	Boreal to tropical, aquatic	Wild-managed, experimental crop
Typha latifolia	Cattail	Northern temperate climates, Africa, Australia	Boreal to subtropical, aquatic	Minor global crop

Sources: Hanelt, *Mansfeld's Encyclopedia of Agricultural and Horticultural Crops*; El Bassam, *Handbook of Bioenergy Crops.*
Note: There are thousands of candidates for biomass grasses. Any sufficiently vigorous species could serve. This table includes species that are currently cultivated or under development as dedicated biomass crops for fuel, paper, and woven items. Also included are grass-like plants and cardoon (which is not related but is a promising new herbaceous biomass crop).

lignocellulosic biofuel feedstock. They are suitable for combustion, densification, gasification, pyrolysis, and, unlike some woody biomass crops, biogas. My primary interest in giant grasses is as a feedstock for materials rather than energy.[49]

Sugarcane (*Saccharum officinarum*) is the most widely grown biomass grass. Many more species are or have been important. Papyrus (*Cyperus papyrus*) is an ancient crop for paper- and mat making. Many other rushes (*Juncus*) and sedges (*Cyperus*) are grown for materials to weave mats and baskets. Totora (*Schoenoplectus californicus* ssp. *tatora*) was domesticated in ancient times in the high Andes for weaving and fodder. Species including pampas grass (*Cortaderia selloana*), sabai grass (*Eulaliopsis binata*), and esparto (*Stipa tenacissima*) are grown as important feedstocks for papermaking. Improved varieties of esparto have been selected for superior paper production.[50] Sabai grass is also being investigated as a binder in biobased composite materials, as are other perennial grasses.[51]

GIANT MISCANTHUS (*MISCANTHUS* × *GIGANTEUS*)

Miscanthus grasses have been popular ornamentals since the 1960s. During the 1980s they attracted the interest of German breeders as a biomass crop. Professor Wolfgang Stander, a miscanthus breeder, is reported to have said, "I have seen the future, I can see grass growing twelve feet high that can fuel power stations."[52] The genus *Miscanthus* hails from eastern Asia and is noted for aggressive naturalization in many areas. Scientists developed sterile hybrids that cannot set viable seed— hybrids that are also extremely vigorous.[53]

Giant miscanthus is a C_4 grass, and an outstanding biomass producer in temperate climates. It needs very little nitrogen fertilizer. Many other *Miscanthus* species

FIGURE 19.10. Sterile hybrid giant miscanthus, an important new perennial biomass grass. Photograph courtesy of Hamsterdancer/ Creative Commons 3.0.

are also in use as biomass crops. Some have even been crossed with sugarcane, helping to create hardy "energy cane" for biomass and ethanol production.[54]

Uses. Most giant miscanthus is used as biofuel, but it could serve as a feedstock for any number of biobased materials.

Harvest and processing. Hybrid miscanthus is highly productive. It can be harvested with large cutter-balers.

Carbon farming applications. Little or nothing is written about miscanthus in agroforestry. It could be a good candidate for contour hedgerow systems and perhaps a fodder bank species as well.

CHAPTER TWENTY

Industrial Starch Crops

Starch is a carbohydrate, formed from a chain of glucose sugar molecules. In plants, it provides energy storage. For humans, it provides a basic source of energy from food—energy that is often also stored. But although we rarely think of it in this way, starch is also the second most commonly used carbohydrate in industrial applications; only cellulose used for papermaking is used more.[1] Unlike many kinds of industrial crops, there is no "synthetic starch" being made from fossil fuels. The situation, however, is not much better. All, or virtually all, industrial starch comes from annual food crops grown in conventional tillage systems. In fact, 17 percent of European grain goes to papermaking every year.[2] So we're using annuals where perennials might do, and we're using food to make cardboard and drywall. This is wasteful and inefficient, and if we want to minimize the use of annuals, we need to find a better way.

Most of the crops reviewed in chapters 12 and 13 could serve as industrial starch feedstocks. Here we'll review the additional category of inedible starch crops.

Bioplastics

In 1976 world plastic production was 50 million tons. By 2010 it was 230 million tons. And by 2015 it is projected to reach 330 million tons.[3] Almost all plastics are made from petroleum or natural gas; 4 to 5 percent of all refined oil and gas goes into plastic manufacturing.[4] Plastics are a clear case where replacing fossil fuels means more than looking at energy. You can't make plastic from solar or wind.

However, people have been making plastics from natural materials since the mid-1800s.[5] In fact, Henry Ford

FIGURE 20.1. Bioplastic cutlery made from cellulose acetate. Photograph courtesy of F. Kesselring/Creative Commons 3.0.

debuted a car mostly made from soy-based plastics in 1941, although World War II distracted the world from this achievement.[6] Bioplastics are "biodegradable plastics whose components are derived entirely or almost entirely from renewable raw materials."[7] They can be made from cellulose, starch, oils, resins, and other plant- and animal-based materials. Today scientists are at work developing biobased, compostable plastics made from renewable feedstocks that can break back down into organic matter.[8] Industry experts think that about 90 percent of current plastics could be made from biobased feedstocks, leaving about 10 percent of specialty plastics that require fossil fuels as raw material.[9] What's missing is an emphasis on perennial, non-destructively harvested feedstocks, especially non-food crops.

Bioplastics are not necessarily biodegradable, nor is their production necessarily non-toxic, although scientists are making headway on both of these aspects of plastic manufacturing and use.[10] Some compostable bioplastics are already available, often simple, starch-based products, although at this point they are mostly from GMO corn. These include the extruded foam packing peanuts you may have received in the mail, as well as agricultural plastics, trash bags, plasticware, and diapers. More durable bioplastics can be created by fermenting starches and other biomaterials. Many more are under development.[11]

When I think of plastic making I imagine giant industrial facilities, so I was surprised and pleased to read E. S. Stevens's *Green Plastics*, which gives recipes to make bioplastics in your kitchen with simple materials like cornstarch, glycerin, and gelatin. I set out one night to make some plastic and found it was just like cooking. In the end, I had a concoction of Korean acorn starch, gelatin, and glycerin simmering on the stove. I poured it into the fruit leather tray of a food dehydrator and the following morning I had a very thin, gummy cellophane-like plastic. It seems worthless for any application, but a good demonstration of using starch from perennial crops for bioplastic on a very small scale.

What does biobased, carbon-sequestering, decentralized, low-tech, socially just plastic production look like? Open Source Ecology, founded in 2003 by Marcin Jakubowski, is working to develop a toolkit of the 50 most important machines for civilization that can be built from scratch using simple technology. Among the items in their "Global Village Construction Set" is a bioplastic extruder. I am sometimes asked if bioplastics can work with 3-D printing. They can. Open Source Ecology is developing it.[12]

I've become hopeful about the potential for small-scale, regional bioplastic facilities around the world, providing necessities like irrigation pipes from local, perennial feedstocks. A hectare of farmland can produce feedstock for 2 to 4 tons of bioplastic.[13] We could meet the world's current plastic use, which could and should be halved or more, with 4 to 5 percent of agriculture land.[14] When we factor out bioenergy, meeting human non-food needs with perennial crops starts to look feasible.

INDUSTRIAL STARCH CROP USES

MATERIALS

Starches are important constituents of paper and cardboard, binding to cellulose fibers to strengthen the final product. They are used for their binding properties in textiles.[15] Starches are used, too, for their binding and thickening properties in numerous construction products, such as plasterboard, glues, joint compounds, paints, foams, and ceiling coatings.[16] Starches are also important components of bioplastics. An example already in commercial production is starch-based packing foam, which replaces petroleum-based Styrofoam packing peanuts.[17] During World War II, chemists developed a technique to make synthetic rubber from alcohol derived from starchy grains such as corn, wheat, and potatoes.[18]

CHEMICALS

Starches have binding and stabilizing properties that make them useful in numerous chemical products. For example, they are used in pharmaceuticals, agrochemicals, and other products as binders, coaters, flocculants, coagulants, finishing agents, and stabilizers.[19]

Starches are also used as fermentation substrates for the production of various chemicals. Products include pharmaceuticals, glucose, biopolymers, and "platform chemicals" such as lactic acid, which are used as building blocks in the chemicals industry.[20] Solvents are produced from fermented starch. Among many uses, they are required to extract naturally occurring hydrocarbons from many "petrocrops."[21]

ENERGY

Starches can be used to produce ethanol, although sugar is more commonly used.[22] Ethanol itself is used as fuel and as a feedstock for other industrial processes.

PERENNIAL INDUSTRIAL SUGAR CROPS

Industrial sugar is a massive global industry. The primary product is ethanol. The drive to increase agrofuel production in Brazil has led to many social and ecological problems, the most alarming being the enslavement of laborers.

MATERIALS

The residual by-product from sugarcane extraction is called bagasse and is an important biomass feedstock around the world. Other sugar crops such as sugar palms do not produce much residue.

CHEMICALS

Industrial sugars are used to make solvents. In the first half of the 20th century, chemurgists developed techniques to derive many products, including rubber, from ethanol.[23]

ENERGY

The primary use of industrial sugar is as an ethanol feedstock. Industrial sugar crops are the same species as edible sugar crops. The sugar is produced from pods, fruits, extracted sugarcane juice, and saps (from palms, agaves, and maples).

Perennial Crop Types

Non-destructively harvested perennial starches include nuts, grains, woody pods, starchy fruits, starchy resprouting trunks, and aerial tubers. Edible perennial starch crops can be used for ethanol production and industrial starch uses, although food should come first. More interestingly, the need for non-destructively harvested perennial starch crops for industrial purposes offers a somewhat novel and intriguing use for a class of plants that has, until now, been largely neglected: plants producing poisonous or non-edible carbohydrates, such as inedible nuts and starchy fruits. Why not just use edible

perennial carbohydrates? Because broadening the biodiversity of crops we grow can help to reduce pest and disease pressure, gives us more elements to integrate in polycultures, and lets us grow industrial starch in more diverse soils and climates. To my knowledge, industrial use of inedible starches is still hypothetical.

This would be a great use for the inedible forms of air potato, which are already abundant in areas where they have naturalized and are native to almost half the world's humid tropics. This might also provide a use for toxic nuts that have traditionally required extensive processing before eating, including horse chestnuts, cycad nuts, and Moreton Bay chestnuts. It would, of course, be crucial to distinguish between edible and non-edible crops during harvesting and processing.

Breeding oaks for industrial starch would simplify the domestication process. Annual bearing would still be a goal, but there would be no need to breed out the tannins. In fact, tannins themselves are a useful industrial product that could be removed in the processing plant. Betel nut (*Areca catechu*) is a widely grown, addictive, carcinogenic crop chewed for a stimulant effect. It yields 1.5 to 2.5 t/ha of nuts,[24] which might be detoxified and used as a feedstock for industrial starch. The many inedible palm fruits, which are borne so abundantly, should be evaluated for starch content. Overall, this is a difficult category to collect data on because it has not been important until recently. Individual regions of the world could assess their local native species for potential candidates.

CYCADS (DIVISION CYCADOPHYTA)

Although these ancient plants superficially resemble palms, they are part of a lineage that arose long before the origin of flowering plants. They are adapted throughout the tropics and subtropics with species for sun and shade, desert, swamp, and rain forest. These slow-growing but long-lived plants are utterly unrelated to any commercial crops, offering a chance to produce industrial starch without any of the drawbacks of working with existing food crop families. Genera to investigate include *Cycas, Dioon, Encephalartos, Macrozamia, Microcycas,* and *Zamia*.[25] There are native species throughout the tropics.

TABLE 20.1. Inedible Starch Crops

Latin Name	Common Name	Origin	Climate	Cultivation
Aesculus spp.	Horse chestnut	Northern temperate climates	Cold to warm temperate, Mediterranean, semi-arid to humid	Ornamental only
Areca catechu	Betel nut	Asia	Humid lowland tropics	Minor global crop
Caesalpinia digyna	Teri-pod	Asia	Warm temperate to tropical, semi-arid to humid, lowlands and highlands	Regional crop only
Castanospermum australe	Moreton Bay chestnut	Australia, Oceania	Subtropics to tropics, arid to humid	Regional crop
Coix lacryma-jobi	Job's tears	Asia	Humid tropics	Ornamental perennial types
Corynocarpus spp.	Karaka	New Zealand	Warm temperate to subtropical, humid	Wild
Cycas spp.		Africa, Asia	Warm temperate to tropical, arid to humid	Ornamental
Dioscorea bulbifera	Air potato (toxic forms)	Asia, Africa, Australia	Subtropics to tropics, highlands and lowlands, humid	Minor global crop
Encephalartos spp.		Africa	Warm temperate to tropical, arid to humid	Ornamental only
Hyphaene thebaica	Doum palm	Africa	Warm temperate, subtropical, tropical, Mediterranean, semi-arid to humid	Regional crop; young fruits eaten, mature for mannan
Maclura pomifera	Osage orange	North America	Cold to warm temperate, semi-arid to humid	Regional crop
Macrozamia spp.		Australia	Warm temperate to tropical, arid to humid	Ornamental only
Metroxylon amaricum		New Guinea, Oceania	Tropical humid lowlands	Regional trunk starch crop; fruits yield mannan
Microcycas calocoma		Cuba	Tropics, semi-arid to humid	Ornamental only
Phytelephas macrocarpa	Tagua	South America	Tropical humid lowlands	Mannan, important wild-collected
Zamia spp.		Americas	Warm temperate to tropics, arid to humid	Ornamental only

Source: Wink and van Wyk, *Mind-Altering and Poisonous Plants of the World.*
Note: Inedible (or poisonous), non-destructively harvested starch crops of the world. This is a very preliminary table; surely many more species are present throughout the globe.

FIGURE 20.2. Toxic, starchy seeds of the cycad *Macrozamia communis*. Photograph courtesy of AYArktos/Creative Commons 3.0.

Uses. Cycad "nuts" can be produced in fairly large amounts and are high in starch. Most are very toxic, though some species are processed for food during famines.[26]

Carbon farming applications. Most or all cycads fix low amounts of nitrogen through a partnership with blue-green algae in the roots.

HORSE CHESTNUTS AND BUCKEYES (*AESCULUS* SPP.)

This toxic, starchy nut genus in the soapnut family has representatives all over the cold temperate parts of the world, as well as some tropical highland and Mediterranean climates.[27] Given their wide geographic potential, domestication and cultivation of *Aesculus* nuts for industrial starch is worth investigation.

Uses. The nuts closely resemble chestnuts, and they seem to yield almost as well despite no domestication efforts.

Carbon farming applications. Many *Aesculus* species coppice well, and all produce hard, durable nuts. This indicates that they could be investigated for woody agriculture.

Vegetable Ivory: A Starch-Like Product

Several palms produce large, hard seeds that have been used as a substitute for ivory. These include two species of humid tropical wetlands: tagua or ivory nut palm (*Phytelephas macrocarpa*), a sago (*Metroxylon amicarum*), as well as the tropical desert-loving doum palm (*Hyphaene thebaica*). Tagua is the main vegetable ivory of commerce. There are reports of other species as well. Historically, tagua seeds have been important for many industrial uses, particularly for button making. At the highest point of the industry in 1929, Ecuador exported 2,220 tons of seeds.[28] Petroleum-based plastics soon replaced it, although there is a small resurgence today, producing beads and jewelry from wild-harvested nuts.

Like coconuts, the young nuts are full of sweet water, which gradually enters an edible jelly-like state. Finally it hardens into "ivory." This material is made of long-chain storage carbohydrate biopolymers called mannan. Mannan is not quite starch but is similar, composed of long chains of sugars for storing energy.[29] This makes it an intriguing potential feedstock for bioplastics.

FIGURE 20.3. Inedible starchy nuts of horse chestnut (*Aesculus hippocastanum*). Photograph courtesy of Andrew Dunn/Creative Commons 2.0.

FIGURE 20.4. Tagua, a starch-like industrial feedstock, fruiting in Ecuador. Photograph courtesy of Romanofski/Creative Commons 3.0.

Yields of tagua are noteworthy. A tree can produce 45 to 100 kg (100 to 220 pounds) of fruits every year for 50 to 100 years. They reach maturity at 10 to 15 years.[30] If planted at 143 trees per hectare (as oil palms are), they would yield 6.4 to 14.3 t/ha in a monoculture—impressive and highly competitive with many current edible starch bioplastic feedstocks. The ability of these species to produce in wetlands and deserts is also notable. Perhaps they could serve as a foundation for developing regional bioplastic processing industries to meet local needs in tropical areas.

CHAPTER TWENTY-ONE

Industrial Oil Crops

I magine modern civilization without moving metal parts. Industrial machinery and transportation motors require lubricants to function—and today most of these are made from petroleum. In the United States alone 4.2 million liters of industrial lubricants and 5.3 million liters of automotive lubricants are used every year. Together they account for about 1 percent of US petroleum use.[1] Most vegetable oils need to be refined and stabilized to be used for these purposes, although some such as castor oil can be used directly. During World War II and again during the 1970s energy crisis, vegetable oils were utilized more extensively as lubricants.

Interest in biobased lubricants is returning, although most interest in oil crops today is for biodiesel production. Today perennial oil crops such as jatropha and oil palm are grown as agrofuels in massive monoculture plantations to serve as biodiesel feedstocks, at great social and ecological cost. Much of the world's industrial oil also comes from annual crops, notably GMO canola and soy, peanuts, and other edible crops that it would be better to reserve as human food crops if we are looking to minimize annual crops. In fact, if the entire annual US crop of soybeans were converted to biodiesel it would only replace 6 percent of diesel used in the country.[2] I propose that the best use of oil crops is

TABLE 21.1. Inedible Oil Yields Compared

Latin Name	Common Name	Climate	Seed or Fruit Yield t/ha	Oil Yield t/ha
Sapium sebiferum	Chinese tallow tree	Warm temperate, tropical highlands, semi-arid to humid	14[a]	5.4[a]
Jatropha curcas	Jatropha	Tropical lowlands, semi-arid to humid	1–15[b]	0.3–5[c]
Aleurites moluccanus	Candlenut	Tropical humid to semi-arid	16[d]	3[d]
Ricinus communis	Castor bean	Tropics, subtropics, high or lowlands, semi-arid to humid	0.5–5[e]	0.3–2.7[e]
Pongamia pinnata	Pongamia	Subtropics, tropics, lowlands, highlands, humid	5–8[f]	1.8[f]
Prunus pedunculata		Cold to warm temperate, arid to semi-arid	2.5[g]	1.5
Maclura pomifera	Osage orange	Cold to warm temperate, semi-arid to humid		1.5[h]
Simmondsia chinensis	Jojoba	Subtropics, arid to semi-arid	2.2–4.5[i]	0.5–1.1[j]
Azadirachta indica	Neem	Tropics, humid to semi-arid		0.5[k]

Note: Organized in order of maximum yield potential. These yields compare favorably with the edible oils from chapters 15 and 16.

[a] Duke, *Handbook of Nuts*, 264.
[b] Sharma and Kuman, "*Jatropha curcas*," 202
[c] At 37% oil, as reported in Duke, *Handbook of Nuts*, 179.
[d] Ibid., 13.
[e] Duke, *Handbook of Energy Crops*.
[f] Kesari and Rangan, "Development of *Pongamia pinnata* as an alternative biofuel crop," 132.
[g] Chu, Xu, and Zhang, "Production and properties of biodiesel produced from *Amygdalus pedunculata* Pall," 374.

[h] Moser et al., "Preparation of fatty acid methyl esters from Osage orange (*Maclura pomifera*) oil and evaluation as biodiesel," 1870.
[i] Duke, *Handbook of Nuts*, 274–75; El Bassam, *Handbook of Bioenergy Crops*, 216.
[j] Speculated. Duke, *Handbook of Nuts*, 274–25; El Bassam, *Handbook of Bioenergy Crops*, 95.
[k] El Bassam, *Handbook of Bioenergy Crops*, 252.

as feedstocks for bioplastics and lubricants for industry and transportation—necessary products that, unlike energy, can't be made from sunlight, wind, geothermal, or hydropower energy.

Perennial Industrial Oil Crop Types

Because the focus here is on industrial oils, I'll cover only inedible oils. The many edible oils profiled in chapters 15 and 16 can also serve the same functions, although it seems a shame to run a diesel tractor on olive oil! Edible or not, oils are almost exclusively pressed from the seeds or fruits of trees, shrubs, and herbs, making them particularly well suited to non-destructive harvest. The presscake, or oilcake, that is left behind has various new and traditional uses, including livestock feed and the full range of biomass products.

Most of the best producers in this category are woody plants in the Euphorbiaceae family, including Chinese tallow tree, jatropha, candlenuts, and castor bean. Not all oils are equal, and each has a different composition. Some lend themselves especially well to certain industrial purposes. In fact, some are being bred to improve oil profiles for superior biodiesel and other feedstock suitability.

CASTOR (*RICINUS COMMUNIS*)

Castor is a rare example of a global, domesticated perennial industrial crop. A small tree or large shrub, it was already an important crop 6,000 years ago in Egypt, where it was grown for lamp oil and medicine.[3] It has spread from its origin in East Africa and India to become a pantropical weed, popular ornamental, and important feedstock. Like many industrial crops, castor is in the Euphorbiaceae family.

There were 1.7 million hectares (4.2 million acres) of castor in production in 2012. They produced 1.9 million tons of seed worth $1.1 billion US.[4] This major world crop is suited to the tropics and subtropics, highlands and lowlands, in arid to humid conditions and is also a salt-tolerant halophyte. As a perennial, castor is usually grown for two to four years.[5] The Global Invasive Species Database designates it as an invasive species.[6]

INDUSTRIAL OIL CROP USES

MATERIALS

Vegetable oils are important components of bioplastics. Glycerin, a by-product of biodiesel production from oils, is particularly important for this purpose. The residues, or "presscakes," left behind after extracting oils have a variety of uses, from livestock feed to biodigester biomass feedstock.

CHEMICALS

Vegetable oils are used as industrial lubricants, from motor oil to grease to hydraulic fluid and are also used in soaps, surfactants, solvents, paints, varnishes, and agrochemicals. Some platform chemicals and other feedstocks are derived from plant-based oils.

ENERGY

Vegetable oils are the main feedstock for biodiesel. Diesel motors can be modified to use straight vegetable oil as fuel.

Uses. The seeds (erroneously called "beans") are 43 to 53 percent oil. They contain the extremely toxic poison ricin. A small handful of seeds is enough to kill an adult human. Castor oil is a high-quality lubricant, still in use today. It is used to make varnishes, paints, waxes, and even a nylon-like fiber called rilson. The oil is high in ricinoleic acid; its unusually polar chemical structure makes it an especially suitable feedstock for chemical preparations. The oil can be used directly as a transportation lubricant and is used today in jet engines and race cars. The drawback of raw, unprocessed castor oil is that it forms gums over time, requiring occasional rebuilding of engines. Today castor is being grown as a feedstock for biodiesel. Castor oil has many traditional and modern medicinal uses (many poisonous plants are medicinal in small amounts, though in this case the poison seems to be removed with the presscake). The presscake itself is used as a high-nitrogen fertilizer or can be detoxified and fed to livestock.[7]

TABLE 21.2. Inedible Oil Crops

Latin Name	Common Name	Origin	Climate	Cultivation
Aleurites moluccanus	Candlenut	Malay archipelago	Tropical humid to semi-arid	Regional
Anacardium occidentale	Cashew	Brazil	Tropical lowland, semi-arid to humid	Global
Azadirachta indica	Neem	South Asia	Arid to humid tropical lowlands	Minor global
Caesalpinia digyna	Teri-pod	Asia	Warm temperate to tropical, semi-arid to humid, lowlands and highlands	Regional
Ceiba pentandra	Kapok	Tropical Americas, Africa	Tropics, semi-arid to humid	Minor global
Croton tiglium	Croton oil plant	South and Southeast Asia	Arid to humid subtropics and tropics, lowlands	Regional
Cuphea micropetala	Achanclan	Mesoamerica	Warm temperate to tropical, humid	Regional
Hevea brasiliensis	Para rubber	South America	Humid tropical lowlands	Global rubber crop, minor oil
Jatropha curcas	Jatropha	Tropical Americas	Tropical lowlands, semi-arid to humid	Minor global
Licania rigida	Oiticica	Americas	Tropics, semi-arid	Regional
Litsea calophylla	Tagutugan	Asia	Humid tropical lowlands	Regional
Litsea glutinosa	Maida lakri	Asia	Humid tropical lowlands	Regional
Melia azedarach	Chinaberry	Asia	Warm temperate to tropics, lowlands and highlands, semi-arid to humid	Minor global
Parinari curatellifolia	Mabo cork-tree	Central Africa	Semi-arid tropics	Regional
Pongamia pinnata	Pongam	India, Southeast Asia, Australia	Subtropics, tropics, lowlands, highlands, humid	New crop
Prunus armeniaca	Apricot	Central Eurasia	Cold to warm temperate, semi-arid to humid	Minor global
Prunus armeniaca var. *mandschurica*	Manchurian apricot	East Asia	Boreal, cold temperate, semi-arid to humid	Regional
Prunus pedunculata		East Asia	Cold to warm temperate, arid to semi-arid	Wild
Prunus sibirica	Siberian apricot	East Asia	Boreal, cold temperate, semi-arid to humid	Regional
Prunus spinosa	Sloe	Europe	Warm and cold temperate, humid	Regional
Rhus succedanea	Wax tree	China	Warm and cold temperate, humid	Regional
Ricinus communis	Castor bean	Africa or India	Tropics, subtropics, highlands or lowlands, arid to humid	Global
Sapindus saponaria	Soapnut	Americas	Semi-arid to humid tropics	Minor global
Sapium sebiferum	Chinese tallow tree	Asia	Warm temperate, tropical highlands, semi-arid to humid	Minor global
Semecarpus anacardium	Indian marking nut tree	India	Semi-arid to humid tropics	Regional
Simarouba amara	Paradise tree	Mesoamerica, Caribbean	Tropical humid lowlands	Experimental
Simmondsia chinensis	Jojoba	Southwest America, Mesoamerica	Subtropics, arid to semi-arid	New crop
Terminalia bellirica	Celeric myrobalan	India, Southeast Asia	Semi-arid tropical lowlands	Regional
Toxicodendron sylvestre	Yame-haze	East Asia	Warm and cold temperate	Regional
Toxicodendron vernicifluum	Lacquer tree	China	Warm temperate, humid	Regional
Trichilia emetica	Mafura butter	Africa, Middle East	Semi-arid to humid tropical lowlands	Regional
Vernicia fordii	Tung oil tree	China	Subtropics, semi-arid to humid	Minor global
Vernicia montana	Mu-oil tree	Asia	Warm temperate to tropical, semi-arid to humid	Regional
Virola sebifera	Virola nut	Americas	Humid tropical lowlands	Regional

Note: These crops, combined with the edible oil crops from chapters 15 and 16, can provide oil feedstock for a perennial biobased industry. Many are already important industrial crops.

Figure 21.1. Castor beans, an ancient perennial industrial crop, at Las Cañadas.

Yield. Yields range from 1 to 3 t/ha, with 0.3 to 0.4 t/ha in semi-arid areas without irrigation.[8] Average global yield is 1.16 t/ha.[9]

Harvest and processing. In the tropics, castor is typically grown as a perennial. It yields the first year and continues to do so for 10 to 15 years. In colder areas, it is grown as an annual. Cultivars have been developed for multiple variables, including annual and perennial types, hand- and machine-harvested forms, and strains with high content of particular desired chemicals.[10]

Carbon farming applications. Castor is quite "sociable" and commonly grown in polyculture systems. Often it is intercropped with annuals when planted. Later it can be used as a shade provider for understory crops such as ginger, turmeric, and sugarcane. Castor is often grown at field borders and can be used as a windbreak. It is also grown as a host for lac insects and silkworms.[11] This species might be adapted as a short-lived woody agriculture crop.

CHINESE TALLOW TREE (*SAPIUM SEBIFERUM*)

Chinese tallow tree is another tree in the Euphorbiaceae family that produces industrial oils. It is native to eastern Asia and has naturalized, sometimes aggressively, in other semi-arid to humid, warm temperate, and subtropical regions of the world. It is a minor global crop, cultivated on 400,000 hectares (988,000 acres) in 2012. World yield was 985,000 tons.[12] The Global Invasive Species Database designates it as an invasive species.[13]

Uses. Tallow tree produces two different oils. The waxy outer layer of the fruits is called tallow and is

FIGURE 21.2. The fruit of a Chinese tallow tree, a very high-yielding temperate oil crop. Photograph courtesy of KENPEI/ Creative Commons 3.0.

melted off the seeds with hot water or steam in processing. The tallow is used for candles, soaps, and fuel. At times, the tallow is also used to prepare an edible replacement for lard. The oil from the seed is called stillingia oil and has many uses as varnish, paint, machine oil, and illuminant. Both oils have medicinal uses, and both have potential for bioplastics and biodiesel.[14]

Yields. Yields are high, with 14 t/ha of fruits yielding 2.6 t/ha of oil and 2.8 of tallow, for a combined oil yield almost as high as the world champion African oil palm—but suited to much colder climates.[15] There is a lot of interest in developing *Sapium* as a biodiesel feedstock. Global average in yield in 2012 was 2.4t/ha.[16] Trees can yield until they are 70 to 100 years old.[17]

Harvest and processing. The seeds are fairly easily removed from the fruit for separate processing of the tallow and stillingia oil.[18]

Carbon farming applications. Tallow tree is intercropped with annual crops in southwest China.[19] Tea grown under tallow tree yields better, particularly in dry years.[20] It might be suited to woody agriculture.

CANDLENUT, TUNG, AND MU-OIL TREES (*ALEURITES* AND *VERNICIA* SPP.)

The genus *Aleurites* (recently split into *Aleurites* and *Vernicia*), which belongs to the Euphorbiaceae family, is native to Southeast Asia and the Western Pacific. Multiple species are in use as industrial crops. These are minor global and regional crops.

Candlenut (*A. moluccanus*) was domesticated in the Malay archipelago and carried throughout the Pacific in ancient times. Today it is the state tree of Hawaii. Author Craig Elevitch describes candlenut as "one of the great domesticated multipurpose trees of the world."[21] Unlike other *Aleurites* species, the roasted and pounded seeds are used as a condiment in small amounts, with at least one variety ('Maewo') edible and apparently non-toxic.[22] If a number of candlenuts are placed on a skewer or string, they can be burned like a candle, with each nut in turn burning for about three minutes. Yields are high, up to 16 t/ ha with about 3 t/a of oil.[23] Candlenuts grow in tropical humid and semi-arid climates. They are considered invasive species in the Global Invasive Species Database.[24]

Tung (*V. fordii*) is similar to candlenut but more cold tolerant. It requires a hot, humid summer and at least 300 to 400 "chilling hours" of low, near-freezing temperatures. Today it remains an important finish for high-quality furniture and musical instruments. At one time, a tung oil industry was established in the United States, notably in Florida. Hurricanes and extreme cold events put an end to this effort after several decades.[25] The global average yield is 2.8 t/ha, and 184,000 hectares (455,000 acres) are planted to tung globally. This land yields 518,000 tons, worth $166 million US.[26]

Mu-oil trees (*V. montana*) are similar to tung but have a more cosmopolitan range, able to produce in warm temperate through tropical conditions. They are beginning to replace tung in areas where its climate sensitivity is an impediment to production.[27]

Figure 21.3. Candlenuts, a multipurpose industrial feedstock, at Fruit and Spice Park in Florida.

Figure 21.4. A jatropha plantation in Hawaii. Photograph courtesy of Forest and Kim Starr/Creative Commons 2.0.

Uses. The nuts are pressed for oil, which is used as a furniture finish, varnish, paint, illuminant, linoleum feedstock, and lubricant. In most cases the nuts and oil are toxic. They have application as bioplastic and biodiesel feedstocks. Medicinal uses have a long history as well. Improved varieties have been developed.[28]

Carbon farming applications. These species could be tested for woody agriculture.

JATROPHA (*JATROPHA CURCAS*)

Jatropha is a Euphorbia family shrub from Mesoamerica, growing in semi-arid tropical lowlands. I remember when, as a student at the Institute for Social Ecology in 1994, I took a field trip to the workshop of appropriate technology engineer Carl Bielenberg. Carl lit a jatropha nut, and I was amazed to see it burn with a bright blue flame. At the time, biodiesel was a wild and crazy new idea. The decades that followed saw rapid expansion of jatropha planting for agrofuel biodiesel in Africa and Asia. India in particular has selected jatropha as a key element of its energy program, with 10.9 million hectares (26.9 million acres) set aside for jatropha production.

Much of the excitement about jatropha was the idea that it could bear on marginal land.[29] Unfortunately, this led to jatropha becoming one of the most oversold and underdelivering "miracle" crops of the 21st century thus far. While jatropha grows fine on marginal land, it does not yield well there. In one study, it yielded 20 percent of what was expected. This has led to the need for irrigation and fertilization and undermined the economic and non-food competition arguments for jatropha production. Those "marginal lands" have been occupied in many cases by subsistence farmers and pastoralists, who have become displaced by industrial monocultures. Despite this, more jatropha expansion is on the horizon.[30]

There are valuable lessons here. First, any crop—and jatropha is a fine one with enough water and decent soils—can be a nightmare if grown in monoculture. Second, don't spread a crop rapidly until it has been tested in the conditions where you will be growing it. Third, if you plant out large tracts of undomesticated crops, like jatropha, you may be disappointed. The story of jatropha tempers my enthusiastic visions of expansive plantings of mesquite, edible acacias, and other perennials in the world's drylands. Jatropha still has many useful applications. With sufficient moisture and fertility, in integrated farming systems, it could provide bioplastic or lubrication feedstock, or even some biodiesel for local use.

Uses. Jatropha's primary use today is as an inedible oil for biodiesel. It has traditional uses in Mesoamerica, including use as an illuminant, for soap, and as a candlenut. Some forms even have edible nuts, which taste to me something like pistachios.[31]

Yield. Jatropha yields 1 to 15 t/ha of nuts, usually on the lower end of that range.[32] At 37 percent oil, that would produce 0.3 to 5 t/ha of oil.[33]

Harvest and processing. Currently most jatropha is hard-harvested. Researchers and farmers are looking at mechanical harvest techniques. The challenge is that the fruits do not ripen all at once, spreading the harvest over several months.[34]

Carbon farming applications. Jatropha makes a fine living fence, growing easily from live stakes. Because the foliage is toxic to livestock, it is an effective green corral or livestock excluder.[35] It is also used for crop shade and can be grown as a host plant for lac insects. Jatropha is worth testing as a woody agriculture species.[36]

JOJOBA (*SIMMONDSIA CHINENSIS*)

Jojoba is an inconspicuous shrub of the arid and semi-arid southwestern United States and adjacent Mexico. Its oil is actually a liquid wax, similar to the now unavailable spermaceti oil formerly "destructively harvested" from sperm whales. Jojoba has received a lot of attention as a new industrial crop, although it has yet to take off as a major feedstock, perhaps due to competition from petroleum-based products.

In a post-fossil-fuel scenario, jojoba could become an important feedstock crop. Its adaptability to arid, alkaline soils offers an exciting crop to some of the least fertile regions of the world. This crop is suited to tropical and subtropical arid and semi-arid areas.

Uses. This oil, like castor, needs no refinement before being used as a lubricant. It also has many uses in the pharmaceuticals and cosmetics. The seeds were (and probably still are) eaten and used for medicine by indigenous people, although researchers today consider them inedible.[37] I've eaten a few and found them waxy and interesting.

Yield. Jojoba seed yields are up to 2.2 to 4.5 t/ha, with 0.5 to 1.1t/ha of "oil."[38]

Harvest and processing. Mechanical harvesters have been modified from blueberry and grape pickers in

FIGURE 21.5. Fruits of jojoba, a wax crop for arid lands. Photograph courtesy of Michael Wolf/Creative Commons 3.0.

the United States and Israel. Vacuum pickers have been trialed to pick up dropped fruits.[39]

Carbon farming applications. Jojoba has been intercropped with asparagus, a perennial vegetable. The asparagus served as a windbreak for the establishing jojoba shrubs. Some annual crops have also been trialed. The primary difficulty with intercropping is competition for water in the desert environments where jojoba is grown.[40]

PONGAM (*PONGAMIA PINNATA*)

The poor results of jatropha cultivation have led some researchers to look for inedible oilseeds that can be grown more successfully in poor soils. Pongam, though undomesticated, is a promising candidate. It bears seed heavily every year, oil content of the seeds is 30 to 45 percent, it has few pest and disease problems, it tolerates poor and salty soils and long dry seasons, and it is a fast-growing nitrogen-fixing tree.[41] Pongam is native from India through northern Australia and grows in sub-humid to humid tropical and subtropical lowlands. The pods float, and it can disperse effectively if planted near bodies of water, making it a potential invasive species outside of its ample native range.[42]

This perennial, nitrogen-fixing crop has most promising potential for a range of industrial uses. Researchers and agrofuel companies hope to see pongam planted in large plantations of hundreds or thousands of hectares. They do acknowledge that this will increase pest and disease pressure.[43] Of course, I'd recommend integrated food and feedstock systems.

Uses. Pongam has multiple biofuel uses. It is a high-quality, coppicing firewood. The oil can be used for biodiesel, and the whole pods can be burned, with a calorie value comparable to low- or medium-grade coal.[44] The oil could also serve as a raw material for bioplastics, lubricating oils, and other products. Pongam oil is used today in India for fuel, soaps, and lubrication and as a pesticide. The seed presscake left over after oil extraction is applied to the soil as an organic nematocide (a pesticide for killing pest nematodes).[45]

Yields. Pongamia yields 5 to 8 t/ha of beans, yielding 1.8 t/ha of oil.[46] Domestication could substantially improve the yields.

Harvest and processing. Large-scale techniques are still under development.

Carbon farming applications. Pongam is used for crop shade, windbreak, and fodder bank.[47] As a

FIGURE 21.6. Pongam is a promising new oil crop and a multipurpose nitrogen-fixing tree. Photograph courtesy of Forest and Kim Starr/Creative Commons 3.0.

nitrogen-fixing plant it would make a promising intercrop for other species. It could potentially be produced using woody agriculture techniques.

CHAPTER TWENTY-TWO

Hydrocarbon Crops

Although they are the primary cause of climate change and other social and economic catastrophes, fossil fuels are remarkable substances. Their utility derives from their high concentration of hydrocarbons, chemical compounds constructed solely from hydrogen and carbon. Imagine my surprise when I learned that thousands of plants naturally and sustainably produce hydrocarbons and some have been cultivated for this purpose for millennia.

Hydrocarbons are compounds made only of carbon and hydrogen. Some, like methane (CH_4), are short, while others are long chains. In plants they provide energy storage, protection from herbivores, and protective coatings for wounds.[1] Hydrocarbons are also the main ingredient of crude oil and, in that form, the essential source of materials, chemicals, and energy for today's global civilization. Hydrocarbons you may recognize include methane, propane, and butane.

As I researched industrial crops for this book, no other category set my imagination afire like hydrocarbon crops. These remarkable plants produce many of the same compounds found in oil, coal, and natural gas. Although the potential of "petrocrops" is unfulfilled, they offer the promise that we don't need to abandon all of the technological breakthroughs that fossil fuels have allowed, at least in the sectors of material and chemical production. As we transform civilization to respond to climate change, we face resistance from the fossil fuel industries. This class of crops lets us extend an olive branch to at least some small percentage of our opponents. With some reskilling of technicians and retrofitting of infrastructure, we can put their remarkable knowledge and technology to work in a carbon-sequestering solution that still provides essential needs of modern humanity, if on a much smaller scale.

Some hydrocarbon crops have been used since ancient times and today represent important world commodities: notably rubber, turpentine, and resins. These crops offer cautionary tales about the potential for industrial-scale production to lead to environmental destruction and human suffering. On the hopeful side, however, particularly in the case

TABLE 22.1. Hydrocarbon Crop Yields Compared

Latin Name	Common Name	Climate	Yield/ha
Hevea brasiliensis	Para rubber	Humid tropics, lowlands	2.5 tons latex[a]
Parthenium argentatum	Guayule	Arid and semi-arid subtropics	1.5–1.9 tons latex[b]
Euphorbia tirucalli	Petroleum plant	Arid to semi-arid tropics	1.8 tons latex[c]
Pinus spp.	Turpentine pines	Warm temperate to tropical, semi-arid to humid	0.8 tons resin[d]
Asclepias syriaca	Common milkweed	Cold to warm temperate, semi-arid to humid	0.7 tons latex[e]
Pittosporum resiniferum	Petroleum nut	Humid tropical highlands	0.4 barrels of oil equivalent[f]

[a] Singh and Singh, "Natural rubber," 367.
[b] Ray et al., "Guayule," 392.
[c] 2,200 liters; a barrel of oil has 159 liters and weighs 138kg (304 pounds). Loke, Mesa, and Franken, "*Euphorbia tirucalli* Bioenergy Manual," 56.
[d] FAO, "Turpentine from pine resin," 7.
[e] 13 t/ha biomass @ 5.7% latex for undomesticated plants; from Duke, *Handbook of Energy Crops.*
[f] Calvin, "Fuel oils from higher plants," 155.

of *Hevea* rubber, we see the production potential of a hydrocarbon crop realized: great increases in yield through breeding and non-destructive harvesting of a long-lived tree crop. What is still almost completely missing is production for regional processing and use in integrated polycultures instead of vast corporate export monocultures.

There is not enough land to produce enough hydrocarbons to replace the vast amounts of fossil fuels, which represent the accumulated photosynthesis of hundreds of millions of years.[2] It is not difficult to imagine efforts to sow monocultures of these plants on vast tracts of land, contributing to social and environmental problems in a desperate and hopeless attempt to maintain the current level of consumption of our unsustainable societies. Even more than most industrial crops, the use of these species should be guided by the principles of democracy, decentralization, integrated production systems, and industrial sovereignty.

Hydrocarbon crops come in two primary types. Latex is produced by thousands of species around the world and is the basis for natural rubber production. Resins are also produced by thousands of species and have long been used for turpentine, tar, pitch, medicinal, and ceremonial uses. There are several types of non-destructive harvest. Tapping is common for both resin- and latex-bearing species. Tapping can be benign or fatal to the tree, depending on the species and the practices used. Extraction techniques remove hydrocarbons from biomass using various solvents, sometimes toxic ones. The biomass is typically harvested by coppicing or haying. Steam-distillation of *Eucalyptus globulus* foliage, a non-coppicing species, yields terpene hydrocarbons abundant enough to excite Nobel Prize–winning chemist Melvin Calvin. Harvesting hydrocarbons from leaves of coppiced, fast-growing trees might be far more labor-efficient than other methods like tapping rubber trees.[3] A few species, such as stinking toe (*Hymenaea* spp.) and petroleum tree (*Pittosporum* spp.), produce hydrocarbons in their fruits, making for the best-case scenario of non-destructive harvest.

Depending on the types of hydrocarbons present in a given species, they may be best suited to particular uses. Any plant hydrocarbons could likely be broken down ("cracked") and recombined to assemble virtually any material, chemical, or energy desired, just as fossil hydrocarbons are. The energy and expense required to do so could be considerable, and as of the time of writing this book, this is still only being done on an experimental basis. It is probably easiest to use the hydrocarbons in the form or forms they come in, just as petroleum companies have found with crude oil.[4]

The Potential of Petrofarming

It wasn't long into the petroleum age that researchers began trying to use hydrocarbon crops to replace fossil fuels. In the mid-1930s Italians attempted to extract fuels from *Euphorbia abyssinica* in Ethiopia. The French soon followed with *E. resinifera* in Morocco. Promising though it was, the world wars largely directed research into other applications.[5]

Again during the oil crisis of the 1970s, researchers, notably Nobel Prize–winning chemist Melvin Calvin—who is credited with co-discovering the Calvin cycle—began working on this project. Calvin published a series of articles such as "Petroleum plantations for fuel and materials" and was enthusiastic about the potential for *Euphorbia* and milkweed family plants to replace dwindling fossil fuels and forestall global climate change. He called for aggressive breeding and production efforts and established a research farm producing a range of potential hydrocarbon crops.[6] Although other scientists, including the great James Duke, think Calvin overestimated the potential yields, Calvin deserves great credit as the modern ambassador of the hydrocarbon crop concept.[7]

In the 1970s Professor R. A. Buchanan and a team of researchers tested several hundred wild midwestern North American species to evaluate their potential as rubber and hydrocarbon crops. They found some surprises, including high yields of rubber from mints, milkweed, goldenrods, and sumac—all abundant, weedy species.[8] Other scientists have continued this assessment of the native and naturalized floras of their regions, including western North America, Mexico, India, and Greece. Extraction techniques are under development.[9]

HYDROCARBON CROP USES

MATERIALS

- **Rubber** is an essential component of a technological civilization. Every motor or engine with moving parts requires rubber gaskets and hoses. Rubber is also essential to modern medicine. Of course, it is also necessary for the tires of cars, trucks, buses, and airplanes.[10]
- **Plastics** represent a major use of petroleum today. Although bioplastics can be made from starches, plant oils, and even biomass, hydrocarbon crops could provide a fantastic sustainable feedstock that might require minimal changes to contemporary plastic-making technology and skills.[11]
- **Asphalt** is made from the least desirable by-products of petroleum refining. Although plants could never provide the vast amounts of asphalt used today, they could produce a fraction of what we currently use.[12] Intriguingly, scientists have recently developed techniques to make asphalt from hog manure.[13] Shredded tires, a waste product from another hydrocarbon material, can also be used to make asphalt.[14]

CHEMICALS

- **Naval stores** include turpentine and rosin. Once essential for producing tar and pitch to waterproof wooden ships, these materials are still of global importance.[15]
- Although most **lubricants** can be made from oil crops, hydrocarbon crops could provide a contribution to this essential ingredient of a modern civilization.[16]
- As a source of **chemical feedstocks**, hydrocarbon crops have enormous potential. Currently chemists coax thousands of products from the basic platform chemicals refined from fossil fuels. These platform chemicals are the basic building blocks of virtually the entire chemical and pharmaceutical industries.[17] Although most of this infrastructure should shift to chemicals isolated from biomass, hydrocarbon crops could provide an opportunity to retain some skills and technology from the original equipment.

ENERGY

- Although they cannot remotely provide the amount of energy humanity uses today, these crops can be refined like petroleum to create the full range of hydrocarbon fuels used today, from propane to gasoline and jet fuel.[18] Some, like diesel tree (*Copaifera* spp.), can even be tapped for resin that can be used directly in a diesel engine.[19] Petrocrops can also be feedstock for methane digesters.[20]

In the 1970s and '80s Brazil almost began the age of petrofarming as they considered domesticating their native diesel tree (*Copaifera* spp.) for massive agrofuel production. Instead they chose ethanol from sugarcane, an already fully developed and studied crop.[21] Thus, unlike rubber and turpentine, "petroleum" production from plants remains largely hypothetical. Given its great potential, it should be a high priority for research and funding. Initial results are interesting, with one effort in Japan yielding the equivalent of 25 to 50 barrels of oil per hectare from *Euphorbia* species.[22]

Calvin estimated in 1979 that even with domesticated hydrocarbon crops it would take hundreds of millions of acres to meet the world's demand.[23] He thought that producing the 1976 US gasoline needs would require an area the size of Arizona (roughly 29 million hectares/72 million acres) dedicated to hydrocarbon production.[24] Consumption has only increased since then, and the food-versus-fuel debate has shown us the foolishness of attempting to meet global energy needs from plants. But a small portion of energy and a lot of materials and chemicals could be possible.

Calvin developed a hypothetical budget for a processing facility to convert the product of 121 hectares (299 acres) of *E. lathyris* to fuel oil and ethanol. Total cost was $1 million (1985 US dollars). He calculated an annual cost (including repaying construction loans) at $1.25 million with an annual income ranging from $1.23 to $1.51 million. He estimated annual production at 9,000 to 12,000 barrels of oil and 12–15,000 barrels of ethanol from the 121-hectare planting.[25] Calvin calculated a cost of about $100 per barrel in 1985 dollars, roughly equivalent to $221 in 2014 dollars.[26] The price of oil is $40 as I write.[27] Carbon taxes or other prices on fossil carbon could make petrofarming cost-effective.

Rubber and Latex

Rubber is a hydrocarbon that comes in the form of a milky latex. It is produced in at least seven plant families, 300 genera, and by more than 7,500 species.[28] As of 2009, 70 percent of rubber was synthesized from petroleum. Of the remaining 30 percent, almost all comes from *Hevea brasiliensis*, a tropical domesticated crop with a number of challenges.[29] Before I began writing this book, I'd scarcely given a thought to rubber and where it comes from. Besides gazing out the windows of a bus passing a plantation of hule rubber (*Castilla ulei*) once in Guatemala, I don't think I've ever even seen any part of the rubber production process.

Yet I use rubber every day and indirectly benefit from it to an even greater degree. More than 40,000 products are made from it. Rubber is used in transportation, for tires and belts. It is used in industry, for conveyor belts, hoses, sealers, and gloves. Consumer uses include clothing and mats. Rubber has more than 400 medical uses including surgical and inspection gloves, contraceptives, and syringes. The bottom line is that we need it. Imagine our civilization without engines or motors. Both are impossible without rubber gaskets, hoses, and belts. In fact, even synthetic rubber won't do for many uses; scientists and engineers have been unable to match the quality of natural rubber.[30]

Rubber production has come at a great cost, however. During the industrial and colonial periods, atrocities were committed to ensure and increase the flow of this important feedstock. In his book *1493*, Charles Mann dedicates a chapter to the sordid colonization history of "black gold," describing the enslavement of indigenous people in Brazil and the deaths of tens of thousands of immigrants from malaria and other diseases.[31] King Leopold II of Belgium used the now Democratic Republic of the Congo as his source of a vast personal rubber fortune; the destructive harvest of *Landolphia* vines

and forests for *gutta* rubber was an ecological disaster, but far worse was the unspeakable exploitation of the indigenous population. Rubber harvesters who did not meet quotas had their hands cut off. As many as 10 million people died in the Belgian Congo, one of the great crimes against humanity in human history.[32] Today *Hevea* rubber is grown on vast plantations in Southeast Asia, displacing indigenous people's food production for an inedible export commodity.[33]

Surely relying on one small region to provide the world's rubber is unsustainable. Synthetic rubber from petroleum actually relieves the situation in many ways, yet we need to drastically reduce use of fossil fuels to address climate change. Replacing synthetic rubber must come through reducing consumption and domesticating additional rubber species so that each region can produce much of its own, or the social and environmental cost of greatly expanded *Hevea* rubber production in Southeast Asia could be enormous. The other cultivated rubber crops are almost all from the humid tropics as well, and none is even close to as domesticated as *Hevea* is.[34]

The challenges of rubber go beyond its production. Rubber degrades poorly, and vulcanized rubber (like most tires) is almost indestructible. Researchers need to develop techniques to upcycle, recycle, or at least compost vulcanized rubber that has outlived its usefulness. Some shredded tires are being used in asphalt with success but without much economic viability.[35] This remains an unsolved challenge for a civilization that depends so heavily on this product.

Around the tropics, many species are grown for rubber as regional crops. These include various species of *Castilla*, *Ficus*, *Manihot*, *Manilkara*, and *Willughbeia*.[36] There have been attempts to domesticate rubber crops for colder and drier regions. Some such efforts are described in Mark Finlay's *Growing American Rubber*, which covers efforts to domesticate the dryland crop guayule, as well as Thomas Edison's attempt to develop a wartime emergency rubber crop for the United States. Edison dedicated the last four years of his life to this project, and his efforts are instructive because even with fame, brilliance, and financial resources, four years is not remotely enough time to develop a new crop. When

World War II arrived and rubber shortages became a crisis, attention shifted to synthetic rubber from petroleum. (Chemists also developed techniques for synthesizing rubber from ethanol from industrial starch during this time.[37]) While this was in part because of the financial power of the petroleum lobby, the crops were truly not yet ready for serious production.[38] There is a lesson here for us as we prepare for adapting to a changing climate and withdrawing quickly from fossil fuel use in the next few decades: It is difficult if not impossible to bring even a promising crop from the wild state to domestication and large-scale cultivation in just a few years. The US government didn't listen to Edison's repeated entreaties to develop an alternative rubber source *before* it was needed. In the future, we may not have fossil carbon to fall back on. Breeding and development work on industrial crops needs to move to the top of our priority list.

HEVEA OR PARA RUBBER (*HEVEA BRASILIENSIS*)

Hevea rubber is the 25th most widely grown agricultural crop in the world, with 9.8 million hectares (24.2 million acres) in production as of 2012. This global crop amounts to 11.4 million tons annually, at a value of $22.8 billion US.[39] And it is the primary domesticated hydrocarbon crop, representing 30 percent of the world rubber market.[40] Native to Amazonia and preferring humid tropical lowlands, *Hevea* seeds were stolen by a plant explorer and biopirate named Henry Wickham in the 1870s to break a Brazilian monopoly. Their few surviving progeny were brought to Asia and underwent selection and breeding. Meanwhile, back in Amazonia, a devastating disease called South American leaf blight came to devastate plantations. This is why today we rely on Asian production of an Amazonian crop. Those Asian varieties, however, are all descended from Wickham's

FIGURE 22.1. *Hevea* rubber tapping. This species is the primary cultivated rubber crop. Photograph courtesy of CIAT/Creative Commons 2.0.

TABLE 22.2. Hydrocarbon Latex Crops

Genus	Common Name	Origin	Climate	Cultivation
Apocynum cannabinum	Dogbane hemp	North America	Temperate, Mediterranean, semi-arid to humid	Minor global
Artemisia ludoviciana	White sagebrush	Western North America, Mesoamerica	Boreal to subtropical, semi-arid to humid	Experimental
Artocarpus altilis	Breadfruit	New Guinea, Pacific	Lowland humid tropics	Minor global
Artocarpus integer	Champedak	Southeast Asia, New Guinea	Tropical humid lowlands	Regional
Asclepias curassavica	Curassavian swallow-wort	Tropical Americas	Tropics, subtropical, warm temperate, humid	Regional
Asclepias erosa	Desert milkweed	Southwestern North America, Baja California	Subtropical, arid to semi-arid	Experimental
Asclepias incarnata	Swamp milkweed	North America	Cold to warm temperate, semi-arid to humid	Experimental
Asclepias speciosa	Showy milkweed	Western North America	Cold temperate, warm temperate, semi-arid	Experimental
Asclepias subulata	Desert milkweed	Mesoamerica, southwestern North America	Warm temperate to subtropical, arid to semi-arid	Experimental
Asclepias syriaca	Common milkweed	North America	Cold temperate, warm temperate, semi-arid to humid	Experimental
Asclepias tuberosa	Butterfly weed	North America	Cold temperate, warm temperate, subtropical, humid to semi-arid	Experimental
Brosimum alicastrum	Ramon, Mayan breadnut	Mesoamerica	Tropical lowlands, humid to semi-arid	Regional
Brosimum utile	Milk tree, cow tree	Americas	Tropical lowlands, humid to semi-arid	Regional
Calotropis gigantea	Akon	South and Southeast Asia	Tropical arid to humid	Regional
Calotropis procera	French cotton	Africa through India	Tropical arid to humid	Regional
Castilla elastica	Panama rubber tree	Americas	Humid tropical lowlands	Regional
Castilla ulei	Hule rubber	Americas	Humid tropical lowlands	Regional
Chiococca alba	West Indian milkberry	Tropical Americas	Tropical, subtropical, humid to semi-arid	Experimental
Chondrilla ambigua	Chondrilla	Northern middle Asia	Cold temperate, warm temperate, semi-arid to humid	Experimental
Cnidoscolus elasticus	Highland chilte	Mesoamerica	Arid, semi-arid tropical highlands	Experimental
Couma macrocarpa	Leche caspi	Amazonia	Humid lowland tropics	Experimental
Couma utilis	Sorva	South America	Humid tropical lowlands	Regional
Croton tiglium	Croton oil plant	South and Southeast Asia	Arid to humid subtropics and tropics, lowlands	Regional
Cryptostegia grandiflora	India rubber vine	Madagascar	Humid lowland tropics	Minor global
Dyera costulata	Lutong	Southeast Asia	Tropical humid lowlands	Regional
Ericameria nauseosa	Rubber rabbitbrush	Western North America	Boreal to warm temperate, arid to semi-arid	Experimental
Eucommia ulmoides	Chinese rubber tree	Asia	Cold to warm temperate, semi-arid to humid	Regional
Euphorbia abyssinica	Desert candle	Africa	Arid to semi-arid tropics to warm temperate	Experimental
Euphorbia esula	Leafy spurge	Eurasia	Boreal to subtropical, arid to humid	Experimental
Euphorbia intisy	Intisy	Madagascar	Tropics, semi-arid	Regional
Euphorbia tirucalli	Petroleum plant	Africa	Tropics, arid to semi-arid	Minor global
Ficus annulata	Panggang	India, Southeast Asia	Humid tropical lowlands	Regional
Ficus elastica	India rubber tree	Southeast Asia	Tropical humid lowlands	Minor global
Ficus racemosa	Cluster fig	India, Southeast Asia, Australia	Semi-arid to humid, tropical lowlands and highlands	Regional
Funtumia elastica	West African rubber tree	West Africa	Tropical humid lowlands	Regional
Hancornia speciosa	Mangaba rubber tree	Eastern South America	Semi-arid to humid, lowland tropics	Regional

Genus	Common Name	Origin	Climate	Cultivation
Hevea benthamiana		South America	Humid tropical lowlands	Regional
Hevea brasiliensis	Para rubber	South America	Humid tropical lowlands	Global
Landolphia heudelotii	Landolphia rubber	West Africa	Humid tropical lowlands	Regional
Landolphia kirkii	Zanzibar rubber	East Africa	Humid tropical lowlands	Regional
Manihot caerulescens	Mandioca de viado	Brazil	Humid tropical lowlands	Regional
Manihot carthaginensis ssp. *glaziovii*	Ceara rubber	Brazil	Humid tropical lowlands	Regional
Manihot dichotoma	Jequie manicoba	Brazil	Humid tropical lowlands	Regional
Manihot heptaphylla	Manicoba de Sao Francisco	South America	Humid tropical lowlands	Regional
Manilkara bidentata	Balata	Caribbean	Tropical lowlands	Regional
Manilkara huberi	Cow tree	South America	Tropical humid lowlands	Regional
Palaquium gutta	Gutta percha tree	Southeast Asia	Humid tropical lowlands	Regional
Parthenium argentatum	Guayule	Southwest North America, Mesoamerica	Arid to semi-arid, subtropics	Minor global
Payena leerii	Getah sundek	Southeast Asia	Humid tropical lowlands	Regional
Pouteria guianensis		Tropical South America	Humid tropical lowlands	Regional
Rhus copallinum	Winged sumac	North America	Cold to warm temperate, semi-arid to humid	Experimental
Solidago leavenworthii	Leavenworth's goldenrod	Southeast North America	Warm temperate to subtropical, semi-arid to humid	Experimental
Solidago nemoralis	Gray goldenrod	Eastern North America	Cold to warm temperate, semi-arid to humid	Experimental
Willughbeia coriacea	Akar gerit-gerit besi	Asia	Humid tropical lowlands	Regional
Willughbeia edulis	Gedraphol	India, Southeast Asia	Tropical lowlands	Regional

Sources: Hanelt, *Mansfeld's Encyclopedia of Agricultural and Horticultural Crops*; Finlay, *Growing American Rubber.*
Note: This is a small sample of the 7,500 latex-producing species of the world. Those listed here include species that are currently or historically cultivated for rubber as regional crops (mostly humid tropical species), as well as some that have been experimentally cultivated or identified as potential sources for colder and drier climates.

few seedlings—and are fully susceptible to the blight. Grown in monocultures as they are, they are even less resistant. As a result, the world's current supply of natural rubber is highly vulnerable; sooner or later the blight will arrive.[41]

Uses. *Hevea* rubber is the main source of commercial natural rubber. It is of high quality and used in tires, medical products, and other applications that require superior material. In recent decades many people have developed allergies to *Hevea* latex. This has been a factor in the development of alternative crops such as guayule. *Hevea* flowers make excellent honey, which is often an important secondary product for rubber farmers. The seeds are high in oil and could be developed as an industrial oil crop. When rubber production declines, the trees are harvested for timber.

Yield. Para rubber breeding is the gold standard for hydrocarbon crops. Wickham's unselected trees yield 225kg (496 pounds) per hectare. By the 1920s yields were 650kg (1,433 pounds) per hectare. By the 1950s they had increased to 1,600kg (3,527 pounds) per hectare. In the 1990s they had increased to 2,500kg (5,512 pounds) per hectare.[42] This 10-fold increase in a century provides a good benchmark for other potential hydrocarbon crops. Average global yield in 2012 was 1.16t/ha.[43]

Harvest and processing. Para rubber trees are a moderately long-lived perennial crop. They begin to produce 5 to 10 years after planting. Peak production comes in years 12 to 15, and the production declines to non-commercial levels by years 25 to 30.[44] *Hevea* rubber tapping and collection is labor-intensive. When I think of the many species that are laboriously tapped for hydrocarbons, I wonder if the tubing systems developed by sugar maple producers might be used to centralize collection from multiple trees into a series of tanks or barrels. Unfortunately, in this case I imagine that the latex might literally "gum up the works."

Carbon farming applications. Annual and short-lived crops are often grown between the young trees. Indonesian farmers have developed a unique "jungle rubber" agroforestry system that intercrops a main crop of rubber with timber, fruits, rattan, and medicinal crops. This system is practiced on 2.8 million hectares (6.9 million acres) by smallholders.[45] In China young rubber trees are interplanted with cereals, hemp, tea, coffee, black pepper, pineapples, sugarcane, bananas, medicinal shade crops, and more.[46] Shade-tolerant nitrogen-fixing ground covers are often grown in the understory. The dense evergreen shade cast by solid stands of Para rubber trees is the primary challenge to intercropping.[47] Surely contour rows or bands of *Hevea* rubber could be intercropped with food-producing tropical trees such as breadfruit or peach palm with integrated livestock or shade crops.

GUAYULE (*PARTHENIUM ARGENTATUM*)

Guayule is a shrubby species from semi-arid desert plateaus in northern Mexico and southern Texas. Guayule has attracted attention because it can grow in dry subtropical areas, unlike Para rubber and most rubber crops. It is currently a minor global crop. Efforts to develop guayule as a crop go back more than 100 years. The International Rubber Company produced 9,500 tons of rubber in 1910, although much of this was from destructively harvested wild plants.[48]

During World War II there was a major effort to jump-start production in the western United States. Major breeding and production advances were made by Japanese-Americans held at the Manzanar internment camp. Mark Finlay says of these skilled nursery experts: "This small group of Japanese-American scientists, working on shoestring budgets in makeshift laboratories and on tiny research plots, had more success than the large and well-funded research teams of the USDA."[49]

Uses. This species is currently the only crop being scaled up to meet the need for non-allergenic rubber production. With allergies to *Hevea* rubber latex on the rise, the economic incentive has been present to further domesticate this promising species.[50] Guayule processing provides some interesting additional products, including a resin used in waterproofing. The spent biomass is used to create termite-resistant composite boards and has other biomass feedstock uses.[51]

Yield. The species presents some breeding challenges because of its biology, but some forms have seen the yield increase up to 300 percent from breeding. Yields of elite lines currently range from 1.5 to 1.9 t/ha, competitive with what Para rubber was about 50 years ago.[52]

Harvest and processing. Guayule used to be harvested by digging up the whole plants after five years since some of the rubber is in the roots. Many contemporary producers coppice the aboveground biomass but leave the roots in place for future harvests.[53]

Carbon farming applications. Perhaps guayule might be intercropped in contour strips alongside crops such as jojoba, wattleseed acacias, sisal, moringa, and mongongo.

Crop development. Guayule is the subject of ongoing breeding work.

FIGURE 22.2. Guayule, an up-and-coming desert rubber crop. Photograph by Edwin Remsberg.

COMMON MILKWEED
(*ASCLEPIAS SYRIACA*)

Common milkweed is an herbaceous species, spreading underground to form large colonies in grasslands, urban lots, and abandoned areas. It is native to North America and naturalized in Eurasia centuries ago. It tolerates cold to warm temperate humid and semi-arid areas. Milkweed bast (stem) fibers have long been used for cordage by indigenous people. The fluffy seed fibers also have a long history and were collected in the United States during World War II to fill sailors' life vests. It is also an excellent vegetable. However, milkweed offers more than food and fiber. It produces a hydrocarbon-rich latex, chemically similar to the naphtha fraction of crude petroleum. Economic botanist James Duke proposed milkweed as a new "botanochemical crop" in 1981. In his *Handbook of Energy Crops*, he outlined a diversity of products, including rubber, edible seed oil, livestock feed, floss for insulation, bast fibers for premium papermaking and cordage, chemical intermediates from the latex and polyphenol hydrocarbons, and more.[54]

Little follow-up was done on this promising research until the recent biofuels boom. Should humanity decide to replace fossil fuels in part with solar-powered plant products, breeding and research attention to this roadside wildflower could at last bring out its capacity to meet a wide range of human needs in a sustainable way. There are many other *Asclepias* species (and other related genera in the Apocynaceae family) with similar uses.

Uses. Milkweed produces latex for rubber and hydrocarbon feedstock. It serves as a dual fiber and vegetable and has many other industrial uses. Milkweed seed oil is being investigated as a feedstock for cosmetics such as sunscreen and conditioner.[55]

Yield. As a hydrocarbon crop milkweed has been estimated to yield 0.7 t/ha of latex per hectare per year.[56]

Harvest and processing. As an experimental hydrocarbon crop, milkweed is harvested with hay equipment and chopped. Solvents are added to extract the latex.

Carbon farming applications. Milkweed might make a fine intercrop for other temperate perennial food and industrial crops. It is a frequent weed in orchards today.

Crop development. Though it is considered an agronomic weed, one of the main challenges in milkweed domestication is deliberately getting it to grow on a large scale. Once this is worked out, breeding work could proceed.

PETROLEUM PLANT
(*EUPHORBIA TIRUCALLI*)

Petroleum plant is a member of the genus *Euphorbia*, in the Euphorbiaceae family, one of the most important hydrocarbon plant families. It is a shrub to medium-sized tree native to arid and semi-arid, low-elevation regions of tropical Africa. It is adapted to poor soils and extended dry seasons. It also tolerates salty soils and brackish irrigation water. Widely grown around the world as an ornamental—I have seen plants in Mexico, Cuba, and other Caribbean islands—in recent decades it has been investigated as a promising energy crop.

The major drawback of this species is its toxicity. Touching the plants can burn the skin and eyes, and it is toxic to ingest. Even the runoff from plants can contaminate well water, a major consideration in siting a planting. It has naturalized to some degree outside of its native range. These factors combine to make it a species to plant with caution outside sub-Saharan Africa.[57] Petroleum plant has an extremely efficient respiration

FIGURE 22.3. Milkweed—a multipurpose hydrocarbon-fiber-vegetable plant ready for domestication—in Vermont.

FIGURE 22.4. Petroleum plant in Hawaii. Photograph courtesy of Forest and Kim Starr.

strategy, using ordinary C₃ photosynthesis during the rainy season and switching to water-conserving crassulacean acid metabolism (CAM) in the dry season.[58]

Uses. Petroleum plant is a source of biomass, latex, and hydrocarbons. It can also be burned directly, pressed into briquettes, or made into charcoal. It has potential as a feedstock for bioplastics and biorefinery operations, as well.

Yield. Petroleum plant yields 1.8 tons per hectare of latex annually, the equivalent of 14 barrels of oil.[59] Biomass yields have ranged from 20 to 150 t/ha fresh weight (3 to 22.5 t/ha dry).[60] With sufficient equipment it can be converted to biodiesel or high-octane gasoline, producing the equivalent of 220 liters per hectare fuel oil annually. In the developing world, it has excellent potential as a biodigester feedstock for rural areas. (Using this appropriate technology, the coppiced biomass gives off methane, burned as fuel.) In one experiment this approach produced 11,880 kWh of electricity per hectare, compared with 2,330 kWh for jatropha and 5,850 kWh for African oil palm. An added benefit of biodigestion is that only hydrogen and carbon are removed; the remaining composted biomass can be returned to the field as a fertility input.[61]

Harvest and processing. Petroleum plant is coppiced and harvested for latex extraction. Harvesters and processors need to wear protective gear to prevent burns.

Carbon farming applications. Petroleum plant is often used as a toxic, goat-proof living fence. It has been intercropped with jatropha.[62]

LEAFY SPURGE (*EUPHORBIA ESULA*)

Leafy spurge is an herb whose native range runs from Britain to Siberia. It has naturalized in the United States and Canada, where it is considered a noxious weed of pastures. It ranges from boreal to subtropical, and arid to humid, tolerating temperatures as low as −45°C (−49°F) and annual rainfalls as low as 180 millimeters (7 inches). Currently leafy spurge is not cultivated as a hydrocarbon crop.

Uses. Leafy spurge is reasonably high in hydrocarbon latex. It is worth investigating its breeding potential as an extremely cold-hardy hydrocarbon species, at least for its broad native range and perhaps for other areas it has already overtaken. Leafy spurge hay burned with four times the energy of wheat straw in one experiment.[63] Although it is toxic to cattle,

FIGURE 22.5. Leafy spurge is a hated weed in the Americas, with potential as a latex crop in cold, dry climates. Photograph courtesy of Matt Lavin/Creative Commons 2.0.

leafy spurge is a fine high-protein fodder for sheep and goats, offering the intriguing prospect of pastures providing meat, milk, wool, and plastic! According to Melvin Calvin, all *Euphorbia* species contain latex. Of the annual *E. lathyris*, he writes, "The terpenes can be cracked like crude oil and the sugar can be fermented like sucrose, whiles the lignocellulose can be used in a way similar to the bagasse of sugarcane, or, better yet, compressed into more efficiently burning pellets."[64]

Yield. Unknown.

Harvest and processing. Presumably leafy spurge would be harvested with haymaking equipment.

Resins

Resins are the other primary type of hydrocarbons in plants. In *Plant Resins: Chemistry, Evolution, Ecology, and Ethnobotany*, Jean Langenheim points out that many different plant exudates have been called resins but are often substances with very different chemistries. She lays out a definition of plant resin that I've used as my guide here: "A lipid-soluble mixture of volatile and non-volatile terpenoid and/or phenolic secondary compounds that are 1) usually secreted in specialized structures located either internally or on the surface of the plant, and 2) of potential significance in ecological interactions."[65]

Resins come in two primary types. The first, which are true hydrocarbons, are terpenoid resins, or terpenes. More than 30,000 types of terpene compounds have been found in plants. They are believed to defend plants from herbivores and diseases. Volatile terpenes are more fluid and are used for fuel and turpentine. Non-volatile terpenes, sometimes called copals or dammars, are more viscous and are used for varnishes. The second type of resin is phenolic resins, which, while interesting, are not hydrocarbons and won't be discussed here.[66]

Resins have been used for millennia. Amber, a fossil resin, has been used as a ceremonial and ornamental gemstone since prehistoric times. One of the earliest recorded industrial uses of resin was to caulk and waterproof wooden ships, as well as waterproofing ropes and tarps for shipboard use. Naval stores, as these pine-resin-derived products are called, were essential strategic materials during the age of wooden ships. Pitch, or cooked resin, was used as an early form of napalm in ancient warfare. Today pine resins remain important in the form of turpentine and rosin and their many products. Turpentine, the volatile percentage ("fraction") of pine resin, is used as a paint thinner, solvent, cleaner, and fragrance. Rosin, the non-volatile fraction, has hundreds of uses, including adhesive tape, ink, paper, and rubber.[67]

Historically many tropical resins were used as illuminants for torches and fuel for lanterns. Solid resin pieces are still used as highly flammable fire starters. Many resins have been, and some are still, used as lacquers and varnishes to beautify and preserve wood and other materials. Resins have been used as medicine, cosmetics, and incense for millennia. In fact, the famous frankincense and myrrh from the Old Testament are both resins.[68]

Modern science has developed additional uses for plant resins. Linoleum was originally made by blending linseed oil (from flaxseeds), *Agathis* resin, and cork.[69] By swapping out flax for a perennial oil, linoleum could be made from three non-destructively harvested perennials! Today linoleum is made from petroleum, which has also replaced most traditional uses of resin from varnishes to paint thinners, although a lot of pine resin is currently used as a feedstock for the chemical industry. Thanks to visionaries like Melvin Calvin, we can envision domesticated plant resins providing a renewable and carbon-sequestering alternative to fossil hydrocarbons, with a nearly limitless range of potential products.[70]

Like rubber, resin production has been associated with a lot of social and ecological devastation. Tapping trees can be performed in ways that doesn't harm the trees, but it has often been highly destructive, particularly where wild resin trees were highly abundant. Like rubber, copal was also extracted from Congolese forests under King Leopold II's vicious reign.[71] Turpentine production is hard work, and in the United States it was performed first by enslaved people and then by exploited poor laborers.[72]

TABLE 22.3. Hydrocarbon Resin Crops

Genus	Common Name	Origin	Climate	Cultivation
Agathis australis	Kauri	New Zealand	Subtropical to tropical humid	Regional
Cistus ladanifer	Rockrose	North Africa, Mediterranean	Semi-arid warm temperate to subtropical	Regional
Copaifera langsdorffii	Diesel tree	Amazonia	Tropical humid lowlands	New crop
Copaifera multijuga		South America	Humid tropical lowlands	Regional
Copaifera officinalis		Americas	Humid tropical lowlands	Regional
Daniellia oliveri	African copaifera	Africa	Humid tropical lowlands	Regional
Dipterocarpus alatus		Southeast Asia	Humid tropical lowlands	Wild
Dipterocarpus gracilis		India and Southeast Asia	Humid tropical lowlands	Wild
Dipterocarpus grandiflorus		India and Southeast Asia	Humid tropical lowlands	Wild
Dipterocarpus kerrii		India to Southeast Asia	Humid tropical lowlands	Wild
Dyera costulata	Lutong	Southeast Asia	Tropical humid lowlands	Regional
Eremophila fraseri	Burra	Western Australia	Warm temperate to subtropical, arid to semi-arid	Wild
Ericameria nauseosa	Rubber rabbitbrush	Western North America	Boreal to warm temperate, arid to semi-arid	Experimental
Eucalyptus globulus	Bluegum	Australia	Warm temperate to subtropical, arid to humid	Minor global
Garcinia hanburyi	Siam gamboge	Southeast Asia	Humid tropical lowlands	Regional
Garcinia xanthochymus	Gamboge	Southeast Asia	Tropical and subtropical lowlands and highlands, humid	Regional
Grindelia hirsutula	Gumweed	Western North America	Arid to semi-arid, cold to warm temperate	Experimental
Grindelia squarrosa	Gumweed	Western North America	Arid to semi-arid cold to warm temperate	Wild
Guibourtia copallifera	Sierra Leone gum copal	Africa (Congo)	Humid tropical lowlands	Regional
Hymenaea courbaril	Stinking toe	Tropical Americas	Humid tropical lowlands	Regional
Hymenaea verrucosa		Africa	Humid tropical lowlands	Regional
Larix occidentalis	Western larch	North America	Boreal to warm temperate, humid	Regional
Larrea tridentata	Creosote bush	Southwestern North America, Mesoamerica	Arid to semi-arid, warm temperate to subtropical	Experimental
Parthenium argentatum	Guayule	Southwestern North America, Mesoamerica	Arid to semi-arid subtropics	Minor global
Pinus elliottii	Slash pine	North America	Warm temperate to tropical, semi-arid to humid	Global
Pinus halepensis	Aleppo pine	Mediterranean, Eurasia	Mediterranean	Regional
Pinus jeffreyi	Jeffrey pine	North America	Cold temperate to warm temperate, arid to humid	Regional
Pinus massoniana	Horsetail pine	Asia	Warm temperate to tropical	Regional
Pinus merkusii	Sumatran pine	Asia	Tropical lowlands and highlands, semi-arid to humid	Regional
Pinus oocarpa	Pino amarillo	Mesoamerica	Tropics, subtropics	Regional
Pinus palustris	Longleaf pine	North America	Warm temperate to subtropical, semi-arid to humid	Minor global
Pinus pinaster	Cluster pine	Africa, Mediterranean	Mediterranean, warm temperate	Regional
Pinus ponderosa	Ponderosa pine	North America	Cold temperate to subtropics, semi-arid to humid	Regional
Pinus radiata	Monterey pine	North America, Mesoamerica	Warm temperate, Mediterranean	Regional
Pinus sylvestris	Scots pine	Eurasia	Boreal to cold temperate, humid	Regional
Pinus taeda	Loblolly pine	North America	Warm temperate to subtropics, humid	Regional

Genus	Common Name	Origin	Climate	Cultivation
Pistacia lentiscus	Chios mastic tree	Mediterranean	Warm temperate to subtropical, Mediterranean, arid to semi-arid	Regional
Pistacia terebinthus	Terebinth tree	Mediterranean	Warm temperate to subtropical, Mediterranean, arid to semi-arid	Regional
Pittosporum resiniferum	Petroleum nut	Philippines	Humid tropical highlands	Experimental
Pittosporum undulatum	Australian cheesewood	Australia	Subtropics to tropics, Mediterranean, semi-arid	Experimental
Shorea javanica		Asia	Humid tropics	Experimental
Silphium laciniatum	Rosinweed	North America	Cold to warm temperate, semi-arid to humid	Wild

Sources: Hanelt, *Mansfeld's Encyclopedia of Agricultural and Horticultural Crops*; Langenheim, *Plant Resins*.
Note: These species produce terpene and heptane hydrocarbon resins. It is a challenge to develop a short list of the most promising resin crops from the great number of resin-bearing species. Some are cultivated, but few have been domesticated. And in some cases, it is difficult to ascertain which are destroyed by harvesting. I encourage botanists, chemists, agronomists, foresters, and agroecologists to prioritize research and development of this group of plants. For a more thorough review see Langenheim's *Plant Resins*.

In China, as early as 4000 BC, the lacquer tree (*Toxicodendron verniciﬂuum*) was already being tapped. While lacquered products are beautiful, the raw resin is highly toxic. The lacquer tree is closely related to poison oak and ivy and, like them, contains urushiol, the allergenic chemical present in many members of the cashew family. Harvesters and processors are often afflicted with terrible rashes.[73] (I'm highly allergic to urushiol, and it's hard for me to imagine cultivating trees to extract the urushiol-laden resin on purpose!)

Not all resin crops are tapped, however. Some, like guayule, rosinweed, and creosote bush, have resin in their leaves. Others, like some *Dyera*, *Hymenaea*, and *Pittosporum* species, produce fruits that are high in resin. A lot of shellac is still made from the resinous exudates of lac insects, farmed on host fodder trees.

Like most crops, there is nothing inherently bad about resin-producing species. The monoculture plantations, destructive harvests, and labor exploitation are products of human greed and oppression, not products of the plants per se. A cooperative of turpentine tappers in Honduras has developed an equitable community-based forestry system.[74]

Agroforestry systems sometimes include resin plants. One such system intercrops young *Agathis* with short-lived food crops and nitrogen-fixing *Leucaena* trees. Shade crops such as coffee and black pepper are grown under *Shorea* resin trees. Many timber trees are tapped to provide extra income for farmers while they wait for the timber yield.[75]

To my knowledge, of all resin crops only a few turpentine pines have received any breeding attention. Typically resin yields seem to be lower than those of latex crops, at least the domesticated ones. Developing some high-yielding, non-destructively harvested resin crops could be an important strategy for keeping some hydrocarbon feedstocks available in the post-fossil-fuel future we need to embrace. For more information, I recommend Langenheim's *Plant Resins*. At more than 500 pages, it has a far more thorough treatment of the subject than is possible here.

DIESEL TREE (*COPAIFERA* AND *GUIBOURTIA* SPP.)

These are large trees of the tropical lowlands, originating in Amazonia (*Copaifera* spp.) and Africa (*Guibourtia* spp.). These are regional crops, new crops, or wild-harvested, suited to humid or semi-arid tropical lowlands. Although they are legumes they probably do not fix nitrogen. The diesel tree (*C. langsdorffii*) was one of the "petrocrops" that most excited Melvin Calvin.[76] Perhaps most remarkable is that the resin can be passed through a filter and poured directly into a diesel tank to run motors and engines. Many *Copaifera* and *Guibourtia* species in Amazonia and tropical Africa are cultivated or wild-tapped for their similar resins.[77]

Diesel trees are great examples of potential hydrocarbon crops that could have great impact without the need for expensive biorefineries. They have enormous, if still undeveloped, potential to provide liquid fuel for rural villages in the humid tropics, for tractors, pickup trucks, grain grinders, oil presses, and other small-scale equipment. (They have promise as a larger-scale hydrocarbon feedstock as well.) Should domestication efforts be fruitful, I imagine a grove of diesel trees in every village and farm cluster in the humid tropics. This is an absolutely appropriate use of local biofuel. That said, I also fear that a domesticated diesel tree might be grown in vast corporate agrofuel monocultures the way palm oil and sugarcane are today, and I reiterate that even with remarkable trees like this, there is not nearly enough land to meet our current liquid fuel needs *and* produce food. Food must come first, and almost all energy must come from solar, wind, and water. Still, I can envision a future for domesticated, perhaps neo-hybrid diesel trees, grown in integrated polycultures, serving as a supply of local liquid fuel and a feedstock for regional medicine, bioplastics, and a wide range of other uses.

Some large-scale planting has already begun. Farmers in tropical Queensland, Australia, had planted 20,000 diesel trees as of 2008. They are hoping that a hectare of diesel trees will meet the fuel needs of a farm.[78] We may not know until those trees reach maturity if they were of improved varieties or just random seedlings. I suspect it is much too early for plantings of this scale since yields are still unpredictable.

Uses. Indigenous people of the Amazon have used diesel tree resin for millennia. Forest people use "copaiba oil" for lamp oil and in cosmetics, varnish, and paint. The tree is known as the "antibiotic of the forest" for its many medicinal uses.[79] The resin can be filtered and used to run diesel engines. It also has many other industrial applications.

Yield. Although it is exciting, diesel tree is far from ready for wide cultivation. Currently most resin is obtained from wild trees in managed forests. Many wild trees yield no resin at all. Others range from 1 to 60 liters per year of resin. Few wild trees are high producers. Trees need to be 15 to 20 years old before they can be tapped, Some have a productive life of 70 years of twice-annual tapping. Other wild trees show diminished yields after only a few years of tapping. A

FIGURE 22.6. The remarkable diesel tree, whose tapped resin can be used in diesel engines after a bit of filtering. Photograph courtesy of Mauro Halpern/Public Domain.

program of selecting the best wild trees and breeding is needed.[80]

Harvest and processing. The resin can only be stored for about three months before degrading. However, trees can be tapped throughout the year to potentially provide a steady supply.[81]

Carbon farming applications. These species have been managed in their natural forest settings for millennia. They could serve many functions in agroforestry systems.

Crop development. There has been little or no published work on domesticating diesel trees or selecting improved varieties with superior resin production.

TURPENTINE PINES (*PINUS* SPP.)

Many pine species are tapped for turpentine. They range from temperate to tropical, and semi-arid to humid. Pines were the critical suppliers of naval stores during the wooden ship era, and turpentine remains an important product today. There are different methods to extract turpentine from the raw oleoresin, only some of which are non-destructive. Destructive practices include use of aged stumps (wood naval stores) and extraction from pulped wood (sulfate naval stores).

Turpentine from tapped trees is called gum turpentine and can also be destructive or non-destructive.

FIGURE 22.7. Tapped turpentine pines in China. Photograph courtesy of Brock Dolman.

With proper techniques trees are barely bothered. Trees in pine timber plantations are often tapped for turpentine, and the impact has been closely studied. Some trees can be tapped for 20 years and live a long, healthy life afterward. Poor practices, however, can increase the potential for pests and diseases to attack the trees through the cutting or tapping wounds. Global production of gum turpentine is about 100,000 tons. This is about a third of the total of global biobased turpentine production. The remainder comes from destructively harvested trees.[82]

Millions of hectares of pine plantations around the world are not tapped for turpentine, providing an opportunity to quickly ramp up biobased hydrocarbon production if needed. This might provide a transitional strategy while we wait for other newly planted hydrocarbon crops to come into production. It also bears mentioning that techniques for grazing cattle in pine plantations have been well established, which could add productivity and reduce understory management

needs to this once and future hydrocarbon production system. The Gaviotas community in the Colombian savanna region have planted millions of Caribbean pines (*P. caribaea*) in a highly successful restoration effort, resulting in regeneration of native forest and income from resin products.[83]

Uses. Turpentine is the distilled resin of pines and a few other species. It is primarily composed of terpene hydrocarbons. It is used as a cleaning agent and solvent, and as the basis for producing disinfectants, cleaning agents, and pharmaceuticals. Like other hydrocarbon plant products, it has the potential to be used in a post-fossil-fuel substitute for petroleum and coal in producing materials and chemicals. It has a strong history of use in ships, evidenced by the alternate name, *naval stores*. Rosin, a by-product of turpentine production, is itself used for many industrial purposes, such as soldering, ink, varnish, and waxes.[84]

Yield. A typical turpentine pine will yield 2 to 3 kilograms (4.4 to 6.6 pounds) per year (4kg/8.8 pounds in some tropical species). Approximately 300 to 500 trees are needed to produce a ton of oleoresin. Typically about 200 trees are planted per hectare, producing 0.8 t/ha of resin. The resin is distilled to produce 150kg (331 pounds) of turpentine and 650 to 700kg (1,433 to 1,543 pounds) of gum rosin.[85]

Harvest and processing. Use of good tapping practices need not diminish the health of the tree. Processing is relatively low-tech with inexpensive technology and few advanced skills required.

Carbon farming applications. Some turpentine pines are used as windbreaks.[86] Pines are well suited to silvopasture systems, although collecting equipment needs to be protected from livestock damage.

Crop development. There is yield variation between and within species, indicating potential for breeding. Indeed, hybrids developed for the wood industry, such as *P. elliottii × P. caribaea*, have higher turpentine yields. Slash pine has had a bit of breeding work for turpentine yield. Otherwise, there appears to have been little to no breeding for increased turpentine production.[87] Perhaps someday neohybrid turpentine pines will be a common part of the landscape with selections for climates from extremely cold arid mountains to tropical lowlands.

Jeffrey Pine (*Pinus jeffreyi*)

Jeffrey pine (*Pinus jeffreyi*) is a North American pine with a limited distribution in the mountains of California and Oregon. It tolerates temperatures as low as −31°C (−24°F), arid conditions, and infertile soils. When tapped like turpentine pines, this species produces highly explosive heptane rather than turpentine. Heptanes are an ingredient in gasoline, and few plants are known to produce them. Heptane is used as the zero point in setting the octane rating scale for gasoline engines as its explosive nature causes "knocks" (tiny explosions) in internal combustion engines.[88] At one time botanists did not distinguish Jeffrey pine from ponderosa pine (*P. ponderosa*), a similar species tapped for turpentine. When Jeffrey pine resin ended up in the distilleries, it caused deadly explosions. Today's

FIGURE 22.8. Jeffrey pines in Nevada. In the foreground is rubber rabbitbrush, another wild hydrocarbon plant.

botanists distinguish Jeffrey pine in the field by its vanilla-scented needles.[89]

Uses. The heptane-rich resin could be used as a hydrocarbon source.

Yield. Unknown.

Harvest and processing. Intentional harvest and processing of Jeffrey pine heptanes would need to proceed carefully to avoid explosions!

Kauri (*Agathis australis*)

Kauri is a New Zealand conifer in the Araucariaceae family. The trees can tower up to 50 meters (164 feet) high. They are impressive resin producers. Resin is tapped, but it is also often found in the soil at the base of trees. European colonization of New Zealand was in large part driven by the market for kauri resin, which had accumulated in vast quantities—with some chunks up to 270 kilograms (595 pounds) in size—and which was used in Europe for linoleum and varnish production. Between 1850 and 1950, 450 million kilograms (992 million pounds) were exported. Destructive tapping practices were largely responsible for the demise of 99.5 percent of wild kauris. Today kauri and other *Agathis* species are cultivated as timber and resin trees.[90]

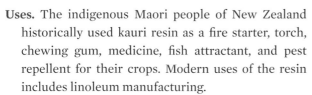

Figure 22.9. Kauri pine, once an important feedstock for linoleum production. Photograph courtesy of J. Brew/Creative Commons 2.0.

Figure 22.10. Petroleum nut, whose fruit is high in explosive heptane resin. Photograph courtesy of Dick Culbert/Creative Commons 2.0.

Uses. The indigenous Maori people of New Zealand historically used kauri resin as a fire starter, torch, chewing gum, medicine, fish attractant, and pest repellent for their crops. Modern uses of the resin includes linoleum manufacturing.

Harvest and processing. Today kauri resin is tapped commercially using less destructive methods.

PETROLEUM NUT
(*PITTOSPORUM RESINIFERUM*)

Petroleum nuts are native to "elfin" tropical highland forests in the Philippines, where they grow among stunted oaks and pines covered with ephiphytes. Unlike most petrocrops, which require tapping or biomass extraction with solvents, *Pittosporum resiniferum* concentrates its product in the easily harvested nuts. This experimental crop is suited to humid tropical highlands and mid-elevation areas.[91] The related Australian *P. undulatum* has similar resin content.[92]

Uses. These nuts, which are full of heptanes, have long been used as torches in the Philippines. They have many other potential applications as hydrocarbon crops.

Yield. Melvin Calvin estimated 6.5 to 10 barrels of oil per acre (0.4 hectare) per year equivalent.[93]

CHAPTER TWENTY-THREE

Fiber Crops

It's easy to overlook fiber crops, but we are surrounded by them every day: clothing, rugs, sheets and blankets, upholstered and stuffed furniture, and much more. Fiber crops are among the earliest domesticates, with several species of cotton, for example, going back thousands of years in both the New and Old Worlds. They are used for more than just textiles; archaeologists have found 12,000-year-old ceramic pottery reinforced with hemp fibers.[1]

Plant fibers are made of cellulose, a long linear chain (or polymer) of glucose. Cotton fibers are essentially long, long chains of sugar. Different plants produce fibers with different characteristics. Some species produce fibers around their seeds, like cotton and milkweed. Some produce bast fibers in the stems, including ramie, flax, and hemp. Palms produce a range of fibers in their leaves and fruits. Other crops, such as sisal, have long fibers in their succulent leaves. Several banana relatives are cultivated for the fibers in their trunks.

That said, relatively few plants produce fibers that are useful to human civilization. The perennial ones tend to be clustered in several important families: the Malvaceae or mallow family (cotton, kapok), Asparagaceae or asparagus family (sisal, henequen, bowstring hemp), Apocynaceae or milkweed family (milkweed, kendyr), the palms (coconut coir, raffia), and Urticaceae or nettle family (ramie, nettle). Fiber crops are generally less heavy feeders than food crops and are less damaging to the soil. They are often intercropped with food plants and represent a good model of industrial crops grown in polycultures.

Currently almost none of our fibers are produced from perennial crops. Many textiles, including nylon, spandex, and polyester, are actually made from petroleum. Some are made from synthetics from biobased feedstocks, such as rayon—a plastic made from biomass. Although this is better than petroleum, synthesizing fibers from biobased feedstocks is energy-intensive. In 2012 world fiber production volume totaled 88.5 million tons. Of those, 56.0 million tons were petroleum-based synthetics. Natural fibers accounted for 32.5 million tons produced. Cellulosic synthetics such as rayon, derived from wood and bamboo, accounted for 5.2 million tons.[2]

Of those natural fibers, in 2009, 73 percent were cotton. Globally, annual cotton production is more than 24 million tons. Non-destructively harvested perennials are far behind, with coconut coir at 1.03 million tons.[3] In 2012 global production of sisal was 428,000 tons, ramie was 83,000 tons, abacá was 168,000 tons, and kapok was 161,000 tons.[4] Global wool production is 1.22 million tons.[5]

Cotton's excellence as a fiber crop accounts for its dominance, but it comes at great cost. Cotton is grown on 2.4 percent of the world's cropland but accounts for 11 percent of all crop protection chemicals (fungicides, insecticides, and herbicides). Cotton alone accounts for 25 percent of all insecticide use. One study estimated that an incredible 20,000 people die every year from pesticide poisoning from cotton.[6] Cotton production is responsible for salting soils on irrigated lands and other forms of land degradation. For example, in Uzbekistan, the Aral Sea has shrunk by half due to withdrawal of irrigation water for cotton, with consequent salting of half of Uzbekistan's irrigated cropland.[7] Interestingly, kendyr, a salt- and

TABLE 23.1. Fiber Crop Yields Compared

Latin Name	Common Name	Climate	Avg. Global Yield t/ha[a]	Yield t/ha Dry Fiber
Cocos nucifera	Coconut	Tropical lowlands, semi-arid to humid		8.6–10.7 coir[b]
Corchorus olitorius	Jute	Warm temperate to tropical humid	2.1	2–5 bast[c]
Agave sisalana	Sisal	Tropics, subtropics, semi-arid to humid	0.5	2–4.5 lamina[c]
Gossypium hirsutum	Annual cotton	Warm temperate to tropical humid as annual	2.0[a]	1–4 seed fiber[d]
Boehmeria nivea	Ramie	Warm temperate to tropical, humid	1.8	1.5–3.5 bast[d]
Sansevieria spp.	Bowstring hemp	Tropics, subtropics, semi-arid to humid		3.1 lamina[d]
Cannabis sativa	Hemp	Cold temperate to tropical, semi-arid to humid		1–3 bast[d]
Asclepias syriaca	Common milkweed	Cold temperate, warm temperate, semi-arid to humid		2.6 (1.3 bast, 1.3 seed fiber)[e]
Linum usitatissimum	Flax	Boreal to warm temperate, semi-arid to humid	1.1	1.5–2.5 bast[d]
Urtica dioica	Stinging nettle	Boreal to subtropics, humid		1.7 bast[f]
Musa textilis	Abaca	Tropical lowlands and highlands, humid	0.6	0.5–1.5 banana[d]
Ceiba pentandra	Kapok	Tropics semi-arid to humid		0.4–0.7 seed fiber[d]
Apocynum venetum	Kendyr	Cold to warm temperate, arid to semi-arid		0.5–0.6 bast, plus seed fiber[e]

Note: These are arranged in order from highest to lowest yield. These figures represent the yield of extracted and dried fiber only; the total biomass yields of fiber crops are much higher. All of these are cultivated crops with the exception of common milkweed. Annual crops provided for comparison are marked with an asterisk (*).

[a] FAO Statistics Division online.
[b] 1,000 coconut husks yield 80kg (176 pounds) of coir. 18 to 24t/ha of nuts is 108,000 to 144,000 nuts/ha (at 3,000 nuts per ton), yielding 8.6 to 10.7 tons of coir per hectare per year. Calculated from: Duke, *Handbook of Nuts*, 105.
[c] Plant Resources of Tropical Africa database.
[d] Plant Resources of Tropical Africa database.
[e] Duke, *Handbook of Energy Crops*.
[f] Bacci et al., "Fiber yields and quality of fiber nettle (*Urtica dioica* L.) cultivated in Italy," 480–84.
[e] Yields 5 to 6 t/ha at 10% fiber. Calculated from Thevs et al., "*Apocynum venetum* L. and *Apocynum pictum* Schrenk (Apocynaceae) as multi-functional and multi-service plant species in Central Asia," 159–67.

drought-tolerant perennial fiber crop, is being grown on some of the land damaged by cotton.[8] Although perennial cotton forms exist and produce high-quality fiber, they are not amenable to today's mechanized production, so almost all cotton production today is from annual plants.[9]

Cotton production has been linked to atrocities for centuries. The high labor demands of cotton drove the growth of slavery around the world. Sharecropping followed slavery in the United States, with terrible conditions for low-income farmers. In fact, cotton was pivotal in both the industrial revolution and the development of capitalism itself, for good and for ill.[10] Cotton's dominance is not due to superior production. Perennial fiber crops such as coconut and sisal yield more, and several others are competitive. It is cotton's high quality that accounts for its ubiquity.

Thus we have an industry dominated by fossil fuel feedstocks with an unsustainable annual crop as the primary alternative. Where cotton is the only fiber

for the job, sustainable production of annual cotton seems like a good priority use of annual cropland, particularly with the addition of some perennial agroforestry support species. Cotton is perennial in the tropics, and perhaps larger-scale perennial production systems can be developed. It could be a candidate for woody agriculture; it coppices and has a durable fruit crop.

High-yielding perennial fiber crops can fill in the gap, serving coarser purposes that don't require cotton, or used in cotton blends. They might be especially useful for non-textile fiber uses such as composites and insulation. Wool also has an important role to play in replacing cotton and fossil-fuel-based textiles. Sheep and other wool-producing livestock such as llamas can be produced on perennial pastures and silvopastures with managed grazing for a positive carbon impact. As in every industrial crop category, we need to reduce consumption. Clothing today is too often made to fall apart in a year, to be replaced

FIBER CROP USES

Materials

- **Traditional fiber materials** have been made for thousands of years, since long before the dawn of agriculture. These products include rope, knitted goods, nets, mats, carpets, cordage, and twine.
- **Modern textiles** include woven and non-woven fabrics with products that include clothing, upholstery, sails, rugs, curtains, and many more.
- Many fiber plants are used for making **paper**. Abacá is particularly favored for this and is used to make the currency of several countries.[11]
- **Geotextiles** are an interesting category. These meshes, blankets, and logs are used for erosion control. They eventually degrade after sites are rehabilitated. I have used these products extensively in wetland restoration, and I can attest to their utility. Current international use amounts to 20,000 tons per year, but much more could be used in efforts to restore degraded lands.[12]
- **Composite** materials combine fibers with rubber, plastics, or other materials. Many molded car parts such as door and instrument panels are being made with natural fibers. In recent decades, glass fibers have been used for this kind of reinforcing material, but glass is heavier and more energy-intensive. Switching to natural fibers is greener and allows for lighter vehicles with reduced fuel needs. Fiber crops and their residues are also used in composite panels such as particleboard.[13]
- Finally, fibers have some uses in **construction**. Insulation is a critical part of improving energy efficiency in buildings, which is an important aspect of reducing fossil fuels. Fiber crops are processed to serve as insulation in the forms of fleeces, felts, and panels. These can also be used for soundproofing. Fibers are being used as reinforcement in cement, notably for lightweight roofing tiles.[14]

Chemicals and Energy

- Chemical and energy uses of fibers crops primarily work with the residues left after removing the fibers. These residues are used like any biomass feedstock.

with the next season's fashions. Durability must again become a virtue to address the world textile feedstock situation.

There has been little modern breeding work on fiber crops with the exception of cotton. Most fiber crops except cotton are grown by small farmers. Often they lack processing equipment and must use intensive hand labor to separate the fibers. Investment in perennial fiber crop breeding and appropriate-scale processing technologies should be a priority. There is room to produce significantly more coir; only a few countries in Southeast Asia and the Pacific currently process the husks into fiber.[15]

Seed Fibers and Fruit Hairs

These plant fibers grow on and surround the seeds. Cotton is the best-known seed fiber crop. This category also includes various members of the milkweed family, as well as the tree fiber kapok.

Perennial Cottons (*Gossypium* spp.)

Few people today know that there are cultivated, perennial cottons and that annual cottons were bred from perennial wild ancestors. Perennial cotton fiber is apparently superior and highly desirable due to longer fibers, but annual cottons are grown because they are adapted to mechanical harvest and can be grown outside the tropics.[16] Perennial cottons are suited to arid to somewhat humid tropical conditions.

Cultivated perennial cottons are typically grown on a small scale but are found throughout the world's tropics. Many were more important before the dawn of industrial agriculture. Perennial cultivated "races" include *G. arboreum* ssp. *burmanicum*, *G. arboreum* ssp. *indicum*, *G. arboreum* ssp. *soudanense*, *G. barbadense* f. *brasiliense*, *G. herbaceum* var. *acerifolium*, *G. hirsutum* var. *marie-galante*, and *G. hirsutim* var. *taitense*.[17] Given the terrible problems caused by annual cotton production and the great need for cotton fibers, researching perennial cotton varieties and production systems should be a high priority.

TABLE 23.2. Perennial Seed and Fruit Hair Fiber Crops

Latin Name	Common Name	Origin	Climate	Cultivation
Apocynum cannabinum	Dogbane hemp	North America	Temperate, Mediterranean, semi-arid to humid	Minor global
Apocynum pictum	Kendyr	Central Asia	Cold to warm temperate, arid to semi-arid	Regional
Apocynum venetum	Kendyr	Central Asia	Cold to warm temperate, arid to semi-arid	Regional
Asclepias curassavica	Curassavian swallow-wort	Tropical Americas	Tropics, subtropical, warm temperate, humid	Regional
Asclepias incarnata	Swamp milkweed	North America	Cold to warm temperate, humid	Experimental
Asclepias syriaca	Common milkweed	North America	Cold temperate, warm temperate, semi-arid to humid	Experimental
Bombax ceiba	Paina	Asia, Australia	Tropics, humid, lowlands	Regional
Calotropis gigantea	Akon	South and Southeast Asia	Tropical arid to humid	Regional
Calotropis procera	French cotton	Africa through India	Tropical arid to humid	Regional
Ceiba aesculifolia	Pochote	Mesoamerica	Tropical semi-arid to humid	Regional
Ceiba pentandra	Kapok	Tropical Americas, Africa	Tropics semi-arid to humid	Minor global
Funtumia elastica	West African rubber tree	West Africa	Tropical humid lowlands	Regional
Gossypium arboreum ssp. *burmanicum*	Tree cotton	Old World tropics	Tropics, semi-arid to humid	Regional
Gossypium arboreum ssp. *indicum*	Tree cotton	Old World tropics	Tropics, semi-arid to humid	Regional
Gossypium arboreum ssp. *soudanense*	Tree cotton	India	Tropics, semi-arid to humid	Regional
Gossypium barbadense f. *brasiliense*	Sea Island cotton	South America	Tropics, semi-arid to humid	Regional
Gossypium barbadense f. *darwinii*	Galapagos cotton	Galapagos Islands	Tropics, semi-arid to humid	Perennial wild relative
Gossypium herbaceum var. *acerifolium*	Levant cotton	Ethiopia	Semi-arid to humid tropics	Regional
Gossypium herbaceum spp. *africanum*	Levant cotton	South Africa	Semi-arid to humid tropics	Perennial wild relative
Gossypium hirsutum var. *marie-galante*	Upland cotton	Tropical Americas	Tropics, arid to humid	Regional
Gossypium hirsutum f. *punctatum*	Upland cotton	Tropical Americas	Tropics, arid to humid	Historic
Gossypium hirsutum ssp. *taitense*	Upland cotton	Pacific, Australia	Tropics, arid to humid	Regional
Tilia europaea	Linden, lime tree	Europe	Cold to warm temperate, humid	Historic
Trachomitum lancifolium	Kendyr	Central Asia	Boreal to warm temperate, semi-arid to humid	Minor global
Trachomitum sarmatiense	Kendyr	Central Asia	Boreal to warm temperate, semi-arid to humid	Regional
Trachomitum scabrum	Kendyr	Eurasia	Cold to warm temperate, arid to semi-arid	Regional
Typha angustifolia	Cattail	Europe, North America	Boreal to tropical, aquatic	Wild-managed
Typha latifolia	Cattail	Northern temperate climates, Africa, Australia	Boreal to subtropical, aquatic	Minor global

Source: Hanelt, *Mansfeld's Encyclopedia of Agricultural and Horticultural Crops*.

Uses. Commercial cotton is the fiber surrounding the seeds in the cotton "fruit." The fibers are almost pure cellulose and are easily spun. These ideal fiber qualities, and the relatively high yields, have made cotton a critical part of the world economy. Cotton seeds are also pressed for oil.

Yield. Annual cotton yields 1 to 4 tons per hectare of fiber.[18] Yield data on perennial cottons is unavailable.

Harvest and processing. Currently perennial cottons are harvested by hand. In Brazil and Cameroon, the plants are coppiced or "rattooned" annually.[19]

Carbon farming applications. Moco cotton (*G. hirsutum* var. *marie-galante*) is still grown as a perennial on a small scale in Brazil, where it is coppiced annually. Moco grows in semi-arid areas and is often intercropped with edible cacti.[20] This system is labor-intensive and low yielding, but worthy of further investigation. Perennial cottons are susceptible to all the pest and disease issues of annual cotton, which may make multi-year monoculture production impossible. Perhaps strips of tree cotton intercropped with other crops and support species could be viable. Because perennial cottons are woody plants that coppice and produce a fruit-borne crop not overly sensitive to bruising, they might be adapted to the Badgersett woody agriculture system, offering a chance to mechanize perennial cotton production.

KAPOK (*CEIBA PENTANDRA*)

Kapok or silk cotton tree is a tree from the American and West African semi-arid and humid tropics, now cultivated throughout the tropics on three continents as a minor fiber crop. Annual production of kapok in 2012 was 161,000 tons on 161,000 hectares (398,000 acres), worth $8 million US.[21] Several related species are minor regional crops, including *Bombax ceiba*, as well as *C. aesculifolia*, cultivated for fiber used to insulate refrigerators.[22]

Uses. Kapok seed fibers are not good for spinning but make good padding, stuffing, and insulation. The seeds are pressed for inedible oil. Young leaves and fruits are edible.[23]

Yield. Kapok produces 0.4 to 0.7 t/ha of floss.[24]

Harvest and processing. Fruits are knocked down from trees and harvested.

Carbon farming applications. Kapok is a minor component of various agroforestry systems. It can be coppiced and might be suitable for woody agriculture.

FIGURE 23.1. Wild perennial cotton, presumably *G. barbadense*, in Barbados.

FIGURE 23.2. Floss of kapok, a tree fiber crop. Photograph courtesy of Craig Morey/Creative Commons 3.0.

Palm Fibers

Many palms produce useful fibers. These are often visible on the trunk, appearing like woven burlap. By far the most important palm fiber is coir, which is extracted from coconut husks. Other important palm fibers, like raphia, can be non-destructively extracted from leaves or other parts of the plant. Some edible palm crops are secondarily used for fiber. These edibles with additional fiber benefits include *Acrocomia aculeata, Arenga pinnata, Borassus flabellifer, Caryota urens, Corypha utan, Elaeis guineensis, Mauritia flexuosa, Metroxylon sagu,* and *Phoenix dactylifera*.

Coconut (*Cocos nucifera*)

Coconut is not only an important food plant, but the world's most important perennial fiber crop. The fiber, called coir, is made from the husks of the nuts and is a productive use of this abundant crop residue. About 1 million tons of coir are produced annually.[25] Coir is only made in a few countries, with the majority of the world's coconut hulls unprocessed and representing a potential fiber resource.[26]

Uses. Traditional uses of coir include rope, mats, and carpets. Its salt resistance has made it an important maritime cordage material for millennia. Today coir is an important material in geotextiles, notably coir

Figure 23.3. Processing of coir fiber from coconut husks. Photograph courtesy of Fotokannan/Creative Commons 3.0.

"logs," which are widely used in erosion control. Coir is also rubberized (sprayed with molten rubber latex) and used in mattresses, pillows, and cushions for vehicles, including cars and trains.[27] "White" coir, of a higher quality for textiles, is extracted from the shells of green (young) coconuts, like those used for coconut water.[28]

Yields. Coconuts can produce an estimated 8.6 to 10.7 tons per hectare of coir.[29]

TABLE 23.3. Palm Fiber Crops

Genus	English Name	Origin	Climate	Cultivation Status
Astrocaryum vulgare	Tucuma palm	South America	Tropical humid lowlands	Regional
Attalea funifera	Piassava palm	Tropical Americas	Tropical humid lowlands	Regional
Chamaerops humilis	Dwarf fan palm	North Africa, Mediterranean	Mediterranean, warm temperate to subtropics, semi-arid to humid	Regional
Cocos nucifera	Coconut palm	Asia	Tropical humid lowlands	Global
Leopoldinia piassaba	Piassava palm	South America	Tropical humid lowlands	Regional
Oenocarpus bataua	Pataua palm	South America	Tropical humid lowlands	Experimental
Raphia farinifera	Madagascar raffia palm	Africa	Humid tropics, subtropics	Minor global
Raphia hookeri	Ivory Coast raphia palm	Africa	Tropical humid lowlands	Regional
Raphia vinifera	Bamboo palm	Africa	Tropical humid lowlands	Regional
Trachycarpus fortunei	Chinese windmill palm	Asia	Warm temperate to subtropics, humid	Regional

Source: Hanelt, Mansfeld's Encyclopedia of Agricultural and Horticultural Crops.

Stem Fibers (Bast and Bark)

Known as bast fibers, these are stem fibers that are "retted" out of the stem through fermentation. Important annual bast fibers include hemp (*Cannabis sativa*), kenaf (*Hibiscus cannabinus*), and flax (*Linum usitatissimum*). Bast production is one area where perennial fiber crops shine with a number of high-yielding, high-quality species. This category also includes some bark fibers.

RAMIE (*BOEHMERIA NIVEA*)

Ramie is an ancient herbaceous perennial fiber crop from Asia. There are two primary varieties of ramie, the tropical var. *tenacissima* and the surprisingly cold-hardy var. *nivea*. This hardy form can overwinter in warm temperate climates such as Washington, DC, with temperatures as low as −17°C (1.4°F), although it does not yield well in areas that cold. Ramie is a minor global perennial fiber crop with world production at 83,000 tons on 83,000 hectares (205,000 acres), worth $233 million US.[30]

Uses. Ramie fibers are of very high quality; they are long, silky, and tough.

Yield. Ramie yields 1.5 to 3.5 t/ha of fiber.[31]

Harvest and processing. A gummy resin makes extraction of the fibers difficult for small-scale producers, although a modern industrial process has been developed. Development of small-scale decentralized processing technology would be a boon to growers.

Carbon farming applications. The leaves of this (non-stinging) member of the nettle family are used as high-protein pig fodder in Asia. Perhaps they could also be useful as a source of leaf protein concentrate for human consumption.

FIGURE 23.4. Ramie, a perennial bast fiber, at Las Cañadas.

TABLE 23.4. Bast and Bark Fiber Crops

Latin Name	Common Name	Origin	Climate	Cultivation
Abelmoschus esculentus	Perennial okra	Africa	Tropical lowlands, semi-arid to humid	Regional
Abroma augusta	Perennial Indian hemp	Southeast Asia, Australia	Tropics, lowlands, humid	Regional
Adansonia digitata	Baobab	Africa	Tropical lowlands, arid to semi-arid	Important wild fiber
Adenanthera pavonina	Condori wood	South Asia through Australia	Tropical humid lowlands	Regional
Apocynum cannabinum	Dogbane hemp	North America	Cold to warm temperate, Mediterranean, semi-arid to humid	Minor global
Apocynum pictum	Kendyr	Central Asia	Cold to warm temperate, arid to semi-arid	Regional
Apocynum venetum	Kendyr	Central Asia	Cold to warm temperate, arid to semi-arid	Regional
Artocarpus altilis	Breadfruit	New Guinea, Pacific	Lowland humid tropics	Minor global
Asclepias curassavica	Curassavian swallow-wort	Tropical Americas	Tropics, subtropical, warm temperate, humid	Regional
Asclepias incarnata	Swamp milkweed	North America	Cold to warm temperate, humid	Experimental
Asclepias syriaca	Common milkweed	North America	Cold temperate, warm temperate, semi-arid to humid	Experimental
Bixa orellana	Annatto	Tropical Americas	Tropics and subtropics, semi-arid to humid	Minor global
Boehmeria nivea	Ramie	Asia	Warm temperate to tropical, humid	Minor global
Broussonetia papyrifera	Paper mulberry	Asia	Cold temperate to tropics, humid	Regional
Calotropis gigantea	Akon	South and Southeast Asia	Tropical arid to humid	Regional
Calotropis procera	French cotton	Africa through India	Tropical arid to humid	Regional
Clappertonia ficifolia	Bolo bolo	Africa	Tropics, subtropics, lowlands, humid	Regional
Cryptostegia grandiflora	India rubber vine	Madagascar	Humid lowland tropics	Minor global
Edgeworthia tomentosa	Paper bush	Asia	Warm temperate, subtropics, humid	Regional
Firmiana simplex	Gum karaya	South Asia	Subtropics, semi-arid	Regional
Gnetum gnemon	Jointfir, bago	Asia	Humid tropical lowlands	Regional
Hibiscus radiatus	Clavelina	Asia	Tropics and subtropics, humid	Minor global
Hibiscus tilliaceus	Beach hibiscus	Pantropical	Tropics to subtropics, lowlands, humid	Minor global
Linum hybrids	Perennial flax	Hypothetical	Boreal to warm temperate, semi-arid to humid	Under development
Pavonia spinifex	Gingerbush	North America, tropical Americas	Tropics, subtropics, warm	Regional
Pueraria montana var. *lobata*	Kudzu	Asia	Cold to warm temperate humid	Minor global
Pueraria phaseoloides	Tropical kudzu	Asia	Humid tropics	Minor global
Reynoutria japonica	Japanese knotweed	Asia	Cold to warm temperate, humid	Experimental
Senna auriculata	Tanner's cassia	Asia	Tropics, lowlands, semi-arid to humid	Minor global
Sida rhombifolia	Arrowleaf sida	Tropical Americas	Tropics, subtropics, lowlands and highlands, semi-arid to humid	Minor global

Latin Name	Common Name	Origin	Climate	Cultivation
Spartium junceum	Spanish broom	Mediterranean, North Africa	Mediterranean	Regional
Trachomitum lancifolium	Kendyr	Central Asia	Boreal to warm temperate, semi-arid to humid	Regional
Trachomitum sarmatiense	Kendyr	Central Asia	Boreal to warm temperate, semi-arid to humid	Regional
Trachomitum scabrum	Kendyr	Eurasia	Cold to warm temperate, arid to semi-arid	Regional
Trema orientalis	African elm	Asia	Humid tropical lowland	Regional
Wissadula periplocifolia	White velvetleaf	Mesoamerica	Tropical, subtropical, humid	Regional

Source: Hanelt, *Mansfeld's Encyclopedia of Agricultural and Horticultural Crops.*

STINGING NETTLE (*URTICA DIOICA*)

This species seems like an unlikely candidate for producing soft fibers since the leaves and stems, when green, possess stinging hairs that can be painful to touch. But nettles have been cultivated on a small scale for their high-quality bast fibers for centuries. They produce a fine linen called *Nesseltuch* in Germany, where during World War I nettles were cultivated on a large scale.[32] More recent experiments in Germany demonstrate continued interest in this multipurpose perennial. Nettles may also have been cultivated as a

FIGURE 23.5. Stinging nettle, a minor global perennial fiber. Photo courtesy of Harry Rose.

fiber crop in ancient Egypt and in North America before the arrival of Europeans. They are suited to cold temperate through subtropical humid climates.

Uses. Nettles produce a high-quality bast fiber.

Yield. Stinging nettle yields 1.7 t/ha of fiber.[33]

Harvest and processing. I suspect that enclosed tractors are used for harvest, or else stem fiber harvest comes after frost has safely killed the stinging hairs.

Carbon farming applications. Intercropping efforts would have to take into account the impact of this weedy, stinging crop on harvest and management of neighboring species. Nettles are already a common weed in forest edges and in orchards, indicating potential use as an understory species for partial shade.

DOGBANE-MILKWEED FAMILY FIBER CROPS (*ASCLEPIAS, APOCYNUM, CALOTROPIS,* AND *TRACHOMITUM* SPP.)

These crops are all members of the new expanded dogbane-milkweed family (Apocynaceae). Around the world they have served as industrial crops for millennia, with a fair number of species in active cultivation as regional crops today.[34]

Uses. All of these crops are dual-purpose fibers, offering bast fibers from the stem and seed fiber or "floss" in the fruit pods. Many have also been identified as potential hydrocarbon crops due to their high latex content.[35] Some have edible and medicinal uses as well.

FIGURE 23.7. A sisal plantation in Tanzania. Photograph courtesy of Dom Pates/Creative Commons 2.0.

cement-sisal tiles already in use. It is cultivated in China for use in metal polishing.[41]

Yield. Sisal yields 2 to 4.5 t/ha of lamina fibers, making it one of the highest-yielding fibers in the world.[42]

Harvest and processing. Lack of access to mid-scale sisal fiber processing technology is a challenge for many tropical farmers. The first harvest is not ready until two to five years after planting. Often shorter-lived crops are intercropped during this period. Sisal plants live for 7 to 15 years. After flowering and dying, they produce multiple basal suckers for propagation.[43]

Carbon farming applications. Agave species are often used as thorny living fences.

BOWSTRING HEMP (*SANSEVIERIA* SPP.)

Bowstring hemp was once much more widely culti-vated. At least four species remain in production today as regional crops: *S. hyacinthoides*, *S. roxburghiana*, *S. trifasciata*, and *S. zeylanica*.[44] Like sisal, bowstring

FIGURE 23.8. Bowstring hemp growing in shade of coconuts, bananas, gingers, and fruit trees in the Cook Islands. Photograph courtesy of Akos Kokai/Creative Commons 2.0.

hemp species have crassulacean acid metabolism. They are suited to semi-arid tropics and subtropics with rainfall as low as 250 millimeters (10 inches). Bowstring hemp prefers to grow in the shade, a promising characteristic.[45]

Uses. A good-quality lamina fiber.

Yield. Bowstring hemp is recorded as yielding as high as 3.1 t/ha, making it among the highest-yielding perennial fibers.[46]

Harvest and processing. Plants are harvested two to three years after planting and every two years afterward.[47]

Carbon farming applications. Bowstring hemp's suitability to shade and dry climates makes it an excellent understory crop candidate for tree crops such as mesquite, moringa, wattleseed acacias, and castor beans.

Banana Stem Fibers

Several species in the banana family (Musaceae) produce useful fibers. Of these abacá (*Musa textilis*) is by far the most important. The related Japanese fiber banana (*M. basjoo*) is remarkably cold-tolerant; the plants in my Massachusetts garden are 10 years old and resprout each spring after being killed by frost and spending the winter under piles of snow for months at a time. The fibers of enset (*Ensete ventricosum*) are a by-product of starch extraction.

ABACÁ (*MUSA TEXTILIS*)

This relative of bananas was domesticated in the Philippines as a fiber crop. Abacá remains important in the Philippines and is also important in Ecuador, but little

FIGURE 23.9. Processing the fibers of abacá in the Philippines. Photograph courtesy of Ronald Tagra/Creative Commons 2.0.

grown elsewhere. This minor global crop is grown on 168,000 hectares (415,000 acres), yielding 106,000 tons of fiber worth $109 million US.[48] Abacá is not without its challenges. Like bananas, it is vulnerable to pests and diseases, including some that spread through vegetative propagation of the plants. There is also a dearth of small- to mid-scale processing technology, meaning most farmers must tediously remove the fibers by hand. This is a crop for mid-elevation humid tropics.

Uses. The long, strong fibers are useful for textiles as well as specialty papers such as currency and teabags. Recently they are becoming popular as a substitute for glass fibers in composite materials.[49]

TABLE 23.6. Banana Leaf Fiber Crops

Genus	Common Name	Origin	Climate	Cultivation
Ensete ventricosum	Enset	Africa	Tropical highlands, humid	Regional
Musa acuminata	Banana, plantain	Asia, New Guinea	Humid tropical lowlands and highlands	Global
Musa basjoo	Japanese fiber banana	East Asia	Cold temperate to subtropics, humid	Regional
Musa textilis	Abacá	Asia	Tropical lowlands and highlands, humid	Minor global

Source: Hanelt, *Mansfeld's Encyclopedia of Agricultural and Horticultural Crops.*

Yield. Abacá yields 0.5 to 1.5 t/ha of fiber.[50]

Harvest and processing. Abacá plants will bear for about 20 years. The stems are cut and processed for fiber, with new suckers coming up to replace them in a perennial managed multistem system.

Carbon farming applications. Abacá likes 50 percent shade and is typically grown in integrated farming systems in mid-elevation humid tropics. Polyculture companions often include coconut, durian, cacao, coffee, and nitrogen-fixing shade trees.[51]

CHAPTER TWENTY-FOUR

Other Industrial Uses

In many ways perennial biomass, starch, oil, hydrocarbon, and fiber crops are just the beginning. Many other important products, many currently synthesized from petroleum, are or have been produced from perennial crops. This chapter covers a few of these other important categories. They include dyes, cork, tannins, waxes, gums, pesticides, medicines, and soaps.

Before World War II these plants were an important part of the world economy. Today most of the products are synthesized from petroleum. Getting off the fossil carbon train means relearning where the products we use come from, and developing a strategy to reduce consumption and convert to biobased feedstocks—perennial wherever possible.

Dyes

Synthetic dyes and mordants are among the world's worst polluters. However, hundreds of plants are cultivated, and more are wild-harvested, for use as botanical dyes.[1] Many of these are non-destructively harvested perennials. I've chosen to focus here on woody species, although there are many more herbaceous perennials to choose from.

TABLE 24.1. Dye and Mordant Crops

Latin Name	Common Name	Origin	Climate	Cultivation
Acacia catechu	Cutch tree	South and Southeast Asia	Semi-arid to humid tropical lowlands	Regional crop
Bixa orellana	Annatto	Tropical Americas	Tropics and subtropics, semi-arid to humid	Minor global crop
Caesalpinia sappan	Brazilwood	South America	Tropical humid lowlands	Regional
Cotinus coggygria	Smoketree	North America	Cold to warm temperate, humid	Wild
Indigofera suffruticosa	Indigo	Tropical America	Tropical and subtropical humid	Minor global
Indigofera tinctoria	Indigo	India	Tropical and subtopical, humid	Minor global
Juglans nigra	Black walnut	Eastern North America	Cold to warm temperate, humid	Regional
Lawsonia inermis	Henna	Africa, South Asia through Australia	Tropical semi-arid	Regional
Maclura pomifera	Osage orange	North America	Cold to warm temperate, semi-arid to humid	Regional
Rhus copallinum	Winged sumac	North America	Cold to warm temperate, semi-arid to humid	Experimental
Sambucus canadensis	Elderberry	North America, Mesoamerica	Cold temperate to subtropical, humid	Regional
Sambucus nigra	Elderberry	Europe	Cold temperate to subtropical, humid	Minor global
Semecarpus anacardium	Indian marking nut tree	India	Semi-arid to humid tropics	Regional
Senna auriculata	Tanner's cassia	Asia	Tropics, lowlands, semi-arid to humid	Minor global

INDIGO (*INDIGOFERA TINCTORIA, I. SUFFRUTICOSA*)

Indigo is the most celebrated dye plant, for good reason. The beautiful blue color it imparts to textiles doesn't fade with sunlight or washing. A number of legumes and unrelated species produce indoxyl—the chemical that produces indigo—in their dried leaves, but two species of indigo are dominant. *Indigofera tinctoria* is the Old World species and has been grown for thousands of years in India. In the New World, *I. suffruticosa* has a similar history.[2] Both species are medium shrubs in the humid tropics and subtropics and are grown as annuals farther north. Prior to the development of synthetic dyes, indigo was a major item of world trade. It was also a major plantation crop grown by enslaved people in the early American colonies.[3]

Uses. The leaves and twigs are used as the source of a blue dye.

Yields. Indigo is a relatively short-lived perennial when coppiced annually for dye production. It produces 10 to 13 t/ha of fresh biomass.[4]

Carbon farming applications. *I. suffruticosa* is used as a contour hedgerow species.[5] *Indigofera* species fix nitrogen and could have a role in agroforestry systems.

Cork

Cork oak is native to the western Mediterranean including southwestern Europe and North Africa. It is a key component of the semi-arid savannas there. Human-managed cork oak savannas, called *dehesa* in Portugal, cover 2 million hectares (4.9 million acres) in Europe, mostly in Portugal.[6]

CORK OAK (*QUERCUS SUBER*)

Uses. Cork is the inner bark of the tree. It is a remarkable material, sold commercially for bottle stoppers—to such an extent that these bottle stoppers are referred to as "corks." Other uses include insulation, soundproofing, flooring, bulletin boards, gaskets, and shoe cushions.[7]

FIGURE 24.1. Indigo, the most famous dye crop, in Florida.

FIGURE 24.2. Recently harvested cork in a *dehesa* system. Photograph courtesy of Brock Dolman.

Yield. Cork stands produce 5 tons per hectare every 10 years when harvested.[8]

Harvest and processing. The bark is stripped off the trees about every 10 years. It regrows until the trees are 150 to 250 years old.[9]

Carbon farming applications. Cork oak is a key component of *dehesa* silvopasture and silvoarable systems.

Tannins

Tannins are astringent compounds found in many plants that have a long history of use. The term *tannins* comes from their use in tanning hides into leather, but they also have applications as medicines and dyes. I was surprised while researching this book to discover that so many plants are cultivated for tannin production, albeit less commonly than in the past, since synthetic tannins (presumably synthesized from petroleum) dominate the market today. Many crops grown specifically for tannin production include pods (*Acacia farnesiana, Caesalpinia digyna,* and *Senna auriculata*) and coppiced barks (*Acacia catechu, A. tortilis, A. saligna, A. mearnsii, Eucalyptus camaldulensis*). Tannins are a by-product of acorn processing, as well as production of crops that include *Alnus glutinosa* (cones), *Carya illinoinensis* (shells), *Cotinus coggygria* (leaves), and *Juglans regia* (husks). *Acacia mearnsii* may be the most widely cultivated tannin crop, but it is destructively harvested and therefore not profiled here.

TERI-POD (*CAESALPINIA DIGYNA*)

Teri-pod is a viney shrub originating in India and Southeast Asia. It has a wide range, from warm temperate to tropical, and grows in semi-arid to humid conditions.[10] This non-nitrogen-fixing legume is cultivated on a small scale for the tannin-rich pods for tanning and dyeing, and as a spiny hedge. Teri-pod seems ripe for domestication as a multipurpose industrial feedstock species.

Uses. The pods are 50 to 60 percent tannin, making this one of the most concentrated non-destructively

TABLE 24.2. Tannin Crops

Latin Name	English Name	Origin	Climate	Cultivation
Acacia catechu	Cutch tree	South and Southeast Asia	Semi-arid to humid tropical lowlands	Regional crop
Acacia nilotica	Egyptian thorn	Africa	Arid to semi-arid tropical	Regional crop
Acacia saligna	Port Jackson wattle	Australia	Tropical	Minor global crop
Acacia tortilis	Umbrella thorn acacia	Africa	Tropical	Regional crop
Caesalpinia digyna	Teri-pod	Asia	Warm temperate to tropical, semi-arid to humid, lowlands and highlands	Regional crop
Castanea spp.	Chestnuts	Eurasia, North America	Warm to cold temperate, Mediterranean, semi-arid to humid	Global nut, regional tannin
Casuarina equisetifolia	Australian pine	Australia	Tropical and subtropical lowlands, arid to humid	Minor global crop
Eucalyptus camaldulensis	Red River gum	Australia	Warm temperate to tropics, semi-arid to humid	Global pulp and timber crop, regional tannin
Rhus copallinum	Winged sumac	North America	Cold temperate through subtropical, semi-arid to humid	Wild
Quercus spp.	Oaks	Northern Hemisphere	Tropical to boreal, arid to humid	Mostly wild-harvested for tannin
Salix viminalis	Basket willow	Eurasia	Boreal to warm temperate, semi-arid to humid	Regional crop
Senna auriculata	Tanner's cassia	Asia	Tropics, lowlands, semi-arid to humid	Minor global crop
Trema orientalis	African elm	Asia	Humid tropical lowlands	Regional crop

Note: Many tannins are obtained destructively from the bark of logged trees grown for their tannins. Here we profile non-destructively harvested species, which mostly fall into two categories: tannin-rich fruits and coppiced high-tannin barks and woods.

harvested sources of this important compound. The beans inside the pods are also quite useful. They are 25 percent oil, which is used as an illuminant, 40 percent starch, and 14 percent protein. The beans are sometimes used as livestock fodder and are occasionally eaten by humans.[11]

Yield. Yields and other information are missing from the literature.

Carbon farming applications. Teri-pod might be a good candidate for woody agriculture. It is a non-nitrogen-fixing legume.

Waxes

Waxes are water-resistant, malleable substances produced by many different plants, animals, and other organisms. Many different chemicals are lumped together under the term *wax*; some, for instance, are hydrocarbons, while others are not.[12] Waxes are used in polishes and surface treatments, as food additives and lubricants, and in cosmetics and pharmaceuticals.[13] Much of the wax used in commerce today is paraffin, which is made from fossil carbon. The world's most important wax crop is the carnauba palm.

CARNAUBA PALM (*COPERNICIA PRUNIFERA*)

This palm is the source of carnauba wax, the most important plant-based commercial wax. *Copernicia* is a palm of the semi-arid to humid tropical lowlands of northeastern Brazil, where it is extensively wild-harvested and also cultivated on a moderate scale. In 2006 more than 22,000 tons of carnauba were produced.[14]

Uses. Known as the "queen of waxes," carnauba is used in polishes for shoes, vehicles, floors, furniture, and musical instruments. Carnauba wax is also a common ingredient in processed foods.[15]

TABLE 24.3. Wax Crops

Latin Name	Common Name	Origin	Climate	Cultivation
Ceroxylon alpinum	Andean waxpalm	South America	Humid tropical highlands, Mediterranean	Regional
Copernicia prunifera	Carnauba palm	South America	Semi-arid to humid tropical lowlands	Regional
Euphorbia antisyphilitica	Candelillia	Mesoamerica, North America	Warm temperate, subtropical arid to semi-arid	Experimental
Morella cerifera	Wax myrtle	Eastern North America	Warm temperate to tropical, humid	Wild-harvested
Myrica pensylvanica	Bayberry	Eastern North America	Cold to warm temperate, humid	Regional
Raphia farinifera	Madagascar raffia palm	Africa, South Asia, Madagascar	Humid tropics, subtropics	Regional
Rhus succedanea	Wax tree	China	Warm and cold temperate humid	Regional
Ricinus communis	Castor bean	Africa or India	Tropics, subtropics, highlands or lowlands, arid to humid	Global
Saccharum officinarum	Sugarcane	New Guinea	Humid tropical and subtropical lowlands	Global
Simmondsia chinensis	Jojoba	Southwest North America, Mesoamerica	Subtropics, arid to semi-arid	New crop
Stipa tenacissima	Esparto grass	Northwest Africa, Mediterranean	Semi-arid, warm temperate to subtropical	Regional
Syagrus coronata	Ouricury palm	Brazil	Humid tropics and subtropics	Regional
Toxicodendron vernicifluum	Lacquer tree	China	Warm temperate humid	Regional
Trachycarpus fortunei	Chinese windmill palm	Asia	Warm temperate to subtropics, humid	Regional

Source: Hanelt, *Mansfeld's Encyclopedia of Agricultural and Horticultural Crops.*

FIGURE 24.3. Carnauba palm, a commercial source of wax. Photograph courtesy of Tacarijus/Creative Commons 3.0.

Harvest and processing. The leaves are coated with wax and are beaten after harvesting to remove the flakes.

Carbon farming applications. Carnauba seems well suited to intercropping operations of any kind.

Gums

Gums are polysaccharides widely used as thickeners and gelling agents. In foods, they are used as ingredients in processing, glazes in baked goods, and hundreds of other applications we rarely think about—like keeping Popsicles from "sweating" too much. Gums are also used in pharmaceuticals, including denture adhesives. Technical uses include applications in the printing industry.[16] Non-destructively harvested gums come from tapped trees and from seeds. Other gums come from seaweeds (carrageenan), annual crops (guar), destructively harvested roots (tragacanth), and synthetic sources.

GUM ARABIC (*ACACIA SENEGAL*)

Gum arabic is an important world crop. Exports in 2005 totaled 70,000 tons.[17] Considering that the price can be $1,500 to $6,000 US a ton, it represents a significant income for rural people in poor arid and semi-arid regions of Africa and South Asia where gum arabic grows.[18] The gum is wild-collected but is also grown in plantations as a regional crop.

Uses. Gum arabic is an edible gum of the highest quality.

Yield. Yields can be as high as 2 to 3kg (4.4 to 6.6 pounds) per year on the best trees, although they are typically much less.[19]

Harvest and processing. Plants can be tapped in four to five years from seeding. Yields begin to slow at

TABLE 24.4. Gum Crops

Latin Name	Common Name	Origin	Climate	Cultivation
Acacia catechu	Cutch tree	South and Southeast Asia	Semi-arid to humid tropical lowlands	Regional
Acacia saligna	Port Jackson wattle	Australia	Tropical, lowlands, arid to semi-arid	Minor global
Acacia senegal	Gum arabic	Africa, South Asia	Tropics, subtropics, lowlands, highlands, arid to semi-arid	Regional
Acacia seyal		Africa	Tropical, arid to semi-arid	Regional
Bixa orellana	Annatto	Tropical Americas	Tropics and subtropics, semi-arid to humid	Minor global
Ceratonia siliqua	Carob	North Africa, Mediterranean	Mediterranean, arid to semi-arid subtropics	Minor global
Firmiana simplex	Gum karaya	South Asia	Subtropics, semi-arid	Regional
Prosopis juliflora	Mesquite	Central and South America	Tropical lowlands, arid to semi-arid, halophyte	Minor global
Toxicodendron sylvestre	Yame-haze	East Asia	Warm and cold temperate	Regional

FIGURE 24.4. A Sudanese gum arabic producer. Photograph courtesy of Grant Wroe-Street, United Nations Environment Programme.

20 years, at which point the tree can be coppiced for renewed growth and production (as well as firewood). A blade on a long pole called a *sunki* is used to peel back the bark on the branches, which then exude gum. The gum is carefully collected and processed.[20]

Carbon farming applications. Gum arabic is a fantastic multipurpose tree. It fixes nitrogen and coppices well. It is a good firewood. The pods are an excellent fodder for livestock. Gum arabic trees are also used in efforts to reverse desertification and restore degraded land.[21]

Pesticides

Many plants provide natural pesticides to control insects, plant diseases, and other agricultural pest organisms. Some of these, including *Derris* and *Lophophora*, are destructively harvested because the pesticides are in the roots. Producing pesticides on-farm saves money and ensures availability. The table here mostly covers cultivated species with a few wild cold-hardy species included.

NEEM (*AZADIRACHTA INDICA*)

Neem is one of the "miracle trees" that have received a lot of attention in recent decades. It is a remarkable multipurpose species, although no plant can solve humanity's problems on its own. Neem originates in India and Southeast Asia and grows in semi-arid to humid tropical lowlands. I have seen it growing in quite arid regions, although it wants 400 millimeters (16 inches) or more of rainfall for best production. There are an estimated 14 million neem trees in India and 40,000 hectares (99,000 acres) in the Philippines, probably making neem the most widely grown pesticide crop in the world, although it is still a minor global crop.[22] The related chinaberry (*Melia azedarach*) is an alternative for tropical highlands and warm temperate climates.

Uses. Neem-based pesticides are grown, sold, and used around the world. The oil, presscake, leaves, and other parts are utilized as insecticides, fungicides,

TABLE 24.5. Pesticide Crops

Latin Name	Common Name	Origin	Climate	Cultivation
Albizia lebbeck	Lebbek tree	Asia	Tropical, subtropics, lowlands, semi-arid to humid	Minor global
Amorpha fruticosa	False indigo	North America	Boreal through warm temperate, semi-arid to humid	Minor global
Azadirachta indica	Neem	South Asia	Arid to humid tropical lowlands	Minor global
Duboisia myoporoides	Corkwood	Australia	Humid tropical lowlands	Regional
Melia azedarach	Chinaberry	Asia	Warm temperate to tropics, lowlands and highlands, semi-arid to humid	Minor global
Nicotiana glauca	Tree tobacco	South America	Warm temperate to tropical, arid to humid	Regional
Sapindus saponaria	Soapnut	Americas	Semi-arid to humid tropics	Minor global
Sapindus saponaria var. *drummondi*	Western soapberry	Western North America	Cold to warm temperate, semi-arid	Wild
Tanacetum cinerariifolium	Pyrethrum daisy	Eastern Europe	Warm temperate, subtropical, Mediterranean, tropical highlands, semi-arid to humid	Minor global
Tephrosia virginiana	Goat's rue	North America	Cold to warm temperate, humid	Wild
Tephrosia vogelii	Tephrosia	Africa	Tropical humid	Minor global

FIGURE 24.5. Neem with coconut and breadfruit in a Hawaiian homegarden. Photograph courtesy of Forest and Kim Starr/Creative Commons 2.0.

bactericides, amoebacides, and even antivirals.[23] Neem has many medicinal uses. The twigs are used as antibacterial toothbrushes. Neem is antimalarial, spermicidal, antiseptic, febrifuge, contraceptive, and especially effective against Chagas' disease.[24] Neem coppices readily and provides a high-quality firewood. It is often used for charcoal. The seed oil is an important fuel and feedstock. Neem seedcake may slow denitrification in soils, helping preserve precious nitrogen.[25]

Yield. Biomass yields are 10 to 100 t/ha of dry matter, at about 50 percent leaves. Neem firewood and biomass yields are roughly equal to the best such species. Neem fruit yields 0.5 t/ha of pesticidal oil.[26]

Harvest and processing. For leaf production, neem is coppiced. For oil production, the trees are not coppiced, and mostly harvested by hand. Trees begin to flower at three to five years of age.[27]

Carbon farming applications. The trees are fast growing and are often used in agroforestry, in reforestation efforts, and as shade trees and windbreaks.[28]

Medicinal Plants

The international medicinal plant consumer market is $18 billion US. There are many times more medicinal plants than edible ones. Most pharmaceuticals today

TABLE 24.6. Non-Destructively Harvested Medicinal Tree Crops

Genus	Common Name	Origin	Climate	Cultivation
Acacia catechu	Cutch tree	South and Southeast Asia	Semi-arid to humid tropical lowlands	Regional
Acacia senegal	Gum arabic	Africa, South Asia	Tropical	Regional
Azadirachta indica	Neem	South Asia	Arid to humid tropical lowlands and highlands	Minor global
Camptotheca acuminata	Cancer tree	China	Warm temperate	New crop
Cinchona officinalis	Quinine	South America	Tropical humid lowlands	Minor global
Cinchona pubescens	Red cinchona	Tropical Americas	Tropical lowlands, highlands	Minor global
Crataegus pinnatifida	Chinese hawthorn	East Asia	Cold to warm temperate, sermi-arid to humid	Regional
Duboisia myoporoides	Corkwood	Australia	Humid tropical lowlands	Regional
Eucalyptus globulus	Bluegum	Australia	Warm temperate to subtropical, semi-arid to humid	Minor global
Ginkgo biloba	Ginkgo	Asia	Cold temperate to subtropical, humid	Minor global
Lycium chinense	Goji	East Asia	Cold climate to subtropical, tropical highlands, semi-arid to humid	Minor global
Melaleuca alternifolia	Tea tree	Australia	Subtropical to tropical, humid	Minor global
Moringa oleifera	Moringa	India	Tropics and subtropics, lowlands, semi-arid to humid	Minor global
Ricinus communis	Castor bean	Africa or India	Tropics, subtropics, highlands or lowlands, semi-arid to humid	Global
Sambucus nigra	Elderberry	Europe	Cold temperate to subtropical, humid	Regional
Styphnolobium japonicum	Japanese pagoda tree	East Asia	Cold temperate to subtropical, tropical highlands, semi-arid to humid	Regional
Taxus baccata	Yew	Eurasia, Africa	Cold to warm temperate, humid	Regional
Tilia spp.	Linden	Northern temperate climates	Cold to warm temperate, humid	Regional
Trichilia emetica	Mafura butter	Africa, Middle East	Semi-arid to humid tropical lowlands	Regional
Ulmus rubra	Slippery elm	North America	Cold to warm temperate, humid	Regional
Virola sebifera	Virola nut	Americas	Humid tropical lowlands	Regional

Note: In an effort to limit the thousands of medicinal plants, this table focuses on non-destructively harvested woody crops cultivated specifically as medicine. This list could and should be greatly expanded. Many other species profiled in this book have medicinal uses. See appendix A.

are synthesized from petroleum, but many are based on compounds found in plants. In fact, 25 percent of the modern world's medicines are based on plants.[29]

GINKGO (*GINKGO BILOBA*)

At more than 200 million years old, ginkgo is one of the world's most ancient plants.[30] Today it is an important coppiced woody medicinal crop. One US farm, the world's leading ginkgo leaf supplier, produces over 400 hectares (988 acres).[31] Ginkgos are incredibly tough; one famously resprouted after the nuclear blast at Hiroshima.[32] The trees form large lignotubers—woody underground growths from which they resprout. Individual plants can live for thousands of years.[33] I

FIGURE 24.6. Field production of coppiced ginkgo for leaf production. Photograph by Steven Foster.

speculate that plants with lignotubers may be especially good at sequestering carbon.

Uses. Ginkgo leaves have been used medicinally in China for 2,800 years. Today they are used globally for cerebrovascular insufficiency and dementia, along with many other applications.[34] Ginkgos also produce edible nuts, encased in a foul-smelling flesh that causes irritating rashes when it comes in contact with skin. Large amounts, or even small amounts, of raw nuts can be toxic and even fatal.[35] Nonetheless they are widely eaten across Asia.

Yield. Coppiced leaf yields are 20 to 25 t/ha or more.[36] Fruit yields are high but unreported.

Harvest and processing. Ginkgo leaves are machine-harvested in coppiced fields.

Crop development. "Stinkless" ginkgo clones have been selected for nut production.[37] (No word on whether they cause terrible rashes.)

Soaps

Soaps can be made from edible and inedible oils of plant, animal, or petrochemical origin. Several plants manufacture soaps directly; among them, the most important is the non-destructively harvested soapnut. The fruits of several other species are used as soap, although I don't know of any cultivated specifically for this purpose the way *Sapindus* is. These include *Aesculus* spp., *Cucurbita foetidissima*, *Gleditsia* spp. (though not *G. triacanthos*), and *Shepherdia canadensis*.

SOAPNUTS (*SAPINDUS SAPONARIA*)

Soapnut, or soapberry, trees are native to the Americas and Asia. The main species is tropical and subtropical and appears to be an ancient domesticate in the Americas and probably Asia as well. There is a cold-hardy form, the western soapberry (var. *drummondii*), which grows in cold and warm temperate semi-arid to humid regions. Soap nuts represent a great example of appropriate technology. No special equipment is needed to process or use them; they are simply a natural, biodegradable source of soap that

Figure 24.7. Western soapnut, a temperate tree soap, at the Denver Botanic Garden in Colorado.

grows on trees. No farm or village should be without a few trees.

Uses. *Sapindus* fruits are full of soap. They are marketed today as an ecological laundry detergent.[38] I've used them; I simply tie up some dry fruits in a sock and throw them into the laundry machine.

They work well and are safe for use in gray-water systems because they are fully biodegradable. The fruits have also been used as medicine, fish poison, and insecticide. The seeds can be pressed for inedible oil.[39]

Carbon farming applications. Soapnuts would be worth testing in woody agriculture systems.

PART FIVE

Road Map to Implementation

CHAPTER TWENTY-FIVE

A Three-Point Plan to Scale Up Carbon Farming

As I've written this book, I've been envisioning it as a toolkit, packed with useful tools and information on subjects such as perennial staple and industrial crops that are not yet widely understood or implemented. I've tried to resist giving a prescription for allocation of land to different strategies. In part, this is because I lean toward a grassroots, decentralized, democratic model wherein farmers and communities make their own decisions about the best ways to sequester carbon; the goal of the book in that sense is to help inform such efforts by farmers, communities, and policy makers.

There is no one-size-fits-all solution that works for every soil, climate, community, and regional economy. Factors including land tenure, soils and climates, gender equity, levels of poverty, and the desired products and ecosystem services can all influence the selection of carbon farming techniques.[1] Another reason to avoid such a prescription is that it is quite complicated to do so, far beyond the scope of this book. (Recently I've joined Paul Hawken's Project Drawdown team, which *is* developing such scenarios for land use as well as individuals, buildings, communities, businesses, cities, and utilities.[2]) Nor is agricultural sequestration the only such tool we have—restoration of forests and wetlands, and planting of timber trees, are also necessary to successful mitigation.

That said, one of the challenges with the approach I've chosen is that the information can seem overwhelming and it can be hard to know where to start—in terms of both on-farm modifications and the sorts of policy measures that might enable them. And because I am sometimes asked what my ideas would look like if they were put into practice on the broad, global scale I'm calling for, I inevitably spend time envisioning that and what it would take to get us there. So I have, in fact, developed a "prescription," of sorts, although it is focused on broad implementation, not farmland itself. Like so many aspects of carbon farming, these three points are interrelated and interdependent; moving forward on any one of them will impact the others. Ignoring any one of them will also impact on the others.

A Three-Point Plan to Scale Up Carbon Farming

1. Support Farmers and Farming Organizations to Make the Transition
2. Effectively Finance Carbon Farming Efforts
3. Remove National and International Policy Barriers

Most of the rest of part 5 will explore these points in greater detail. But before elaborating, I think it's important to briefly revisit why it matters so much. We're currently living in what some refer to as "Decade Zero"—our last chance for massive, meaningful, and effective action against climate change. The chief economist for the International Energy Association, Fatih Birol, asserts: "The door to two degrees is about to close. In 2017 it will be closed forever."[3] The IPCC declares that the "adaptation and mitigation choices in the near term will affect the risks of climate change throughout the 21st century."[4] To avoid disastrous warming of 4 or 6°C (7 to 11°F), the wealthy nations need to peak emissions in 2015 and reduce emissions by 10 percent every year thereafter. This is a virtually unprecedented rate of reduction; during the collapse of the Soviet Union, emissions fell by 5 percent.[5] Only Cuba has experienced a greater emissions reduction. (We'll explore the Cuban story later in this chapter.)

If we had initiated a meaningful response in 1988, for example, when James Hansen first raised broad awareness about climate change during his congressional testimony, we wouldn't be looking at such drastic action. But we didn't, and now we're paying the price for political inaction, due in large part to stonewalling and disinformation campaigns by fossil fuel companies.[6]

In other words, the time is now. Fortunately, a veritable army of experts has been hard at work answering the technical questions about climate change mitigation. We are already well down the road to a climate solution. The harder part is that we have to challenge some incredibly powerful assumptions and even more powerful people and industries. The main barrier to climate stabilization—in terms of both reducing emissions and sequestering carbon—isn't our understanding of perennial plant breeding, the need to develop more sophisticated carbon measurement tools, or the development of clean energy. It's the lack of a strong international movement and the political will to make such a transformation happen.[7] According to journalist Naomi Klein, at this point we won't get back down to 350ppm without drastic changes and an overhaul to our political and economic systems.[8] (For a deep analysis of the relationship between climate and capitalism, I highly recommend Klein's *This Changes Everything*.)

Many people around the world have already devoted countless hours, and even their lives and careers, to bringing about the kind of revolution we so desperately need in order to stave off catastrophic climate change. And we even have some political leaders who acknowledge the urgency and want to act but realistically can't without a powerful popular movement. They need a clear and unequivocal message from their constituencies that the time is now. Climate change calls for a rise in citizen power and a weakening of corporate power.[9]

FIGURE 25.1. Most of the technical issues in mitigating climate change have been worked out. What's needed is to grow the international movement that insists on the necessary changes. Photograph courtesy of South Bend Voice/Creative Commons 2.0.

We need to put a political price on climate change denial and resistance, meaning politicians have to know that failing to act will cost them their positions.[10] A global mass movement has emerged and must grow until it is unstoppable. And the sequestration potential of agriculture—in addition to a reduction in fossil fuel emissions—needs to be front and center in that conversation.

Rattan Lal, one of the foremost experts in the world on carbon sequestration, argues that sequestering carbon in soil is the most cost-effective mitigation strategy, and that it can even be a negative cost (a net gain) because of the many economic benefits of increased production and environmental services.[11] The IPCC has compared the costs of mitigation strategies and found that agriculture is cost-competitive with energy-, transportation-, and forestry-based options.[12]

Rather than launching brand-new (untested and expensive) carbon farming projects and initiatives, it makes more sense to provide greater support to the thousands of farmers and organizations that are already developing perennial crop and cropping systems and other carbon-sequestering practices. In many cases these efforts have other focuses—food security, building local economies, reversing desertification or land degradation. That's part of what makes carbon farming so exciting. We don't need to invest huge amounts of money starting this effort from scratch. We need to study the efforts that are already under way, do a better job supporting them, and incorporate carbon sequestration as an important, and aligned, goal. Almost every practice profiled in this book, and many of the crops, has networks and organizations devoted to it. These groups are local, regional, national, and international. Directing climate funds to these groups is a low-hanging fruit of mitigation. We need to help them maintain and scale up their efforts.

And since carbon farming is such a multifunctional climate solution, we are presented with the opportunity to partner with strong movements and organizations. In fact, we must if this effort is to succeed. Some logical partner movements include food security, food sovereignty, agroecology, sustainable development, industrial ecology, climate justice, and the broader environmental, climate, and social justice movements, as well as socially responsible businesses and social entrepreneurship programs.

Cuba: Agriculture After Fossil Fuels

One country has already gone farther down the road to carbon farming transition than any other. Cuba's story provides insight, inspiration, and cautions. In December 2013 I attended the International Permaculture Convergence in Cuba. The organizers felt that Cuba was the ideal location for the convergence because attendees would see firsthand what the transition to post-petroleum agriculture can look like. With the collapse of the Soviet Union in the late 1980s, Cuba lost its primary trading partner and essentially learned overnight to farm without fossil fuels. The country's strategy was to maintain a commitment to food security, to view food as a national security issue, and to rebuild its agricultural system along organic and low-input lines. Since the early 1990s Cuba has made the world's largest-scale national conversion to organic and sustainable practices.[14] The Cuban government and various Cuban NGOs have won multiple international sustainability awards, including the Right Livelihood Award and the Goldman Prize (the "Green Nobel").[15] As part of the convergence, I had the opportunity to visit multiple farms in rural and urban settings and speak to many Cuban producers about how this political experience affected them.

LEAKEY'S 12 PRINCIPLES

Roger Leakey is the author of *Living with the Trees of Life*, former head of the World Agroforestry Centre, and developer of the participatory breeding system I describe in chapter 10. His article "Twelve Principles for Better Food and More Food from Mature Perennial Agroecosystems" provides an approach based on his extensive experience to the question of carbon farming implementation at every scale from the farm to international policy. Leakey's article is essential reading on implementation strategy.

1. Ask farmers what they want, do not tell them what they should do.
2. Provide appropriate skills and understanding, not unsustainable infrastructure.
3. Build on local culture, tradition, and markets.
4. Use appropriate technology and indigenous perennial species.
5. Encourage species and genetic diversity.
6. Encourage gender and age equity.
7. Encourage farmer-to-farmer dissemination.
8. Promote new business and employment opportunities.
9. Understand and solve underlying problems —the big picture.
10. Rehabilitate degraded land and reverse social deprivation—close the yield gap.
11. Promote multifunctional agriculture for environmental, social, and economic sustainability and relief of hunger, malnutrition, poverty, and climate change.
12. Encourage integrated rural development.[13]

Prior to 1990, Cuban agriculture was highly industrialized, almost like a tropical version of California's Central Valley.[16] Exported sugarcane accounted for 60 percent of Cuban cropland.[17] Petroleum and other agricultural inputs were almost entirely imported, mostly from Cuba's Soviet bloc trading partners since

FIGURE 25.2. This urban farm outside Havana, Cuba, provides produce to the city and income for the members of its grower cooperative. Soil organic matter is increased with worm castings produced from the urban waste stream.

the US embargo created extremely tight constraints. Cuba imported 97 percent of its livestock feed, 98 percent of its herbicides, and 94 percent of its fertilizers, as well as tractors and other agricultural equipment.[18] With the collapse of the Soviet Union, imports of agricultural fuel fell by 50 percent, fertilizer imports by 77 percent, food imports by 50 percent, and pesticide imports by 60 percent. This island nation found itself in the position of needing to double food production and cut its inputs in half.[19]

This was the beginning of the "Special Period in Peacetime," a period of austerity in which the country took on the challenge of food self-reliance as a national security issue. It was not a smooth or easy transition.

Food production on the island collapsed. National tuber production fell by 96 percent, rice by 68 percent, beans by 62 percent, milk by 53 percent, beef by 48 percent, pork by 52 percent, vegetables by 64 percent, and fruit by 75 percent.[20] Daily protein and calorie intakes fell by 30 percent, and the average Cuban lost 9kg (20 pounds) in the six years between 1989 and 1995.[21] It was not until 2000 that nutrition returned to its 1989 levels.[22] By 2003 Cuba met the UN Millennium Goals for food security.[23]

Prior to the commencement of the Special Period, however, many Cuban scientists were already promoting an agroecological approach to food production. President Fidel Castro was himself a longtime advocate of sustainable agriculture, so there was a foundation in place for sustainable agriculture and, perhaps as important, food sovereignty—or "the right of a country to define its own agricultural and food policy"—before the disaster hit.[24] Cuba responded to the crisis by prioritizing food as a national security issue. On-island food production became a "non-negotiable priority."[25]

It would have been extremely difficult for Cuba to take this stance if it were part of the international capitalist economy. It had no IMF loans with structural adjustment policy requirements and was not subject to WTO trade court. The government could subsidize food production and encourage local, sustainable production without fear of bring sued. Cuba participates in alternative international trade networks such as ALBA (Alianza Bolivariana para los Pueblos de Nuestra América), an economic integration organization of social democratic and socialist nations in the tropical Americas. ALBA provides opportunities for credit and fair trade among member countries.[26] And Cuban farmers are not threatened by "dumping" of cheap foreign agricultural surpluses.[27]

The Cuban government also subsidizes food production to ensure food access for its people and instituted some free market reforms, such as support for local farmers markets, to make food sovereignty possible.[28] Access to land and land tenure have also been important components of Cuba's national agricultural policy. The country had already been through land reform, but during the Special Period a lot of land was abandoned or left fallow. Unused land (much of it government-owned) was made available to individuals and cooperatives, typically through 10- to 25-year leases, through a series of policy reforms, although these farmers often still have difficulty obtaining the equipment they need to start and operate their farms.[29]

Critically, Cuba decentralized its decision making, and decisions about land redistribution, production, marketing, and distribution are increasingly decentralized.[30] The urban agriculture program, in particular, is "a decentralized, participatory grassroots program, but with strong government and institutional support."[31] A strong farmer-to-farmer agroecology movement has impacted 100,000 farms, including farmer-led breeding and seed-saving efforts.[32]

A particularly interesting example is "popular rice." Across the country, ordinary citizens planted rice in any unused land, from fields to ditches, in an effort to increase availability of this important staple. They didn't wait for their government to solve their problems, but took action to feed their communities; eventually the government and institutions supported its citizens with information and resources.[33] Today popular rice accounts for 35 percent of the rice consumed in Cuba.[34]

Excluding sugarcane, Cuba now devotes only 3.6 percent of its cropland to permanent crops, mostly fruits, coconuts, coffee, cacao, and fibers such as sisal.[35] I spoke to farmers and agroecologists there who would love to introduce new woody staple crops like tropical nuts but have been stymied by the US economic blockade. Perhaps now as US–Cuban relations appear to be normalizing, Cuban growers will have access to the many excellent perennial staples for tropical humid and semi-arid areas.

Cuba's efforts in urban agriculture have captured the attention of people worldwide. Promotion and support of community gardens and urban and peri-urban farms increased access to fresh vegetables by 1,000 percent. By 2007 these farms and gardens were responsible for 75 percent of the nation's fresh vegetables.[36] Localizing production in this way is a critical strategy for reducing transportation-related emissions.

The country has also built soil fertility through reduced tillage, crop rotations and green manures, composting, and extensive earthworm farming.[37] In 2007 Cuban farmers produced 16 million tons of worm castings, and compost was applied on half a million hectares (1.2 million acres).[38] Other key components of the country's farming system include the use of organic pesticides, managed grazing, animal traction (instead of tractors), and planting in time with seasonal rains to reduce irrigation.[39] A reforestation effort restored 55,000 hectares (136,000 acres) of forest, bringing the island to 25 percent forest cover.[40]

By 2009, 600,000 hectares (1.5 million acres; 40 percent of Cuba's cultivated land) had been improved with living barriers, hedgerows, contour rainfall infiltration ditches, compost, or worm castings.[41] If we assign a low carbon sequestration rate of 1 to 2 t/ha/yr for these agricultural practices, we can estimate that these lands will have sequestered 12 to 24 million tons of carbon in 20 years.

That said, Cuban agriculture today is far from a local, organic paradise. Food imports are on the rise.[42] Potatoes, which are beloved but not suited to the Cuban climate, and sugarcane, which is still an important export crop though on a far smaller scale than during the Soviet period, are grown with chemical-intensive practices. As food security has improved, urban food production has declined, and debate on introducing GMO crops is under way.[43] Like most people, Cubans have been reluctant to change their diets to revolve around the foods their country can sustainably produce, even when living in a time of food scarcity.[44] Nonetheless Cuba remains our best example of a country that has transformed agriculture on this scale, and both rapidly and widely implemented carbon-sequestering practices such as reforestation and increases in soil organic matter.

What can this teach us about carbon farming? One lesson for me when I visited Cuba was that while the world needs to transition to a climate-friendly model of agriculture, doing so in the middle of a hunger crisis like the Special Period would be hugely challenging. Fortunately, we have the opportunity to plan ahead rather than retool in the middle of a hunger crisis—but only if we take advantage of that opportunity. I was also struck by the lack of hunger and poverty I witnessed. Many Cubans are dissatisfied with the austerity of their lives, but as an American living in a city where many people's needs go visibly unmet, I was struck to see a place where the basics of food, housing, education, and health care are provided to all. I was also amazed by the power of farmer- and citizen-led movements, which are actually supported, instead of resisted, by government and powerful institutions. This should be the model we work toward.

More specific to perennial crops, I realized that we need to respect how difficult it is for people to change their basic diets—even undernourished people in the midst of a food crisis. For example, Cubans seem uninterested in reducing the amount of meat they eat, suggesting to me that perennial crops might have their first and greatest impact on livestock feed. Many people have speculated about how relevant the Cuban example truly is to other countries given the unique circumstances surrounding its transition to such self-sufficient agriculture. On the other hand, May Ling Chan and Eduardo Francisco Feyre Roach conclude in their book *Unfinished Puzzle: Cuban Agriculture: The Challenges, Lessons, and Opportunities*, "If a small country like Cuba can achieve international recognition for its sustainable agriculture, [surely] countries with far greater resources could obtain substantial results."[45]

The year 2015 witnessed the beginning of some big changes in Cuba's relationship with the United States. It remains to be seen whether the future will bring the end of the blockade, and how that might impact Cuba's agriculture.

CHAPTER TWENTY-SIX

Support Farmers and Farming Organizations to Make the Transition

Agricultural climate mitigation relies on millions of farmers to make changes to the way they farm. Most of the world's farmers, be they market farmers or subsistence farmers, work very hard and are frequently close to the edge financially. Even though most carbon farming practices increase yields and resilience to extreme weather, these impacts are not felt right away. In fact they tend to take several years to implement and typically result in several years of reduced yield and income before benefits are seen.[1] Perennial crops in particular require years (sometimes decades) of care before yielding. (These are key reasons that secure land tenure is essential.) Why take the risk if you might not be farming that land next year?

FIGURE 26.1. A group of carbon farmers associated with the Riba Agroforestry Resource Center in Cameroon. Photograph courtesy of the World Agroforestry Centre.

There are actions that can help farmers overcome this initial period of difficulty. For carbon farming to be successful on a large scale, we'll need plans to address these needs. In the chapters that follow we'll look at land tenure and financing, two key issues to enable adoption of carbon farming practices.

In a 2011 paper, researchers Nancy McCarthy, Leslie Lipper, and Giacomo Branca identified five major challenges facing farmers as they transition to "climate-smart agriculture."[2]

1. **Investment costs** represent the money or labor needed to initially implement practices. They include the purchase of trees or special equipment, renting or hiring equipment to install rainwater catchment, or the added extra time of managed grazing instead of extensive grazing. The costs of implementing carbon-friendly practices on the farm range from $12 to $600 US per hectare.[3]
2. **Variable and maintenance costs** are the costs needed to keep the practice going on an ongoing basis. This would include annual labor and the purchase of inputs such as cover crop seed.[4]
3. **Opportunity costs** represent the lost land, time, and money that could have gone to the previous farm operation. For example, taking land out of annual crop production to plant trees may mean several years of reduced income or food security. Some tree crops takes decades before they yield.[5]
4. **Transaction costs** look at the time and money it takes to learn about practices (including training), track down the information and resources to implement them, and negotiate and report on any carbon finance arrangement. Most perennial staple crops, for example, are extremely difficult to acquire.[6]
5. **Risk costs** reflect the serious potential of losing income or even the farm itself. There is usually a period of several years of reduced yields and income, even when transitioning to well-known practices.[7] Risks are even higher with more experimental crops and practices, or unknown or unstable markets.

Farmers are more likely to adopt practices if they help them to address existing risks or challenges. For

TABLE 26.1. Widely Used Carbon Farming Practices Developed Since the 1970s and Their Scale

Practice	Region	Scale
Conservation agriculture	Global	124 million ha[a]
No-till annual cropping (also included in conservation agriculture)	Global	100 million ha[b]
Holistic Management grazing	Global	12 million ha[c]
System of Rice Intensification	Global	10 million farmers[d]
Organic annual cropping	Global	6.3 million ha[e]
Farmer-managed natural regeneration	Niger	4.8 million ha[f]
Paulownia strip intercropping	China	3 million ha[g]

Note: This table references carbon farming practices that are documented to have spread over 1 million hectares since 1970.

[a] Farooq and Siddique, "Conservation agriculture," 8.
[b] Lal, "Managing soils and ecosystems for mitigating anthropogenic carbon emissions and advancing global food security," 710.
[c] Neely and De Leeuw, "Home on the range," 337.
[d] Vidal, "India's rice revolution."
[e] IFOAM, *The World of Organic Agriculture*, 61.
[f] Garrity et al., "Evergreen agriculture," 205.
[g] Li, *Agro-Ecological Farming Systems in China*, 107.

example, farmers on sloping land, in drier regions, and in unpredictable climates are more likely to adopt agroforestry. Farmers are more likely to adopt multifunctional practices that provide direct yields than purely conservation (or carbon sequestration) efforts.[8] In some regions, factors beyond a farmer's control such as brush fires and free-roaming livestock can be serious challenges to establishing agroforestry practices. Many grazers are working on communal land, making decision making particularly difficult.[9]

It will be important for successful implementation of carbon farming to look at past successes and failures in development and diffusion of crops and farming practices and tree-planting projects. Of particular relevance is the wave of sustainable farming practices that was developed in response to the environmental crisis since the early 1970s. Some of these techniques have spread widely and quickly—often with little government or institutional support. Those profiled in table 26.1 have spread to more than 1 million hectares (2.5 million acres) since their development, though many others have remained marginal or spread more slowly. There remains significant room for each of these to grow. They stand ready to be scaled up further to address the climate crisis. Studying why these have been successful can offer important insights for use in scaling up carbon farming.

TABLE 26.2. Scale of Selected Indigenous and Traditional Carbon Farming Practices

Practice	Region	Scale
Indigenous land management	Australia prior to colonization	769 million ha, historic[a]
Indigenous land management	United States prior to colonization	493 million ha, historic[b]
Indigenous land management	Amazonia	21–200 million ha, historic but some persists[c]
Complex swidden rotations	Thailand, Laos, Cambodia, Vietnam, Myanmar, Malaysia, Indonesia	25.2–39.6 million ha[d]
Shea nut parkland	Semi-arid West Africa	22.9 million ha[e]
Managed bamboo	Mostly Asia, tropical Americas	22 million ha[f]
Silvopasture	Central America	9.2 million ha[g]
Tropical homegardens	Indonesia, India, Sri Lanka, Bangaladesh	7.9 million ha[h]
Cacao agroforests	Humid tropics	7.8 million ha[g]
Amazonian terra preta soils	Amazonia	5.5 million ha[i]
Dehesa fodder silvopasture	Spain and Portugal	5.5 million ha[e]
Tropical homegardens	Philippines	70% of households[h]
Coffee under diverse shade canopy	Humid tropics	2.4 million ha[j]
Streuobst	Europe	1 million ha[k]

Note: I could add many more practices—terraces including metepantli, rainwater harvesting, and al-Hima grazing—to this list. Unfortunately, data on the scale of these systems is hard to come by. Even for the examples here, it's important to note that these only represent specific instances from available scientific literature; swidden, for example, is practiced much more widely than just in Southeast Asia. Only practices that are documented to have over 1 million hectares in production are listed here.

[a] Gammage, *The Biggest Estate on Earth*, 173.
[b] Rough estimate based on half of area of contemporary United States.
[c] Mann, *1491*.
[d] Schmidt-Vigt et al., "An assessment of trends in the extent of swidden in Southeast Asia," 272.
[e] Boffa, *Agroforestry Parklands in Sub-Saharan Africa*.
[f] Henley and Yiping, "The Climate Change Challenge and Bamboo."
[g] Zomer et al., *Trees on Farm*, 3.
[h] Nair and Kumar, "Introduction," 3.
[i] Mann, "The real dirt on rainforest fertility," 920.
[j] Jha et al., "Shade coffee."
[k] Mosquera-Losada et al., "Definitions and components of agroforestry practices in Europe," 9.

A major aspect of promoting carbon farming is acknowledging and understanding the value of traditional and indigenous knowledge. Carbon farming practices are—for the most part—not new, marginal, or experimental; rather, it is thinking of them in terms of sequestering carbon in the soil and aboveground biomass that is new. Many of these practices and crops are already practiced widely, on hundreds of millions of hectares of farmland; while writing this book I was struck over and over by the vast scale of traditional carbon farming and carbon-friendly land management systems. In most cases, these systems are rapidly losing ground to industrial agriculture, and in some cases they are almost gone.

We need to fully appreciate that indigenous and traditional communities are, in many cases, practicing sophisticated management of farms, forests, and grasslands—and that in many cases this management *is* carbon farming. These practices often maintain a balance between productivity for humans on the one hand and biodiversity and ecosystem function on the other. We need policies that will protect these communities and their anthropogenic ecosystems and encourage their spread and further development. This is another aspect of supporting carbon farmers—to help them continue to store the carbon in their systems and protect them from land grabs and efforts to displace and replace them.

The World Agroforestry Centre promotes a climate-smart landscape approach, which looks at more than just individual farms, and acknowledges that tracts of land can and do have more than one function.[10] Indeed, with increasing competition for limited land, multifunctionality is essential to meeting the goals, including food security, carbon sequestration, and ecosystem services.[11] "Mechanisms that allow co-investments of public and private interests and resources are needed to catalyze actions in landscapes, guided by landscape democracy and transparency."[12]

The World Agroforestry Centre also promotes landscape democracy, the use of "democratic and good governance principles (such as transparency, account-ability, participation, legitimacy, and coordination) in multi-stakeholder processes at the landscape level." Landscape democracy values "self-determination (autonomy in decision-making); co-determination and participation (ensuring common good); impartiality and respect for arguments . . . , and procedures and multi-order impartiality (rules for deliberation that ensure respect and equity)."[13] There's no carbon farming without farmers. An understanding of the challenges faced by farmers, and best practices in adoption, lets us develop implementation programs with the greatest chance of success.

When it comes to carbon sequestration, some crops and practices are clearly more effective. That doesn't mean those are the right practices in any given situation, but it does mean that for any landscape, preference could and should be given to practices that have a better climate impact wherever possible. For example, chapter 3 makes clear that systems that incorporate woody plants sequester more carbon per hectare on an annual basis and also have greater lifetime soil carbon storage capacity. Though they come with many challenges, as described in chapter 11, scaling up of these systems will have a more powerful mitigation impact on a per-area basis than improved annual cropping or managed grazing alone.

The ideal carbon farming system performs multiple functions—carbon sequestration, climate change adaptation, and high yields that can compete with the baseline carbon-unfriendly practices in the region. Almost all of the practices in chapters 6 through 9 meet all three criteria. All of the perennial crops in chapters 12 through 24 satisfy the first two criteria by virtue of being perennial, and many them satisfy the third. (This is especially true in the tropics and subtropics.) Even the perennials that yield less than their annual counterparts in monoculture comparison can be inter-cropped or otherwise incorporated into a polyculture system, resulting in an overyield. In fact, Roger Leakey reports that many scientists think it is possible to double our global food output on the same land using less water, land, energy, and fertilizer with these kinds of techniques.[14] Maximizing carbon sequestration and

multifunctionality to the greatest possible degree is a key part of the recipe for a successful climate mitigation strategy through carbon farming.

Specifically, we need to widely implement the practices and perennial crops that have competitive yields and are ready to scale up. We can begin this by backing the existing efforts of the countless farmer groups, NGOs, and others who have been practicing and promoting these strategies for decades, often with little recognition. Among the best of these tools are:

- Improved strategies for growing annual crops such as the System of Rice Intensification, regenerative organic production of annual crops, reduced tillage, cover cropping, green manures, and nutrient management.
- Agroforestry and perennial–annual intercropping systems including strip intercropping, FMNR, ever-green agriculture, windbreaks, and pasture cropping.
- Livestock systems like crop–livestock integration, managed grazing and pasture management, silvopasture and intensive silvopasture, and perennial feeds for livestock.
- Perennial cropping systems such as multistrata agroforestry; perennial crops; bamboo, coppice, and SRC systems; biomass grasses; and aquaforestry.
- Other techniques including rainwater harvesting, biochar, and protecting and restoring indigenous and traditional land management systems.

We should also invest heavily in research and development of practices and perennial crops that have great potential and could be ready in a few decades. Perennial grain development, for example, should be a high priority because it has such incredible potential for widespread adoption and replacement of annual grains. Practices such as woody agriculture and other cold-climate agroforestry techniques and crops should also be a priority since cold-climate crops and techniques currently lag behind and account for much of the world's population and agricultural production. Some practices, such as keyline, may be shovel-ready—the practice is fully fleshed out and ready for farmers to start using today—but their carbon impact is not yet well understood. Researching the carbon impact of these practices should be a high priority.

Crop Production

- Perennial crops are the winners in terms of sequestering carbon. How far can we push agriculture in this direction? How much annual cropland might be converted to perennials without sacrificing productivity? In the many cases where perennial crops are already yielding as well as or better than annuals, conversion should proceed.

- Multistrata agroforestry systems represent the highest level of carbon sequestration in food production. How much more widely can they be expanded? How much farmland might be converted to this model? There is a need for development of multistrata production systems for non-tropical climates.

- Efforts to perennialize the annual grains we depend on should be fully funded and fast-tracked. Likewise, efforts to develop new perennial systems such as woody agriculture should be top funding priorities.

- The priority areas to switch to perennial crops and systems should be sloping and degraded farmland. Let's stop growing annuals on sloping and degraded lands unless agroforestry, terracing, pasture cropping, or other regenerative practices ensure a carbon-friendly production model. Flat or gently sloping lands are where we should focus on producing annuals.

- We should think of annual crops and arable cropland as precious resources, to be used wisely.

- These precious annual croplands should be managed using the best agroecological techniques, such as reduced tillage, green manures and cover crops, and mulching or preservation of crop residues. Wherever they equal or increase overall yields, woody elements like windbreaks, FMNR, *Faidherbia* evergreen agriculture, and strip intercropping systems should be incorporated.

- Priority for annual cropland should be on the things that are difficult to produce with perennials. This includes grains, root crops, annual vegetables, and cotton.

- Rainwater harvesting should once again be a central part of farmland planning and management.

- Biochar's potential to improve soils and yields, especially in degraded and infertile tropical conditions, should be thoroughly explored. Implementation should follow as appropriate.

The remarkable *Lost Crops of Africa* book series has put much thought into potential partners for crop development. For each of their crops, they identify relevance to potential partners in the areas of nutrition,

TABLE 26.3. Perennial Staple Crop-by-Crop Partnerships as Proposed in *Lost Crops of Africa*

Latin Name	Common Name	Nutrition	Food Security	Rural Development	Sustainable Land Care
Adansonia digitata	Baobab	***	***	***	***
Balanites aegyptiaca	Balanites	***	***	**	***
Dacryodes edulis	Butterfruit	***	***	***	***
Ensete ventricosum	Enset	*	***	*	**
Irvingia gabonensis	Dika	**	*	***	***
Lablab purpureus	Lablab bean	**	**	***	***
Moringa oleifera	Moringa	***	**	***	**
Parkia biglobosa	Locust bean	**	***	**	***
Sclerocarya birreya	Marula	***	***	***	***
Sphenostylis stenocarpa	Yambean	***	**	*	***
Tylosema esculentum	Marama bean	*	*	*	*
Vitellaria paradoxa	Shea	*	**	***	***

Source: Adapted from *Lost Crops of Africa*, vols. 2 and 3.
Note: Three asterisks is the highest score.

food security, rural development, and sustainable landcare. Perennial staples from their assessment are featured in table 26.3.

Livestock Production

- I propose we stop feeding annual grains that are fit for human consumption to livestock. This could free up a third of our annual cropland for perennial systems. It will probably mean less meat consumption in the Global North.
- Livestock and crop production should be reintegrated to solve many of industrial agriculture's environmental problems and improve mitigation and adaptation.
- Ruminants such as cattle and sheep can be quite happy on pasture, hay, and fodder trees. Any need to increase nutrition, such as for highly productive dairy animals, could be met with perennial staples, processed crop residues like oil presscake, or silage and hay from perennials.
- Silvopasture is a proven practice and should be more widely implemented. Fodder tree silvopasture in particular is worthy of broad-scale expansion.
- Scientists and farmers need to work together to determine what it is exactly about managed grazing that sequesters carbon under specific conditions of soil, climate, livestock species, scale, and intensity of management. The hundreds of thousands of successful, carbon-sequestering managed grazing operations should be studied carefully to develop science-based guidelines for grazing in their areas.
- Intensive silvopasture should be much more widely practiced in suitable climates. Development of similar techniques outside of the humid tropics should be prioritized.
- Monogastric livestock such as hogs, poultry, and fish should be fed perennial feeds and convert the food waste stream to additional meat and eggs. Insects should be investigated as feed for livestock and as human food where desired.
- Biochar should be thoroughly explored as a method to reduce methane emissions from livestock manure and ruminant digestion.

Carbon Farming Research Centers

The Consortium of International Agricultural Research Centers (CGIAR) is a global partnership. It runs 15 research centers including the World Agroforestry Centre, the International Rice Research Institute, which pioneered work on perennial rice, and Bioversity International.

We need a Carbon Farming Research Center for each of the world's agricultural climates (arid tropics, humid temperate, and so on). These might be CGIAR projects, or UN-funded, or otherwise operated. Their mandates would be to research and promote practices and crops for their climate. For example, an international tropical humid highlands institute might research and distribute the most promising crops for that climate, such as enset, New Guinean pandanus nuts, and Andean tree beans. We also need regional research centers—Sahel, *cerrado*, Pacific Islands, cold dry western United States—with the goal of regionalizing this work and trialing useful native species.

10,000 Nurseries

At this time, many of the world's most promising crops are difficult or impossible to acquire, especially outside their native range. Improved and elite varieties are even harder to come by. We need thousands of nurseries to rise to this challenge around the world. These could include state-run nurseries, but I'm primarily advocating for private nurseries—from the backyard micronursery to the huge international suppliers. I worked for many years in the nursery world; I have a lot of faith in it and find that private enterprises can often get things done more quickly.

The bureaucracy around international shipping of plants is a major obstacle. It's difficult, and sometimes illegal, to move plants between countries. Some of these protections are to prevent the spread of pests and diseases of crops, which I support. Others are to control the spread of invasive species, a more controversial issue I address in chapter 10. The bottom line is that international plant trade laws need to be streamlined to permit more rapid transit of elite varieties of important perennial crops.

Simple Carbon Measuring Tools

The research gap that most impacts implementation, certification, and financing of carbon farming is carbon monitoring. We need a standardized set of tools to measure on-farm carbon sequestration. They must be affordable and easy to use. As of 2014, no such system exists, although many are under development.[15] There are a number of carbon calculators for more conventional farming practices. Of 18 reviewed by FAO, only half measured soil carbon, with a far greater emphasis on emissions reduction than carbon sequestration.[16]

Currently agroforestry researchers use many different methods, which makes comparing and developing predictive models very difficult. For example, soil organic carbon may be measured at 10cm, 30cm, 60cm, 100cm, or deeper. Some researchers don't even say how deep they tested. This results in data that is difficult to synthesize into broader understanding. Without being able to offer standard sequestration rates by practice, climate, and soil type, it will be difficult to attract large-scale carbon financing to carbon farming practices.[17]

This is a particular barrier for smallholders in the tropics. The Kenya Agricultural Carbon Project has an interesting methodology to address this. They are developing standard sequestration rates for various farming practices for their climates and soils through research. These will then be occasionally spot-checked on farms without requiring robust and expensive annual testing.[18] The Cool Farm Tool and Climate Yardstick are tools that help larger, Western-style farmers estimate emissions and sequestration.[19]

We'll never transition all of the world's farmland to carbon farming practices. So getting the biggest carbon impact we can on the farmland that can be converted is essential. The much higher numbers reported for some of the crops and practices in this book could greatly increase total agricultural mitigation potential if widely implemented.

CHAPTER TWENTY-SEVEN

Effectively Finance Carbon Farming

Climate scientist James Hansen, who was among the first people to raise awareness of climate change during his 1988 congressional testimony, writes: "A reward system for improved agricultural and forestry practices that sequester carbon could remove the current CO_2 overshoot."[1] There are several such efforts underway, and although most of them leave a lot to be desired, it's worth being aware of where we stand. Even if we haven't hit upon the right incentives and programs yet, I strongly believe that climate change mitigation is an effort that needs to be financed and that we won't get far until we figure out how to address it.

International Climate Finance Mechanisms

Climate finance refers to the "financial resources paid to cover the costs of transitioning to a low-carbon global economy and to adapt to, or build resilience against, current and future climate change impacts."[2] These funds come from a complex mix of aid agencies, governments, and financial institutions.[3]

According to the Climate Policy Initiative, approximately $331 billion US was spent on carbon finance efforts around the world in 2013. This spending was split almost exactly in half between developed and developing countries and was mostly invested in the country of origin.[4] But $331 billion is far less than what we currently need, and the need is growing. This spending compares poorly with the additional $5 trillion US

that the International Energy Agency projects will be needed in clean energy investments alone by 2020 to limit warming to 2°C (3.6°F).[5]

Ninety-one percent ($302 billion) of the $331 billion went toward mitigation, primarily to clean energy projects. Of this $302 billion, roughly 2 percent ($6 billion) went to agriculture, forestry, land use, and livestock management.[6] This is a disproportionate percentage given the fact that agriculture is responsible for 23 to 30 percent of anthropogenic emissions.[7] We need to direct far greater spending to agriculture mitigation efforts.

Within this tiny allocation for agricultural mitigation, almost all of it was directed at emissions reduction—primarily from more efficient use of nitrogen fertilizer, reducing methane emissions, the use of biofuels, and reduced crop residue combustion. According to the international advisory firm Climate Focus, between 2010 and 2012, 1 to 9 percent of agricultural climate finance went to the widely applicable but minimally carbon-sequestering practice of no-till (estimated at 0.3 t/ha/yr). Less than 1 percent per year went to agroforestry.[8] That's 0.02 percent of the already insufficient mitigation funding for agroforestry.

Compared with the $302 billion (91 percent) spent on total mitigation efforts in 2013, only $25 billion (7 percent) was spent on adaptation efforts. Most of these adaptation funds went to water supply and management while only $2 billion (8 to 9 percent of total adaptation funds) was directed toward agriculture, forestry, land use, and livestock management. About 1 percent of total carbon finance funding in 2013 went to projects that both mitigate and adapt.[9] That said, adaptation, as opposed to mitigation, occupies a growing percentage of carbon finance funding within agriculture, representing the majority of agricultural carbon finance funding in 2012.[10]

Clearly these numbers are only a drop in the bucket of what's needed to attain 350ppm through carbon farming. Focusing on projects that mitigate *and* adapt seems like a good approach for future agricultural funding.

The Clean Development Mechanism (CDM)—one of the "flexibility" mechanisms defined in the Kyoto Protocol that provides for emissions reduction projects that generate Certified Emissions Reduction units, which can then be used in emissions trading schemes— acknowledges only a few agricultural practices: manure management, biogas, and biomass energy from agricultural residues.[11] It's no wonder then that only 2.5 percent (174) of the 6,989 CDM-funded projects are agricultural.[12] Likewise, the European Emission Trading Scheme—the first large greenhouse gas emissions trading scheme in the world—did not acknowledge agricultural or forestry projects as of 2014.[13]

This represents a serious failure in focus. More funding needs to be available to support carbon farming, especially for practices that both mitigate climate change and adapt to it. But most of the governments, NGOs, and financial institutions involved in carbon finance do not fund agriculture or land use at all. I'm not suggesting that agriculture merits a larger share of current carbon finance pie; I'm not advocating that we redirect funding from renewable energy, for example, to agriculture. I'm arguing that we need to spend far more money on carbon mitigation and adaptation, with substantial additional funding for agriculture.

Carbon finance has been criticized for more than just insufficient funding. Among these criticisms are:

- Many are based in carbon offsets, which (a) have frequently been shown to be ineffective, and (b) don't actually reduce total greenhouse gases (unless a "shrinking cap" is present, which is rarely the case, though California is considering implementing one).[14]
- Many carbon funds do not fund agriculture projects at all.

- Most carbon farming practices are not covered by carbon financing mechanisms. One study found that only 25 percent of basic practices were eligible.[15] Bamboo and agroforestry often fall into the bureaucratic cracks between forestry and agriculture, or are not acknowledged at all.[16]
- Despite the higher productivity of small farms, it is extremely difficult for small farmers to get access to carbon finance because most programs require huge amounts of land to qualify. This means that small-holders need to aggregate into larger groups, which introduces a number of financial and organizational problems.[17] One possible partial solution would be for carbon finance institutions to recognize existing farmer cooperatives and groups.[18]
- Carbon finance record-keeping and paperwork requirements are well above the capacity level of most grassroots farmer organizations.[19]
- Many carbon funds have been criticized for lack of transparency and other forms of corruption.[20] This appears to be a major vulnerability in offset-based systems.[21]

Despite the many shortcomings, it is good that there are institutions working to fund climate mitigation and adaptation through agriculture. We need to reform salvageable programs, replace those that are unworkable, and greatly ramp up carbon farming funding.

Small-Scale, Grassroots, and Alternative Carbon Farming Finance

As of 2014 very few individual farmers around the world have access to international carbon finance tools that would allow them to make the transition to carbon farming. And there are many other limitations to international carbon finance programs that need to be worked out before they're truly effective. But there are also alternative financing options available to at least some farmers, and there is a huge amount of opportunity for individuals, organizations, and businesses to invest their money in a better future for us all. Increasing access

CARBON FARMING CERTIFICATION

We can't shop our way out of climate change, but while we wait for pressure on governments to create carbon farming policies, people who have the means can help early-innovating carbon farmers. Organic certification has provided premium prices to farmers, enabling a wide-scale transition to organic farming. I saw the enthusiasm among both farmers and consumers for organic food long before it received support from governments and academic institutions.

I suspect that it will be some time before a majority of national governments get behind carbon farming. Meanwhile an independent certification system could promote carbon farming and provide price incentives to help farmers make the transition (and to reward those already practicing it). Perhaps such a system might even track emissions from processing and transport on the way to market.

The Sustainable Agriculture Network is an international body that certifies organic and other practices on farms. They have developed a Climate Module, although it currently looks at reduced on-farm emissions rather than carbon sequestration. Two hundred thousand farms on 1 million hectares (2.5 million acres) in Latin America and Africa have received the certification, though there is not yet a premium price market available for it as of 2014.[22] Nonetheless the SAN Climate Module represents a promising step forward.

to financing is an important task for lenders, investors, and others who want to make a difference with their money in the decades to come. We should explore every possibility, both conventional and alternative, to come up with a finance toolkit so that no legitimate carbon farming project is stymied for lack of funding.

Especially in the Global North, banks and agricultural lenders can play an important role, in offering low-interest loans to carbon farmers. Fair trade and socially entrepreneurial businesses looking to source carbon-friendly materials may offer trade credit to farmer-suppliers.[23] Groups like Slow Money, Investors' Circle, and Solari Investment Groups represent investors who want to put their money in projects like carbon farming. More work is needed to formalize those connections.

Financing is a greater challenge in the Global South, where interest rates for farmers are extremely high.[24] And many of the powerful new tools for financing, such as crowdsourcing and peer-to-peer lending, are also difficult for these farmers to utilize due to lack of Internet access. Community and microfinance lenders and rural financial coops can provide financing, although in the Global South world microfinance loans are usually of shorter duration (one to two years) than is needed for most carbon farming efforts.[25] There are a lot of opportunities to further develop alternative finance mechanisms, especially those that could be put to work in the Global South, such as partnerships that would connect farmers with fossil fuel divestment investors in wealthy countries.

Universities, NGOs, and foundations can have multiple impacts. Of course they can provide philanthropic grants to farmers or farmer groups. Even more powerfully, many are starting to perform "program-related investment." Foundation grants are usually only given from the annual income from investments made with the endowment. Today hundreds of foundations in the United States are investing with a portion of their endowment, providing loans to businesses (including agricultural businesses) that are in line with their mission.[26]

The NGO Heifer International has a particularly interesting philanthropic model that turns recipients into donors. Farming families receive livestock or plants with agroecological training. Within a few years they "pass on the gift" and give the same livestock or plants to another farm family, transforming them from recipients to donors themselves.[27]

There is a very exciting movement for institutions (and individuals) to divest the money they have invested in fossil fuel companies (I might add that some agribusiness companies should perhaps be on the divestment list as well). What is needed is a more robust system to help invest that divested wealth in clean energy, green industry based on perennial feedstocks, and of course carbon farming. The IPCC notes that serious mitigation requires major changes in investment patterns.[28] Building this new financial infrastructure is a major challenge for our decade.[29]

Putting a Price on Carbon

Climate change impacts are one of the "externalized" costs of our global economy. Many activists, scientists, and policy makers think a price on carbon emissions is an essential part of climate change mitigation.

The IPCC estimates that agricultural production could reduce emissions by 7.1 to 10.6 billions tons of CO_2 a year by 2030 with a price on carbon dioxide emissions of $100 US per ton. At $20/ton, about a third of that impact could be achieved.[30] James Hansen states that achieving 350ppm requires a rising price on emissions.[31] A $50/ton tax on carbon could raise $450 billion a year for integrated mitigation, adaptation, and development efforts.[32] Clearly these strategies are important components of an overall climate policy and could greatly facilitate carbon farming adoption. Carbon taxes and cap-and-trade systems are the two primary strategies currently at play.

Cap and trade provides to industry licenses to emit greenhouse gas pollutants. The "cap" is the ceiling on total carbon emissions, which must be set by governments. Rather than reduce their own emissions, a company could trade to another company that can clean up their act less expensively.[33]

Carbon offsets, part of carbon trading, enable greenhouse gas polluters, including individuals, to pay someone else to reduce emissions or sequester carbon to cancel out their own emissions. The UN Clean Development Mechanism is such a system. The slippery nature of such estimation has meant a net increase in emissions.[34] Offsets and carbon trading do nothing to reduce emissions (again, without the rarely seen and elusive shrinking cap). Rather they shift responsibility to others. The carbon trading component that actually reduces emissions is the cap—but capping efforts have been weak and ineffectual at best.[35]

In 2013 a coalition of 130 NGOs called for the abolition of the European Union Emissions Trading System. They claimed it had failed to reduce emissions—which in fact increased under its watch—and that it had to go to "make room for climate measures that work."[36]

While carbon taxes are very unpopular politically, the *New York Times* reports, "Cap and trade, by contrast, is almost perfectly designed for the buying and selling of political support through the granting of valuable emissions permits to favor specific industries and even specific Congressional districts."[37] That carbon trading schemes have failed to have much impact is no surprise: "While commodity prices can do many things, one thing that they have never achieved is to solve problems that require structural change in so many fundamental areas of industry and agricultural practices."[38] The IPCC states that although cap-and-trade systems could work in theory, they have a poor track record and limited effect due to "loose caps" and caps that fail to constrain emissions.[39] On the other hand the IPCC notes that taxes on emissions are effective.[40]

Carbon taxes are less complicated and opaque than the financial manipulations of carbon trading. They tax carbon at the source—usually a fossil fuel extraction site. Their primary drawback is that higher costs can get passed on to the consumer, creating hardship for the limited-resource people who did the least to cause the problem in the first place.[41] James Hansen proposes a carbon tax with 100 percent of the proceeds going back to the public.[42]

While neither strategy is perfect, cap-and-trade and offset efforts have largely failed. It's time to implement carbon taxes. Funds from carbon taxes could be used to support farmers to make the transition to carbon farming, through grants, loans, training, and more. Money from a carbon tax also can and should be used to build a world where low-carbon living is not only a lifestyle choice for the affluent. Choices in transportation, energy, and food need to be made easy through public programming.

Other Market-Based Strategies

There is a critical role here for social entrepreneurs. Markets, especially markets with good prices, are necessary for farmers to succeed. Businesses that build relationships with groups of carbon farmers help to finance transition (and help farmers maintain their carbon farming practices). Fair trade and other models can be part of making carbon farming viable, especially before national and international policies are in effect.

An example is the Brazilian automobile factory that uses industrial crops to make biocomposites. These are purchased from associations of small-scale agroforestry producers, who supply rubber, castor oil, cashew nutshell oil, indigo, and perennial fibers such as coir, sisal, and ramie. In addition to providing a market for perennial products grown in diverse polycultures, the company funds village development, including clean water and education. In return the company has lower manufacturing costs (due to use of natural products) and builds its social and environmental reputation.[43]

CHAPTER TWENTY-EIGHT

Remove National and International Policy Barriers

If measures are taken to restructure agriculture and the larger food system based on food sovereignty, small-scale farming, agroecology and local markets, global GHG emissions could be cut by half within a few decades. There is no need for carbon markets or techno-fixes. What is needed are the right policies and programmes that bring about a shift from the current industrial food system to a sustainable, equitable and truly productive one. —GRAIN[1]

C urrently there are few incentives to reward on-farm sequestration, and few if any consequences for emissions from farming. The bottom line for effective climate change mitigation is that we need to remove the incentives for unsustainable farming and establish incentives that support carbon farming. This needs to be a multidisciplinary effort at international, national, and regional levels. And it needs to be integrated into government policies and programs. These changes mean substantial changes to national agriculture department budgets. The emphasis should be on practices that push the agenda of mitigation, adaptation, and development.[2] Without a major overhaul of policy at all levels, it will be extremely difficult to scale up carbon farming to the levels needed for effective climate mitigation.

The MAD Challenge: Mitigation, Adaptation, and Development

Many climate leaders have pointed out that multifunctional climate change mitigation efforts will be more successful and popular. They can win more allies and funding and help resolve many of the world's toughest problems.[3] The IPCC states: "Understanding how to enhance positive feedbacks between mitigation, adaptation, and sustainable development (e.g., win–win and triple-win interventions) while minimizing potential trade-offs between them . . . is an essential part of planning for and pursuing climate-resilient pathways."[4] In fact, policies that integrate adaptation and mitigation are more effective than addressing either alone.[5]

Climate funding can provide the opportunity for the Global South to build truly sustainable infrastructure that mitigates and adapts to climate change while creating opportunities to improve the basic human well-being of the world's majority.[6] MAD opportunities are not limited to these populations but should be targeted there.

Roger Leakey writes in *Living with the Trees of Life*, "In thinking about the future of tropical agriculture, I think we should be much more observant about what developing-country farmers say they want, and in some cases what they are actually doing. These farmers don't have a voice in setting the development agenda—what they do for themselves is the only way they can illustrate their needs and their actions speak louder than words . . . One of the things they are doing is expressing continued interest in, and need for, the indigenous trees producing fruits and nuts . . ."[7]

The carbon farming strategy outline in the book, including the potential of perennial feedstocks for industry, offers abundant opportunities to meet the

PAYING CARBON DEBT

Carbon debt is the notion that the wealthy countries, who are responsible for burning most of the carbon, owe a debt to the countries of the Global South, who are suffering and will continue to suffer the most from climate change. Due to some accidents of biology and history, carbon farming happens to be a near-perfect strategy to address climate justice and carbon debt.

There are several key reasons for this. One is that some of the most effective practices are best developed in the tropics and have been little explored in temperate climates. This is true for agroforestry in general, with examples including multistrata agroforests, intensive silvopasture, FMNR, and *Faidherbia* evergreen agriculture. Two, lots of tropical and subtropical perennial crops yield as well as or better than their annual competitors, which is not the case in colder climates (see parts 3 and 4). Finally, trees in the tropics have the least undesirable albedo impact, making the region the most strategic place for edible afforestation.[8] Thus here we have a strategy that sequesters carbon and feeds the people most vulnerable to climate change while providing an opportunity for wealthy nations to pay off some of their climate debt. That it also serves as a powerful adaptation strategy is icing on the cake.

MAD challenge. There have already been some instructive successes with food and industrial crops. Brazil has a strong movement of farmers producing industrial crops in integrated agroforestry systems. Their products are purchased by Mercedes and used in making cars, with train and truck parts in the planning stage.[9] Food multinational Unilever has developed an agreement with producers of *Allanblackia* oil. This fair trade agreement—which respects indigenous intellectual property—won a Union for Ethical Biotrade Award.[10]

International Agriculture and Trade Policy Reform

At the largest scale, most current international (and national) agriculture and trade policy is a barrier to the development and successful implementation of carbon farming programs. The World Trade Organization (WTO)—the intergovernmental organization that regulates international trade among participating countries—pushes farmers in the Global North toward industrial agriculture—exactly where we don't want them to be. For example, Lim Li Ching writes that international trade policy "[prompts] producers toward greater specialization . . . monocropping, increased mechanization and utilization of chemicals . . . higher dependence on external inputs . . . and enhanced scales of production."[11] International trade rules currently limit the ability of countries in the Global South to create incentives for carbon farming practices such as Payment for Environmental Services and financing for carbon farming transition.[12] Climate-friendly trade policy would remove subsidies for industrial agriculture and permit countries to protect and support their small, sustainable farmers. For an overview of these issues I recommend the United Nations Conference on Trade and Development publication *Wake Up Before It Is Too Late: Make Agriculture Truly Sustainable Now for Food Security in a Changing Climate.*

Market agreements such as the Trans-Pacific Partnership need to be overhauled as well, because sustainable and small producers are marginalized there, too. For example, because loopholes are provided for subsidies in the Global North, but not the Global South, tropical farmers are forced to compete against subsidized crops.[13] We also need "adequate regulation of agricultural markets to shield vulnerable agricultural producers against dumping—the export of surplus food at highly subsidized prices, impacting farmers in target countries—and price volatility."[14]

Massive transnational corporations have far too much control over research, trade, and public policies. Although many policy experts have proposed changes to make agriculture more sustainable and democratic, powerful corporations and the governments that do their bidding through trade alliances won't let it happen.[15] We need "targeted and strategic interventions" to democratize control of agriculture.[16]

Speculation on food commodities serves no useful purpose beyond profit. It results in highly volatile prices that impact farmers, especially the smallest, whose intensive operations are critical to climate mitigation.[17] We should instead redirect these funds into carbon farming transition, as part of the climate finance effort described in chapter 27.

Some barriers are mostly bureaucratic. For example, national and international policy frequently has no mechanism to understand or categorize agroforestry. Is it agriculture or is it forestry? This question matters because agricultural policy differs from forestry policy. Agroforestry is often managed by forestry ministries rather than agriculture ministries, or has no true home anywhere. As a result, farmers find themselves having to pay the higher loan rates that are applied to forestry programs, for example. After this problem was identified in France, changing the policy resulted in more farmers adopting agroforestry practices.[18] In a similar example, bamboo production is not currently eligible for the Reducing Emissions from Deforestation and Degradation (REDD+) program—a United Nations initiative that offers incentives for developing countries to reduce emissions from forested lands and invest in low-carbon paths to sustainable development—and other carbon forestry programs.[19] These problems are apparently due to an inability of bureaucracies to conceive of land uses outside of a narrow mind-set, even when farmers and progressive employees advocate for change. In a similar vein, the US Department of Agriculture recognizes only five forms of agroforestry, despite the great diversity of practices encompassed within the 19 basic building blocks identified by P. K. Nair.[20]

In addition to the on-farm challenges of carbon farming there are many potential pitfalls and unintended consequences to carbon farming initiatives—especially when we raise the discussion to the level of national and international trade policy and the interests of the massive corporations that pull the strings of those policies. This international effort must be handled with sound judgment. It is too easy for good ideas to run amok when they're scaled up and put into practice. Negative outcomes could include loss of land tenure or displacement of farmers and indigenous groups; increased

THE QUICKEST ROUTE TO SUSTAINABLE INTENSIFICATION: SUPPORT WOMEN FARMERS

Women produce half the world's food but receive only 7 percent of investment and aid. A 2011 report from the World Agroforestry Centre stated that if female farmers had access to the resources that male farmers have, global food production could increase by 10 to 30 percent. This act alone could reduce the number of hungry people by up to 17 percent.[21] Although this strategy is about intensification rather than mitigation, it is worthy of policy focus.

FIGURE 28.1. A North Korean farmer shows her plans to implement agroforestry on her farm using the Sloping Agricultural Land Technique. Photograph courtesy of The World Agroforestry Centre.

inequality and land conflict; impacts on animal welfare; land grabbing (aka "large-scale land acquisition"); and concentration of economic benefits that would "marginalize local stakeholders."[22]

Not only would these outcomes prove disastrous for many individuals, communities, and small nations, but a poorly managed international effort would also likely fail to mitigate climate change to the necessary level. According to the IPCC's report *Climate Change 2014: Mitigation of Climate Change,* "A large-scale drive towards mitigation without inclusion of key stakeholder communities involved would likely not be greeted favorably."[23] In other words, people won't get on board—and people need to get on board if this is going to work. Massive projects (strongly backed by European Union policy) to produce biofuels from oil palm and sugarcane should be a wake-up call as we move forward. Proposed large-scale biochar projects that displace small farmers and herders from their land should also be a cautionary tale.

Just because a massive global effort to mitigate climate change through carbon farming has the potential to run amok if poorly managed or taken over by corporations and investors with dollar signs in their eyes, that doesn't mean we can back off. It means the effort has to be managed correctly, smartly, and with our eyes on the ultimate goal—climate change mitigation for a livable future. This can't happen exclusively from the top down. Community control and real benefits to local people and ecosystems need to be baseline goals of carbon farming finance efforts.[24] And we know this isn't impossible. There are many successful examples of carbon farming put into practice with incredibly positive, beneficial results. We need to study and implement the best practices of sustainable development for this effort to work, like the many projects profiled in Leakey's *Living with the Trees of Life*.

Finally, although this book is focused on "supply-side" agricultural mitigation, there is so much potential for significant impact if we make changes to our diets and reduce our food waste. Today about a third of all the food grown worldwide is wasted. This represents a huge opportunity to feed far more of the world's seven billion people without clearing more forest for farmland.[25] These and other emissions reduction strategies are essential counterparts to agricultural carbon sequestration.

National Agricultural Policy Reform

Several countries have started to reform their agricultural policies for the better. For example, Kenya's new constitution requires farmers to grow trees on 10 percent of their land.[26] Brazil has a national program offering low-interest loans to help farmers transition to sustainable agriculture.[27] In 2014 India announced a new national agroforestry policy with the goal of increasing tree cover from 25 percent to 33 percent—representing 26 million hectares (64 million acres)—through greater implementation of agroforestry on existing farms. In 2013 an estimated $60 billion US of national climate finance funding was invested in-country from public budgets around the world.[28] Advocates have called for a national carbon farming grants program in the United States.[29] Imagine the impact it would have if the National Research Foundation began administering such a program. The IPCC argues that carbon farming is a cost-effective abatement strategy compared with other approaches, such as geoengineering.[30]

In 2011 Australia implemented a national Carbon Farming Initiative (CFI), funded through a cap-and-trade system, to provide rewards to farmers and landholders for reducing greenhouse gas emission and storing carbon on their land. Farmers and landholders earned credits by implementing specific practices—including forestry plantations, regeneration of native vegetation, biochar application, biodigestion of manure, early dry-season savanna burning, and forest protection—and then selling those credits as carbon offsets.[31] Unfortunately, the new conservative government is in the process of gutting Australia's entire suite of carbon emissions reduction policies, including substantial changes to the CFI.[32]

Many countries already use Payment for Environmental Services (PES) to encourage farmers to adopt or continue practices that maintain water quality, prevent erosion, and support other important ecosystem functions. It's basically a way to thank farmers for practices that may cost them money but benefit society. In the United States, the Environmental Quality Incentives Program (EQUIP)—administered by the US Department of Agriculture's Natural Resources Conservation Service (NCRS)—is an example of PES and could be an excellent model for incentivizing on-farm carbon sequestration. In the wealthy countries, farmland is already eligible for PES, whereas in the developing world only forestry typically is. That needs to change. PES or similar incentives are necessary for the world's

1.3 to 1.5 billion smallholders who cannot invest to make the needed changes due to lack of funding.[33]

Courtney White, the author of *Grass, Soil, Hope* and coalition builder between ranchers and environmentalists, proposes an interesting model. Rather than tying PES funds to specific practices, he suggests, "offer to pay land owners to double the carbon content of their soil, then stand back as they choose from the regenerative toolbox, knowing that no matter what methods they choose they'll be creating a cascade of cobenefits . . ."[34] Others suggest that a system that pays by the ton of carbon sequestered instead of by the practice adopted would be more efficient and effective.[35] Farmers would appreciate the freedom and limited bureaucracy of such a program.

The United States has recently unveiled a national carbon farming initiative. This includes a set of regional Climate Hubs, addressing mitigation and adaptation. The USDA aims to sequester 32.6 million tons of carbon by 2025 through their new climate initiatives.[36] This includes the new COMET-Planner, a tool to help farmers calculate sequestration impacts of various practices, most of which are already eligible for cost sharing by Payment for Environmental Services through the NRCS. This cost sharing is currently based on other environmental benefits, not carbon sequestration. The NRCS is working with the state of California to implement a new offset-based carbon farming market.[37]

Nations have a lot of options for incentivizing carbon farming and deincentivizing unsustainable agricultural practices. The IPCC has asserted that financial disincentives for poor practices including fines for emissions, taxes on emissions, and taxes on emission-related farm inputs such as nitrogen fertilizers are effective.[38] They assert that financial incentives for positive practices are effective as well. "Economic incentives (e.g., special credit lines for low-carbon agriculture, sustainable agriculture and forestry practices, tradable credits, payment for ecosystem services) and regulatory approaches (e.g., enforcement of environmental law to reduce deforestation, set-aside policies, air and water pollution control reducing nitrate load and N_2O emissions) have been effective in different cases. Investments in research, development, and diffusion (e.g., improved fertilizer

use efficiency, livestock improvement, better forestry management practices) could result in positive and synergistic impacts for adaptation and mitigation."[39]

All of these are tools in a national agricultural policy reform toolkit. As nations (hopefully) begin to address climate change more directly and effectively, agriculture must be a critical part of that conversation. These changes could have multiple benefits for farmers and rural economies around the world.

Land Reform and Tenure

La Via Campesina argues that "small farmers cool the planet."[40] This is not because small farms sequester more carbon, but because their intensive methods yield more per unit area, exemplifying "sustainable intensification." Most farms in Asia and Africa are 2 hectares (5 acres) or less.[41] Even the US Department of Agriculture acknowledges that smaller farmers produce more food per hectare, whether measured in weight, sales, or calories.[42]

Globally small farmers represent 92 percent of all farmers but occupy only 25 percent of the land—and their share of farmland is shrinking, in a trend referred to as "reverse agrarian reform." Yet according to a 2014 report from GRAIN, small farmers produce 80 percent of the food in developing countries, and more than their share almost everywhere else.[43] Small farms are generally more productive than larger farms. For example, in nine European Union countries, smaller farms are twice as productive as large farms. If all Kenyan farms were as productive as its small farms, production would double; it would almost triple in Central America, Ukraine, Hungary, and Tajikistan; and in Russia, productivity would increase sixfold.[44] High productivity means less deforestation—a critical factor in climate change mitigation. For so many reasons—including climate change, food security, food sovereignty—we need a bold new wave of land reform. The bottom line is that we need to get land back into the hands of small farmers who reliably put it to the best possible use. The World Bank economists now acknowledge that global land redistribution would increase food production and overall yields.[45]

Long-term land tenure is one of the key limits to the adoption of carbon farming practices such as agroforestry. Implementing carbon farming practices can be expensive and may not "bear fruit" for many years. As we saw in chapter 26 there are many risks and expenses involved in a transition to carbon farming—risks that don't pencil out without secure land tenure. Tree tenure is a set of rights to plant, use, and manage trees. This may refer to the right to use trees on land you do not own or rent and also can constrain your ability to use trees on land you own or rent. Tree tenure is a widespread practice in much of the world and can be a major limitation to carbon farming adoption.[46] Changes in Forest Code tree tenure in Niger removed the key barrier to rapid spread of farmer-managed natural regeneration. Initially farmers faced harsh regulations limiting the use of trees on their farms, understandably dimming the appeal of this agroforestry practice.[47] The Food and Agriculture Organization (FAO) has developed *Voluntary Guidelines on the Responsible Governance of Tenure of Land, Fisheries, and Forests* to ensure pro-poor and climate-adaptive rural development.[48]

New Economic Priorities

The IPCC writes: "Some climate change management objectives require large-scale transformations in human societies" and states that successful mitigation of climate change requires "decoupling human well being from economic growth."[49] It may be that certain institutions, even some governments and economic systems, are just not compatible with addressing climate change. For example, drastic emissions reduction is fundamentally in conflict with the wildly profitable fossil fuel industry. Stepping back behind the tipping point requires a rethinking of economic priorities.[50]

Some core components of capitalism as we know it are not compatible with climate change mitigation:

- Our addiction to constant economic growth is in conflict with the very real limits of nature. Nor is it compatible with the requirement to drastically reduce consumption in wealthy countries.[51]
- Placing profit above all else has driven us to the brink of catastrophic climate change. Care for people and

care for the environment have to be moved to the center of economic priorities.
- Many of the social and ecological costs of our economy are not paid for by those who cause them. Economists call these consequences "externalities." We need to put an end to the externalization of climate costs. A new economy must be rigorous about accounting for emissions and sequestration.[52]

Today the world's richest 7 percent are responsible for 50 percent of emissions.[53] Meanwhile the poorest half of the world's population have almost zero emissions today, and many lack basic human rights, like access to clean water, health care, and sufficient food.[54]

Scientists Anderson and Bows conclude their report on emissions reduction scenarios: "(Extremely) dangerous climate change can only be avoided if economic growth is exchanged, at least temporarily, for a period of planned austerity within Annex 1 [Global North] nations and a rapid transition away from fossil-fueled development within non-Annex 1 nations [Global South]."[55]

Addressing climate change thus requires a rapid strategy of degrowth and reduced consumption among the world's wealthy countries and people.[56] Wealthy nations must cut their emissions by 8 to 10 percent a year.[57] This doesn't mean a return to Stone Age living. Scientists say that for wealthy countries, it means more of a return to the standard of living of the 1970s—not such a terrible sacrifice.[58]

Climate-Smart Agriculture

The FAO defines *climate-smart agriculture* as "agriculture that sustainably increases productivity, resilience (adaptation), reduces/removes greenhouse gases (mitigation), and enhances the achievement of national food security and development goals."[59] The concept of climate-smart agriculture embodies multifunctionality and represents an important rallying point.

In 2014 a partnership of companies, NGOs, governments, researchers, and farming organizations launched the Global Alliance for Climate-Smart Agriculture (GACSA). Members include governments of the Global North including the United States, the UK,

Ireland, and France; and the Global South including Malawi, Nigeria, and Mexico. It also includes CGIAR, the Environmental Defense Fund, Food and Agriculture Organization of the United Nations, The Nature Conservancy, the Organic Consumers Association, and the World Bank.[60]

The emergence of GACSA is, in some ways, a cause for celebration because it signals broad acknowledgment across sectors and at various scales of critically important intersection of agriculture and climate change. But there are also some troubling aspects to the coalition. This collaborative promotes efforts that focus on any one of the three goals, rather than requiring a focus on all three, meaning it's all too easy to interpret "sustainable intensification" as ramped-up industrial agriculture with higher yields while ignoring adaptation and mitigation altogether. Real climate-smart agriculture would return to the FAO definition, focused on strategies that address all three goals at once (as so many practices that are profiled in this book do).[61]

A coalition of more than 100 NGOs has publicly criticized the climate-smart agriculture movement for promoting land grabs and GMOs.[62] This opposition movement argues that any "climate-smart" agriculture movement that includes McDonald's, Walmart, and the world's biggest seed and fertilizer companies—as GACSA does—is little more than a carbon farming greenwash that will direct yet more resources at industrial agriculture under the guise of climate mitigation and adaptation.[63] La Via Campesina, which has denounced "climate-smart agriculture" as a continuation of Green Revolution ideologies that decimated numerous peasant economies, also points out that although GACSA focuses its efforts on developing countries, carbon farming is needed everywhere.[64]

That said, the idea of an international coalition for carbon farming is an excellent one, and many of the organizations that are part of it have strong track records of carefully developing sound projects that have actual positive impacts on the communities in which they work. It remains to be seen whether GACSA will revise its policies or if it will need to be replaced with a coalition that promotes truly "climate-smart agriculture."

CHAPTER TWENTY-NINE

Strategic Next Steps

We can't expect to implement carbon farming strategies at the scale they're needed until people know what carbon farming is and understand its potential. We need to see examples of carbon farming in action—from journalists, from NGOs, and in our own neighborhoods. And we need to equate carbon farming with a broad set of strategies, the most effective of which are currently the least discussed. (It is currently the strategies with the weakest sequestration potential—no-till, organic annual cropping, and managed grazing—that are the most widely promoted and discussed.) We need a broad promotion and education campaign, to the point that the public equates carbon farming with clean energy as a climate solution strategy.

No individual, movement, government, or sector alone can achieve the monumental task of transforming global agriculture and bringing atmosphere carbon back down to 350ppm. We need a collective international movement—governments, politicians, policy makers, individuals, farmers, universities, activists, NGOs, international development agencies, businesses, philanthropists, investors, and more. In other words, everyone can, and should, have a role—because everyone is a stakeholder. Not everyone will agree on the many aspects of this global effort, and not everyone needs to. But we all do need to agree that climate change is serious and that it is a priority. We need a broad set of shared goals, and some level of coordination and communication among a great diversity of efforts. Below are action items that I think various sectors should focus on. Many of these action items cut across sectors. And it's not an exclusive list; at this point there is no end to the useful action we can undertake.

Governments, Politicians, Policy Makers

Not all governments have completely failed to respond to climate change as the critical issue it is (Australia and India are examples of some that have made strides). But it's certainly true of most, especially in wealthy Western countries that contribute the most carbon to the atmosphere but whose citizens are more insulated from the consequences than people in the rest of the world. We need more concerted and effective action—a level of mobilization we know politicians and governments are capable of because we've seen them muster it to combat terrorism and wage war. The points below are specifically geared toward the centralized governments of Western nations—those that have been the slowest to act but, ironically, are also the best positioned to do so—but could also be applied to other scales and areas of governance.

- Get with the program. Preventing catastrophic climate change is the central challenge for humanity in the 21st century.
- Stop bowing to pressure from fossil fuel lobbyists.
- Develop and sign a binding global agreement that would reduce carbon in the atmosphere to 350ppm.
- Set a high price on carbon, preferably through a tax on fossil carbon at the source.
- Remove subsidies for unsustainable farming. Develop robust incentives for carbon farming, including Payment for Environmental Services (PES) programs.
- Fund research, development, and implementation of carbon farming practices.

- Increase citizen control and decrease corporate influence on government.
- Subsidize climate-smart buildings, transportation, energy, industry, and agriculture.
- Support and protect traditional and indigenous carbon farming and land management practices.
- Streamline regulations for international movement of plants.
- Implement land reform to get more small farmers on land with secure tenure.
- Implement austerity plans to reduce consumption in wealthy countries.

Individuals

When it comes to climate, there is a tension about whether the actions of an individual can really be meaningful in the face of a problem so large—especially in the absence of political will. Admittedly, there's something absurd about the idea of changing your lightbulbs being enough to do the job. But individual action can also build to the size of a movement, and it can foster a sense of empowerment in the face of an intractable problem. Why not both? There's no reason that individual action and political pressure for policy change need to be mutually exclusive. Here are some ideas on how I think an individual can best direct his or her efforts.

- Act. Don't wait for governments or other leaders to get it together.
- Put a political price on climate change denial and failure to act. Vote out or throw out politicians who are not part of the solution.
- Dramatically reduce your consumption.
- Purchase carbon farming "products." You can't shop your way to a livable climate, but your support can help make it possible for farmers to transition to carbon farming practices and to stay the course.
- Eat like you care about the planet. Adopt a diet rich in carbon-sequestering foods with short supply chains.
- Divest from fossil fuels. Put your money into carbon farming, clean energy, and other climate-smart practices.

- Educate. No one knows everything, so learn what you can—what you're passionate about—and share that knowledge with others.
- Plant a food garden, the more perennial the better.

Farmers

Farmers are on the front lines of climate change and carbon farming. They are in a position to transition to carbon farming or to implement certain practices to make their farms more climate-smart. But they are also often the first people to see and understand the real impacts of climate change because they are engaged with the land at a deep level on a daily basis. And unfortunately, they are among the most vulnerable because they rely on the land for their living.

- Carbon farm.
- Share knowledge farmer-to-farmer and within farmer organizations you belong to.
- Start (or join) a carbon farming producers association or cooperative to increase access to seeds, equipment, and markets.
- Build regional, national, and international carbon farmers movements to advocate for policy change.
- Connect with alternative carbon finance and marketing entities as they emerge.

Researchers, Students, and Universities

As I was working on this book, I wanted to include a table that would lay out all the areas of carbon farming that need additional research. It quickly became clear that such an endeavor would delay the book indefinitely because the need for further research is so profound. The bad news about that is that we still have a long way to go before we'll understand, in a deep and specific way, the carbon sequestration potential of various crops and practices. The good news is that there are countless areas of study for science students and researchers to investigate, and that the fruits of that labor could make a meaningful contribution to our fight for a livable future. In addition to identifying and pursuing specific areas of research, here are some more general ideas for students, researchers, and universities that rise to the surface in my mind.

- Work with farmers to perform research and development on carbon farming crops and systems and perennial industrial feedstocks, especially for regions that are not well served.
- Develop economic models that are in line with the constraints given to us by climate scientists, that de-externalize costs and narrow the gap of inequality within and among nations.
- Divest endowments from fossil fuels and invest in a post-fossil carbon future.
- Develop standardized carbon monitoring tools.

Climate Activists

Climate activists have forced the climate conversation and put pressure on political leaders not only to acknowledge that climate change is one of the most critical issues of our time, but even to acknowledge that it's real and happening. Their work is essential—carbon farming alone without it will fail to attain 350ppm. Activists are up against wealthy, powerful, and merciless opponents who have a stake in business as usual. It is not an easy fight, nor is it easy to remain positive and motivated day in and day out when our political leaders have been so pathetic in their non-response. My hope is that climate activists will take up the mantle of carbon farming and acknowledge and advocate for the very real, positive impact that intelligently practiced agriculture can have on our climate. Here are some ideas for putting that in motion.

- Support carbon farming the way you support clean energy. Understand it. Fight for it. Put it at the center of your climate solutions platform.
- Make it about people, not just polar bears. Carbon farming is not only about sequestration. It is about justice. Make this connection.
- Build alliances. Carbon farming is about multifunctional solutions. Find common ground, and develop shared strategies with your counterparts in movements for food sovereignty, peasant farmers, and other allies. Your goals overlap; work toward them side by side.
- Press governments to act. From local to global, don't let up. Educate leaders, and press for the development of policy that puts carbon farming into practice.

NGOs

NGOs are well positioned to push forward a carbon farming agenda because they are already on the ground, in communities worldwide, working with farmers and other individuals who may already be practicing carbon farming. Because carbon farming is inseparable from other needs that NGOs address, such as strengthening local economies and empowering women, these groups can integrate carbon farming into existing projects and focus on it in the development of new projects. NGOs also provide a bridge between individuals and policy makers and can offer an important perspective on what works, what doesn't work, what might be possible, and what support is needed to make the possible actually happen. Because carbon farming is, in many ways, farther along in the more impoverished countries around the world, NGOs can serve as a conduit to pass on this knowledge and experience to the wealthier nations that have further to go.

- Develop a carbon farming certification system.
- Advocate, build coalitions, coordinate carbon farming implementation.
- Retool climate-smart agriculture to require mitigation, adaptation, and intensification.
- Serve as bridge between farmers and governments, to advocate for farmers and their needs.
- Serve as a voice for farmers, climate justice activists, and others to the wider public.

International Agencies

International agencies include CGIAR (including the World Agroforestry Centre), UN groups like FAO, economic groups like WTO and the World Bank, and of course the IPCC. These groups have great power and influence that could be directed to carbon mitigation. Many branches of these agencies are at the leading edge of climate work, while others remain part of the problem. These organizations should place climate mitigation at the top of their priorities.

- Set up carbon farming research and demonstration centers for each climate.
- Strengthen the focus on perennial crops, especially at the FAO and IPCC.
- Support traditional and indigenous carbon farming and land management practices.
- Promote scenarios that look at 350ppm and not 450, 500, or more (IPCC).
- Reform or replace WTO to promote rather than discourage (or make illegal) policies that promote local production and sustainable practices.
- Prioritize agriculture in carbon finance mechanisms.

Businesses

Businesses have an important role in mitigation, be they global or micro-scale. While I don't believe that business alone can get us to 350ppm, changing aspects of enterprises in the manufacturing, service, transportation, energy, and food sectors is clearly essential. There's a need for tens of thousands of new carbon-friendly enterprises, as well as major changes to existing businesses.

- Develop fair trade sourcing from carbon farm producers and producer associations.
- Replace current use of fossil carbon and annual food grains as feedstock with non-destructively harvested perennial biobased feedstocks wherever possible.
- Develop industry in the countries and regions where climate-smart raw materials are grown.
- Develop equipment for efficient planting, harvest, and processing.
- Shorten supply chains.

Funders and Philanthropists

Many foundations are already on board with climate change mitigation and agriculture. They have funded critical research and implementation for decades. This sector has the resources to mark a new trail that the rest of society will follow.

- Fund what governments don't yet: research, breeding, and grant-financing farmer transition.
- Support continuation and expansion of existing successful carbon farming programs.
- Help farmers get over the establishment hump with grants.

Lenders and Investors

Access to financing is a critical limitation for farmers who wish to transfer to carbon farming practices. Redirecting investment to carbon farming and other mitigation-adaptation-development efforts is an essential effort for the 21st century. There are vast gaps today in the mechanisms for connecting investors with carbon farmers and their associations that need to be filled in.

- Finance carbon farms and the perennial feedstock industry.
- Assist farmers to get over the establishment hump and learning curve for transition to carbon farming and perennial crops.
- Create a robust network to link divested funds to carbon farming, clean energy, and other climate solutions.

Whoever you may be, envision your group's role in the complete transformation of your society. This includes overhauling governance, economics, finance, transportation, the built environment, industry, energy, and, of course, agriculture. Doing so provides the additional chance to right many long-standing wrongs. This great human project of the 21st century has a role for you. I guarantee it will present sufficient challenge and reward to keep you engaged for many years to come.

ACKNOWLEDGMENTS

In 2008 I read Tim Flannery's *Now or Never: Why We Must Act Now to End Climate Change and Create a Sustainable Future*. Short and to the point, it was my personal wake-up call around climate change. I also realized that the perennial crops I had spent my adult life investigating (and cultivating) were a very promising potential solution that was not being discussed. I left my job as an urban farm project manager in 2009 to begin this project.

I pursued multiple options for financing the effort. I applied for a fellowship, inquired about a PhD program, submitted article pitches to major magazines, and even got an agent to shop the idea around to book publishers. Those few who even wrote me back were unable to provide the funds required to actually write a book.

The *Permaculture Activist* magazine gave me the chance to write two articles, one on carbon farming and the other on perennial staple crops. Writing these allowed me to begin a deeper exploration of the concepts. Opportunities to teach at the Carbon Farming Course and keynote the Northeast Organic Farming Association conference gave other venues to explore and communicate these ideas.

I decided to crowdsource the funds with Kickstarter, as I knew that there was strong interest from the public. Elizabeth Ü provided excellent coaching through the process of launching a campaign. Thanks to permaculturenews.org, International Permaculture Day, both thepermaculturepodcast.com and permaculturepodcast.org, permies.com, and Celsias for posting content related to the campaign, as well as everyone who helped on Facebook and YouTube.

Without my Kickstarter backers, you simply would not be reading this book. I'm deeply grateful for their support. That goes for those who gave $1 all the way to my $1,000 patrons.

In 2013 I also received word that I was appointed as a lecturer at Yale University, where I will teach my third class on Carbon-Sequestering Multifunctional Agroforestry this fall (2015) in the Masters of Forestry and Environmental Science program. Jonah Adels, who had been my student in a carbon farming course, organized Yale students to bring me there to teach. He died as a result of a car accident on the way home from field research at Badgersett Research Corporation. The Yale course, and this book, are a testament to his vision and organizing ability.

Thank you to those who have blazed the trail—particularly P. K. Nair, Rattan Lal, the Land Institute, and the World Agroforestry Centre. A special appreciation to the team at the Institute for Plant Genetics and Crop Plant Research, whose six-volume, 3,644-page *Mansfeld's Encyclopedia of Agricultural and Horticultural Crops* was my essential plant reference.

For the first time I've been able to hire research assistants. Their work was extremely helpful. Thank you to Emma Cutler, Connor Stedman, Sarah Tolbert, Shane Feyers, Mary Johnson, and the extremely capable Ruth Metzel, also teaching fellow for my Yale class.

Throughout the process my friend and colleague Rafter Ferguson has challenged me to consider myself a citizen scientist and use the scientific literature as my guide and my standard for this book. He has provided assistance in navigating the world of peer-reviewed articles, constant thoughts on the organization and development of these ideas, and encouragement to teach the controversy where it exists. He also used his impressive gifts for the visual representation of data to help me develop the tables in chapter 3 around which the argument of the book is framed (along with other figures). I was fortunate to have his companionship on the research trips to Cuba and Mexico as well.

Many individuals were gracious enough to answer questions in person or via phone or email. Thanks to Ken Asmus of Oikos Tree Crops; Danny Blank of the Hunger Education and Resource Training Institute (HEART), formerly of Educational Concerns for Hunger Organization (ECHO); Brock Dolman of the Occidental Arts and Ecology Center (OAEC) and the Water Institute; Richard Felger of the University of Arizona herbarium and the Sky Island Alliance; Rob Fetter, formerly of Equitable Origin eco-social petroleum certification; Bart Fredericks of the FACT Foundation; Christine Jones of Amazing Carbon; Neil Logan of Integrated Living Systems Design; Owen Hablutzel of Whole Systems Transformations; Mark Krawczyk, coauthor of the forthcoming *Coppice Agroforestry*; Ricardo Romero of Las Cañadas; Phil Rutter of Badgersett Research Corporation; Martin Price, founder of the Educational Concerns for Hunger Organization (ECHO); Lincoln Smith of Forested; Rachel Smolker of Biofuel Watch; Rachel Steele of USDA Climate Hubs; Adam Chambers of USDA NRCS; Rattan Lal of the Carbon Management and Sequestration Center; and David Van Tassel of the Land Institute.

Thanks to the many individuals and organizations who provided the photos that so enhance the readability of this book. All are credited. In particular I'd like to thank the assemblers of the Flickr image banks of the World Agroforestry Centre and AgForward, and everyone who shares their photography with the world through open source licenses on Flickr and Wikimedia Commons. Thanks as well to the many others who shared excellent photos that I was unable to include for reasons of space.

Many people reviewed drafts and manuscript chapters. Brian Tokar in particular has gone above and beyond the call of duty, offering thoughts, contacts, and reviews from the very beginning of this project. Ricardo Romero reviewed the introduction. Reviewers for part 1 include Anne Toensmeier (chapters 1–4), Brian Tokar (1), Jonathan Bates (2), Pandora Thomas (2), Craig Hepworth (3 and 4), and Mario Yanez (5). Part 2 reviewers were Rafter Ferguson (7), David Van Tassel (9, 10), Meghan Giroux (6–10), and Vida Chavez-Garcia (9). Part 3 was reviewed by Craig Hepworth (11–17) and David Van Tassel (11, 13, 16). Reviewers for part 4 include Sarah Seitz (18–24), Brian Tokar (18), Brad Ward (18), and Chaia Heller (18). Part 5 was reviewed by Diego Angarita (28, 29), Brian Tokar (28), Rick Burnette (26), and Pandora Thomas (29). Thanks to all of them for taking the time from their busy lives to read and offer their thoughts and expertise. Dr. Cheikh Mbow, of the World Agroforestry Centre and IPCC panelist, performed a very helpful review of the entire manuscript and had many useful suggestions.

I have had great experiences working with the team at Chelsea Green. Publisher Margo Baldwin has been on board since the beginning about the need for lots of beautiful photographs. My editor Brianne Goodspeed really understands how to bring out the best in my writing. She has been a critical anchor to this book, helping to frame the structure to best communicate the ideas.

Thanks to all the friends and counselors whose support kept me going at a steady pace. Diego Angarita, Amy Calendrella, Kara Nye, and Daniel Ross have been particularly helpful in this regard.

Thanks also to Marikler and Daniel for allowing me to finish this writing marathon in a 40-day sprint. I'm so very lucky to call them my family.

I have surely forgotten to mention some of the many people who have been of assistance in this project. Thanks to all of you and my apologies—if you are not mentioned here your work was no less helpful. Finally, final responsibility for any and all errors lies with me and not my many reviewers, contributors, interviewees, and editors.

Global Species Matrix

Tables like this are notoriously difficult to use efficiently. Electronic versions, more easily searchable, will be maintained at www.perennialsolutions.org, as will lists of species for each climate type.

Genus and **Species** reflect the Latin or scientific name of the crop. **Common Name** is only one of usually many names used. Where possible local language names from the region of origin are used. **Family** indicates the botanical family. All scientific names and families are up to date as of 2014 in line with the Kew Botanical Gardens website ThePlantList.org.[1] Appendix D permits searching for outdated names.

Origin reports the region where the crop grows wild or where it is cultivated or was domesticated. **Invasive** indicates if the species is reported as listed by the Global Invasive Species Database.[2] **Climate** and **Humidity** refer to the categories described in chapter 10.

Cultivation regards the level of domestication and active production of the species, as described in chapter 10. **Form** and **Management** describe the type of plant and the non-destructive management systems that are used in cultivation as described in chapter 4.

Staple Crop lists uses for the six categories described in chapter 11, while **Industrial Crop** does the same for the six categories from chapter 18.

Agroforestry Services and **Fodder** list agroforestry support and livestock fodder uses. **Other Systems** lists agroforestry and production systems the crop is reported to be used in.

Primary sources for this table include *Mansfeld's Encyclopedia of Agricultural and Horticultural Crops*, P. K. Nair's *Introduction to Agroforestry*, the references from parts 3 and 4, and Dave's Garden, a user-generated resource that proved remarkably helpful in determining cold-hardiness of many species.[3]

Genus	Species	Common Name	Family	Origin	Invasive	Climate	Humidity
Abelmoschus	*esculentus*	perennial okra	Malvaceae	Africa		tropical	semi-arid to humid
Abroma	*augusta*	perennial Indian hemp	Malvaceae	SE Asia, Australia		tropical	humid
Acacia	*angustissima*	timbre	Fabaceae	Mesoamerica, North America		subtropical to tropical, tropical highlands	semi-arid to humid
Acacia	*catechu*	cutch tree	Fabaceae	South & SE Asia		tropical	semi-arid to humid
Acacia	*colei*	Cole's wattle	Fabaceae	Australia, Asia		tropical	semi-arid
Acacia	*cowleana*		Fabaceae	Australia		tropical	arid to semi-arid
Acacia	*holosericea*		Fabaceae	Australia		tropical	semi-arid to humid
Acacia	*koa*	koa	Fabaceae	Hawaiian Islands		tropical, tropical highlands	humid
Acacia	*koaia*	koai'a	Fabaceae	Hawaiian Islands		tropical, tropical highlands	semi-arid to humid
Acacia	*leucophloea*		Fabaceae	Asia		tropical	semi-arid to humid
Acacia	*murrayana*		Fabaceae	Australia		warm temperate to subtropical	arid to semi-arid
Acacia	*nilotica*	Egyptian thorn	Fabaceae	Africa	yes	tropical	arid to semi-arid
Acacia	*saligna*	Port Jackson wattle	Fabaceae	Australia	yes	tropical	arid to semi-arid
Acacia	*senegal*	gum arabic	Fabaceae	Africa, South Asia		subtropical to tropical, tropical highlands	arid to semi-arid
Acacia	*seyal*		Fabaceae	Africa		tropical	arid to semi-arid
Acacia	*tortilis*	umbrella thorn acacia	Fabaceae	Africa		tropical	arid to semi-arid
Acacia	*victoriae*	elegant wattle	Fabaceae	Australia		warm temperate to tropical	arid to semi-arid
Acanthocereus	*tetragonus*	pitahaya	Cactaceae	Colombia	yes	subtropical to tropical	semi-arid to humid
Acanthosicyos	*horridus*	nara melon	Cucurbitaceae	Africa		subtropical to tropical	arid to semi-arid
Acer	*saccharinum*	silver maple	Sapindaceae	North America		boreal to warm temperate	humid
Acer	*saccharum*	sugar maple	Sapindaceae	North America		boreal to cold temperate	humid
Acrocomia	*aculeata*	macauba palm	Arecaceae	tropical Americas		tropical	humid
Adansonia	*digitata*	baobab	Malvaceae	Africa		tropical	arid to semi-arid
Adenanthera	*pavonina*	condori wood	Fabaceae	South Asia through Australia	yes	tropical	humid
Aechmea	*magdalenae*	ixtle	Bromeliaceae	tropical Americas		tropical	humid
Aesculus	spp.	horse chestnut	Sapindaceae	northern temperate		cold to warm temperate, Mediterranean	semi-arid to humid
Afraegle	*paniculata*	Nigerian powder-flask fruit	Rutaceae	West Africa		tropical	humid
Agathis	*australis*	kauri	Araucariaceae	New Zealand		subtropical to tropical	humid
Agave	*americana*	maguey	Asparagaceae	Mesoamerica	yes	warm temperate to tropical	arid to semi-arid
Agave	*cantala*	cantala	Asparagaceae	Mesoamerica		subtropical to tropical	arid to humid
Agave	*fourcroydes*	henequen	Asparagaceae	Mesoamerica		subtropical to tropical	semi-arid to humid
Agave	*lechuguilla*	ixtle	Asparagaceae	Mesoamerica		subtropical to tropical	arid to humid
Agave	*murpheyi*	Hohokam agave	Asparagaceae	SW North America		warm temperate, tropical highlands	arid to semi-arid
Agave	*parryi*	Parry's agave	Asparagaceae	Mexico & SW North America		boreal to warm temperate	arid to semi-arid
Agave	*salmiana*	maguey	Asparagaceae	Americas		warm temperate to tropical	arid to semi-arid
Agave	*sisalana*	sisal	Asparagaceae	Mesoamerica	yes	subtropical to tropical	semi-arid to humid
Agave	*tequilana*	tequila agave	Asparagaceae	Americas		subtropical, tropical highlands	arid to semi-arid
Agave	*vivipara*	Mescal casero	Asparagaceae	Mesoamerica		subtropical to tropical	semi-arid to humid
Albizia	*julibrissin*	mimosa, silk tree	Fabaceae	Asia	yes	cold to warm temperate, Mediterranean	semi-arid to humid
Albizia	*lebbeck*	lebbek tree	Fabaceae	Asia	yes	subtropical to tropical	semi-arid to humid
Albizia	*lucidior*	tapria siris	Fabaceae	South & SE Asia		tropical	humid

Cultivation	Form	Management	Staple Crop	Industrial Crop	Agroforestry Services	Fodder	Other Systems
regional crop	shrub	standard, coppice	protein-oil	fiber			
regional crop	tree	coppice		fiber			
regional crop	tree	standard, coppice		biomass	nitrogen, alley crop		
regional crop	tree	coppice		tannin, medicinal, gum, dye	nitrogen	insect	
historic wild staple, new crop	tree	standard, coppice	protein		nitrogen, windbreak		FMAFS
historic wild staple, new crop	tree	standard, coppice	protein		nitrogen, windbreak		FMAFS
historic wild staple, new crop	tree	standard, coppice	protein		nitrogen, windbreak		FMAFS
regional crop	tree	standard			nitrogen, crop shade		
regional crop	tree	standard			nitrogen, crop shade		
regional crop	tree	standard			nitrogen	pod	
historic wild staple, new crop	tree	standard, coppice	protein		nitrogen, windbreak		FMAFS
regional crop	tree	standard		tannin	nitrogen, alley crop, windbreak	pod	
minor global crop	tree	standard, coppice		biomass, tannin, gum	nitrogen, windbreak	pod, bank	
regional crop	tree	coppice		biomass, gum, medicinal	nitrogen	pod, bank	
regional crop	tree	standard		gum	nitrogen	pod	
regional crop	tree	standard, coppice		biomass, tannin	nitrogen, windbreak, crop shade	pod, bank	parkland
historic staple, new crop	tree	standard, coppice	protein		nitrogen, windbreak		FMAFS
regional crop	cactus	standard			living fence		
wild staple	shrub	standard	protein-oil				
experimental	tree	standard, coppice	sugar	biomass	crop shade	bank	
regional crop	tree	standard	sugar		crop shade	bank	
regional crop only	palm	standard	oil	fiber			
new crop	tree	standard, coppice	protein-oil	fiber		bank	parkland, FMAFS
regional crop only; beans toxic raw	tree	standard, coppice	protein-oil	fiber	nitrogen, alley crop, crop shade	pod	
regional crop	succulent	standard		fiber			
hypothetical	tree	standard		starch, medicinal			
regional crop only	tree	standard	oil	glue			
regional crop	tree	standard		hydrocarbon, medicinal			
regional crop only	succulent	standard	basic starch, sugar				
regional crop only	succulent	standard	basic starch	fiber	living fence		
minor global crop	succulent	standard		fiber	living fence		
regional crop	succulent	standard		fiber	living fence		
historic crop	succulent	standard	basic starch				
historic crop	succulent	standard	basic starch				
regional crop only	succulent	standard	basic starch		living fence		
minor global crop	succulent	standard		fiber	living fence		
regional crop only	succulent	standard	basic starch, sugar		living fence		
regional crop	succulent	standard		fiber	living fence		
regional crop	tree	standard, coppice		biomass	nitrogen, alley crop, contour hedgerow, crop shade	bank, insect	
minor global crop	tree	standard, coppice		biomass, pesticide	nitrogen, crop shade	bank	
regional crop only	tree	standard	protein		nitrogen		

Genus	Species	Common Name	Family	Origin	Invasive	Climate	Humidity
Albizia	saman	monkeypod	Fabaceae	tropical Americas	yes	tropical	semi-arid to humid
Aleurites	moluccanus	candlenut	Euphorbiaceae	Malay archipelago	yes	tropical	semi-arid to humid
Allanblackia	floribunda	allanblackia	Clusiaceae	West Africa		tropical	humid
Allanblackia	parviflora	vegetable tallow tree	Clusiaceae	West Africa		tropical	humid
Allanblackia	stuhlmannii	mkanye	Clusiaceae	East Africa		tropical	humid
Alnus	acuminata	aliso	Betulaceae	tropical Americas		subtropical to tropical, tropical highlands	humid
Alnus	glutinosa	grey alder	Betulaceae	Europe	yes	boreal to warm temperate	humid
Alnus	jorullensis	alum	Betulaceae	Americas		warm temperate	humid
Alnus	nepalensis	alder	Betulaceae	Asia		tropical highlands	humid
Alnus	rubra	red alder	Betulaceae	western North America		cold to warm temperate	humid
Alocasia	macrorrhizos	bore	Araceae	Asia, Pacific		tropical, tropical highlands	humid
Amorpha	fruticosa	false indigo	Fabaceae	North America		boreal to warm temperate	semi-arid to humid
Amphicarpaea	bracteata	ground bean	Fabaceae	eastern North America		cold to warm temperate	semi-arid to humid
Amphicarpaea	bracteata subsp. edgeworthii	saekong	Fabaceae	East Asia		cold to warm temperate	humid
Anacardium	occidentale	cashew	Anacardiaceae	Brazil		tropical	semi-arid to humid
Ananas	comosus	pineapple	Bromeliaceae	tropical Americas		tropical	semi-arid to humid
Apios	americana	groundnut	Fabaceae	North America		cold temperate to subtropical	humid
Apocynum	cannabinum	dogbane hemp	Apocynaceae	North America		warm temperate, Mediterranean	semi-arid to humid
Apocynum	pictum	kendyr	Apocynaceae	Central Asia		cold to warm temperate	arid to semi-arid
Apocynum	venetum	kendyr	Apocynaceae	Central Asia		cold to warm temperate	arid to semi-arid
Arachis	glabrata	forage peanut	Fabaceae	South America		subtropical to tropical	humid
Arachis	pintoi	pintoi peanut	Fabaceae	South America		subtropical to tropical, tropical highlands	humid
Araucaria	angustifolia	pinheiro	Araucariaceae	South America		subtropical, tropical highlands	humid
Araucaria	araucana	monkey puzzle	Araucariaceae	Chile & Argentina		warm temperate	humid
Araucaria	bidwillii	bunya bunya	Araucariaceae	Australia		Mediterranean, subtropical to tropical	humid
Areca	catechu	betel nut	Arecaceae	Asia		tropical	humid
Arenga	pinnata	sugar palm	Arecaceae	Asia		tropical	humid
Argania	spinosa	argan	Sapotaceae	Morocco		subtropical to tropical	arid to semi-arid
Artemisia	ludoviciana	white sagebrush	Asteraceae	western North America, Mesoamerica		boreal to subtropical	semi-arid to humid
Artocarpus	altilis	breadfruit	Moraceae	New Guinea, Pacific		tropical	humid
Artocarpus	camansi	breadnut	Moraceae	New Guinea, Pacific		tropical	humid
Artocarpus	heterophyllus	jakfruit	Moraceae	India		subtropical to tropical	humid
Artocarpus	integer	champedak	Moraceae	SE Asia, New Guinea		tropical	humid
Artocarpus	mariannensis	dugdug	Moraceae	Pacific		tropical	humid
Arundinaria	spp.	running bamboo	Poaceae	Asia		cold to warm temperate	humid
Arundinaria	gigantea	giant cane	Poaceae	North America		cold to warm temperate	humid
Arundo	donax	giant reed	Poaceae	Europe, Asia	yes	cold temperate to tropical, tropical highlands	semi-arid to humid
Asclepias	curassavica	Curassavian swallow-wort	Apocynaceae	tropical Americas		warm temperate to tropical	humid
Asclepias	erosa	desert milkweed	Apocynaceae	SW North America, Baja California		subtropical	arid to semi-arid
Asclepias	incarnata	swamp milkweed	Apocynaceae	North America		cold to warm temperate	semi-arid to humid
Asclepias	speciosa	showy milkweed	Apocynaceae	western North America		cold to warm temperate	semi-arid

Cultivation	Form	Management	Staple Crop	Industrial Crop	Agroforestry Services	Fodder	Other Systems
regional crop	tree	standard	balanced carb		nitrogen, crop shade, living trellis	pod, bank, insect	
regional crop	tree	standard		oil, medicinal			
new crop	tree	standard	oil				
new crop	tree	standard	oil				
new crop	tree	standard	oil				
regional crop	tree	standard, coppice		biomass	nitrogen, crop shade, living trellis		
minor global crop	tree	standard, coppice		biomass, tannin	nitrogen		SRC
regional crop	tree	standard, coppice			nitrogen, crop shade		
regional crop	tree	standard, coppice			nitrogen, crop shade	bank	
regional crop	tree	standard, coppice			nitrogen, alley crop	bank	
minor global crop	herb	standard				bank	
minor global crop	shrub	coppice		pesticide	nitrogen, alley crop, contour hedgerow, windbreak	bank	
historic crop	herb	standard, fodder	protein		nitrogen, understory legume		
regional crop only	herb	standard, fodder	protein		nitrogen, understory legume	pasture	
global crop	tree	standard	protein-oil	oil	living fence		homegarden
global fruit crop, minor fiber	succulent	standard		fiber	contour hedgerow, living fence		
new crop for tubers, also edible beans	herb vine	standard	protein		nitrogen		
minor global crop	herb	hay		hydrocarbon, fiber			
regional crop	herb	hay		fiber			
regional crop	herb	hay		fiber			
minor global crop	herb	hay, fodder			nitrogen, understory legume		
minor global crop	herb	hay, fodder			nitrogen, understory legume	pasture	
regional crop only	tree	standard	balanced carb				
regional crop only	tree	standard	balanced carb				
regional crop only	tree	standard	balanced carb				
minor global crop		standard		starch			
minor global crop	palm	standard	sugar	fiber			multistrata
new crop	tree	standard	oil				
experimental	shrub	coppice		hydrocarbon			
minor global crop	tree	standard	basic starch	hydrocarbon, fiber	windbreak, crop shade	fruit	homegarden, multistrata
minor global crop	tree	standard	balanced carb				homegarden, multistrata
minor global crop	tree	standard	basic starch, balanced carb			fruit	homegarden, multistrata
regional crop only	tree	standard	basic starch, balanced carb	hydrocarbon			
regional crop only	tree	standard	basic starch, balanced carb				
minor global crop	bamboo	managed multistem		biomass		bank	
wild	bamboo	managed multistem		biomass		bank	
minor global crop	giant grass	hay		biomass		bank	
regional crop	herb	hay		hydrocarbon, fiber			
experimental	herb vine	hay		hydrocarbon			
experimental	herb	hay		hydrocarbon, fiber			
experimental	herb	hay		hydrocarbon			

Genus	Species	Common Name	Family	Origin	Invasive	Climate	Humidity
Asclepias	*subulata*	desert milkweed	Apocynaceae	Mesoamerica, SW North America		warm temperate to subtropical	arid to semi-arid
Asclepias	*syriaca*	common milkweed	Apocynaceae	North America		cold to warm temperate	semi-arid to humid
Asclepias	*tuberosa*	butterfly weed	Apocynaceae	North America		cold temperate to subtropical	semi-arid to humid
Astragalus	*glycyphyllos*	licorice milkvetch	Fabaceae	Eurasia		cold to warm temperate	semi-arid to humid
Astrebla	*squarrosa*	Bull Mitchell grass	Poaceae	Australia		subtropical to tropical	semi-arid
Astrocaryum	*vulgare*	tucuma palm	Arecaceae	NE South America		tropical	humid
Atriplex	*canescens*	fourwing saltbush	Amaranthaceae	North America		cold to warm temperate	arid to semi-arid
Atriplex	*nummularia*	old-man saltbush	Amaranthaceae	Australia		warm temperate to subtropical	arid to semi-arid
Atriplex	*semibaccata*	Australian saltbush	Amaranthaceae	Australia		warm temperate to tropical	arid to semi-arid
Atriplex	*truncata*	Utah saltbush	Amaranthaceae	North America—halophyte		cold to warm temperate	arid to semi-arid
Attalea	*funifera*	piassava palm	Arecaceae	tropical Americas		tropical	humid
Attalea	*maripa*	maripa palm	Arecaceae	northern South America		tropical	humid
Attalea	*speciosa*	babassu palm	Arecaceae	northern South America		tropical	humid
Azadirachta	*indica*	neem	Meliaceae	South Asia		tropical	arid to humid
Bactris	*gasipaes*	peach palm	Arecaceae	tropical Americas		tropical	humid
Balanites	*aegyptiaca*	balanites	Zygophyllaceae	Africa, Asia		subtropical to tropical	arid to semi-arid
Bambusa	*balcooa*	clumping bamboo	Poaceae	Asia		subtropical to tropical, tropical highlands	humid
Bambusa	*bambos*	clumping bamboo	Poaceae	Asia		subtropical to tropical, tropical highlands	humid
Bambusa	*blumeana*	clumping bamboo	Poaceae	Asia		subtropical to tropical, tropical highlands	humid
Bambusa	*chungii*	clumping bamboo	Poaceae	China		subtropical to tropical	humid
Bambusa	*heterostachya*	clumping bamboo	Poaceae	Asia		subtropical to tropical, tropical highlands	humid
Bambusa	*nutans*	clumping bamboo	Poaceae	Asia		subtropical to tropical, tropical highlands	humid
Bambusa	*oldhamii*	clumping bamboo	Poaceae	Asia		subtropical to tropical, tropical highlands	humid
Bambusa	*pervariabilis*	clumping bamboo	Poaceae	Asia		subtropical to tropical, tropical highlands	humid
Bambusa	*polymorpha*	clumping bamboo	Poaceae	Asia		subtropical to tropical, tropical highlands	humid
Bambusa	*textilis*	clumping bamboo	Poaceae	Asia		warm temperate to tropical	humid
Bambusa	*tulda*	clumping bamboo	Poaceae	Asia		subtropical to tropical, tropical highlands	humid
Bambusa	*vulgaris*	clumping bamboo	Poaceae	Asia	yes	warm temperate to tropical	humid
Barringtonia	*edulis*	vutu kana	Lecythidaceae	Fiji		tropical	humid
Barringtonia	*novae-hiberniae*	cutnut	Lecythidaceae	New Guinea, West Pacific		tropical	humid
Barringtonia	*procera*	cutnut	Lecythidaceae	New Guinea, SW Pacific		tropical	humid
Bauhinia	*petersiana*	Kalahari white bauhinia	Fabaceae	Africa		subtropical to tropical	semi-arid
Bauhinia	*thonningii*		Fabaceae	Africa		tropical	semi-arid to humid
Bertholletia	*excelsa*	Brazil nut	Lecythidaceae	South America		tropical	humid
Bixa	*orellana*	annatto	Bixaceae	tropical Americas		subtropical to tropical	semi-arid to humid
Blighia	*sapida*	akee	Sapindaceae	West Africa		tropical	humid
Boehmeria	*nivea*	ramie	Urticaceae	Asia		warm temperate to tropical	humid
Bombax	*ceiba*	paina	Malvaceae	Asia, Australia		tropical	humid
Borassus	*aethiopum*	African palmyra palm	Arecaceae	tropical Africa		tropical	semi-arid to humid
Borassus	*flabellifer*	palmyra palm	Arecaceae	Asia		subtropical to tropical	semi-arid to humid

Cultivation	Form	Management	Staple Crop	Industrial Crop	Agroforestry Services	Fodder	Other Systems
experimental	herb	hay		hydrocarbon			
experimental	herb	hay		hydrocarbon, fiber			
experimental	herb	hay		hydrocarbon			
minor global crop	herb	hay, fodder			nitrogen, understory legume	pasture	
historic crop	grass	standard	balanced carb				
regional crop only	palm	standard	oil				
minor global crop	shrub	coppice, fodder				bank	
regional crop	shrub	coppice, fodder			windbreak	bank	
minor global crop	shrub	coppice, fodder				bank	
regional crop	shrub	coppice, fodder				bank	
regional crop only	palm	standard	oil	fiber			
regional crop	palm	standard	oil				
regional crop only	palm	standard	oil	oil			
minor global crop	tree	standard, coppice		biomass, oil, pesticide	windbreak		parkland
regional crop only	palm	managed multistem	balanced carb, oil		windbreak, crop shade, living trellis		homegarden, multistrata
regional crop only	tree	standard, coppice	protein-oil	biomass		bank	parkland
minor global crop	bamboo	managed multistem		biomass			strip intercrop, multistrata
minor global crop	bamboo	managed multistem		biomass			strip intercrop, multistrata
minor global crop	bamboo	managed multistem		biomass			strip intercrop, multistrata
regional crop	bamboo	managed multistem		biomass			
minor global crop	bamboo	managed multistem		biomass			strip intercrop, multistrata
minor global crop	bamboo	managed multistem		biomass			strip intercrop, multistrata
minor global crop	bamboo	managed multistem		biomass		bank	strip intercrop, multistrata
minor global crop	bamboo	managed multistem		biomass			strip intercrop, multistrata
minor global crop	bamboo	managed multistem		biomass			strip intercrop, multistrata
minor global crop	bamboo	managed multistem		biomass		bank	strip intercrop, multistrata
minor global crop	bamboo	managed multistem		biomass			strip intercrop, multistrata
minor global crop	bamboo	managed multistem		biomass			strip intercrop, multistrata
regional crop only	tree	standard	balanced carb				
regional crop only	tree	standard	balanced carb				
regional crop only	tree	standard	balanced carb				
wild staple	tree	standard	protein-oil				
regional crop	tree	standard	balanced carb		nitrogen	pod	
minor global crop, mostly wild-collected	tree	standard	protein-oil				multistrata
minor global crop	tree	standard		fiber, dye, gum	living fence, windbreak, crop shade, living trellis		
regional crop only; deadly when unripe	tree	standard	protein-oil				parkland, homegarden
minor global crop	herb	hay, fodder	protein	fiber		bank	
regional crop	tree	standard		fiber			
regional crop only	palm	standard	oil, sugar				parkland
regional crop only	palm	standard	oil, sugar	fiber			

Genus	Species	Common Name	Family	Origin	Invasive	Climate	Humidity
Brassica	carinata	Ethiopian kale	Brassicaceae	Africa		tropical, tropical highlands	semi-arid to humid
Brassica	oleracea	perennial kales and collards	Brassicaceae	Mediterranean		warm temperate to subtropical, Mediterranean	humid
Bromelia	karatas	karata	Bromeliaceae	tropical Americas		tropical	semi-arid to humid
Bromelia	pinguin	Caraguata	Bromeliaceae	South America		tropical	semi-arid to humid
Bromelia	serra	Caraguata	Bromeliaceae	South America		tropical	semi-arid to humid
Brosimum	alicastrum	ramon, Mayan breadnut	Moraceae	Mesoamerica		tropical	semi-arid to humid
Brosimum	parinarioides	leite de amapa	Moraceae	Americas		tropical	humid
Brosimum	utile	milk tree, cow tree	Moraceae	Americas		tropical	semi-arid to humid
Broussonetia	papyrifera	paper mulberry	Moraceae	Asia		cold temperate to tropical	humid
Buchanania	cochinchinensis	almondette	Anacardiaceae	India		tropical	semi-arid to humid
Bursera	simaruba	gumbo limbo	Burseraceae	Americas		tropical	semi-arid to humid
Caesalpinia	digyna	teri-pod	Fabaceae	Asia		warm temperate to tropical	semi-arid to humid
Caesalpinia	sappan	Brazilwood	Fabaceae	South America		tropical	humid
Cajanus	cajan	pigeon pea	Fabaceae	India		subtropical to tropical, tropical highlands	semi-arid to humid
Calamus	spp.	rattan palm	Arecaceae	Asia, Australia		tropical	humid
Calliandra	calothyrsus	red calliandra	Fabaceae	Mesoamerica		tropical, tropical highlands	semi-arid to humid
Calopogonium	caeruleum	bejuco culebra	Fabaceae	Mexico		tropical	semi-arid to humid
Calopogonium	mucunoides	calopo	Fabaceae	Mesoamerica		tropical	humid
Calotropis	giganteus	akon	Apocynaceae	South & SE Asia		tropical	arid to humid
Calotropis	procera	French cotton	Apocynaceae	Africa through India		tropical	arid to humid
Camptotheca	acuminata	cancer tree	Cornaceae	China		warm temperate	humid
Canarium	indicum	ngala	Burseraceae	New Guinea, West Pacific		tropical	humid
Canarium	luzonicum	Elmi canary tree	Burseraceae	Philippines		tropical	humid
Canarium	ovatum	pili nut	Burseraceae	Philippines		tropical	humid
Canarium	schweinfurtii	African elemi	Burseraceae	tropical Africa		tropical	humid
Canavalia	gladiata	sword bean	Fabaceae	Africa, Asia		tropical, tropical highlands	semi-arid to humid
Caragana	arborescens	Siberian pea shrub	Fabaceae	Asia		boreal to cold temperate	semi-arid to humid
Caragana	microphylla	littleleaf peashrub	Fabaceae	East Asia		boreal to warm temperate	arid to semi-arid
Carex	meyeriana		Cyperaceae	East Asia		boreal to warm temperate	aquatic
Carex	morrowii	iwashiba	Cyperaceae	East Asia		boreal to warm temperate	aquatic
Carya	hybrids	hybrid and neohybrid hickories	Juglandaceae	hybrid origin		cold temperate to subtropical, tropical highlands	humid
Carya	cathayensis	Chinese hickory	Juglandaceae	China		cold to warm temperate	humid
Carya	illinoiensis	pecan	Juglandaceae	North America, Mesoamerica		cold temperate subtropical, tropical highlands	humid
Carya	laciniosa	shellbark hickory	Juglandaceae	eastern North America		cold to warm temperate	humid
Carya	ovata	shagbark hickory	Juglandaceae	North America, Mesoamerica		cold temperate subtropical, tropical highlands	humid
Caryocar	brasiliense	pequi	Caryocaraceae	Brazil, Bolivia, Paraguay		tropical	semi-arid
Caryocar	nuciferum	souari-nut	Caryocaraceae	S. America		tropical	humid
Caryocar	villosum	piquia	Caryocaraceae	Brazil, French Guiana		tropical	humid
Caryodendron	orinocense	inche	Euphorbiaceae	NW South America		tropical	humid
Caryota	urens	fishtail palm	Arecaceae	Asia		tropical, tropical highlands	humid
Cassia	fistula	golden shower tree	Fabaceae	South & SE Asia		tropical	humid
Cassia	grandis		Fabaceae	tropical Americas		tropical	humid

Cultivation	Form	Management	Staple Crop	Industrial Crop	Agroforestry Services	Fodder	Other Systems
minor global crop	herb	standard	protein, oil				
minor global crop	herb	standard	protein				
regional crop	succulent	standard		fiber	living fence		
regional crop	succulent	standard		fiber	living fence		
regional crop	succulent	standard		fiber	living fence		
regional crop only	tree	standard, coppice	balanced carb, protein	hydrocarbon		bank	
wild staple		standard	protein				
regional crop	tree	standard	balanced carb, protein	hydrocarbon			
regional crop	tree	coppice					
regional crop only	tree	standard	protein-oil				
regional crop	tree	coppice		biomass	living fence		
regional crop only	shrub	standard	protein-oil	starch, oil, tannin		pod	
regional crop	tree	coppice		dye			
global crop as annual, new perennial crop	shrub	standard, coppice	protein		nitrogen, alley crop, windbreak, crop shade	bank, insect	perennial fallow, homegarden
regional crop	palm vine	coppice					multistrata
minor global crop	tree	coppice		biomass	nitrogen, alley crop, contour hedgerow, windbreak, crop shade	bank, insect	
minor global crop	herb	fodder			nitrogen, understory legume	pasture	
minor global crop	herb	fodder			nitrogen, understory legume	bank	
regional crop	shrub	standard, coppice		hydrocarbon, fiber			
regional crop	shrub	standard, coppice		hydrocarbon, fiber			
new crop	tree	coppice		medicinal			
regional crop only	tree	standard	oil				
regional crop only	tree	standard	oil				
regional crop	tree	standard	oil				
regional crop only	tree	standard	oil				
regional crop	vine	standard	protein				
minor global crop	shrub	standard, coppice	protein	biomass	nitrogen, living fence, windbreak		
regional crop	shrub	standard, coppice			nitrogen, contour hedgerow, windbreak	bank	
regional crop	grass-like	hay		biomass			
regional crop	grass-like	hay		biomass			
minor global crops	tree	standard	oil				
regional crop only	tree	standard	oil				
global crop	tree	standard	oil				strip intercrop
regional crop only	tree	standard	oil				
regional crop	tree	standard	oil				
regional crop only	tree	standard	oil				
regional crop only	tree	standard	oil				
wild staple, experimental crop	tree	standard	oil				
experimental, very promising	tree	standard	protein-oil				
regional crop only	palm	standard	sugar	fiber			
regional crop only	tree	standard	sugar			pod	
regional crop only	tree	standard	sugar		living fence, crop shade	pod	

Genus	Species	Common Name	Family	Origin	Invasive	Climate	Humidity
Castanea	spp.	hybrid and neohybrid chestnuts	Fagaceae	hybrid origin		cold to warm temperate	humid
Castanea	crenata	Japanese chestnut	Fagaceae	Asia		cold to warm temperate	humid
Castanea	dentata	American chestnut	Fagaceae	North America		cold to warm temperate	humid
Castanea	mollissima	Chinese chestnut	Fagaceae	Asia		cold to warm temperate	humid
Castanea	sativa	European chestnut	Fagaceae	Europe		Mediterranean, warm temperate	semi-arid to humid
Castanopsis	acuminatissima		Fagaceae	Asia, New Guinea		tropical highlands	humid
Castanopsis	inermis	berangan	Fagaceae	Malacca, Sumatra		tropical	humid
Castanopsis	javanica		Fagaceae	Java		tropical	humid
Castanopsis	tribuloides		Fagaceae	SE Asia		tropical highlands	humid
Castanospermum	australe	Moreton Bay chestnut	Fabaceae	Australia, Oceania		subtropical to tropical	arid to humid
Castilla	elastica	Panama rubber tree	Moraceae	Americas	yes	tropical	humid
Castilla	ulei	hule rubber	Moraceae	Americas		tropical	humid
Casuarina	equisetifolia	Australian pine	Casuarinaceae	Australia	yes	subtropical to tropical	arid to humid
Casuarina	oligodon		Casuarinaceae	New Guinea		tropical highlands	humid
Catharanthus	roseus	Madagascar periwinkle	Apocynaceae	Madagascar		tropical	semi-arid to humid
Ceiba	aesculifolia	pochote	Malvaceae	Mesoamerica		tropical	semi-arid to humid
Ceiba	pentandra	kapok	Malvaceae	tropical Americas, Africa		tropical	semi-arid to humid
Centrosema	pubescens	centro	Fabaceae	Mesoamerica		tropical	semi-arid
Cephalostachyum	pergracile	clumping bamboo	Poaceae	Asia		subtropical to tropical	humid
Ceratonia	siliqua	carob	Fabaceae	North Africa, Mediterranean		Mediterranean, subtropical	arid to semi-arid
Cercidium	microphyllum	paloverde	Fabaceae	SW North America, Mesoamerica		subtropical	arid to semi-arid
Cercocarpus	ledifolius	curl-leaf mountain mahogany	Rosaceae	western North America		boreal to warm temperate	arid to semi-arid
Cereus	hildmannianus		Cactaceae	South America		subtropical to tropical	arid to semi-arid
Cereus	repandus	Peruvian apple cactus	Cactaceae	South America		subtropical to tropical	arid to semi-arid
Ceroxylon	alpinum	Andean waxpalm	Arecaceae	South America		tropical, Mediterranean	humid
Chamaecrista	rotundifolia		Fabaceae	Mesoamerica		tropical, tropical highlands	semi-arid
Chamaerops	humilis	dwarf fan palm	Arecaceae	Mediterranean		Mediterranean	semi-arid
Chenopodium	spp.	perennial quinoa	Amaranthaceae			cold temperate to subtropical, tropical highlands	humid
Chenopodium	bonus-henricus	good King Henry	Amaranthaceae	Europe		cold to warm temperate	humid
Chiococca	alba	West Indian milkberry	Rubiaceae	tropical Americas		subtropical to tropical	semi-arid to humid
Chondrilla	ambigua	chondrilla	Compositae	northern middle Asia		cold to warm temperate	semi-arid to humid
Chrysobalanus	icaco	cocoplum	Chrysobalanaceae	Americas	yes	tropical	humid
Chrysopogon	zizanioides	vetiver	Poaceae	India		warm temperate to tropical	semi-arid to humid
Cinchona	officinalis	quinine	Rubiaceae	South America		tropical	humid
Cinchona	pubescens	red cinchona	Rubiaceae	tropical Americas	top 100	tropical, tropical highlands	humid
Cinnamomum	camphora	camphor	Lauraceae	Asia	yes	tropical	humid
Cistus	ladanifer	rockrose	Cistaceae	North Africa, Mediterranean		warm temperate to subtropical	semi-arid
Citrullus	colocynthis	perennial egusi	Cucurbitaceae	Africa through South Asia		subtropical to tropical	arid to semi-arid
Clappertonia	ficifolia	bolo bolo	Malvaceae	Africa		subtropical to tropical	humid
Clitoria	ternatea	butterfly pea	Fabaceae	Asia		tropical	semi-arid to humid
Cnidoscolus	aconitifolius	chaya	Euphorbiaceae	Mesoamerica		tropical, tropical highlands	arid to humid
Cnidoscolus	elasticus	highland chilte	Euphorbiaceae	Mesoamerica		tropical highlands	arid to semi-arid
Cocos	nucifera	coconut palm	Arecaceae	Asia		tropical	semi-arid to humid
Coix	lacryma-jobi	Job's tears	Poaceae	Asia		subtropical to tropical	humid

Cultivation	Form	Management	Staple Crop	Industrial Crop	Agroforestry Services	Fodder	Other Systems
new crop	tree	standard, coppice	balanced carb	biomass, tannin			woody agriculture
regional crop only	tree	standard	balanced carb	tannin			
historic regional crop, disease-resistant under development	tree, shrub	standard	balanced carb	tannin			
minor global crop	tree	standard	balanced carb	biomass, tannin			strip intercrop
global crop	tree	standard	balanced carb	biomass, tannin			
regional crop only	tree	standard	balanced carb				
regional crop only	tree	standard	balanced carb				
regional crop only	tree	standard	balanced carb				
regional crop only	tree	standard	balanced carb				
regional crop	tree	standard		starch			
regional crop	tree	standard		hydrocarbon	crop shade		
regional crop	tree	standard		hydrocarbon			
minor global crop	tree	standard		tannin	nitrogen, windbreak		
minor global crop	tree	standard			nitrogen		
hypothetical	herb	hay					
regional crop	tree	standard		fiber			
minor global crop	tree	standard		oil, fiber			
minor global crop	herb	fodder			nitrogen, understory legume	pasture	
regional crop	bamboo	managed multistem		biomass			
minor global crop	tree	standard	balanced carb, sugar	gum	windbreak	pod	irreg. intercrop
wild staple	tree	standard	protein				
regional crop	tree	standard			nitrogen	bank	
regional crop	cactus	standard			living fence		
regional crop	cactus	standard			living fence		
regional crop	palm	standard		wax			
minor global crop	shrub	standard, coppice			nitrogen, understory legume	pasture	
regional crop	palm	standard		fiber			
hypothetical	herb	standard					
historic grain crop	herb	standard	balanced carb				
experimental	shrub	coppice		hydrocarbon			
experimental	herb	hay		hydrocarbon			
regional crop only	shrub	standard	protein-oil		living fence		
minor global crop	grass	standard, hay		biomass	contour hedgerow		
minor global crop	tree	coppice		medicinal			
minor global crop	tree	coppice		medicinal			
minor global crop	tree	coppice			windbreak, crop shade		
regional crop	shrub	coppice		hydrocarbon			
regional crop only	herb	standard	protein-oil				strip intercrop
regional crop	shrub	coppice		fiber			
minor global crop	vine	hay	protein		nitrogen		
minor global crop	shrub	standard, coppice	protein		living fence, crop shade		homegarden
experimental	shrub	coppice		hydrocarbon			
global crop	palm	standard	oil, sugar	fiber	crop shade		homegarden, multistrata
hypothetical	grass	standard	balanced carb			bank	

Genus	Species	Common Name	Family	Origin	Invasive	Climate	Humidity
Colophospermum	mopane	mopane	Fabaceae	Africa		subtropical to tropical	arid to semi-arid
Copaifera	langsdorffii	diesel tree	Fabaceae	Amazonia		tropical	humid
Copaifera	multijuga		Fabaceae	South America		tropical	humid
Copaifera	officinalis		Fabaceae	Americas		tropical	humid
Copernicia	prunifera	carnauba palm	Arecaceae	South America		tropical	semi-arid to humid
Cordeauxia	edulis	yeheb	Fabaceae	Africa		tropical	arid to semi-arid
Cordyline	australis	dracaena palm	Asparagaceae	New Zealand		warm temperate to subtropical	humid
Cornus	sericea	silky dogwood	Cornaceae	North America		boreal to warm temperate	semi-arid to humid
Cortaderia	selloana	pampas grass	Poaceae	South America		warm temperate to tropical	semi-arid to humid
Corylus	hybrids & neohybrids	hybrid and neohybrid hazel	Betulaceae	hybrid		boreal to warm temperate, Mediterranean	semi-arid to humid
Corylus	avellana	hazelnut	Betulaceae	Europe		warm temperate, Mediterranean	humid
Corylus	colurna	Turkish tree hazel	Betulaceae	Central Eurasia		warm temperate, Mediterranean	semi-arid to humid
Corylus	heterophylla	Siberian hazel	Betulaceae	NE Asia		boreal to warm temperate	semi-arid to humid
Corylus	maxima	giant filbert	Betulaceae	Central Eurasia		cold to warm temperate, Mediterranean	semi-arid to humid
Corylus	sieboldiana	Manchurian hazel	Betulaceae	NE Asia		boreal to warm temperate	humid
Corymbia	citriodora	lemon-scented gum	Myrtaceae	Australia		subtropical to tropical, tropical highlands	semi-arid to humid
Corynocarpus	spp.	karaka	Corynocarpaceae	New Zealand		warm temperate to subtropical	humid
Corypha	utan	buri palm	Arecaceae	Asia		tropical	humid
Cotinus	coggygria	smoketree	Anacardiaceae	North America	yes	cold to warm temperate	humid
Coula	edulis	African walnut	Olacaceae	Africa		tropical	humid
Couma	macrocarpa	leche caspi	Apocynaceae	Amazonia		tropical	humid
Couma	utilis	sorva	Apocynaceae	South America		tropical	humid
Crataegus	pinnatifida	Chinese hawthorn	Rosaceae	East Asia		cold to warm temperate	semi-arid to humid
Crescentia	alata	jícaro	Bignoniaceae	Mesoamerica		tropical	semi-arid
Crescentia	cujete	calabash tree	Bignoniaceae	Mesoamerica		tropical	semi-arid
Croton	tiglium	croton oil plant	Euphorbiaceae	South & SE Asia		subtropials to tropical	arid to humid
Cryptostegia	grandiflora	India rubber vine	Apocynaceae	Madagascar	yes	tropical	humid
Cucurbita	digitata	coyote gourd	Cucurbitaceae	SW North America, Mesoamerica		subtropical	arid to semi-arid
Cucurbita	ficifolia	figleaf gourd	Cucurbitaceae	Mesoamerica		tropical highlands	humid
Cucurbita	foetidissima	buffalo gourd	Cucurbitaceae	western North America		cold temperate to subtropical	arid to semi-arid
Cuphea	micropetala	achanclan	Lythraceae	Mesoamerica		warm temperate to tropical	humid
Cycas	spp.		Cycadaceae	Africa, Asia		warm temperate to tropical	arid to humid
Cylindropuntia	spp.	cholla	Cactaceae	Mesoamerica		cold temperate to tropical	arid to semi-arid
Cynara	cardunculus	cardoon	Asteraceae	Mediterranean, North Africa	yes	warm temperate, Mediterranean	semi-arid
Cyperus	cephalotes		Cyperaceae	South Asia through Australia		warm temperate to tropical	aquatic
Cyperus	giganteus	capim de esteira	Cyperaceae	Mexico to Argentina		warm temperate to tropical	aquatic
Cyperus	papyrus	papyrus	Cyperaceae	North Africa		Mediterranean, subtropical	aquatic
Cyperus	textilis		Cyperaceae	South Africa		Mediterranean, warm temperate	aquatic
Cytisus	proliferus	tagasaste	Fabaceae	Canary Islands		Mediterranean, warm temperate	semi-arid
Cytisus	scoparius	Scotch broom	Fabaceae	Europe		cold to warm temperate, Mediterranean	semi-arid to humid
Dacryodes	edulis	safou, Africado, butter fruit	Burseracaeae	West Africa		tropical	semi-arid to humid
Daniellia	ogea		Fabaceae	Africa		tropical	humid
Daniellia	oliveri	African copaifera	Fabaceae	Africa		tropical	humid
Dendrocalamus	asper	clumping bamboo	Poaceae	Asia		subtropical to tropical	semi-arid to humid
Dendrocalamus	brandisii	clumping bamboo	Poaceae	Asia		subtropical to tropical	humid

Cultivation	Form	Management	Staple Crop	Industrial Crop	Agroforestry Services	Fodder	Other Systems
new crop	tree	standard, coppice		biomass		bank, insect	
new crop	tree	standard		hydrocarbon, medicinal			
regional resin crop	tree	standard		hydrocarbon, medicinal			
regional crop	tree	standard		hydrocarbon, medicinal			
regional crop	palm	standard		wax			
wild staple, under development	shrub	standard	balanced carb		nitrogen	bank	
minor global crop	succulent	standard		fiber			
regional crop	shrub	coppice		biomass	living fence	bank	
minor global crop	grass	hay		biomass	windbreak		
new crop	shrub	standard, coppice	protein-oil				woody agriculture
global crop	shrub	standard, coppice	protein-oil	biomass			
regional crop only	tree	standard	protein-oil				
regional crop only	shrub	standard, coppice	protein-oil				strip intercrop
regional crop only	tree	standard, coppice	protein-oil				
regional crop only	shrub	standard, coppice	protein-oil				
minor global crop	tree	standard, coppice		biomass			SRC
wild	tree	standard		starch			
regional crop only	palm	standard	sugar	fiber			
wild	tree	coppice		dye, tannin			
regional crop	tree	standard					
experimental	tree	coppice		hydrocarbon			
regional crop	herb	hay		hydrocarbon			
minor global crop	tree	standard		medicinal			
new crop	tree	standard	protein-oil				
new crop	tree	standard	protein-oil				
regional crop	tree	standard		oil, hydrocarbon			
minor global crop	vine	coppice		hydrocarbon, fiber			
new crop	herb	standard	protein-oil				
regional crop only	herb	standard	protein-oil				
new crop	herb	standard	protein-oil				
regional crop	herb	standard		oil			
hypothetical	cycad	standard		starch			
minor global crop	cactus	standard			living fence		
minor global vegetables	herb	standard, hay	oil	biomass			
regional crop	grass-like	hay		biomass			
regional crop	grass-like	hay		biomass			
regional crop, historic	grass-like	hay		biomass			
regional crop	grass-like	hay		biomass			
minor global crop	shrub	coppice			nitrogen, alley crop	bank	
minor global crop	shrub	coppice			nitrogen		
regional crop	tree	standard	protein-oil				homegarden, multistrata
regional crop	tree	standard		hydrocarbon			
regional crop	tree	standard		hydrocarbon			
minor global crop	bamboo	managed multistem		biomass		bank	strip intercrop, multistrata
minor global crop	bamboo	managed multistem		biomass			strip intercrop, multistrata

Genus	Species	Common Name	Family	Origin	Invasive	Climate	Humidity
Dendrocalamus	giganteus	clumping bamboo	Poaceae	Asia		subtropical to tropical	humid
Dendrocalamus	hamiltonii	clumping bamboo	Poaceae	Asia		subtropical to tropical	humid
Dendrocalamus	hookeri	clumping bamboo	Poaceae	Asia		subtropical to tropical	humid
Dendrocalamus	latiflorus	clumping bamboo	Poaceae	Asia		subtropical to tropical	humid
Dendrocalamus	membranaceus	clumping bamboo	Poaceae	Asia		subtropical to tropical	humid
Dendrocalamus	strictus	clumping bamboo	Poaceae	Asia		subtropical to tropical, tropical highlands	humid
Desmanthus	illinoensis	Illinois bundleflower	Fabaceae	North America		cold to warm temperate	semi-arid to humid
Desmodium	cinereum	rensoni	Fabaceae	Asia		tropical highlands	humid
Desmodium	intortum	pega-pega	Fabaceae	Mesoamerica		tropical, tropical highlands	semi-arid to humid
Desmodium	paniculatum	panicled tick trefoil	Fabaceae	eastern North America		cold to warm temperate	semi-arid to humid
Desmodium	triflorum		Fabaceae	Asia		tropical	humid
Desmodium	uncinatum		Fabaceae	Americas		tropical	humid
Dioscorea	bulbifera	air potato	Dioscoreaceae	Asia, Africa, Australia	yes	subtropical to tropical, tropical highlands	humid
Dioscorea	polystachya		Dioscoreaceae	East Asia	yes	cold to warm temperate	humid
Dipterocarpus	alatus		Dipterocarpaceae	SE Asia		tropical	humid
Dipterocarpus	gracilis		Dipterocarpaceae	India & SE Asia		tropical	humid
Dipterocarpus	grandiflorus		Dipterocarpaceae	India & SE Asia		tropical	humid
Dipterocarpus	kerrii		Dipterocarpaceae	India to SE Asia		tropical	humid
Distichlis	palmeri	nipa	Poaceae	Mesoamerica		tropical	humid, aquatic
Duboisia	myoporoides	corkwood	Solanaceae	Australia		tropical	humid
Dyera	costulata	lutong	Apocynaceae	Southeast Asia		tropical	humid
Echinochloa	polystachya	aleman grass	Poaceae	Americas		warm temperate to tropical	humid
Edgeworthia	tomentosa	paper bush	Thymelaeaceae	Asia		warm temperate to subtropical	humid
Elaeagnus	angustifolia	Russian olive	Elaeagnaceae	Eurasia	yes	boreal to warm temperate	semi-arid
Elaeagnus	rhamnoides	sea buckthorn, seaberry	Elaeagnaceae	Eurasia		boreal to warm temperate	semi-arid
Elaeagnus	umbellata	autumn olive	Elaeagnaceae	Asia	yes	cold to warm temperate	semi-arid to humid
Elaeis	guineensis	African oil palm	Arecaceae	West Africa	yes	tropical	humid
Elaeis	oleifera	American oil palm	Arecaceae	tropical Americas		tropical	humid
Elymus	hispidus	Kernza	Poaceae	Asia		cold to warm temperate	semi-arid to humid
Encephalartos	spp.		Zamiaceae	Africa		warm temperate to tropical	arid to humid
Ensete	ventricosum	enset	Musaceae	Africa		tropical highlands	humid
Entada	phaseoloides	gogo bean	Fabaceae	Africa & Asia		tropical	humid
Enterolobium	cyclocarpum	guanacaste	Fabaceae	Mesoamerica		subtropical to tropical	semi-arid to humid
Eremophila	fraseri	burra	Scrophulariaceae	West Australia		warm temperate to subtropical	arid to semi-arid
Erharta	stipoides	weeping rice grass	Poaceae	Australia		Mediterranean, warm temperate	semi-arid
Ericameria	nauseosa	rubber rabbitbrush	Asteraceae	western North America		boreal to warm temperate	arid to semi-arid
Erythrina	edulis	chachafruto	Fabaceae	Andes		tropical highlands	semi-arid to humid
Erythrina	poeppigiana		Fabaceae	South America		tropical	humid
Erythrina	sandwicensis	wiliwili	Fabaceae	Hawaii		tropical	semi-arid
Erythrina	subumbrans	coral beans	Fabaceae	Asia		tropical	humid
Escontria	chiotilla		Cactaceae	Mesoamerica		tropical highlands	arid to semi-arid
Eucalyptus	brassiana	Cape York red gum	Myrtaceae	Australia, New Guinea		tropical	humid
Eucalyptus	camaldulensis	Red River gum	Myrtaceae	Australia		subtropical to tropical	arid to humid

Cultivation	Form	Management	Staple Crop	Industrial Crop	Agroforestry Services	Fodder	Other Systems
minor global crop	bamboo	managed multstem		biomass		bank	strip intercrop, multistrata
minor global crop	bamboo	managed multstem		biomass			strip intercrop, multistrata
minor global crop	bamboo	managed multstem		biomass			strip intercrop, multistrata
minor global crop	bamboo	managed multstem		biomass			strip intercrop, multistrata
minor global crop	bamboo	managed multstem		biomass			strip intercrop, multistrata
minor global crop	bamboo	managed multstem		biomass			strip intercrop, multistrata
under development	herb	standard	protein		nitrogen		
minor global crop	herb	fodder			nitrogen, contour hedgerow, understory legume	bank	
minor global crop	herb	fodder			nitrogen, understory legume	pasture	
wild	herb	fodder			nitrogen, understory legume	bank, pasture	
minor global crop	herb	fodder			nitrogen, understory legume	pasture	
minor global crop	herb	fodder			nitrogen, understory legume	pasture	
minor global crop	vine	standard	basic starch	starch			
regional crop	vine	standard	basic starch				
wild	tree	standard		hydrocarbon			
wild	tree	standard		hydrocarbon			
wild	tree	standard		hydrocarbon			
wild	tree	standard		hydrocarbon			
wild staple, under development	grass	standard, fodder	balanced carb				
regional crop	tree	standard		pesticide, medicinal			
regional crop	tree	standard		hydrocarbon			
new crop	grass	hay		biomass			
regional crop	shrub	coppice		fiber			
minor global crop	tree	standard, coppice		biomass	nitrogen, windbreak	bank	
minor global crop	shrub	standard, coppice	oil	biomass	nitrogen, alley crop, windbreak	bank	strip intercrop
minor global crop	shrub	standard, coppice		biomass	nitrogen, windbreak	bank	
global crop	palm	standard	oil	fiber			parkland, multistrata
regional crop only	palm	standard	oil				
under development	grass	standard	balanced carb				
hypothetical	cycad	standard		starch			
regional crop only	giant herb	managed multistem	basic starch	fiber			
regional crop only	woody vine	standard	protein-oil		nitrogen		
regional crop	tree	standard			nitrogen, evergreen ag	pod, bank	
wild	herb	hay		hydrocarbon			
under development	grass	standard	balanced carb				
experimental	shrub	coppice		hydrocarbon			
regional crop only	tree	standard	protein		nitrogen, living fence	pod, bank	multistrata
minor global crop	tree	standard, coppice			nitrogen, alley crop, living fence, crop shade	bank	
regional crop	tree	standard, coppice			nitrogen, living fence, windbreak, crop shade, living trellis	bank	
regional crop	tree	standard, coppice			nitrogen, living fence, crop shade, living trellis	bank	
regional crop	cactus	standard			living fence	bank	
minor global crop	tree	standard, coppice		biomass			SRC
global crop	tree	standard, coppice		biomass, tannin	windbreak		SRC

Genus	Species	Common Name	Family	Origin	Invasive	Climate	Humidity
Eucalyptus	*globulus*	bluegum	Myrtaceae	Australia		warm temperate to subtropical	arid to humid
Eucalyptus	*gomphocephala*	tuart	Myrtaceae	Australia		Mediterranean, subtropical to tropical, tropical highlands	humid
Eucalyptus	*grandis*	flooded gum	Myrtaceae	Australia		subtropical to tropical, tropical highlands	humid
Eucalyptus	*microtheca*	coolabah	Myrtaceae	Australia		subtropical to tropical	arid to semi-arid
Eucalyptus	*tereticornis*	forest red gum	Myrtaceae	Australia, New Guinea		subtropical to tropical	semi-arid to humid
Eucalyptus	*urophylla*	ampupu	Myrtaceae	Indonesia		tropical, tropical highlands	humid
Eucommia	*ulmoides*	Chinese rubber tree	Eucommiaceae	Asia		cold to warm temperate	semi-arid to humid
Eugeissona	*utilis*	nanga palm	Arecaceae	Asia		tropical	humid
Eulaliopsis	*binata*	sabai grass	Poaceae	South to SE Asia		subtropical to tropical, tropical highlands	humid
Euphorbia	*abyssinica*	desert candle	Euphorbiaceae	Africa		warm temperate to tropical	arid to semi-arid
Euphorbia	*antisyphilitica*	candelillia	Euphorbiaceae	Mesoamerica, North America		warm temperate to subtropical	arid to semi-arid
Euphorbia	*esula*	leafy spurge	Euphorbiaceae	Eurasia	yes	boreal to subtropical	arid to humid
Euphorbia	*intisy*	intisy	Euphorbiaceae	Madagascar		tropical	semi-arid
Euphorbia	*lactea*		Euphorbiaceae	Africa		tropical	arid to humid
Euphorbia	*neriifolia*		Euphorbiaceae	South & SE Asia		tropical	semi-arid to humid
Euphorbia	*tirucalli*	petroleum plant	Euphorbiaceae	Africa		tropical	arid to semi-arid
Euryale	*ferox*	gorgon fruit	Nymphaeaceae	Asia		tropical	aquatic
Fagopyrum	spp.	perennial buckwheat	Polygonaceae	Eurasia		boreal to cold temperate	arid to semi-arid
Faidherbia	*albida*	apple-ring acacia	Fabaceae	Africa, Middle East		tropical	semi-arid to humid
Ficus	*annulata*	panggang	Moraceae	India, Southeast Asia		tropical	humid
Ficus	*elastica*	India rubber tree	Moraceae	Southeast Asia		tropical	humid
Ficus	*racemosa*	cluster fig	Moraceae	India, Southeast Asia, Australia		tropical, tropical highlands	semi-arid to humid
Firmiana	*simplex*	gum karaya	Malvaceae	South Asia		subtropical	semi-arid
Flemingia	*macrophylla*	flemingia	Fabaceae	Asia		tropical, tropical highlands	humid
Fouquieria	*splendens*	ocotillo	Fouquieriaceae	Mesoamerica		warm temperate to tropical	arid to semi-arid
Funtumia	*elastica*	West African rubber tree	Apocynaceae	West Africa		tropical	humid
Furcraea	*acaulis*	cocuiza	Asparagaceae	South America		tropical highlands	semi-arid to humid
Furcraea	*andina*	pacpa	Asparagaceae	South America		tropical highlands	semi-arid to humid
Furcraea	*cabuya*	cabuya	Asparagaceae	tropical Americas		tropical highlands	semi-arid to humid
Furcraea	*foetida*	Mauritius hemp	Asparagaceae	tropical Americas	yes	tropical highlands	semi-arid to humid
Furcraea	*hexapetala*	fique	Asparagaceae	tropical Americas		tropical highlands	semi-arid to humid
Garcinia	*hanburyi*	Siam gamboge	Clusiaceae	SE Asia		tropical	humid
Garcinia	*xanthochymus*	gamboge	Clusiaceae	SE Asia		subtropical to tropical, tropical highlands	humid
Gevuina	*avellana*	Chilean hazelnut	Proteaceae	Chile		warm temperate, Mediterranean	semi-arid to humid
Gigantochloa	*albociliata*	clumping bamboo	Poaceae	Asia		subtropical to tropical	humid
Gigantochloa	*apus*	clumping bamboo	Poaceae	Asia		subtropical to tropical	humid
Gigantochloa	*atroviolacea*	clumping bamboo	Poaceae	Asia		subtropical to tropical	humid
Gigantochloa	*balui*	clumping bamboo	Poaceae	Asia		subtropical to tropical	humid
Gigantochloa	*hasskarliana*	clumping bamboo	Poaceae	Asia		subtropical to tropical	humid
Gigantochloa	*levis*	clumping bamboo	Poaceae	Asia		subtropical to tropical	humid
Gigantochloa	*verticillata*	clumping bamboo	Poaceae	Asia		subtropical to tropical	semi-arid to humid
Ginkgo	*biloba*	ginkgo	Ginkgoaceae	Asia		cold temperate to subtropical	humid
Gleditsia	*triacanthos*	honey locust	Fabaceae	North America		boreal to warm temperate, Mediterranean	semi-arid to humid
Gliricidia	*sepium*	madre de cacao	Fabaceae	Mesoamerica		tropical	semi-arid to humid

Cultivation	Form	Management	Staple Crop	Industrial Crop	Agroforestry Services	Fodder	Other Systems
minor global crop	tree	standard, coppice		hydrocarbon, medicinal			SRC
minor global crop	tree	standard, coppice		biomass			SRC
global crop	tree	standard, coppice		biomass			SRC
global crop	tree	standard, coppice		biomass			SRC
minor global crop	tree	standard, coppice		biomass			SRC
minor global crop	tree	standard, coppice		biomass			SRC
regional medicinal crop	tree	coppice		hydrocarbon			
regional crop only	palm	managed multistem	basic starch				
regional crop	grass	hay		biomass			
experimental	shrub	coppice		hydrocarbon			
experimental	shrub	coppice		wax			
experimental	herb	hay					
regional crop	shrub	coppice					
minor global crop	shrub	coppice			living fence		
minor global crop	shrub	coppice			living fence		
minor global crop	shrub	coppice		biomass, hydrocarbon	living fence		
regional crop only	herb	standard	balanced carb				
hypothetical	herb	standard	balanced carb				
regional crop	tree	standard	balanced carb		nitrogen	pod, bank	evergreen ag, parkland, FMNR
regional crop	tree	standard		hydrocarbon			
minor global crop	tree	standard		hydrocarbon		insect	
regional crop	tree	standard		hydrocarbon	crop shade	insect	
regional crop	tree	standard		fiber, gum			
minor global crop	shrub	coppice			nitrogen, alley crop, contour hedgerow	bank	
experimental	shrub	standard					
regional crop	tree	standard		hydrocarbon, fiber			
regional crop	succulent	standard		fiber	living fence		
regional crop	succulent	standard		fiber	living fence		
regional crop	succulent	standard		fiber	living fence		
regional crop	succulent	standard		fiber	living fence		
regional crop	succulent	standard		fiber	living fence		
regional crop	tree	standard		hydrocarbon			
regional crop	tree	standard		hydrocarbon			
new crop	tree	standard	oil				
regional crop	bamboo	managed multistem		biomass		bank	
regional crop	bamboo	managed multistem		biomass			
regional crop	bamboo	managed multistem		biomass			
regional crop	bamboo	managed multistem		biomass			
regional crop	bamboo	managed multistem		biomass			
regional crop	bamboo	managed multistem		biomass			
regional crop	bamboo	managed multistem		biomass		bank	
minor global crop	tree	standard, coppice	balanced carb	medicinal			
minor global crop	tree	standard, coppice	balanced carb	biomass	living fence	pod, bank	
minor global crop	tree	standard, coppice		biomass	nitrogen, alley crop, contour hedgerow, living fence, windbreak, crop shade, living trellis	bank	

Genus	Species	Common Name	Family	Origin	Invasive	Climate	Humidity
Glyceria	fluitans	manna grass	Poaceae	Eurasia, Africa		boreal to warm temperate, Mediterranean	humid
Glycine	spp.	perennial soybean	Fabaceae	cultivated origin		hypothetical	semi-arid to humid
Gnetum	africanum	eru	Gnetaceae	Africa		tropical	humid
Gnetum	buchholzianum	eru	Gnetaceae	Africa		tropical	humid
Gnetum	gnemon	jointfir, bago	Gnetaceae	Asia		tropical	humid
Gossypium	arboreum ssp. burmanicum	tree cotton	Malvaceae	South & SE Asia		tropical	semi-arid to humid
Gossypium	arboreum ssp. indicum	tree cotton	Malvaceae	Old World tropics		tropical	semi-arid to humid
Gossypium	arboreum ssp. soudanense	tree cotton	Malvaceae	India		tropical	semi-arid to humid
Gossypium	barbadense var. brasiliense	Sea Island cotton	Malvaceae	South America		tropical	semi-arid to humid
Gossypium	barbadense f. darwinii	Sea Island cotton	Malvaceae	Galapagos Islands		tropical	semi-arid to humid
Gossypium	herbaceum var. acerifolium	Levant cotton	Malvaceae	Ethiopia		tropical	semi-arid to humid
Gossypium	herbaceum ssp. africanum	Levant cotton	Malvaceae	South Africa		tropical	semi-arid to humid
Gossypium	hirsutum var. marie-galante	upland cotton	Malvaceae	tropical Americas		tropical	arid to humid
Gossypium	hirsutum ssp. punctatum	upland cotton	Malvaceae	tropical Americas		tropical	arid to humid
Gossypium	hirsutum var. taitense	upland cotton	Malvaceae	Pacific, Australia		tropical	arid to humid
Grindelia	hirsutula	gumweed	Asteraceae	western North America		cold to warm temperate	arid to semi-arid
Grindelia	squarrosa	gumweed	Asteraceae	western North America		cold to warm temperate	arid to semi-arid
Guadua	angustifolia	clumping bamboo	Poaceae	South America		subtropical to tropical	humid
Guazuma	ulmifolia	guasimo	Malvaceae	tropical Americas		tropical	humid
Guibourtia	coleosperma	African rosewood	Fabaceae	Africa, South Asia		subtropical to tropical	semi-arid
Guibourtia	copallifera	Sierra Leone gum copal	Fabaceae	Africa—Congo		tropical	humid
Gustavia	superba	membrillo	Lecythidaceae	tropical Americas		tropical	humid
Gynerium	sagittatum	wild cane	Poaceae	tropical Americas		tropical	humid
Haloxylon	spp.	saxaul	Amaranthaceae	Central Asia		cold to warm temperate	arid to semi-arid
Hancornia	speciosa	mangaba rubber tree	Apocynaceae	eastern South America		tropical	semi-arid to humid
Helianthus	hybrids	perennial sunflower	Asteraceae	hybrid		boreal to warm temperate	semi-arid to humid
Helianthus	cusickii	Cusick's sunflower	Asteraceae	North America		cold to warm temperate	arid to semi-arid
Helianthus	maximiliani	Maximilian sunflower	Asteraceae	North America		boreal to warm temperate	semi-arid to humid
Helianthus	pauciflorus	stiff sunflower	Asteraceae	North America		boreal to warm temperate	semi-arid to humid
Helianthus	tuberosus	sunchoke	Asteraceae	North America		tropical highlands	semi-arid to humid
Hesperaloe	funifera	New Mexico false yucca	Asparagaceae	North America, Mesoamerica		warm temperate to subtropical	arid to semi-arid
Hesperoyucca	whipplei	chapparal yucca	Asparagaceae	North America, Mesoamerica		Mediterranean, subtropical	arid to semi-arid
Hevea	benthamiana		Euphorbiaceae	South America		tropical	humid
Hevea	brasiliensis	Para rubber	Euphorbiaceae	South America		tropical	humid
Hibiscus	radiatus	clavelina	Malvaceae	Asia		subtropical to tropical	humid
Hibiscus	tilliaceus	beach hibiscus	Malvaceae	pantropical		subtropical to tropical	humid
Hordeum	vulgare	perennial barley	Poaceae	under development		boreal to warm temperate, Mediterranean	semi-arid to humid
Hymenaea	courbaril	stinking toe	Fabaceae	tropical Americas		tropical	humid

Cultivation	Form	Management	Staple Crop	Industrial Crop	Agroforestry Services	Fodder	Other Systems
historic staple	grass	standard	balanced carb				
hypothetical	herb	standard	protein-oil				
new crop, important wild for leaves	vine	coppice	balanced carb				
wild staple, new crop	vine	coppice					
regional crop	tree	standard	balanced carb	fiber			
regional crop	shrub	standard, coppice	oil	fiber			
regional crop	shrub	standard, coppice	oil	fiber			
regional crop	shrub	standard, coppice	oil	fiber			
regional crop	shrub	standard, coppice	oil	fiber			
wild	shrub	standard, coppice	oil	fiber			
regional crop	shrub	standard, coppice	oil	fiber			
wild	shrub	standard, coppice	oil	fiber			
regional crop	shrub	standard, coppice	oil	fiber			strip intercrop
regional crop	shrub	standard, coppice	oil	fiber			
regional crop	shrub	standard, coppice	oil	fiber			
experimental	herb	hay		hydrocarbon			
wild	herb	hay		hydrocarbon			
regional crop	bamboo	managed multistem		biomass		bank	
regional crop	tree	standard, coppice			living fence	mast, bank	
wild staple	tree	standard	protein-oil				
regional crop	tree	standard		hydrocarbon, medicinal			
regional crop only	tree	standard	oil				
regional crop	giant grass	hay		biomass		bank	
regional crop	shrub	coppice				bank	
regional crop	tree	standard		hydrocarbon			
under development	herb	standard	protein-oil				
under development	herb	standard	protein-oil				
under development as seed crop	herb	standard	protein-oil				
under development	herb	standard	protein-oil				
minor global tuber crop	herb	standard	protein-oil				
under development	succulent	standard		fiber			
regional crop	succulent	standard		fiber			
regional crop	tree	standard		hydrocarbon			
global crop	tree	standard		oil, hydrocarbon	crop shade		strip intercrop, multistrata
minor global crop	shrub	coppice		fiber			
minor global crop	tree	coppice		biomass, fiber	living fence, crop shade, living trellis		
under development	grass		balanced carb				
regional crop	tree	standard		hydrocarbon	nitrogen		

Genus	Species	Common Name	Family	Origin	Invasive	Climate	Humidity
Hymenaea	*verrucosa*		Fabaceae	Africa		tropical	humid
Hyphaene	*thebaica*	doum palm	Arecaceae	Africa		warm temperate to tropical, Mediterranean	semi-arid to humid
Indigofera	*suffruticosa*	indigo	Fabaceae	tropical America		subtropical to tropical	humid
Indigofera	*tinctoria*	indigo	Fabaceae	India		subtropical to tropical	humid
Inga	*edulis*	ice cream bean	Fabaceae	South America		tropical	semi-arid to humid
Inga	*vera*	ice cream bean	Fabaceae	Caribbean		tropical	humid
Inocarpus	*fagifer*	Tahitian chestnut	Fabaceae	Oceania		tropical	humid
Irvingia	*gabonensis*	dika nut	Irvingiaceae	West Africa		tropical	humid
Iryanthera	*laevis*	cumala	Myristicaceae	South America		tropical	humid
Jatropha	*curcas*	jatropha	Euphorbiaceae	tropical Americas		tropical	semi-arid to humid
Jubaea	*chilensis*	Chilean wine palm	Arecaceae	Chile		warm temperate to tropical, Mediterranean	semi-arid
Juglans	*ailanthifolia var. cordiformis*	heartnut	Juglandaceae	East Asia		cold to warm temperate	humid
Juglans	*cinerea*	butternut	Juglandaceae	eastern North America		boreal to warm temperate	humid
Juglans	*neotropica*	Andean walnut	Juglandaceae	South America		subtropical, tropical highlands	humid
Juglans	*nigra*	black walnut	Juglandaceae	eastern North America		cold to warm temperate	humid
Juglans	*olanchana*	nogal	Juglandaceae	tropical Americas		tropical	humid
Juglans	*regia*	Persian walnut	Juglandaceae	Central Asia		warm temperate, Mediterranean	semi-arid to humid
Juncus	*effusus*	mat rush	Juncaceae	East Asia, North America		cold to warm temperate	aquatic
Lablab	*purpureus*	lablab bean	Fabaceae	Africa		tropical, tropical highlands	semi-arid to humid
Landolphia	*heudelotii*	landolphia rubber	Apocynaceae	West Africa		tropical	humid
Landolphia	*kirkii*	Zanzibar rubber	Apocynaceae	East Africa		tropical	humid
Larix	*occidentalis*	western larch	Pinaceae	North America		boreal to warm temperate	humid
Larrea	*tridentata*	creosotebush	Zygophyllaceae	SW North America & Mesoamerica		warm temperate to subtropical	arid to semi-arid
Lawsonia	*inermis*	henna	Lythraceae	Africa, South Asia through Australia		tropical	semi-arid
Lecythis	*corrugata*	guacharaco	Lecythidaceae	Amazonia		tropical	humid
Lecythis	*minor*	coco de mono	Lecythidaceae	northern South America		tropical	humid
Lecythis	*ollaria*	sabucaia	Lecythidaceae	Venezuela		tropical	humid
Lecythis	*pisonis*	castanha e sapucaia	Lecythidaceae	Amazonia		tropical, tropical highlands	humid
Lecythis	*zabucajo*	monkey pot	Lecythidaceae	northern South America		tropical	humid
Leopoldinia	*piassaba*	piassava palm	Arecaceae	tropical South America		tropical	humid
Lepidium	spp.	perennial lepidium	Brassicaceae	northern temperate		boreal to warm temperate	semi-arid to humid
Lepironia	*articulata*	cicao	Cyperaceae	South Asia to Australia, Pacific		tropical	aquatic
Lespedeza	*bicolor*	bush clover	Fabaceae	NE Asia		cold to warm temperate	semi-arid to humid
Lesquerella	hybrids	perennial lesquerella	Brassicaceae	SW North America		warm temperate to subtropical	semi-arid
Leucaena	*diversifiolia*	ipil-ipil	Fabaceae	Central America		tropical highlands	humid
Leucaena	*leucocephala*	leucaena	Fabaceae	tropical Americas	yes	tropical	semi-arid to humid
Leymus	*arenarius*	beach wildrye	Poaceae	standard		boreal to warm temperate	semi-arid to humid

Cultivation	Form	Management	Staple Crop	Industrial Crop	Agroforestry Services	Fodder	Other Systems
regional crop	grass	standard		hydrocarbon			
regional crop	palm	standard		vegetable ivory	crop shade		
minor global crop	shrub	coppice		dye	nitrogen, contour hedgerow		
minor global crop	shrub	coppice		dye	nitrogen		
regional crop	tree	standard, coppice			nitrogen, alley crop, living fence, crop shade		
minor global crop	tree	standard, coppice		biomass	nitrogen, alley crop, living fence, crop shade		
regional crop	tree	standard	balanced carb		nitrogen, crop shade		multistrata
regional crop only	tree	standard	oil				
wild-harvested and sold	tree	standard	oil				
minor global crop	shrub	standard	protein-oil	oil	living fence, living trellis	insect	
regional crop only	palm	standard	oil				
regional crop only	tree	standard	protein-oil				strip intercrop
regional crop	tree	standard	protein-oil				
regional crop only	tree	standard	protein-oil				
regional crop only	tree	standard	protein-oil	dye			strip intercrop
regional crop only	tree	standard	protein-oil				
global crop	tree	standard	protein-oil				irreg. intercrop, strip intercrop
regional crop (Asia), wild-managed	grass-like	hay		biomass			
minor global crop	vine	standard	protein		nitrogen, understory legume	pasture	
regional crop	vine	coppice		hydrocarbon			
regional crop	vine	coppice		hydrocarbon			
regional timber	tree	standard		hydrocarbon			
experimental	shrub	coppice		hydrocarbon			
regional crop	shrub	coppice		dye			
regional crop only	tree	standard	protein-oil				
regional crop only	tree	standard	protein-oil				
regional crop, somewhat toxic in selenium soils	tree	standard					
regional crop	tree	standard	protein-oil				
regional crop	tree	standard	protein-oil				
regional crop only	palm	standard	oil	fiber			
under development	herb	standard	oil				
regional crop	grass-like	hay					
minor global crop	shrub	coppice, fodder	protein	biomass	nitrogen	bank	
new crop as annual, perennials under development	herb	standard	oil				
minor global crop	tree	standard, coppice			nitrogen, alley crop, contour hedgerow, crop shade	bank	
minor global crop	tree	standard, coppice		biomass	nitrogen, alley crop, contour hedgerow, living fence, windbreak, crop shade, living trellis	bank, intensive silvopasture	
wild-collected as food, grown for beach stabilzation	grass		balanced carb				

Genus	Species	Common Name	Family	Origin	Invasive	Climate	Humidity
Licania	rigida	oiticica	Chrysobalana-ceae	Americas		tropical	semi-arid
Linum	hybrids	perennial flax	Linaceae	hypothetical		boreal to warm temperate	semi-arid to humid
Litsea	calophylla	tagutugan	Lauraceae	Asia		tropical	humid
Litsea	glutinosa	maida lakri	Lauraceae	Asia	yes	tropical	humid
Lupinus	polyphyllus	Washington lupine	Fabaceae	western North America		boreal to warm temperate	semi-arid to humid
Macadamia	hildebrandii	Celebes nut	Proteaceae	Sulawesi		tropical	humid
Macadamia	integrifolia	macadamia	Proteaceae	East Australia		tropical	semi-arid to humid
Macadamia	ternifolia	macadamia nut	Proteaceae	East Australia		subtropical, tropical highlands	semi-arid to humid
Maclura	pomifera	Osage orange	Moraceae	North America		cold to warm temperate	semi-arid to humid
Macrozamia	spp.		Zamiaceae	Australia		warm temperate to tropical	arid to humid
Madhuca	longifolia	butter tree	Sapotaceae	India		subtropical	semi-arid
Manihot	caerulescens	mandioca de viado	Euphorbiaceae	Brazil		tropical	humid
Manihot	carthaginensis ssp. glaziovii	ceara rubber	Euphorbiaceae	Brazil		tropical	humid
Manihot	dichotoma	jequie manicoba	Euphorbiaceae	Brazil		tropical	humid
Manihot	heptaphylla	manicoba de Sao Francisco	Euphorbiaceae	South America		tropical	humid
Manilkara	bidentata	balata	Sapotaceae	Caribbean		tropical	humid
Manilkara	huberi	cow tree	Sapotaceae	South America		tropical	humid
Mauritia	flexuosa	buriti palm	Arecaceae	tropical South America		tropical	humid
Medicago	sativa	alfalfa	Fabaceae	Central Asia		boreal to warm temperate, tropical highlands	semi-arid to humid
Melaleuca	alternifolia	tea tree	Myrtaceae	Australia		subtropical to tropical	humid
Melia	azedarach	chinaberry	Meliaceae	Asia	yes	warm temperate to tropical	semi-arid to humid
Melocanna	baccifera	clumping bamboo	Poaceae	Asia		tropical	humid
Mesua	ferrea	surli nut	Calophyllaceae	South & SE Asia		tropical	humid
Metroxylon	amicarum		Arecaceae	New Guinea, Oceania		tropical	humid
Metroxylon	sagu	sago palm	Arecaceae	New Guinea		tropical	humid
Microcycas	calocoma		Zamiaceae	Cuba		tropical	semi-arid to humid
Mimosa	pudica	morivivir	Fabaceae	Brazil		tropical	semi-arid to humid
Miscanthus	sacchariflorus	Amur silver grass	Poaceae	East Asia		cold to warm temperate	humid
Miscanthus	sinensis	eulalia	Poaceae	East Asia		cold temperate to subtropical	humid
Miscanthus	× giganteus	giant miscanthus	Poaceae	hybrid of Asian species		cold temperate to subtropical	humid
Morella	cerifera	wax myrtle	Myricaceae	eastern North America		warm temperate to tropical	humid
Moringa	oleifera	moringa	Moringaceae	India		subtropical to tropical	semi-arid to humid
Moringa	peregrina	ben tree	Moringaceae	Middle East, North Africa		tropical	arid to semi-arid
Moringa	stenopetala	African moringa	Moringaceae	Africa		subtropical to tropical, tropical highlands	semi-arid to humid
Morus	alba	white mulberry	Moraceae	Asia	yes	cold temperate to tropical, tropical highlands	semi-arid to humid
Musa	acuminata	banana	Musaceae	Asia, New Guinea		tropical, tropical highlands	humid
Musa	basjoo	Japanese fiber banana	Musaceae	East Asia		cold temperate to subtropical	humid
Musa	textilis	abaca	Musaceae	Asia		tropical, tropical highlands	humid
Myrica	pensylvanica	bayberry	Myricaceae	eastern North America		cold to warm temperate	humid
Nelumbo	lutea	American lotus	Nelumbonaceae	North America		cold temperate to tropical	aquatic
Nelumbo	nucifera	Chinese water lotus	Nelumbonaceae	Asia, Australia, Middle East		warm temperate to tropical	aquatic
Neololeba	atra	clumping bamboo	Poaceae	Asia		subtropical to tropical, tropical highlands	humid

Cultivation	Form	Management	Staple Crop	Industrial Crop	Agroforestry Services	Fodder	Other Systems
regional crop	tree	standard		oil			
under development?	herb	standard, hay	oil	fiber			
regional crop	tree	standard		oil			
regional crop	tree	standard		oil		bank	
new crop	herb	standard	protein		nitrogen		
new crop	tree	standard	oil				
global crop	tree	standard	oil		crop shade		multistrata
minor global crop	tree	standard	oil				multistrata
regional crop	tree	standard, coppice	protein-oil	biomass, starch, dye	living fence	insect	
hypothetical	cycad	standard		starch			
regional crop only	tree	standard	oil				
regional crop	tree	standard		hydrocarbon			
regional crop	tree	standard		hydrocarbon	crop shade		
regional crop	tree	standard		hydrocarbon			
regional crop	tree	standard		hydrocarbon			
regional crop	tree	standard		hydrocarbon			
regional crop	tree	standard	protein	hydrocarbon			
regional crop only	palm	standard	oil	fiber			
global crop	herb	hay	protein				
regional crop	tree	standard, coppice		medicinal			
minor global crop	tree	standard, coppice		biomass, oil, pesticide			
regional crop	bamboo	managed multistem		biomass			
regional crop	tree	standard	oil				
regional crop	palm	managed multistem	basic starch	vegetable ivory			
regional crop	palm	managed multistem	basic starch	fiber	windbreak	insect	
hypothetical	cycad	standard		starch			
regional crop	herb	fodder, hay			nitrogen, understory legume	pasture	
minor global crop	grass	hay		biomass			
minor global crop	grass	hay		biomass			
new crop	grass	hay		biomass		bank	
wild harvested	shrub	standard		wax			
minor global crop	tree	standard, coppice	protein, oil	medicinal	alley crop, contour hedgerow, living fence, crop shade, living trellis	bank	homegarden, FMAFS
regional crop only	tree	standard	oil				
regional crop only	tree	standard, coppice	oil				
minor global crop	tree	standard, coppice		biomass	contour hedgerow, living fence	bank, insect	dike-pond
global crop	giant herb	managed multistem	basic starch	fiber	crop shade		dike-pond, homegarden, multistrata
regional crop	giant herb	managed multistem		fiber			
minor global crop	giant herb	managed multistem		fiber			multistrata
regional crop	shrub	standard		wax	nitrogen		
historic crop	herb		balanced carb				
minor global crop	herb		balanced carb				
minor global crop	bamboo	managed multistem		biomass			strip intercrop, multistrata

Genus	Species	Common Name	Family	Origin	Invasive	Climate	Humidity
Newtonia	*buchananii*		Fabaceae	Africa		tropical	humid
Nicotiana	*glauca*	tree tobacco	Solanaceae	South America		warm temperate to tropical	semi-arid to humid
Nopalea	*cochenillifera*		Cactaceae	Mesoamerica	yes	tropical, tropical highlands	arid to semi-arid
Nypa	*fruticans*	mangrove palm	Arecaceae	Asia	yes	tropical	humid, aquatic
Ochlandra	spp.	clumping bamboo	Poaceae	South Asia		tropical	humid
Oenocarpus	*bataua*	pataua palm	Arecaceae	South America		tropical	humid
Oenocarpus	*distichus*	bacaba palm	Arecaceae	tropical South America		tropical	semi-arid to humid
Olea	*europaea*	olive	Oleaceae	Mediterranean		Mediterranean, tropical	semi-arid
Opuntia	*dillenii*		Cactaceae	Mesoamerica		tropical	semi-arid
Opuntia	*ficus-indica*	nopale cactus	Cactaceae	Mesoamerica		warm temperate to tropical, tropical highlands	arid to humid
Opuntia	*tomentosa*		Cactaceae	Mesoamerica		tropical, tropical highlands	arid to semi-arid
Oryza	hybrids	perennial rice	Poaceae	hypothetical hybrid		cold temperate to tropical, tropical highlands	humid, aquatic
Oryza	*longistaminata*	African perennial rice	Poaceae	Africa		tropical	humid
Oryzopsis	*hymenoides*	Indian ricegrass	Poaceae	North America		cold to warm temperate	arid to humid
Oxytenanthera	*abyssinica*	African clumping bamboo	Poaceae	Africa		subtropical to tropical	semi-arid to humid
Pachira	*glabra*	pachira nut	Malvaceae	Central America		tropical	semi-arid to humid
Pachycereus	*hollianus*		Cactaceae	Mesoamerica		subtropical to tropical	arid to semi-arid
Palaquium	*gutta*	gutta percha tree	Sapotaceae	SE Asia		tropical	humid
Pandanus	*conoideus*	marita	Pandanaceae	New Guinea, Moluccas		tropical, tropical highlands	humid
Pandanus	*julianettii*	karuka	Pandanaceae	New Guinea		tropical highlands	humid
Panicum	*decompositum*	native millet	Poaceae	Australia		Mediterranean, subtropical to tropical	arid to semi-arid
Panicum	*maximum*	Guinea grass	Poaceae	Africa, Middle East		tropical, tropical highlands	humid
Panicum	*turgidum*	afezu	Poaceae	Africa through South Asia		tropical	arid to semi-arid
Panicum	*virgatum*	switchgrass	Poaceae	North America		boreal to warm temperate	semi-arid to humid
Parajubaea	*cocoides*	coquillo palm	Arecaceae	South America		tropical	semi-arid to humid
Parinari	*curatellifolia*	Mabo cork-tree	Chrysobalana-ceae	Central Africa		tropical	semi-arid
Parkia	*biglobosa*	African locust bean	Fabaceae	Africa		tropical	semi-arid
Parkinsonia	*aculeata*	Jerusalem thorn	Fabaceae	tropical Americas		subtropical to tropical	arid to semi-arid
Parthenium	*argentatum*	guayule	Asteraceae	SW North America, Mesoamerica		subtropical	arid to semi-arid
Pavonia	*spinifex*	gingerbush	Malvaceae	North America, tropical Americas		warm temperate to tropical	semi-arid
Payena	*leerii*	getah sundek	Sapotaceae	SE Asia		tropical	humid
Pennisetum	*purpureum*	Napier grass	Poaceae	Africa		warm temperate to tropical	semi-arid to humid
Pennisetum	*setaceum*	fountain grass	Poaceae	North Africa, Middle East		warm temperate to subtropical, Mediterranean	semi-arid
Pentaclethra	*macrophylla*	African oil bean	Fabaceae	Africa		tropical	humid
Pereskia	*sacharosa*	guayapa	Cactaceae	South America		tropical	arid to semi-arid
Persea	*americana*	avocado	Lauraceae	Mesoamerica, Caribbean		tropical, tropical highlands	humid
Phalaris	*arundinacea*	reed canary grass	Poaceae	Europe, Asia, North Africa, North America	yes	cold to warm temperate	semi-arid to humid
Phaseolus	*coccineus*	runner bean	Fabaceae	Mesoamerica		warm temperate, tropical highlands	humid
Phaseolus	*coccineus* ssp. *polyanthus*	cache bean	Fabaceae	Mesoamerica		tropical highlands	semi-arid
Phaseolus	*lunatus*	lima bean	Fabaceae	tropical Americas		tropical, tropical highlands	semi-arid to humid

Cultivation	Form	Management	Staple Crop	Industrial Crop	Agroforestry Services	Fodder	Other Systems
regional crop						pod	
regional crop				pesticide			
regional crop	cactus				living fence	bank, insect	
regional crop	palm		sugar		windbreak		
regional crop	bamboo	managed multistem		biomass			
experimental new crop	palm		oil	fiber			
regional crop	palm		oil, sugar				
global crop			oil				woody agriculture, irreg. intercrop, strip intercrop
regional crop	cactus				living fence	bank, insect	
minor global crop	cactus				living fence	bank	
regional crop	cactus				living fence	bank, insect	
under development	grass		balanced carb				
wild-harvested	grass		balanced carb				
new crop	grass		balanced carb				
regional crop	bamboo			biomass		bank	
experimental oil crop			protein-oil				
regional crop	cactus				living fence		
regional crop	tree	standard		hydrocarbon			
regional crop only	palmlike	standard	oil				
regional crop only	palmlike	standard	oil				
historic crop	grass	standard	balanced carb				
minor global crop	grass	hay	protein		contour hedgerow	bank, pasture	
wild staple	grass	standard	balanced carb				
regional crop	grass	hay		biomass			
regional crop	palm	standard	oil, sugar				
regional crop	tree	standard	oil	oil			
regional crop only	tree	standard	protein-oil		nitrogen	pod	parkland
minor global firewood crop	tree	standard	protein	biomass		pod, bank	
minor global crop	shrub	coppice		hydrocarbon			
regional crop	shrub	coppice		fiber			
regional crop	tree	standard		hydrocarbon			
global crop	giant grass	hay	protein	biomass	contour hedgerow	bank, pasture	
regional crop	grass	standard	balanced carb				
regional crop	tree	standard	protein-oil		nitrogen		
regional crop	cactus	standard			living fence		
global crop	tree	standard	oil	medicinal		fruit	homegarden, multistrata
new crop	grass	hay	protein	biomass			
minor global crop	vine	standard	protein		nitrogen		
regional crop	vine	standard	protein		nitrogen		
minor global crop	vine	standard	protein		nitrogen		

Genus	Species	Common Name	Family	Origin	Invasive	Climate	Humidity
Phaseolus	*polystachios*	thicket bean	Fabaceae	North America, Mesoamerica		cold temperate to tropical	humid
Phleum	*pratense*	timothy	Poaceae	Eurasia		boreal to warm temperate	semi-arid to humid
Phoenix	*canariensis*	Canary Island date palm	Arecaceae	Canary Islands	yes	Mediterranean	semi-arid
Phoenix	*dactylifera*	date palm	Arecaceae	North Africa, Middle East, India		subtropical to tropical	semi-arid
Phormium	*tenax*	New Zealand flax	Xanthorrhoeaceae	New Zealand	yes	warm temperate, Mediterranean, tropical highlands	semi-arid to humid
Phragmites	*australis*	common reed	Poaceae	Europe, Asia, Africa, N. America	yes	cold temperate to tropical	humid, aquatic
Phyllostachys	*bambusoides*	running bamboo	Poaceae	Asia		cold to warm temperate	humid
Phyllostachys	*edulis*	running bamboo	Poaceae	Asia		cold to warm temperate	humid
Phyllostachys	*glauca*	running bamboo	Poaceae	Asia		cold to warm temperate	humid
Phyllostachys	*nuda*	running bamboo	Poaceae	Asia		cold to warm temperate	humid
Phytelephas	*macrocarpa*	ivory nut palm	Arecaceae	South America		tropical	humid
Pinus	*brutia*	Calabrian pine	Pinaceae	Mediterranean		Mediterranean	semi-arid
Pinus	*caribaea*	Caribbean pine	Pinaceae	tropical Americas	yes	tropical	semi-arid to humid
Pinus	*cembroides*	Mexican piñon pine	Pinaceae	Mesoamerica		warm temperate to subtropical, tropical highlands	semi-arid
Pinus	*edulis*	Colorado piñon pine	Pinaceae	western North America, Mesoamerica		cold temperate subtropical, tropical highlands	arid to semi-arid
Pinus	*elliottii*	slash pine	Pinaceae	North America		warm temperate to tropical	semi-arid to humid
Pinus	*elliottii × caribaea*		Pinaceae			warm temperate to tropical	semi-arid to humid
Pinus	*halepensis*	aleppo pine	Pinaceae	Mediterranean, Eurasia		Mediterranean	semi-arid
Pinus	*jeffreyi*	Jeffrey pine	Pinaceae	North America		cold to warm temperate	arid to semi-arid
Pinus	*koraiensis*	Korean nut pine	Pinaceae	NE Asia		boreal to warm temperate	humid
Pinus	*massoniana*	horsetail pine	Pinaceae	Asia		warm temperate to tropical	humid
Pinus	*merkusii*	Sumatran pine	Pinaceae	Asia		tropical, tropical highlands	semi-arid to humid
Pinus	*monophylla*	singleleaf piñon	Pinaceae	western North America		cold temperate subtropical, tropical highlands	arid to semi-arid
Pinus	*oocarpa*	pino amarillo	Pinaceae	Mesoamerica		subtropical to tropical	humid
Pinus	*palustris*	longleaf pine	Pinaceae	North America		warm temperate to subtropical	semi-arid to humid
Pinus	*pinaster*	cluster pine	Pinaceae	Africa, Mediterranean	yes	Mediterranean, warm temperate	semi-arid
Pinus	*pinea*	Italian stone pine	Pinaceae	Mediterranean		warm temperate to subtropical, Mediterranean	arid to semi-arid
Pinus	*ponderosa*	ponderosa pine	Pinaceae	North America	yes	cold temperate to subtropical	semi-arid to humid
Pinus	*radiata*	Monterey pine	Pinaceae	North America, Mesoamerica	yes	warm temperate, Mediterranean	semi-arid
Pinus	*roxburghii*	chir pine	Pinaceae	Asia		tropical highlands	humid
Pinus	*sibirica*	Siberian stone pine	Pinaceae	NE Asia		boreal to warm temperate	semi-arid to humid
Pinus	*sylvestris*	Scots pine	Pinaceae	Eurasia	yes	boreal to cold temperate	humid
Pinus	*taeda*	loblolly pine	Pinaceae	North America		warm temperate to subtropical	humid
Pinus	*wallichiana*	blue pine	Pinaceae	Asia		tropical highlands	semi-arid to humid
Pistacia	*lentiscus*	Chios mastic tree	Anacardiaceae	Mediterranean		warm temperate to subtropical, Mediterranean	arid to semi-arid
Pistacia	*terebinthus*	terebinth tree	Anacardiaceae	Mediterranean		warm temperate to subtropical, Mediterranean	arid to semi-arid
Pistacia	*vera*	pistacio	Anacardiaceae	Central Asia		warm temperate to subtropical, Mediterranean	arid to semi-arid
Pithecellobium	*dulce*	Manila tamarind	Fabaceae	Mesoamerica		subtropical to tropical, tropical highlands	semi-arid to humid
Pittosporum	*resiniferum*	petroleum nut	Pittosporaceae	Philippines		tropical highlands	humid
Pittosporum	*undulatum*	Australian cheesewood	Pittosporaceae	Australia		Mediterranean, subtropical to tropical	semi-arid

Cultivation	Form	Management	Staple Crop	Industrial Crop	Agroforestry Services	Fodder	Other Systems
historic crop	vine	standard	protein		nitrogen		
new grain crop; major fodder crop	grass	standard, hay	balanced carb			pasture	
regional crop	palm	standard	sugar				
global crop	palm	standard	sugar	fiber	crop shade		multistrata
minor global crop	succulent	standard		fiber			
minor global crop	giant grass			biomass		bank	
minor global crop	bamboo	managed multistem		biomass			multistrata
minor global crop	bamboo	managed multistem		biomass			multistrata
minor global crop	bamboo	managed multistem		biomass			multistrata
minor global crop	bamboo	managed multistem		biomass			multistrata
regional crop	palm	standard		vegetable ivory			
regional timber	tree	standard		hydrocarbon			
regional timber	tree	standard		hydrocarbon			
regional crop only	tree	standard	protein-oil				
wild staple	tree	standard	protein-oil				
global crop	tree	standard		hydrocarbon	windbreak		
minor global crop	tree	standard					
regional crop	tree	standard		hydrocarbon			
regional timber	tree	standard		hydrocarbon			
minor global crop	tree	standard	protein-oil				irreg. intercrop
regional crop	tree	standard		hydrocarbon			
regional crop	tree	standard		hydrocarbon			
wild staple	tree	standard	protein-oil				
regional timber	tree	standard		hydrocarbon			
minor global crop	tree	standard, coppice		hydrocarbon			
regional crop	tree	standard		hydrocarbon			
minor global crop	tree	standard	protein-oil				
regional timber	tree	standard		hydrocarbon			
regional timber	tree	standard		hydrocarbon			
regional crop	tree	standard		hydrocarbon			
minor global crop	tree	standard	protein-oil				
regional timber	tree	standard		hydrocarbon			
regional timber	tree	standard		hydrocarbon	windbreak		
regional crop	tree	standard		hydrocarbon			
regional crop	tree	standard		hydrocarbon			
regional crop	tree	standard		hydrocarbon			
global crop	tree	standard, coppice	protein-oil				strip intercrop
minor global crop	tree	standard, coppice		biomass, tannin	nitrogen, alley crop, living fence, windbreak, crop shade	pod, bank, insect	
experimental	tree	standard		hydrocarbon			
experimental	tree	standard		hydrocarbon			

Genus	Species	Common Name	Family	Origin	Invasive	Climate	Humidity
Plukenetia	conophora	conophor	Euphorbiaceae	Africa		tropical	humid
Plukenetia	volubilis	sacha inchi	Euphorbiaceae	South America		tropical	humid
Pongamia	pinnata	pongam	Fabaceae	India, SE Asia, Australia		subtropical to tropical, tropical highlands	humid
Populus	hybrids	hybrid poplar	Salicaceae	northern temperate		boreal to warm temperate	semi-arid to humid
Populus	deltoides	eastern cottonwood	Salicaceae	eastern North America		cold to warm temperate	semi-arid to humid
Populus	euphratica	desert poplar	Salicaceae	Eurasia, North Africa		cold temperate to subtropical	semi-arid to humid
Populus	nigra	black poplar	Salicaceae	Eurasia, North Africa		boreal to warm temperate, Mediterranean	semi-arid to humid
Populus	trichocarpa	black cottonwood	Salicaceae	western North America		boreal to warm temperate	semi-arid to humid
Porteresia	coarctata	uri	Poaceae	South Asia		tropical	humid
Pouteria	guianensis		Sapotaceae	tropical South America		tropical	humid
Prinsepia	utilis	cherry prinsepia	Rosaceae	Himalayas		cold to warm temperate	semi-arid to humid
Prosopis	africana		Fabaceae	Africa		tropical	arid to semi-arid
Prosopis	alba	algarrobo blanco	Fabaceae	Central & South America		tropical	arid to semi-arid
Prosopis	chilensis	algarrobo de Chile	Fabaceae	South America		warm temperate to tropical, tropical highlands	arid to semi-arid
Prosopis	cineraria	jandi	Fabaceae	Asia, Middle East		tropical	arid to semi-arid
Prosopis	glandulosa	honeypod mesquite	Fabaceae	North America, Mesoamerica	yes	warm temperate to subtropical, tropical highlands	arid to semi-arid
Prosopis	juliflora	mesquite	Fabaceae	Central & South America	yes	tropical	arid to semi-arid
Prosopis	pallida	mesquite	Fabaceae	South America	yes	tropical	arid to semi-arid
Prosopis	tamarugo	tamarugo	Fabaceae	South America		tropical	arid to semi-arid
Prunus	armeniaca	apricot	Rosaceae	Central Eurasia		cold to warm temperate	semi-arid to humid
Prunus	armeniaca var. mandschurica	Manchurian apricot	Rosaceae	East Asia		boreal to warm temperate	semi-arid to humid
Prunus	dulcis	almond	Rosaceae	Mediterranean		cold temperate to subtropical, Mediterranean	semi-arid to humid
Prunus	pedunculata		Rosaceae	East Asia		cold to warm temperate	arid to semi-arid
Prunus	sibirica	Siberian apricot	Rosaceae	East Asia		boreal to warm temperate	semi-arid to humid
Prunus	spinosa	sloe	Rosaceae	Europe		cold to warm temperate	humid
Psophocarpus	tetragonolobus	winged bean	Fabaceae	New Guinea to India		tropical, tropical highlands	humid
Pueraria	montana var. lobata	kudzu	Fabaceae	Asia	yes	cold to warm temperate	humid
Pueraria	phaseoloides	tropical kudzu	Fabaceae	Asia		tropical	humid
Quercus	spp.	hybrid oaks	Fagaceae	hybrid		boreal to subtropical	arid to humid
Quercus	hybrid	burgambel oak	Fagaceae	natural hybrid in western North America		boreal to warm temperate	semi-arid to humid
Quercus	acutissima	sawtooth oak	Fagaceae	East Asia		warm temperate to subtropical	humid
Quercus	brantii	barro	Fagaceae	Turkey, Iran		warm temperate	semi-arid
Quercus	emoryi	emory oak	Fagaceae	SW North America, northern Mexico		warm temperate to subtropical	arid to semi-arid
Quercus	ilex	encina	Fagaceae	Mediterranean		Mediterranean, warm temperate to subtropical	semi-arid
Quercus	insignis	chicalaba	Fagaceae	Mesoamerica		tropical highlands	humid
Quercus	kelloggii	California black oak	Fagaceae	western North America		cold temperate to subtropical, Mediterranean	semi-arid
Quercus	suber	cork oak	Fagaceae	Mediterranean		warm temperate to subtropical, Mediterranean	semi-arid
Quercus	variabilis	Chinese cork oak	Fagaceae	Asia		cold to warm temperate	humid
Raphia	farinifera	Madagascar raffia palm	Arecaceae	Africa, South Asia, Madagascar		subtropical to tropical	humid

Cultivation	Form	Management	Staple Crop	Industrial Crop	Agroforestry Services	Fodder	Other Systems
regional crop only	vine	standard	protein-oil				
new crop	vine	standard	protein-oil			bank	
new crop	tree	standard, coppice		biomass, oil	nitrogen, windbreak, crop shade	bank	
global crop	tree	standard, coppice	protein	biomass	windbreak, living trellis	bank	SRC, irreg. intercrop, strip intercrop
minor global crop	tree	standard, coppice		biomass			irreg. intercrop, strip intercrop
regional crop only	tree	standard, coppice		biomass			irreg. intercrop, strip intercrop
minor global crop	tree	standard, coppice		biomass	windbreak, living trellis	bank	SRC, irreg. intercrop, strip intercrop
minor global crop	tree	standard, coppice		biomass			SRC, irreg. intercrop, strip intercrop
regional crop	grass	standard	balanced carb				
regional crop	tree	standard		hydrocarbon			
regional crop only	shrub	standard	oil		living fence		
regional crop	tree	standard, coppice			nitrogen	pod	parkland
regional crop	tree	standard, coppice	balanced carb		nitrogen, windbreak	pod	
regional crop	tree	standard, coppice	balanced carb		nitrogen	pod, bank	
regional crop	tree	standard, coppice			nitrogen, windbreak	pod, bank	
historic staple	tree	standard, coppice, coppice	balanced carb	biomass	nitrogen	pod, bank	
minor global crop	tree	standard, coppice	balanced carb	biomass, gum	nitrogen, living fence	pod	SRC
minor global crop	tree	standard, coppice	balanced carb	biomass	nitrogen	pod	SRC
regional crop	tree	standard, coppice			nitrogen	pod, bank	
global fruit, minor nut	tree	standard	protein-oil	oil			
regional crop only	tree	standard	oil	oil			
global crop	tree	standard	protein-oil				
wild	tree	standard		oil			
regional crop only	tree/shrub	standard	oil	oil			
regional crop	tree	standard		oil	living fence, windbreak		
minor global crop	vine	standard	protein-oil		nitrogen		
minor global crop	vine	hay, fodder	protein	fiber	nitrogen	bank, pasture	
minor global crop	vine	hay, fodder		fiber	nitrogen, understory legume	bank, pasture	
minor global crop	tree	standard	balanced carb	biomass		mast	
experimental as food	tree	standard	balanced carb	tannin		mast	
regional crop	tree	standard	balanced carb	tannin		mast	
regional crop	tree	standard	balanced carb	tannin		mast	
wild staple	tree	standard	balanced carb			mast	
regional crop	tree	standard	balanced carb	tannin		mast	dehesa
wild	tree	standard	balanced carb	tannin		mast	
wild staple	tree	standard	balanced carb			mast	
regional crop	tree	standard		cork, tannins		mast	dehesa, strip intercrop
experimental	tree	standard		cork, tannins			
regional crop	palm	standard	oil, sugar	fiber, wax			

Genus	Species	Common Name	Family	Origin	Invasive	Climate	Humidity
Raphia	*hookeri*	Ivory Coast raphia palm	Arecaceae	West Africa		tropical	humid
Raphia	*vinifera*	bamboo palm	Arecaceae	West Africa		tropical	humid
Reynoutria	*japonica*	Japanese knotweed	Polygonaceae	Asia	yes	cold to warm temperate	humid
Rhus	*copallinum*	winged sumac	Anacardiaceae	North America		cold to warm temperate	semi-arid to humid
Rhus	*succedanea*	wax tree	Anacardiaceae	China		cold to warm temperate	humid
Rhynchoryza	*subulata*	arroz bravo	Poaceae	South America		warm temperate to tropical	humid
Ricinodendron	*heudelotii*	djansang	Euphorbiaceae	Africa		tropical	humid
Ricinus	*communis*	castor bean	Euphorbiaceae	Africa or India	yes	subtropical to tropical, tropical highlands	arid to humid
Robinia	*neomexicana*	New Mexico locust	Fabaceae	western North America		cold temperate to subtropical, tropical highlands	semi-arid to humid
Robinia	*pseudoacacia*	black locust	Fabaceae	North America	yes	cold temperate to subtropical, tropical highlands	semi-arid to humid
Rumex	*acetosa*	sorrel	Polygonaceae	Eurasia		boreal to warm temperate, Mediterranean	semi-arid to humid
Rumex	*patientia ×tianschanicus*	hybrid sorrel	Polygonaceae	hybrid		boreal to warm temperate, Mediterranean	semi-arid to humid
Saccharum	hybrids	energy cane	Poaceae	hybrid of Asian species		subtropical to tropical	humid
Saccharum	*officinarum*	sugarcane	Poaceae	New Guinea		subtropical to tropical	humid
Salix	hybrids	hybrid willows	Salicaceae	northern temperate, Africa, South America		tropical highlands	semi-arid to humid, aquatic
Salix	*bonplandiana*	sauce	Salicaceae	Mesoamerica		tropical highlands	aquatic
Salix	*eriocephala*	heartleaf willow	Salicaceae	North America		boreal to warm temperate	semi-arid to humid
Salix	*miyabeana*		Salicaceae	East Asia		boreal to warm temperate	humid, aquatic
Salix	*purpurea*	purple willow	Salicaceae	Europe, North Africa		boreal to warm temperate	humid, aquatic
Salix	*viminalis*	basket willow	Salicaceae	Eurasia		boreal to warm temperate	semi-arid to humid, aquatic
Salvadora	*persica*	salvadora	Salvadoraceae	North Africa through Pakistan		tropical	arid to semi-arid
Sambucus	*canadensis*	elderberry	Adoxaceae	North America, Mesoamerica		cold temperate to subtropical	humid
Sambucus	*nigra*	elderberry	Adoxaceae	Europe		cold temperate to subtropical	humid
Sansevieria	*hyacinthoides*	bowstring hemp	Asparagaceae	Africa	yes	subtropical to tropical	semi-arid to humid
Sansevieria	*roxburghiana*	bowstring hemp	Asparagaceae	Africa		subtropical to tropical	semi-arid to humid
Sansevieria	*trifasciata*	bowstring hemp	Asparagaceae	Africa	yes	subtropical to tropical	semi-arid to humid
Sansevieria	*zeylanica*	bowstring hemp	Asparagaceae	South Asia		subtropical to tropical	semi-arid to humid
Santalum	*acuminatum*	quandong	Santalaceae	Australia		Tropical to subtropical	Arid to semi-arid
Sapindus	*saponaria*	soapnut	Sapindaceae	Americas		tropical	semi-arid to humid
Sapindus	*saponaria* var. *drummondi*	western soapberry	Sapindaceae	western North America		cold to warm temperate	semi-arid
Sapium	*sebiferum*	Chinese tallow tree	Euphorbiaceae	Asia	yes	warm temperate to subtropical, tropical highlands	semi-arid to humid
Saraca	*asoca*		Fabaceae	Asia		tropical	humid
Sarcocornia	*fruticosa*	oilseed glasswort	Amaranthaceae	Old World		Mediterranean	humid
Schinziophyton	*rautanenii*	mongongo nut	Euphorbiaceae	Africa, South Asia		subtropical to tropical	semi-arid
Schizostachyum	spp.	clumping bamboo	Poaceae	Asia		subtropical to tropical	humid
Schleichera	*oleosa*	Macassar oil tree	Sapindaceae	South & SE Asia		tropical	semi-arid to humid
Schoenoplectiella	*mucronata*		Cyperaceae	Eurasia, Australia		warm temperate to subtropical, Mediterranean	aquatic
Schoenoplectus	*californicus* subsp. *tatora*	totora	Cyperaceae	Bolivia, Peru, Chile		tropical highlands	aquatic
Sclerocarya	*birrea*	marula	Anacardiaceae	Africa		subtropical to tropical	semi-arid to humid
Secale	hybrids	perennial cereal rye	Poaceae	recent hybrid origin		boreal to warm temperate, Mediterranean	semi-arid to humid

Cultivation	Form	Management	Staple Crop	Industrial Crop	Agroforestry Services	Fodder	Other Systems
regional crop	palm	standard	oil, sugar	fiber			
regional crop	palm	standard		fiber			
experimental	herb	hay		fiber			
experimental	shrub	coppice		hydrocarbon, tannin, dye			
regional crop	tree	standard		oil, wax			
historic crop	grass	standard	balanced carb				
new crop, major wild staple	tree	standard	protein-oil				
global crop	tree	standard		oil, wax	windbreak, crop shade	insect	multistrata
regional crop	tree	coppice		biomass	nitrogen	bank	
minor global crop	tree	standard, coppice		biomass	nitrogen, windbreak	bank	SRC
minor global crop	herb	standard, hay	balanced carb	biomass			
new crop	herb	standard, coppice	balanced carb	biomass			
new crop	giant grass	hay	sugar	biomass		bank	
global crop	giant grass	hay	sugar	biomass, wax	nitrogen	bank	dyke-pond, homegarden
minor global crop	shrub	coppice		biomass		bank	SRC
regional crop	tree	coppice					chinampas
regional crop	shrub	coppice		biomass		bank	SRC
regional crop	shrub	coppice		biomass		bank	SRC
regional crop	shrub	coppice		biomass	windbreak	bank	SRC
regional crop	shrub	coppice		biomass, tannin		bank	SRC
new crop	tree	standard	protein-oil				
regional crop	shrub	standard		medicinal, dye	contour hedgerow		
minor global crop	shrub	standard		medicinal, dye			
minor global crop	succulent	hay		fiber			
minor global crop	succulent	hay		fiber			
minor global crop	succulent	hay		fiber			
minor global crop	succulent	hay		fiber			
new crop	tree	standard	protein-oil				
minor global crop	tree	standard		oil, soap, pesticide			
wild harvested only	tree	standard		soap, pesticide			
minor global crop	tree	standard, coppice		biomass, oil			SRC, strip intercrop, multistrata
regional crop	tree	standard			nitrogen	pod	
regional crop	succulent	standard	oil				
wild staple	tree	standard	protein-oil				
regional crop	bamboo	managed multistem		biomass			
regional crop only	tree	standard	oil			bank, insect	
regional crop	grass-like	hay		biomass			
regional crop	grass-like	hay		biomass			
wild staple	tree	standard	protein-oil				
under development	grass	standard	balanced carb				

Genus	Species	Common Name	Family	Origin	Invasive	Climate	Humidity
Semecarpus	*anacardium*	Indian marking nut tree	Anacardiaceae	India		tropical	semi-arid to humid
Senna	*auriculata*	Tanner's cassia	Fabaceae	Asia		tropical	semi-arid to humid
Senna	*siamea*	cassod tree	Fabaceae	SE Asia		tropical	semi-arid
Senna	*singueana*		Fabaceae	Africa		tropical	semi-arid
Sequoia	*sempervirens*	coast redwood	Cupressaceae	western North America		warm temperate to subtropical, tropical highlands	humid
Sesbania	*grandiflora*	vegetable hummingbird	Fabaceae	SE Asia to Australia		tropical	humid
Sesbania	*sesban*		Fabaceae	Africa		tropical	semi-arid to humid
Setaria	*palmifolia*	highlands pitpit	Poaceae	New Guinea		tropical highlands	humid
Shorea	*javanica*		Dipterocarpaceae	Asia		tropical	humid
Sida	*rhombifolia*	arrowleaf sida	Malvaceae	tropical Americas		subtropical to tropical, tropical highlands	semi-arid to humid
Silphium	*integrifolium*	prairie rosinweed	Asteraceae	North America		boreal to warm temperate	semi-arid to humid
Silphium	*laciniatum*	rosinweed	Asteraceae	North America		cold to warm temperate	semi-arid to humid
Simarouba	*amara*	paradise tree	Simaroubaceae	Mesoamerica, Caribbean		tropical	humid
Simmondsia	*chinensis*	jojoba	Simmondsiaceae	SW America, Mesoamerica		subtropical	arid to semi-arid
Solidago	*leavenworthii*	Leavenworth's goldenrod	Asteraceae	SE North America		warm temperate to subtropical	semi-arid to humid
Solidago	*nemoralis*	gray goldenrod	Asteraceae	eastern North America		cold to warm temperate	semi-arid to humid
Sorghum	*hybrids*	perennial sorghum	Poaceae	hypothetical hybrid		cold temperate to tropical, tropical highlands	semi-arid to humid
Spartium	*junceum*	Spanish broom	Fabaceae	Mediterranean, North Africa		Mediterranean	semi-arid
Sphenostylis	*stenocarpa*	African yambean	Fabaceae	Africa		tropical, tropical highlands	humid
Sporobolus	*fimbriatus*	sand dropseed	Poaceae	South Africa		subtropical	semi-arid to humid
Stenocereus	*griseus*	Mexican organpipe cactus	Cactaceae	South America		subtropical to tropical	arid to semi-arid
Stenocereus	*stellatus*	organpipe cactus	Cactaceae	Mesoamerica		subtropical to tropical	arid to semi-arid
Stipa	*tenacissima*	esparto grass	Poaceae	NW Africa, Mediterranean		warm temperate to subtropical	semi-arid
Stipagrotis	*scoparia*	drinn	Poaceae	Africa		tropical	arid to semi-arid
Stylosanthes	*guianensis*	estilo	Fabaceae	South America		tropical, tropical highlands	semi-arid to humid
Styphnolobium	*japonicum*	Japanese pagoda tree	Fabaceae	Asia		cold temperate to subtropical, tropical highlands	semi-arid to humid
Syagrus	*coronata*	ouricury palm	Arecaceae	Brazil		subtropical to tropical	humid
Tamarindus	*indica*	tamarind	Fabaceae	Africa		tropical	semi-arid to humid
Tanacetum	*cinerariifolium*	pyrethrum daisy	Asteraceae	Eastern Europe		warm temperate to tropical, Mediterranean	semi-arid to humid
Taxus	*baccata*	yew	Taxaceae	Eurasia, Africa		cold to warm temperate	humid
Taxus	*brevifolia*	western yew	Taxaceae	western North America		cold to warm temperate	humid
Telfairia	*occidentalis*	fluted gourd	Cucurbitaceae	Africa		tropical	humid
Telfairia	*pedata*	oyster nut	Cucurbitaceae	Africa		tropical, tropical highlands	humid
Tephrosia	*virginiana*	goat's rue	Fabaceae	North America		cold to warm temperate	humid
Tephrosia	*vogelii*	tephrosia	Fabaceae	Africa		tropical	humid
Terminalia	*bellirica*	celeric myrobalan	Combretaceae	India, SE Asia		tropical	semi-arid
Terminalia	*catappa*	tropical almond	Combretaceae	Pacific	yes	tropical	semi-arid to humid
Terminalia	*kaernbachii*	okari nut	Combretaceae	New Guinea		tropical	humid
Tetradium	*daniellii*	evodia	Rutaceae	Korea		cold to warm temperate	humid
Theobroma	*cacao*	cacao	Malvaceae	tropical Americas		tropical	humid

Cultivation	Form	Management	Staple Crop	Industrial Crop	Agroforestry Services	Fodder	Other Systems
regional crop	tree	standard		oil			
minor global crop	shrub	standard		fiber, tannin, dye			
minor global crop	tree	standard			nitrogen, alley crop, windbreak, crop shade	bank	
regional crop	tree	standard			nitrogen	pod	
minor global crop	tree	coppice		biomass			
minor global crop	tree	standard, coppice		biomass	nitrogen, living fence, windbreak, crop shade, living trellis	bank	
minor global crop	tree	standard, coppice	protein	biomass	nitrogen, alley crop, living fence, windbreak, crop shade, living trellis	bank	
regional crop	grass	standard, coppice	balanced carb				
experimental	tree	standard		hydrocarbon			
minor global crop	shrub	coppice		fiber			
under development	herb	standard	protein-oil				
wild	herb	standard		hydrocarbon			
experimental	tree	standard		oil			
new crop	shrub	standard		wax			strip intercrop
experimental	herb	hay		hydrocarbon			
experimental	herb	hay		hydrocarbon			
under development	grass	standard, hay	balanced carb, sugar				
regional crop	shrub	coppice		biomass, fiber	nitrogen		
regional crop	vine	standard	protein		nitrogen		
regional crop	grass	standard	balanced carb				
regional crop	cactus	standard			living fence		
regional crop	cactus	standard		biomass	living fence	bank	
regional crop	grass	hay		biomass, wax			
wild staple	grass	standard, coppice	balanced carb				
minor global crop	herb	hay, fodder			nitrogen, understory legume		
minor global crop	tree	standard, coppice		biomass, medicinal	nitrogen, crop shade	bank	
regional crop only	palm	standard	oil	wax			
minor global crop	tree	standard			windbreak, crop shade, living trellis	bank	
minor global crop	herb	hay		pesticide			
regional crop	tree	coppice		medicinal			
regional crop	tree	coppice					
regional crop only	vine	standard	protein-oil				
regional crop only	vine	standard	protein-oil				multistrata
wild	herb	hay		pesticide			
minor global crop	shrub	standard, coppice		pesticide	windbreak, crop shade		
regional crop	tree	standard		oil			
minor global crop	tree	standard	protein-oil			insect	
regional crop only	tree	standard	protein-oil				
regional crop only	tree	standard	oil				strip intercrop, irreg. intercrop
global crop	tree	standard					multistrata

Genus	Species	Common Name	Family	Origin	Invasive	Climate	Humidity
Theobroma	grandiflorum	cupuacu	Malvaceae	tropical Americas, Africa		tropical	humid
Thyrsostachys	siamensis	clumping bamboo	Poaceae	SE Asia		tropical	semi-arid to humid
Tithonia	diversifolia	Mexican sunflower	Asteraceae	Mesoamerica	yes	subtropical to tropical	semi-arid to humid
Torreya	grandis	Chinese kaya	Taxaceae	China		warm temperate to subtropical	humid
Torreya	nucifera	Japanese kaya	Taxaceae	Japan		cold to warm temperate	humid
Toxicodendron	sylvestre	yame-haze	Anacardiaceae	East Asia		cold to warm temperate	humid
Toxicodendron	vernicifluum	lacquer tree	Anacardiaceae	China		warm temperate	humid
Trachomitum	lancifolium	kendyr	Apocynaceae	Central Asia		boreal to warm temperate	semi-arid to humid
Trachomitum	sarmatiense	kendyr	Apocynaceae	Central Asia		boreal to warm temperate	semi-arid to humid
Trachomitum	scabrum	kendyr	Apocynaceae	Eurasia		cold to warm temperate	arid to semi-arid
Trachycarpus	fortunei	Chinese windmill palm	Arecaceae	Asia	yes	warm temperate to subtropical	humid
Treculia	africana	African breadnut	Moraceae	Africa		tropical	humid
Trema	orientalis	African elm	Cannabaceae	Asia		tropical	humid
Trichanthera	gigantea	nacadero	Acanthaceae	South America		tropical highlands	semi-arid to humid
Trichilia	emetica	mafura butter	Meliaceae	Africa, Middle East		tropical	semi-arid to humid
Trifolium	hybridum		Fabaceae	Europe, SW Asia		cold temperate to subtropical, tropical highlands	semi-arid to humid
Trifolium	repens	white clover	Fabaceae	Eurasia		boreal to warm temperate, tropical highlands	semi-arid to humid
Triticum	hybrids	perennial wheat	Poaceae	hypothetical hybrid		boreal to warm temperate, tropical highlands	semi-arid to humid
Tylosema	esculentum	marama bean	Fabaceae	East & South Africa		subtropical	arid to semi-arid
Typha	angustifolia	cattail	Typhaceae	Europe, North America		boreal to tropical	aquatic
Typha	elephantina	bora	Typhaceae	South & SE Asia		subtropical to tropical	aquatic
Typha	latifolia	cattail	Typhaceae	northern temperate, Africa, Australia		boreal to subtropical	aquatic
Ulmus	pumila	Siberian elm	Ulmaceae	NE Asia		boreal to warm temperate	semi-arid to humid
Ulmus	rubra	slippery elm	Ulmaceae	North America		cold to warm temperate	humid
Urochloa	mosambicensis	sabi grass	Poaceae	Africa		tropical	semi-arid to humid
Urtica	dioica	stinging nettle	Urticaceae	Eurasia		boreal to subtropical	humid
Vernicia	fordii	tung oil tree	Euphorbiaceae	China		subtropical	semi-arid to humid
Vernicia	montana	mu-oil tree	Euphorbiaceae	Asia		warm temperate to tropical	semi-arid to humid
Virola	sebifera	virola nut	Myristicaceae	Americas		tropical	humid
Vitellaria	paradoxa	shea butter tree	Sapotaceae	Africa		tropical	semi-arid to humid
Willughbeia	coriacea	akar gerit-gerit besi	Apocynaceae	Asia		tropical	humid
Willughbeia	edulis	gedraphol	Apocynaceae	India, Southeast Asia		tropical	humid
Wissadula	periplocifolia	white velvetleaf	Malvaceae	Mesoamerica		subtropical to tropical	humid
Xanthoceras	sorbifolium	yellowhorn	Sapindaceae	China		cold to warm temperate	semi-arid to humid
Yucca	baccata	banana yucca	Asparagaceae	North America, Mesoamerica		cold to warm temperate	arid to semi-arid
Yucca	elata	soapweed yucca	Asparagaceae	North America, Mesoamerica		cold temperate to subtropical, tropical highlands	arid to semi-arid
Yucca	filamentosa	Adam's needle	Asparagaceae	North America, Mesoamerica		cold temperate to subtropical, tropical highlands	arid to semi-arid
Yucca	gigantea	izote	Asparagaceae	Mesoamerica		warm temperate to subtropical, tropical highlands	semi-arid to humid
Yucca	gloriosa	Spanish dagger	Asparagaceae	North America, Mesoamerica		cold temperate to subtropical, tropical highlands	arid to semi-arid
Zamia	spp.		Zamiaceae	Americas		warm temperate to tropical	arid to humid
Zea	hybrids	perennial corn	Poaceae	hypothetical hybrid		cold temperate to tropical, tropical highlands	semi-arid to humid
Zizania	latifolia	Manchurian water rice	Poaceae	Asia		boreal to warm temperate	humid, aquatic

Cultivation	Form	Management	Staple Crop	Industrial Crop	Agroforestry Services	Fodder	Other Systems
regional crop only	tree	standard					
regional crop	bamboo	managed multistem		biomass			
minor global crop	shrub	coppice			contour hedgerow, living fence	bank	
regional crop only	tree	standard	protein-oil				
regional crop only	tree	standard	protein-oil				
regional crop	tree	standard		oil, gum			
regional crop	tree	standard		oil, wax			
minor global crop	herb	hay		fiber			
regional crop	herb	hay		fiber			
regional crop	herb	hay		fiber			
regional crop	palm	standard		fiber, wax			
new crop	tree	standard	protein		living fence		
regional crop	tree	standard, coppice-coppice, coppice		biomass, fiber	nitrogen, crop shade	bank	
minor global crop	shrub	coppice				bank	
regional crop	tree	standard		oil, medicinal			
minor global crop	herb	hay, fodder			nitrogen, understory legume	pasture	
global crop	herb	hay, fodder	protein		nitrogen, understory legume	pasture	
under development	grass	standard	balanced carb				
wild staple, new crop	prostrate vine	standard	protein-oil				
wild-managed, experimental crop	grass-like	hay	protein	biomass, fiber			
regional crop	grass-like	hay					
minor global crop	grass-like	hay		biomass, fiber			
minor global crop	tree	standard, coppice		biomass	windbreak	bank	
minor global crop	tree	standard, coppice		medicinal			
wild staple, regional pasture crop	grass	standard	balanced carb				
regional crop	herb	hay	protein	fiber			
minor global crop	tree	standard		oil, medicinal			
regional crop	tree	standard		oil, medicinal			
regional crop	tree	standard		oil			
minor global crop	tree	standard	oil				
regional crop	vine	standard		hydrocarbon			
regional crop	vine	standard		hydrocarbon			
regional crop	shrub	coppice		fiber			
regional crop only	tree	standard	protein-oil				strip intercrop
historic crop	succulent	standard		fiber	living fence		
regional crop	succulent	standard		fiber	living fence		
regional crop	succulent	standard		fiber	living fence		
regional crop	succulent	standard			contour hedgerow, living fence		
regional crop	succulent	standard		fiber	living fence		
hypothetical	cycad	standard		starch			
under development	grass	standard	balanced carb				
regional crop	grass	standard	balanced carb				

Clean Dry Weight Yield Calculations

For accurate comparisons of yield between perennial and annual staple crops, it is important to calculate first the edible portion (discounting shells, skins, pits, et cetera), then the dry weight (without water). This table shows the calculations used to provide the dry-weight data at the beginning of each staple crop chapter.

Annual crops are indicated by the presence of an asterisk next to the latin name.

Latin Name	Minimum Fresh Weight t/ha	Average Fresh Weight/ ha/yr in Bearing Years (FAO Stat 2012)	Maximum Fresh Weight t/ha	Percent Inedible Skin, Peel, Pit, Shell or Seed Coat	Dry Weight per Ton	Inedible Portion Reference
BASIC STARCH						
Artocarpus altilis	16	41	50	0	1	n/a
* Colocasia esculenta	5	7.6	30	0.1	0.9	estimated based on yams
* Dioscorea alata	8	11.7	30	0.1	0.9	PROSEA
Dioscorea bulbifera	1		19	0.1	0.9	PROSEA
Ensete ventricosum			5		1	n/a
* Ipomoea batatas	2	13	22	0.03	0.97	estimated
* Manihot esculenta	10	12.8	90	0.15	0.85	PROSEA
Metroxylon sagu	15		25		1	n/a
Musa (banana)	5	20.6	140	0.4	0.6	"Attaining high yield and high quality banana production in Guangxi"
Musa (plantain)		6.8		0.4	0.6	"Attaining high yield and high quality banana production in Guangxi"
BALANCED CARBOHYDRATES						
Artocarpus camansi		11		0.1	0.9	estimated from *Traditional Trees of Pacific Islands*
Bactris gasipaes	20		30	0.2	0.8	estimated
Brosimum alicastrum	7		8		1	n/a
Castanea spp.	3	3.7	5	0.15	0.85	estimated
Ceratonia siliqua	2	2	10	0	1	assuming ground to flour
Cordeauxia edulis	1.6		2.5	0	1	PROTA
Distichlis palmeri			1.2	0	1	n/a
Gleditsia triacanthos			8	0	1	assuming ground to flour
Inocarpus fagifer	3		18	0	1	kernel yield
* Oryza sativa (annual)	3	4.4	15	0	1	n/a
Oryza sativa (perennial)	3.9		7.2	0	1	n/a
Phleum pratense	0.4		0.5	0	1	n/a
Prosopis (hardy)	2		4	0	1	assuming ground to flour
Prosopis (tropical)		10	55	0	1	assuming ground to flour
Quercus ilex	6		15	0.15	0.85	based on chestnut
* Solanum tuberosum	20	19.2	30	0	1	PROSEA
* Sorghum bicolor			3	0	1	PROTA
* Triticum aestivum	3	3.1	9	0	1	n/a
* Zea mays	1	4.9	20	0	1	n/a
PROTEIN						
Acacia colei			1.2		1.056	n/a
* Cajanus cajan (annual)	1	0.8	7.6	0	1	n/a
Cajanus cajan (perennial)			2	0	1	n/a
Canavalia gladiata	1		5	0	1	n/a
Erythrina edulis	12		32	0	1	n/a
Medicago sativa (LPC)						n/a
Phaseolus coccineus	3		5	0	1	n/a
Phaseolus lunatus	3		5	0	1	n/a
* Phaseolus vulgaris	1	0.8	5	0	1	n/a
Sphenostylis stenocarpa			2	0	1	estimated
Treculia africana	25		60	0.15	0.85	estimated
Urtica dioica (LPC)						n/a
* Vigna unguiculata	1	0.5	7	0	1	n/a
PROTEIN-OIL						
Anacardium occidentale	0.8	0.7	1	0	1	n/a
* Arachis hypogaea	1	1.6	5	0.3	0.7	PROTA
Bertholletia excelsa			7.5	0.4	0.6	estimated

% Water	% Water Reference	Proportion of Weight That Is Dry and Edible	Minimum Edible Yield Dry Weight	Average Edible Yield Dry Weight	Maximum Edible Yield Dry Weight
0.62	*Encyclopedia of Fruits and Nuts*	0.38	6.1		19.0
0.7	PROSEA	0.27	1.4	2.1	8.1
0.7	PROSEA	0.27	2.2	3.2	8.1
0.7	PROSEA	0.27	0.3	0.0	5.1
0	*The "Tree Against Hunger"*	1.00	0.0	0.0	5.0
0.7	PROSEA	0.29	0.6	3.8	6.4
0.62	PROSEA	0.32	3.2	4.1	29.1
0	*Traditional Trees of Pacific Islands*	1.00	15.0	0.0	25.0
0.75	*Encyclopedia of Fruits and Nuts*	0.15	0.8	3.1	21.0
0.66	*Encyclopedia of Fruits and Nuts*	0.20	0.0	1.4	0.0
0.61	*Encyclopedia of Fruits and Nuts*	0.35	0.0	3.9	0.0
0.45	*Encyclopedia of Fruits and Nuts*	0.44	8.8	0.0	13.2
0.06	Leung and Flores. *Food Composition Table for Use in Latin America.*	0.94	6.6	0.0	7.5
0.6	USDA Nutrient Database	0.34	1.0	1.3	1.7
0.17	Feedipedia	0.83	1.7	1.7	8.3
0.11	PROTA	0.89	1.4	0.0	2.2
0.13	based on rice from PROTA	0.87	0.0	0.0	1.0
0.25	Feedipedia	0.75	0.0	0.0	6.0
0.76	FAO Pacific Islands Food Composition Tables	0.24	0.7	0.0	4.3
0.13	PROTA	0.87	2.6	3.8	13.1
0.13	PROTA	0.87	3.4	0.0	6.3
0.013	based on rice from PROTA	0.99	0.4	0.0	0.5
0.1	*The* Prosopis juliflora-Prosopis pallida *Complex*	0.90	1.8	0.0	3.6
0.1	*The* Prosopis juliflora-Prosopis pallida *Complex*	0.90	0.0	9.0	49.5
0.4	Characteristics of the acorns selected by free-range Iberian pigs during the *montanera* season	0.51	3.1	0.0	7.7
0.725	PROSEA	0.28	5.5	5.3	8.3
0.09	PROTA	0.91	0.0	0.0	2.7
0.12	PROTA	0.88	2.6	2.7	7.9
0.1	PROTA	0.90	0.9	4.4	18.0
0.11	based on *Phaseolus*	0.88		1.0	
0.1	PROTA	0.90	0.9	0.7	6.8
0.1	PROTA	0.90	0.0	0.0	1.8
0.1	PROTA	0.90	0.9	0.0	4.5
0.84	*Nitrogen Fixing Trees Highlights*	0.16	1.9	0.0	5.1
	n/a			1.2	
0.12	PROTA	0.88	2.6	0.0	4.4
0.11	PROTA	0.89	2.7	0.0	4.5
0.11	PROTA	0.89	0.9	0.7	4.5
0.11	based on *Phaseolus*	0.89			1.8
0.07	*Handbook of Nuts*	0.79	19.8	0.0	47.4
	n/a			0.6	
0.12	PROTA	0.88	0.9	0.4	6.2
0.05	*Encyclopedia of Fruits and Nuts*	0.95	0.8	0.7	1.0
0.06	PROTA	0.66	0.7	1.1	3.3
0.03	*Encyclopedia of Fruits and Nuts*	0.58	0.0	0.0	4.4

Latin Name	Minimum Fresh Weight t/ha	Average Fresh Weight/ ha/yr in Bearing Years (FAO Stat 2012)	Maximum Fresh Weight t/ha	Percent Inedible Skin, Peel, Pit, Shell or Seed Coat	Dry Weight per Ton	Inedible Portion Reference
Blighia sapida	1.2		1.7	0	1	n/a
Caryodendron orinocense	2.4		6	0	1	n/a
Citrullus colocynthis			6.7	0.3	0.7	based on buffalo gourd
Corylus (neohybrid)			1.6	0.57	0.43	based on *C. avellana*
Corylus avellana	2.5	1.5	3.5	0.57	0.43	"Pomological traits and proximate chemical composition of Hazelnut"
Cucurbita foetidissima	2.5		3	0.3	0.7	"The feral buffalo gourd, *Cucurbita foetidissima*"
Dacryodes edulis			10	0.5	0.5	estimated
* *Glycine max*	1	2.3	5	0	1	n/a
* *Helianthus annuus*	1	1.5	6	0	1	n/a
Juglans regia	1	3.4	7.5	0.75	0.25	estimated
Parkia biglobosa	0.3		0.9	0	1	n/a
Pinus koraiensis			0.4	0.3	0.7	based on *P. pinea*
Pinus pinea			9	0.3	0.7	estimated
Pinus sibirica			0.3	0.3	0.7	based on *P. pinea*
Pistacia vera		2	3	0.66	0.34	*Handbook of Nuts*
Prunus dulcis		1.2	6.7	0	1	n/a
Psophocarpus tetragonolobus	1.2		5	0	1	n/a
Schinziophyton rautanenii	0.2		1	0.3	0.7	*Encyclopedia of Fruits and Nuts*
Telfairia pedata	1.8		4.2	0.4	0.6	PROTA
Xanthoceras sorbifolium			1.6	0.25	0.75	personal observation
OIL						
* *Brassica napus*	2	1.4	4	0	1	n/a
* *Carthamus tinctorius*	1.5		4.5	0	1	n/a
Carya illinoinensis			3	0.5	0.5	estimated
Caryocar villosum			2.1	0.1	0.9	n/a
Cocos nucifera	2	4.9	6	0	1	n/a
Elaeagnus rhamnoides	10		14	0	1	n/a
Elaeis guineensis	15		38	0	1	n/a
Macadamia spp.			2.2	0	1	*Specialty Crops for Pacific Islands*
Olea europaea	1	1.6	10	0.1	0.9	estimated
Persea americana	5	9	32	0.3	0.7	estimated
Tetradium daniellii	0.3		0.6	0	1	n/a
SUGAR						
Phoenix dactylifera			10	0.15	0.85	estimated

% Water	% Water Reference	Proportion of Weight That Is Dry and Edible	Minimum Edible Yield Dry Weight	Average Edible Yield Dry Weight	Maximum Edible Yield Dry Weight
0.58	PROTA	0.42	0.5	0.0	0.7
0.04	*Encyclopedia of Fruits and Nuts*	0.96	2.3	0.0	5.8
0.05	based on buffalo gourd	0.67	0.0	0.0	4.5
0.05	based on *C. avellana*	0.41	0.0	0.0	0.7
0.05	USDA Nutrient Database	0.41	1.0	0.6	1.4
0.05	*The Buffalo Gourd (Cucurbita foetidissima) HBK*	0.67	1.7	0.0	2.0
0.11	*Encyclopedia of Fruits and Nuts*	0.45	0.0	0.0	4.5
0.08	PROTA	0.92	0.9	2.1	4.6
0.05	PROTA	0.95	1.0	1.4	5.7
0.04	USDA Nutrient Database	0.24	0.2	0.8	1.5
0.07	"Nutritional and sensory analysis of *Parkia biglobosa* (Dawadawa) based cookies"	0.93	0.3	0.0	0.8
0.02	based on *P. pinea*	0.69	0.0	0.0	0.3
0.02	USDA Nutrient Database	0.69	0.0	0.0	0.0
0.02	based on *P. pinea*	0.69	0.0	0.0	0.2
0.04	USDA Nutrient Database	0.33	0.0	0.7	1.0
0.04	USDA Nutrient Database	0.96	0.0	1.2	6.4
0.08	PROTA	0.92	1.1	0.0	4.6
0.05	PROTA	0.67	0.1	0.0	0.7
0.03	PROTA	0.58	1.0	0.0	2.4
0.15	estimated	0.64	0.0	0.0	1.0
0.05	estimated	0.95	1.9	1.3	3.8
0.05	USDA Nutrient Database	0.95	1.4	0.0	4.3
0.03	*Encyclopedia of Fruits and Nuts*	0.49	0.0	0.0	1.5
0.5	*Encyclopedia of Fruits and Nuts*	0.45	0.0	0.0	0.9
0.47	USDA Nutrient Database	0.53	1.1	2.6	3.2
0.8	"Wild edible plants of the Sikkim Himalaya"	0.20	2.0	0.0	2.8
0.1	PROTA (endosperm 35% water, kernel 7% water)	0.90	13.5	0.0	34.2
0.01	USDA Nutrient Database	0.99	0.0	0.0	2.2
0.6	*Encyclopedia of Fruits and Nuts*	0.36	0.4	0.6	3.6
0.73	USDA Nutrient Database	0.19	0.9	1.7	6.0
0.1	estimated	0.90	0.3	0.0	0.5
0.21	*Encyclopedia of Fruits and Nuts*	0.67	0.0	0.0	6.7

Carbon Sequestration Rates

T his table illustrates the differences in annual rates of carbon sequestration between various practices. It is the basis for figure 3.1 in chapter 3.

PRACTICE	LOCATION	T/Ha/Yr	TYPE	Single, Review, or Estimated Range	REFERENCE
ANNUAL CROPPING SYSTEMS					
annual crops alone					
conservation agriculture	global	0.6	SOC	review	Srinivasarao et al., "Conservation agriculture and soil carbon sequestration," 482
conventional no-till	USA	0.1	SOC	expert estimation	USDA NRCS "COMET-Planner," 4.
conventional no-till	global	0.3	SOC	review	Powlson et al., "Limited potential of no-till agriculture for climate change mitigation," 678.
cover cropping	USA	0.1	SOC	expert estimation	USDA NRCS "COMET-Planner," 8.
crop management	global	0.3	SOC	expert estimation	IPCC, *Land Use, Land-Use Change, and Forestry*, 184.
crop rotation	USA	0.1	SOC	expert estimation	USDA NRCS "COMET-Planner," 7.
manure application	USA	2.4	SOC	single study	Seebert-Elverfeldt and Tapio-Bistrom, "Agricultural mitigation approaches for smallholders," 31–34
organic annual cropping	global	0.7–1.4	SOC	expert estimation	Ibid.
regenerative organic	Egypt	0.9	SOC	single study	Rodale Institute, *Regenerative Organic Agriculture and Climate Change*, 11.
regenerative organic	Iran	4.1	SOC	single study	Ibid.
regenerative organic	Thailand	6.3	unknown	single study	Ibid.
regenerative organic	USA	2.3	SOC	single study	Ibid.
silvoarable agroforestry					
convert cropland to agroforestry	tropics	3.1	both	expert estimation	IPCC, *Land Use, Land-Use Change, and Forestry*, 184.
Faidherbia evergreen agriculture	Africa	2.0–4.0	both	expert estimation	Garrity et al., "Evergreen agriculture," 202.
FMNR	Niger	2.0–4.0	both	expert estimation	Ibid.
improved fallows	Africa	1.3–16.5	both	single study	Albrecht and Kandji, "Carbon sequestration in tropical agroforestry systems," 21.
maize with fruit and timber	Philippines	7.8	AGB	single study	Brakas and Aune, "Biomass and carbon accumulation in land use systems of Claveria," 170.
parkland	Sahel	0.4–1.1	AGB	single study	Luedeling et al., "Carbon sequestration potential of agroforestry systems in Africa," 65.
protective systems; e.g. windbreaks, living fences, riparian buffers	arid and semi-arid regions	1.0–8.0	both	expert estimation	Nair, "Climate change mitigation," 53.
riparian buffers	temperate North America	2.6	both	expert estimation	Udawatta and Jose, "Carbon sequestration potential of agroforestry practices in North America," 25.
silvoarable agroforestry	global	0.1–0.2	both	expert estimation	Smith et al., "Greenhouse gas mitigation in agriculture," 795–96.

PRACTICE	LOCATION	T/Ha/Yr	TYPE	Single, Review, or Estimated Range	REFERENCE
strip intercropping	USA	2.5	both	single study	Udawatta and Jose, "Carbon sequestration potential of agroforestry practices in temperate North America," 22.
strip intercropping (trees with annuals)	temperate North America	3.4	both	expert estimation	Udawatta and Jose, "Carbon sequestration potential of agroforestry practices in North America," 25.
strip intercropping (trees with annuals)	USA	0.2–0.5	both	expert estimation	USDA NRCS "COMET-Planner," 28.
tree intercropping systems	global	0.2–0.5	both	expert estimation	IPCC, *Land Use, Land-Use Change, and Forestry*, 184.
tree intercropping systems	humid and subhumid tropics	2.0–5.0	both	expert estimation	Nair, "Climate change mitigation," 53.

LIVESTOCK SYSTEMS not adjusted for methane

grazing and pasture

compost application on rangeland	USA	0.3–1.6	both	single study	Ryals and Silver, "Effects of organic matter amendments on net primary productivity and greenhouse gas emissions in annual grasslands," 46.
convert cropland to pasture	global	0.8	both	expert estimation	IPCC, *Land Use, Land-Use Change, and Forestry*, 184.
grazing land management	global	0.2–0.5	both	expert estimation	Ibid.
managed grazing	global	0.3	both	review	FAO, *Challenges and Opportunities for Carbon Sequestration in Grassland Systems*, 11.
managed grazing	global	0.1–3.0	both	review	Neely et al., *Review of Evidence on Drylands Pastoral Systems and Climate Change*, 15.
managed grazing	global	2.1	both	review	Tennigkeit and Wilkes, *An Assessment of the Potential for Carbon Finance in Rangelands*, 9.
managed grazing	USA	0.1–0.2	both	expert estimation	USDA NRCS "COMET-Planner," 33.
pasture and grazing management	global	0.1–0.2	both	expert estimation	Smith et al., "Greenhouse gas mitigation in agriculture," 795–96.

livestock with trees

convert pasture to silvopasture	tropics	3.1	both	expert estimation	IPCC, *Land Use, Land-Use Change, and Forestry*, 184.
convert pasture to silvopasture	USA	0.2–0.4	both	expert estimation	USDA NRCS "COMET-Planner," 32.
fodder tree blocks	tropics	0.1–0.5	both	expert estimation	Nair, "Climate change mitigation," 53.
intensive silvopasture	Colombia	4.6	both	single study	Naranjo et al., "Balance de gases de efecto invernadero en sistemas silvopastoriles intensivos con *Leucaena leucocephala* en Colombia."
intensive silvopasture plus timber trees	Colombia	9.4	both	single study	Ibid.
silvopasture	global	3.0–10.0	both	expert estimation	Nair, "Climate change mitigation," 53.
silvopasture	temperate North America	6.1	both	expert estimation	Udawatta and Jose, "Carbon sequestration potential of agroforestry practices in North America," 25.

PERENNIAL SYSTEMS

herbaceous monocultures

banana plantation	Philippines	6.2	AGB	single study	Brakas and Aune, "Biomass and carbon accumulation in land use systems of Claveria," 170.
convert annuals to biomass or forage	USA	0.1	both	expert estimation	USDA NRCS "COMET-Planner," 14.
perennial grains	hypothetical	0.2–0.8	both	expert estimation	Rumsey, "Perennial crops and climate change," 14.
switchgrass	USA	6.0	both	single study	Lemus and Lal, "Bioenergy crops and carbon sequestration," 15.

woody biomass monoculture

extensive bamboo	China	8.1		single study	Yiping et al., *Bamboo and Climate Change Mitigation*, 32.
intensive bamboo	China	12.7		single study	Ibid.
leucaena	Puerto Rico	12.0	AGB	single study	Parrotta, "Productivity, nutrient cycling, and succession in single- and mixed-species plantations of *Casuarina equisetifolia*, *Eucalyptus robusta*, and *Leucaena leucocephala* in Puerto Rico," 45–77.
Miombo coppice	Zambia	0.5–0.9	AGB	single study	Luedeling et al., "Carbon sequestration potential of agroforestry systems in Africa," 65.
moso bamboo	China	20.1–34.1	both	single study	Yiping et al., *Bamboo and Climate Change Mitigation*, 32.
SRC black locust	Germany	7.0	SOC	single study	Quinkenstein et al., "Assessing the carbon sequestration in short rotation coppices of *Robinia pseudoacacia L.* on marginal sites in northeast Germany," 201.

PRACTICE	LOCATION	T/Ha/Yr	TYPE	Single, Review, or Estimated Range	REFERENCE
SRC poplar	USA	5.4	SOC	single study	Lemus and Lal, "Bioenergy crops and carbon sequestration," 15.
SRC willow	USA	4.3	SOC	single study	Ibid.
woody crop monoculture					
cacao	Cameroon	7.2	both	single study	Egbe and Tabot, "Carbon sequestration in eight woody non-timber forest species and their economic potentials in southwestern Cameroon," 369–85.
carob	Portugal	4.2	AGB	single study	Geraldo et al., "Carob-tree as CO_2 sink in the carbon market," 119.
citrus	Sicily	5.3–5.5	AGB	single study	Ligouri, Gugliuzza, and Inglese, "Evaluating carbon fluxes in orange orchards in relation to planting density," 637.
coconut	Philippines	0.9	AGB	single study	Brakas and Aune, "Biomass and carbon accumulation in land use systems of Claveria," 170.
coconut	Vanuatu	1.2–5.3	both	single study	Lamade and Bouillet, "Carbon storage and global change," 154.
macadamia	Australia	0.8	AGB	single study	Murphy et al., "Preliminary carbon sequestration modelling for the Australian macadamia industry," 689–98.
mango	Philippines	17.9	AGB	single study	Brakas and Aune, "Biomass and carbon accumulation in land use systems of Claveria," 170.
mongongo	Cameroon	8.4–9.5	both	single study	Egbe and Tabot, "Carbon sequestration in eight woody non-timber forest species and their economic potentials in southwestern Cameroon," 369–85.
oil palm	Indonesia	6.1	both	single study	Lamade and Bouillet, "Carbon storage and global change," 154.
olive	Mediterranean	7.2	both	single study	Sofo et al., "Net CO_2 storage in Mediterranean olive and peach orchards," 17–24.
peach	Mediterranean	4.7	both	single study	Ibid.
peach palm	Brazil	5.1	AGB	single study	Schroth et al., "Conversion of secondary forest into agroforestry and monoculture plantation in Amazonia," 146.
pigeon pea	Tanzania	2.5	AGB	single study	Luedeling et al., "Carbon sequestration potential of agroforestry systems in Africa," 65.
rubber	Brazil	2.5	ABG	single study	Schroth et al., "Conversion of secondary forest into agrofroestry and monoculture plantation in Amazonia," 146.
safao	Cameroon	7.7–8.6	both	single study	Egbe and Tabot, "Carbon sequestration in eight woody non-timber forest species and their economic potentials in southwestern Cameroon," 369–85.
woody polycultures					
agroforestry woodlot polycultures	Puerto Rico	28.7-31.9	both	single study	Parrotta, "Productivity, nutrient cycling, and succession in single- and mixed-species plantations of *Casuarina equisetifolia*, *Eucalyptus robusta*, and *Leucaena leucocephala* in Puerto Rico," 99.
banana with mixed fruits	Philippines	13.6	AGB	single study	Brakas and Aune, "Biomass and carbon accumulation in land use systems of Claveria," 170.
cacao agroforests	Cameroon	5.8	AGB	single study	Nair, "Carbon sequestration in agroforestry systems," 249.
cacao agroforests	Costa Rica	40.6	both	single study	Ibid.
homegardens	Indonesia	8.0	AGB	single study	Nair, "Carbon sequestration in agroforestry systems," 249.
homegardens	Panama	15.7	both	single study	Seebert-Elverfeldt and Tapio-Bistrom, "Agricultural mitigation approaches for smallholders," 31-34.
homegardens	Philippines	9.4	AGB	single study	Brakas and Aune, "Biomass and carbon accumulation in land use systems of Claveria," 170.
multistrata agroforest	Philippines	4.1	AGB	single study	Brakas and Aune, "Biomass and carbon accumulation in land use systems of Claveria," 170.
multistrata systems	tropics	2.0–18.0	both	expert estimation	Nair, "Climate change mitigation," 53.
multistrata with peach palm and Brazil nut	Brazil	3.0–3.8	AGB	single study	Schroth, et al., "Conversion of secondary forest into agroforestry and monoculture plantation in Amazonia," 146.
shade coffee	Togo	6.3	AGB	single study	Nair, "Carbon sequestration in agroforestry systems," 249.

Changes in Latin Names

The Latin names and families in this book are up to date in accordance with Kew Botanical Gardens' *The Plant List: A Working List of All Plant Species*[1] (theplantlist.org) as of 2014. Because botanical nomenclature sometimes changes due to developments in our understanding of taxonomy, I've included a list of species that have changed in the last 5 to 10 years for reference.

Previous Latin Name	Current Latin Name
Agave angustifolia	*Agave vivipara*
Aleurites fordii	*Vernicia fordii*
Aleurites montanus	*Vernicia montana*
Amphicarpaea edgeworthii	*Amphicarpaea bracteata* var. *edgeworthii*
Amygdalus pedunculata	*Prunus pedunculata*
Aristida pungens	*Stipagrostis pungens*
Armeniaca mandschurica	*Prunus armeniaca* ssp. *mandschurica*
Armeniaca pedunculata	*Prunus pedunculata*
Armeniaca sibirica	*Prunus sibirica*
Armeniaca vulgaris	*Prunus armeniaca*
Bambusa atra	*Neololeba atra*
Bromelia fastuosa	*Bromelia pinguin*
Buchanania lanzan	*Buchanania cochinchinensis*
Castanopsis sumatrana	*Castanopsis inermis*
Cereus peruvianus	*Cereus repandus*
Chamaecytisus palmensis	*Cytisus proliferus*
Chamaecytisus proliferus	*Cytisus proliferus*
Cinnamomum zeylanicum	*Cinnamonum verum*
Cnidoscolus chayamansa	*Cnidoscolus aconitifolius*
Copaifera copallifera	*Guibourtia copallifera*
Dioscorea batatas	*Dioscorea polystachya*
Dioscorea opposita	*Dioscorea polystachya*
Dolichos lablab	*Lablab purpureus*
Eucalyptus citriodora	*Corymbia citriodora*
Euphorbia cerifera	*Euphorbia antisyphilitica*
Fallopia japonica	*Reynoutria japonica*
Furcraea humboldtiana	*Furcraea acaulis*

Furcraea macrophylla	*Furcraea hexapetala*
Garcinia tinctoria	*Garcinia xanthochymus*
Gigantochloa pseudoarundinacea	*Gigantochloa verticillata*
Grindelia camporum	*Grindelia hirsutula*
Hippophae rhamnoides	*Elaeagnus rhamnoides*
Jessenia bataua	*Oenocarpus bataua*
Juglans sinensis	*Juglans regia*
Lingnania chungii	*Bambusa chungii*
Manihot glaziovii	*Manihot carthaginensis* ssp. *glaziovii*
Maximiliana maripa	*Attalea maripa*
Microlaena stipoides	*Erharta stipoides*
Mimusops globosa	*Manilkara bidentata*
Myrica cerifera	*Morella cerifera*
Opuntia cochinellifera	*Nopalea cochenillifera*
Orbignya speciosa	*Attalea speciosa*
Phaseolus polyanthus	*Phaseolus coccineus* ssp. *polyanthus*
Phyllostachys pubescens	*Phyllostachys edulis*
Piliostigma thonningii	*Bauhinia thonningii*
Prunus amygdalus	*Prunus dulcis*
Pueraria lobata	*Pueraria Montana* var. *lobata*
Quercus leucotrichophora	*Quercus oblongata*
Rhus sylvestris	*Toxicodendron sylvestre*
Rhus verniciflua	*Toxicodendron vernicifluum*
Samanea saman	*Albizia saman*
Sambucus mexicana	*Sambucus canadensis*
Scirpus californicus var. *tatora*	*Schoenoplectus californicus* ssp. *tatora*
Scirpus mucronatus	*Schoenoplectiella mucronata*
Simarouba glauca	*Simarouba amara*
Sophora japonica	*Styphnolobium japonicum*
Sterculia urens	*Firmiana simplex*
Tetracarpidium conophorum	*Plukenetia conophora*
Thinopyrum intermedium	*Elymus hispidus*
Yucca guatemalensis	*Yucca gigantea*

RECOMMENDED READING

These readings are the key references I returned to again and again. While most are not light reading, they permit a more detailed investigation of the core subjects of this book.

PART 1

FAO, *Climate-Smart Agriculture Sourcebook*

IPCC, *Climate Change 2013: The Physical Science Basis, Summary for Policymakers*

IPCC, *Climate Change 2014: Impacts, Adaptation, and Aspects. Part A: Global and Sectoral Aspects, Summary for Policymakers*

IPCC, *Climate Change 2014: Mitigation of Climate Change, Summary for Policymakers*

Lal, "Abating climate change and feeding the world through soil carbon sequestration"

Lal, "Managing soils and ecosystems for mitigating anthropogenic carbon emissions and advancing global food security"

Lynas, *Six Degrees: Our Future on a Hotter Planet*

Nair et al., "Carbon sequestration in agroforestry systems"

Nair, *Carbon Sequestration Potential of Agroforestry Systems*

Nair, "Methodological challenges in estimating carbon sequestration potential of agroforestry systems"

Oxfam International, *Suffering the Science: Climate Change, People, and Poverty*

Tokar, *Toward Climate Justice: Perspectives on the Climate Crisis and Social Change*

PART 2

Boffa, *Agroforestry Parklands in Sub-Saharan Africa*

Cuartas Cardona et al., "Contribution of intensive silvopastoral systems to animal performance and to adaptation and mitigation of climate change"

Dupraz, *Temperate Agroforestry Systems*

ECHO, *Agricultural Options for the Poor*

FAO, *Challenges and Opportunities for Carbon Sequestration in Grassland Systems: A Technical Report on Grassland Management and Climate Change Mitigation*

Garrity et al., "Evergreen agriculture: A robust approach to sustainable food security in Africa"

Leakey, *Living with the Trees of Life: Towards the Transformation of Tropical Agriculture*

Li, *Agro-Ecological Farming Systems in China*

Nair, *Introduction to Agroforestry*

Nair, *Tropical Homegardens*

Steinfeld et al., *Livestock's Long Shadow*

World Agroforestry Centre, *Participatory Agroforestry Development in DPR Korea*

PART 3

Clay, *World Agriculture and the Environment: A Commodity-by-Commodity Guide to Impacts and Practices*

Duke, *Handbook of Nuts*

FAO, *Perennial Crops for Food Security: Proceedings of the FAO Expert Workshop*

Hanelt, *Mansfeld's Encyclopedia of Agricultural and Horticultural Crops*

Janick and Paull, *The Encyclopedia of Fruits and Nuts*

National Research Council, *Lost Crops of Africa*, vols. 2 and 3

Pasiecznik, *The* Prosopis juliflora–Prosopis pallida *Complex: A Monograph*

Plant Resources of Southeast Asia online

Plant Resources of Tropical Africa online

Wilkinson, *Nut Grower's Guide: The Complete Handbook for Producers and Hobbyists*

PART 4

Anandjiwala and John, *Industrial Applications of Natural Fibres: Structure, Properties, and Technical Applications*

Calvin, "Petroleum plantations for fuel and materials"

Clay, *World Agriculture and the Environment: A Commodity-by-Commodity Guide to Impacts and Practices*

Dillen et al., *Energy Crops*

Duke, *Handbook of Energy Crops*

El Bassam, *Handbook of Bioenergy Crops*

Finlay, *Growing American Rubber: Strategic Plants and the Politics of National Security*

Hanelt, *Mansfeld's Encyclopedia of Agricultural and Horticultural Crops*

Henley and Yiping, "The climate change challenge and bamboo: Mitigation and adaptation"

Koutinas et al., *Introduction to Chemicals from Biomass*

Langenheim, *Plant Resins: Chemistry, Evolution, Ecology, and Ethnobotany*

National Academy of Sciences, *Firewood Crops: Shrub and Tree Species for Energy Production*, vols. 1 and 2

Singh, *Industrial Crops and Uses*

Tokar, "Biofuels and the global food crisis"

Part 5

Chan and Roach, *Unfinished Puzzle: Cuban Agriculture: The Challenges, Lessons & Opportunities*

FAO, *Climate-Smart Agriculture Sourcebook*

IPCC, *Climate Change 2014: Mitigation of Climate Change, Summary for Policymakers*

Klein, *This Changes Everything: Capitalism Versus the Climate*

Leakey, "Twelve principles for better food and more food from mature perennial agroecosystems"

Martinez, Steve. *Local Food Systems: Concepts, Impacts, and Issues*

UNCAD, *Wake Up Before It Is Too Late: Make Agriculture Truly Sustainable Now for Food Security in a Changing Climate*

Wollenberg et al., *Climate Change Mitigation and Agriculture*

NOTES

INTRODUCTION

1. Williams-Linera, *El Bosque de Niebla del Centro de Veracruz*, 50–51, 63.
2. Ibid., 61, 112.
3. Ibid., 168.
4. Ibid., 62.
5. Las Cañadas website, http://www.bosquedeniebla.com.mx.
6. Ibid.
7. Ibid.
8. Marin Carbon Project, "What is carbon farming?"
9. Nath, Lal, and Das, "Managing woody bamboos for carbon farming and carbon trading," 1.
10. Gilbertson and Reyes, "Carbon trading," 11.
11. IPCC, *Climate Change 2014: Mitigation of Climate Change*, 28; Magdoff and Foster, *What Every Environmentalist Needs to Know About Capitalism*, 118.
12. Lal, personal communication.
13. Ibid.
14. Ibid.
15. Smith et al., "Greenhouse gas mitigation in agriculture," 789.

1. CLIMATE REALITIES

1. Anderegg et al., "Expert credibility in climate change," 12107; the study also noted that those few who disagree have less climate expertise.
2. Hansen et al., "Target atmospheric CO_2: Where should humanity aim?," 16.
3. Tokar, *Toward Climate Justice*, 79.
4. Lal, "Managing soils and ecosystems for mitigating anthropogenic carbon emissions and advancing global food security," 708.
5. Ibid.
6. Ibid.
7. Ibid.
8. Ibid.
9. Ibid.
10. Ibid.
11. Ibid., 709.
12. Hoggan, *Climate Cover-Up*.
13. Skeptical Science website, www.skepticalscience.com.
14. Gore, *Our Choice*.
15. IPCC, *Climate Change 2013*, 11.
16. IPCC, *Climate Change 2014: Mitigation of Climate Change*, 8.
17. Anderson and Bows, "Beyond 'dangerous' climate change: Emission scenarios for a new world"; Oxfam International, *Suffering the Science*.
18. Lynas, *Six Degrees*, 22.
19. Flannery, *Now or Never*, 30.
20. Lynas, *Six Degrees*, 270.
21. IPCC, *Climate Change 2014: Impacts, Adaptation and Vulnerability*, 12.
22. Lynas, *Six Degrees*, 82.
23. Oxfam International, *Suffering the Science*, i–ii.
24. IPCC, *Climate Change 2014: Impacts, Adaptation and Vulnerability*, 4–5.
25. Oxfam International, *Suffering the Science*, i–ii.
26. Ibid.
27. Lynas, *Six Degrees*, 231.
28. Ibid., 17.
29. Flannery, *Now or Never*, 34.
30. Lynas, *Six Degrees*, 250.
31. Klein, *This Changes Everything*, 1.
32. Lynas, *Six Degrees*, 48.
33. Hansen et al., "Target atmospheric CO_2: Where should humanity aim?," 12.
34. Flannery, *Now or Never*, 43.
35. GRAIN, "Food, climate change, and healthy soils," 19–20.
36. FAO Statistics Division.
37. Lal, "Abating climate change and feeding the world through soil carbon sequestration," 450.
38. Ibid.
39. Ibid.
40. Ibid., 708.
41. Ibid., 450.
42. Ibid., 443.
43. Ibid., 444.
44. Ibid., 447.
45. Ibid., 450.
46. Ibid., 443.
47. IPCC, *Contribution of Working Group II*, 16.
48. IPCC, *Climate Change 2013*, 70–71.
49. IPCC, *Climate Change 2014: Impacts, Adaptation and Vulnerability*, 12.
50. New et al., "Four degrees and beyond," 8.

51. Klein, "Why #BlackLivesMatter should transform the climate debate," 1.
52. Lynas, *Six Degrees*, 15. Lynas has become something of a lightning rod in the environmental movement due to his endorsement of nuclear power and genetically modified crops as climate strategies. Although I don't agree with him about these so-called "solutions" (see chapters 10 and 18), I do consider his award-winning *Six Degrees* to be an important and reliable resource on the projected impacts of climate change.
53. Oxfam International, *Suffering the Science*, i–ii.
54. Ibid.
55. Lynas, *Six Degrees*, 168.
56. Tokar, *Toward Climate Justice*, 14.
57. Lynas, *Six Degrees*, 66–67.
58. Ibid., 150.
59. Ibid., 255.
60. Ibid., 66–67.
61. Tokar, *Toward Climate Justice*, 24.
62. Klein, *This Changes Everything*, 114.
63. Klein, "Why #BlackLivesMatter should transform the climate debate."
64. Francis, *Laudato Si'*, 49.
65. Tokar, *Toward Climate Justice*, 79.
66. Klein, "Why #BlackLivesMatter should transform the climate debate."
67. Lynas, *Six Degrees*, 45.
68. Ibid., 30.
69. Than, "Causes of California drought linked to climate change," 1.
70. Ibid., 44.
71. Ibid., 125–27.
72. Ibid., 137.
73. Ibid., 172.
74. Ibid., 173.
75. IUCN, *Red List of Threatened Species*.
76. Lynas, *Six Degrees*, 114.
77. IPCC, *Contribution of Working Group II*, 16.
78. Lynas, *Six Degrees*, 54.
79. Ibid., 116.
80. IPCC, *Climate Change 2014: Mitigation of Climate Change*, 845.
81. Ibid., 178.
82. Ibid., 176–77.
83. Ibid., 251–59.
84. Ibid., 58–60.
85. Ibid.
86. IPCC, *Contribution of Working Group II*, 16.
87. Lynas, *Six Degrees*, 78.
88. Ibid., 249.
89. Ibid, 38.
90. Oxfam International, *Suffering the Science*, i–ii.
91. Lynas, *Six Degrees*, 102; Oxfam International, *Suffering the Science*, i–ii.
92. Lynas, *Six Degrees*, 86–92.
93. Hansen et al., "Ice melt, sea level rise and superstorms," 1.
94. Ibid.
95. Lynas, *Six Degrees*, 94.
96. Ibid., 189–90.
97. IPCC, *Climate Change 2014: Impacts, Adaptation, and Vulnerability*, 4.
98. Ibid.
99. Hoffmann, "Agriculture at the crossroads," 3.
100. Lynas, *Six Degrees*, 195.
101. Hoffmann, "Agriculture at the crossroads," 5.
102. Lal, "Abating climate change and feeding the world through soil carbon sequestration," 444.
103. Dar and Gowda, "Declining agricultural productivity and global food security," 2.
104. IPCC, *Climate Change 2014: Impacts, Adaptation, and Vulnerability*, 488.
105. Högy and Fangmeier, "Yield and yield quality of major cereals under climate change," 46.
106. Oxfam International, *Suffering the Science*, i–ii.
107. IPCC, *Climate Change 2014: Impacts, Adaptation, and Vulnerability*, 502.
108. Ibid., 512.
109. Ibid., 489.
110. Ibid., 251.
111. Lynas, *Six Degrees*, 146, 156.
112. Ibid., 161–63.
113. Ibid., 166.
114. Ibid., 187.
115. Ibid.
116. New et al., "Four degrees and beyond," 12.
117. Hoffmann, "Agriculture at the crossroads," 3.
118. Lynas, *Six Degrees*, 187.
119. Ibid.
120. Dar and Gowda, "Declining agricultural productivity and global food security," 6.
121. IPCC, *Climate Change 2014: Impacts, Adaptation, and Vulnerability*, 488.
122. Ibid., 504.
123. Oxfam International, *Suffering the Science*, i–ii.
124. IPCC, *Climate Change 2014: Impacts, Adaptation, and Vulnerability*, 506.
125. Oxfam International, *Suffering the Science*, i–ii.
126. Lynas, *Six Degrees*, 232.
127. Oxfam International, *Suffering the Science*.

128. Hoffmann, "Agriculture at the crossroads," 5.

129. Ibid.

130. Lynas, *Six Degrees*, 174.

131. Oxfam International, *Suffering the Science*, iii.

132. IPCC, *Contribution of Working Group II*, 16.

133. IPCC, *Climate Change 2014: Impacts, Adaptation, and Vulnerability*, 18.

134. IPCC, *Contribution of Working Group II*, 16.

135. Lynas, *Six Degrees*, 80.

136. Oxfam International, *Suffering the Science*, i–ii.

137. Lynas, *Six Degrees*, 180.

138. Ibid., 187.

139. Oxfam International, *Suffering the Science*, i–ii.

140. New et al., "Four degrees and beyond," 12.

141. IPCC, *Climate Change 2014: Impacts, Adaptation, and Vulnerability*, 758–59.

142. Oxfam International, *Suffering the Science*, i–ii.

143. IPCC, *Climate Change 2014: Impacts, Adaptation, and Vulnerability*, 20.

144. Anderson, "Climate change going beyond dangerous," 29.

145. Lynas, *Six Degrees*, 232.

146. Ibid., 236.

147. Oxfam International, *Suffering the Science*.

148. Lal, "Managing soils and ecosystems for mitigating anthropogenic carbon emissions and advancing global food security," 709.

149. Hansen et al., "Target atmospheric CO_2: Where should humanity aim?," 1.

150. Ibid.

151. Ibid., 16.

152. Flannery, *Now or Never*, 82.

153. IPCC, *Climate Change 2013*, 27–28.

2. AGRICULTURAL CLIMATE CHANGE MITIGATION AND ADAPTATION

1. GRAIN, "Food, climate change, and healthy soils," 19.

2. Lal, "Abating climate change and feeding the world through soil carbon sequestration," 450.

3. Martinez, *Local Food Systems*, 48–49.

4. Pelletier et al., "Energy intensity of agriculture and food systems," 226; Chi, MacGregor, and King, *Fair Miles*, 7.

5. Martinez, *Local Food Systems*, 48–49.

6. Pelletier et al., "Energy intensity of agriculture and food systems," 234.

7. Ibid.

8. Mundler and Rumpus, "The energy efficiency of local food systems," 614.

9. Pelletier et al., "Energy intensity of agriculture and food systems," 233.

10. GRAIN, "Food, climate change, and healthy soils," 19–20.

11. Nelson et al., *Food Security, Farming, and Climate Change to 2050*, xv.

12. Lal, "Abating climate change and feeding the world through soil carbon sequestration," 450.

13. Burney, Davis, and Lobell, "Greenhouse gas mitigation by agricultural intensification."

14. Shiva, *Monocultures of the Mind*.

15. UNCTAD, *Wake Up Before It Is Too Late*. According to leading agroecologist Miguel Altieri's website *Agroecology in Action*, "Agroecology is a scientific discipline that uses ecological theory to study, design, manage and evaluate agricultural systems that are productive but also resource conserving . . . [It] is concerned with the maintenance of a productive agriculture that sustains yields and optimizes the use of local resources while minimizing the negative environmental and socio-economic impacts of modern technologies . . . To put agroecological technologies into practice requires technological innovations, agriculture policy changes, socio-economic changes, but mostly a deeper understanding of the complex long-term interactions among resources, people and their environment."

16. Reij, "Adapting to climate change and improving household food security in Africa through agroforestry," 208.

17. Ibid.

18. Pretty et al., "Resource-conserving agriculture increases yields in developing countries."

19. De Schutter, "Agroecology," 34.

20. IPCC, *Land Use, Land-Use Change, and Forestry*.

21. Steinfeld et al., *Livestock's Long Shadow*, 82.

22. Ibid.

23. Verchot, "On forests' role in climate, *New York Times* op-ed gets it wrong."

24. Koutinas et al., "Production of chemicals from biomass," 77–90.

25. Ibid.

26. Nair, "Methodological challenges in estimating carbon sequestration potential of agroforestry systems," 6.

27. Nair et al., "Carbon sequestration in agroforestry systems," 253.

28. Nair, "Methodological challenges in estimating carbon sequestration potential of agroforestry systems," 8.

29. Nair, "Methodological challenges in estimating carbon sequestration potential of agroforestry systems," 5.

30. Jacke and Toensmeier, *Edible Forest Gardens*, vol. 1, 218–19.

31. Kumar, Pandey, and Pandey, "Plant roots and carbon sequestration," 885–86.

32. Jones, "Liquid carbon pathway unrecognized," 1.

33. Leu, "Managing climate change with soil organic matter in organic production systems," 19.

34. Donovan, "The calculation."

35. Nair et al., "Carbon sequestration and agroforestry systems," 253.

36. Soil also contains inorganic carbon in the form of carbonates and bicarbonates, especially in drylands. However, this tends to be in small amounts, with annual sequestration in the range of 2 to 5kg/year (0.02 to 0.05 t/ha/yr)—a quite small fraction of the potential of soil organic carbon. This book is focused on farming practices that increase soil organic carbon. See Lal, "Abating climate change and feeding the world through soil carbon sequestration."

37. Rumpel and Kögel-Knaber, "Deep soil organic matter," 143.

38. USDA NRCS, *Soil Quality Indicators*, 1.

39. Leu, "Managing climate change with soil organic matter in organic production systems," 19.

40. Donovan, "The calculation."

41. Nair, "Methodological challenges in estimating carbon sequestration potential of agroforestry systems," 5–8.

42. USDA NRCS, *Soil Quality Indicators*, 1.

43. IPCC, *Climate Change 2014: Mitigation of Climate Change*, 832.

44. Ibid., 833.

45. Lal, "Intensive agriculture and the soil carbon pool," 64.

46. Lal, "Managing soils and ecosystems for mitigating anthropogenic carbon emissions and advancing global food security," 708.

47. Ibid., 710.

48. Nair et al., "Carbon sequestration in agroforestry systems," 290.

49. Tennigkeit and Wilkes, *An Assessment of the Potential for Carbon Finance in Rangelands*, 8.

50. IPCC, *Land Use, Land-Use Change, and Forestry*.

51. Newell and Vos, "Accounting for forest carbon pool dynamics in product carbon footprints," 2.

52. Lal, "Abating climate change and feeding the world through soil carbon sequestration," 447.

53. Ibid., 444.

54. Ibid.

55. White, *Grass, Soil, Hope*, 220.

56. IPCC, *Climate Change 2014: Mitigation of Climate Change*, 845.

57. Grobe, "Albedo-e hg.png."

58. Global Climate Change Student Guide, "The climate system."

59. Bala et al., "Combined climate and carbon-cycle effects of large-scale deforestation."

60. Betts, "Offset of the potential carbon sink from boreal forestation by decreases in surface albedo."

61. Bala et al., "Combined climate and carbon-cycle effects of large-scale deforestation."

62. Harvey et al., "Climate-smart landscapes," 82.

63. IPCC, *Climate Change 2014: Mitigation of Climate Change*, 846.

64. Ibid., 847.

65. Harvey et al., "Climate-smart landscapes," 79.

66. Ibid.

67. Lal, "Abating climate change and feeding the world through soil carbon sequestration," 444.

68. FAO, *Climate-Smart Agriculture Sourcebook*, 81–104.

69. Harvey et al., "Climate-smart landscapes," 82.

70. Schoeneberger et al., "Branching out."

71. Lal, "Abating climate change and feeding the world through soil carbon sequestration," 444.

72. Harvey et al., "Climate-smart landscapes," 80.

73. Ibid.

74. van Noordwijk et al., "Agroforestry solutions for buffering climate variability and adapting to change," 227.

75. Altieri, "Strengthening resilience of farming systems," 58.

76. Ibid.

77. Harvey et al., "Climate-smart landscapes," 80.

78. van Noordwijk et al., "Agroforestry solutions for buffering climate variability and adapting to change," 227.

79. FAO, *Climate-Smart Agriculture Sourcebook*, 81–104.

80. Jordan et al., "Sustainable development of the agricultural bio-economy," 1570.

81. Evangelista, "The smallholder farmers' secret weapon."

82. Schoeneberger et al., "Branching out."

83. van Noordwijk et al., "Agroforestry solutions for buffering climate variability and adapting to change," 221.

84. FAO, *Climate-Smart Agriculture Sourcebook*, 137.

85. Ibid., 219.

86. Van Noordwijk et al., "Agroforestry solutions for buffering climate variability and adapting to change," 223.

3. CARBON SEQUESTRATION POTENTIALS

1. Calculated from FAO Statistics Division.

2. Seebert-Elverfeldt and Tapio-Bistrom, "Agricultural mitigation approaches for smallholders," 34.

3. Srinivasarao et al., "Conservation agriculture and soil carbon sequestration," 492–499; Lal, "Abating climate change and feeding the world through soil carbon sequestration," 447.

4. Farooq and Siddique, "Conservation agriculture," 8.

5. Nair et al., "Carbon sequestration in agroforestry systems," 283.

6. IFOAM, *The World of Organic Agriculture*, 61.

7. Nair, "Climate change mitigation," 53.

8. Ibid., 47.

9. FAO Statistics Division.

10. Ryals and Silver, "Effects of organic matter amendments on net primary productivity and greenhouse gas emissions in annual grasslands," 46.

11. Lal, "Abating climate change and feeding the world through soil carbon sequestration," 447.

12. White, *Grass, Soil, Hope*, 212.

13. Adams, J. M. "Estimates of preanthropogenic carbon storage in global ecosystem types."

14. IPCC, *Climate Change 2014: Mitigation of Climate Change*, 832.

15. FAO, *Challenges and Opportunities for Carbon Sequestration in Grassland Systems*, 25.

16. Lal, "Abating climate change and feeding the world through soil carbon sequestration," 447; Nair, "Climate change mitigation," 53.

17. Nair, "Climate change mitigation," 53.

18. FAO Statistics Division online.

19. Nair, "Climate change mitigation," 53.

20. Udawatta and Jose, "Carbon sequestration potentials of agroforestry practices in temperate North America," 22.

21. Brakas and Aune, "Biomass and carbon accumulation in land use systems of Claveria, the Philippines," 170.

22. Kumar and Nair, "Introduction," 3.

23. Nair et al., "Carbon sequestration in agroforestry systems," 289; 272.

24. Nair et al., "Carbon sequestration in agroforestry systems," 248.

25. Ibid., 259.

26. Ibid., 248.

27. Eva Wollenberg et al., eds., *Climate Change Mitigation and Agriculture*, 217–83.

28. Nair et al., "Carbon sequestration in agroforestry systems," 281.

29. Ibid., 281.

30. Ibid.

31. Ibid., 276–84.

32. Mosquera-Losada, Freese, and Rigueiro-Rodriguez, "Carbon sequestration in European agroforestry systems," 50.

33. Ibid., 282.

34. Ibid.

35. Ibid., 281.

36. Lemus and Lal, "Bioenergy crops and carbon sequestration," 1–21.

37. Nair et al., "Carbon sequestration in agroforestry systems," 276–84.

38. Ibid., 284.

39. Ibid.

40. Yiping et al., *Bamboo and Climate Change Mitigation*, 32–33.

41. Ibid., 250.

42. Ibid., 272.

43. Montagnini and Nair, "Carbon sequestration," 285.

44. Ibid.; Lal, "Managing soils and ecosystems for mitigating anthropogenic carbon emissions and advancing global food security," 710.

45. IPCC, *Climate Change 2014: Mitigation of Climate Change*, 851.

46. Lal, "Abating climate change and feeding the world through soil carbon sequestration," 444.

47. Smith et al., "Greenhouse gas mitigation in agriculture," 789.

48. IPCC, *Climate Change 2014: Mitigation of Climate Change*, 849.

49. Leakey, *Living with the Trees of Life*, 56.

50. FAO Statistics Division online.

51. IPCC, *Land Use, Land-Use Change, and Forestry*.

4. AGROFORESTRY AND PERENNIAL CROPS

1. Nair et al., "Carbon sequestration in agroforestry systems," 240.

2. Nair et al., "Methodological challenges in estimating carbon sequestration potential of agroforestry systems."

3. Nair, *An Introduction to Agroforestry*, 10.

4. Nair et al., "Carbon sequestration in agroforestry systems," 246.

5. USDA, Census of Agriculture 2012, 558.

6. Nair, "Climate change mitigation," 47.

7. Ibid.

8. Ibid.

9. Ibid., 27.

10. Ibid., 23.

11. Ibid., 23.

12. Ibid., 32.

13. Nair and Kumar, "Introduction," 1.

14. Ibid., 2.

15. Ibid., 7.

16. Ceccolini, "The homegardens of Soqotra Island, Yemen," 107.

17. Nair and Kumar, "Introduction," 2–5.

18. Nair, "Whither homegardens?," 356.

19. Nair and Kumar, "Introduction," 8.

20. Nair, "Whither homegardens?," 362.

21. Brakas and Aune, "Biomass and carbon accumulation in land use systems of Claveria, the Philippines," 170.

22. Kumar, "Carbon sequestration potential of tropical homegardens," 187–92.

23. Nair, "Whither Homegardens?," 358–59.

24. Nair and Kumar, "Introduction," 7.

25. Kumar, "Carbon sequestration potential of tropical homegardens," 186.

26. Nair, "Whither homegardens?," 364–67.

27. Nair and Kumar, "Introduction," 8; Nair, "Whither homegardens?," 356.

28. Kumar, "Carbon sequestration potential of tropical homegardens," 188–90.

29. Li, *Agro-Ecological Farming Systems in China*, 201–09.

30. Ibid., 208.

31. Ferguson and Lovell, "Permaculture for agroecology," 14.

32. FAO, "Turpentine from pine resin."

33. FAO Statistics Division.

34. National Research Council, *Lost Crops of Africa*, vol. 2: *Vegetables*, 86.

35. Ibid., 78.

36. National Research Council, *Lost Crops of Africa*, vol. 3: *Fruits*, 48–49.

37. National Research Council, *Lost Crops of Africa*, vol. 2: *Vegetables*, 76.

38. Sidibe and Williams, *Baobab* (Adansonia digitata).

39. Ibid.

40. Ibid.

5. A MULTIFUNCTIONAL SOLUTION

1. Nair et al., "Carbon sequestration in agroforestry systems," 1.

2. Ferguson and Lovell, "Permaculture for agroecology," 1–3.

3. Ibid., 15.

4. Ibid., 20.

5. Ibid., 16.

6. Mollison and Slay, *Introduction to Permaculture*, 6.

7. Ibid., 8.

8. Renting et al., "Exploring multifunctional agriculture," S112.

9. Keller, Feng, and Oschlies, "Potential climate engineering effectiveness and side effects during a high carbon dioxide-emission scenario," 1.

10. Ibid.

11. Ibid.

12. Goldenberg, "Al Gore says use of geo-engineering to head off climate disaster is insane," 2.

13. Ibid., 1.

14. Keller, Feng, and Oschlies, "Potential climate engineering effectiveness and side effects during a high carbon dioxide-emission scenario," 1.

15. Klein, *This Changes Everything*, 270.

16. Lal, "Managing soils and ecosystems for mitigating anthropogenic carbon emissions and advancing global food security," 719.

17. Costanza et al., "Changes in the global value of ecosystem services," 152.

18. Zimmer, "Putting a price tag on nature's defenses"; Wikipedia. https://en.wikipedia.org/wiki/Gross _world_product.

19. Zimmer, "Putting a price tag on nature's defenses."

20. Schröter et al., "Ecosystem services as a contested concept," 2–4.

21. Leakey, "Addressing the causes of land degradation, food/nutritional insecurity and poverty," 194.

22. Nair, "Whither homegardens?," 362.

23. Jose, "Agroforestry for ecosystem services and environmental benefits," 4–5.

24. Pimentel et al., "Annual vs. perennial grain production," 4.

25. FAO, *Challenges and Opportunities for Carbon Sequestration in Grassland Systems*, 18.

26. Baudry, Bunce, and Burel, "Hedgerows," 13.

27. World Agroforestry Centre, *Treesilience*, 35.

28. Pimentel et al., "Annual vs. perennial grain production," 3.

29. Jordan et al., "Sustainable development of the agricultural bio-economy," 1570.

30. Montagnini and Nair, "Carbon sequestration," 285.

31. Pasternak, "Combating poverty with plants," 17.

32. Mtaita, Manqwiro, and Mphúru, "The role of horticultural plants in combating desertification," 33–35.

33. Harvey et al., "Climate-smart landscapes," 80.

34. Ibid.

35. Pimentel et al., "Annual vs. perennial grain production," 3.

36. Ibid.

37. Nair, "The coming of age of agroforestry," 1616.

38. Eswaran, Lal, and Reich, "Land degradation," 1.

39. Ibid., 2.

40. Ibid., 3; FAO Statistical Service.

41. FAO, *Review of Evidence on Drylands Pastoral Systems and Climate Change*, 7.

42. Eswaran, Lal, and Reich, "Land degradation," 4.

43. Ibid., 2.

44. Ibid.

45. Ibid., 1.

46. Ibid., 2.

47. Lal, "Managing soils and ecosystems for mitigating anthropogenic carbon emissions and advancing global food security," 710.

48. FAO, *Review of Evidence on Drylands Pastoral Systems and Climate Change*, 26; Lal, "Managing soils and ecosystems for mitigating anthropogenic carbon emissions and advancing global food security," 711–15; Albrecht and Kandji, "Carbon sequestration in tropical agroforestry systems," 15–27; Cooper et al., "Agroforestry and the mitigation of land degradation in the humid and sub-humid tropics of Africa," 235–290; Nair, "The Coming of Age of Agroforestry," 1616.

49. White, *Grass, Soil, Hope*, 220.

50. Pimentel et al., "Annual vs. perennial grain production," 3.

51. Nair, "The coming of age of agroforestry", 1617.

52. Wojtkowski, *The Theory and Practice of Agroforestry Design*, 4.
53. Nair, *An Introduction to Agroforestry*, 270.
54. Ibid., 332.
55. Young, *Agroforestry for Soil Conservation*, 76.
56. Ewel, "Designing agricultural ecosystems for the humid tropics," 262.
57. Kumar, "Carbon sequestration potential of tropical homegardens," 194.
58. Cox, Crews, and Jackson, "From genetics and breeding to agronomy to ecology," 159.
59. FAO, *Challenges and Opportunities for Carbon Sequestration in Grassland Systems*, 15; World Agroforestry Centre, "Treesilience," 35.
60. Lal, "Abating climate change and feeding the world through soil carbon sequestration," 444.
61. Schoeneberger et al., "Branching out," 130A.
62. Merriam-Webster Dictionary online.
63. FAO, *Challenges and Opportunities for Carbon Sequestration in Grassland Systems*, 19. In this book, *carbon-friendly* describes practices that sequester carbon and/or reduce emissions.
64. Lal, "Abating climate change and feeding the world through soil carbon sequestration," 444.
65. World Agroforestry Centre, "Treesilience," 36.
66. Ibid., 36.
67. Kime, "Diversification of your operation, why?"
68. Lal, "Abating climate change and feeding the world through soil carbon sequestration," 444.
69. FAO, *Challenges and Opportunities for Carbon Sequestration in Grassland Systems*, 15.
70. Cuartas Cardona et al., "Contribution of intensive silvopastoral systems to animal performance and to adaptation and mitigation of climate change," 9.
71. Ewel, "Designing agricultural ecosystems for the humid tropics," 259.
72. Lefroy, "Perennial farming systems," 173.
73. Mehra and Rojas, *Women, Food Security, and Agriculture in a Global Marketplace*, 1.
74. Kiptot and Franzel, *Gender and Agroforestry in Africa*, viii.
75. Ibid., 2.
76. Nair, "Whither homegardens?," 357.
77. Kiptot and Franzel, *Gender and Agroforestry in Africa*, 2.
78. Ibid., viii.
79. Ibid., ix.
80. Ibid., 27.
81. World Health Organization, "Food security."
82. Mbow et al., "Agroforestry solutions to address food security and climate change challenges in Africa," 61.
83. Ibid., 6.
84. FAO, *Challenges and Opportunities for Carbon Sequestration in Grassland Systems*, 15.
85. FAO, *Review of Evidence on Dryland Pastoral Systems and Climate Change*, 1.
86. Altieri, "Agroecology, small farms, and food sovereignty," 104.
87. Worldwatch Institute, "Leading the fight for food sovereignty: An interview with La Via Campesina's Dena Hoff."
88. Altieri, "Agroecology, small farms, and food sovereignty," 105–08.
89. IPCC, *Contribution of Working Group II*, 16.
90. Nair, *An Introduction to Agroforestry*, 270.
91. Wojtkowski, *The Theory and Practice of Agroforestry Design*, 3.
92. Nair et al., "Carbon sequestration in agroforestry systems," 282.
93. Jackson et al.,"Trading water for carbon with biological carbon sequestration."
94. Barringer, "Water sources for almonds in California may run dry," 1.

6. Annual Cropping Systems

1. FAO Statistical Services online.
2. Farooq and Siddique, "Conservation agriculture," 8. Throughout this book I report the amount of land occupied by various carbon farming techniques. This data is useful to some degree but also quite problematic. Some categories overlap, which could lead to double accounting. Some practices reported here are entirely subsumed within others. Some practices are reported under cropland, while others are reported as forestry. Many are not reported at all. Most are estimates, and even the best data is not to be entirely trusted.
3. Srinivasarao et al., "Conservation agriculture and soil carbon sequestration," 492–99; Lal, "Abating climate change and feeding the world through soil carbon sequestration," 447.
4. Nair, "Climate change mitigation," 47.
5. Ibid., 53.
6. IPCC, *Climate Change 2014: Mitigation of Climate Change*, 830-832.
7. Harvey et al., "Climate-smart landscapes," 79.
8. Gliessman, *Agroecology*, 199.
9. AGROOF, *Agroforesterie* DVD.
10. Nair et al., "Carbon sequestration in agroforestry systems," 250; 272.
11. Miccolis et al., "Oil palm and agroforestry systems."
12. Harvey et al., "Climate-smart landscapes," 80.
13. Altieri, "Strengthening resilience of farming systems," 58.
14. Farooq and Siddique, "Conservation agriculture," 8.
15. Garrity et al., "Evergreen agriculture," 199.
16. Moyer, *Organic No-Till Farming*.

17. Powlson et al., "Limited potential of no-till agriculture for climate change mitigation," 678.

18. Srinivasarao et al., "Conservation agriculture and soil carbon sequestration," 482.

19. Ibid., 681.

20. IPCC, *Climate Change 2014: Mitigation of Climate Change*, 830–32.

21. Vidal, "India's rice revolution."

22. Uphoff, "Agroecological alternatives," 6.

23. Ibid., 8.

24. Tsujimoto et al., "Soil management," 70.

25. Abraham et al., *The System of Crop Intensification*.

26. Tsujimoto et al., "Soil management," 61.

27. IPCC, *Climate Change 2007*, 506–08.

28. FAO, *Climate-Smart Agriculture Sourcebook*, 81–104.

29. Uphoff, "Agroecological alternatives," 8.

30. Abraham et al., *The System of Crop Intensification*, 58–59.

31. Berkhout, Glover, and Kuyenhoven, "On-farm impact of the System of Rice Intensification (SRI)," 1.

32. FAO, "Organic agriculture: What is organic agriculture?," 2015, http://www.fao.org/organicag/oa-faq/oa-faq1/en.

33. Rodale Institute, *Regenerative Organic Agriculture and Climate Change*, 11.

34. Badgley et al., "Organic agriculture and the global food supply," 86.

35. Ponisio et al., "Diversification practices reduce organic to conventional yield gap."

36. Seebert-Elverfeldt and Tapio-Bistrom, "Agricultural mitigation approaches for smallholders," 31–34.

37. Luske and van der Kamp, "Carbon sequestration potential of reclaimed desert soils in Egypt," 25.

38. Rodale Institute, *Regenerative Organic Agriculture and Climate Change*, 11.

39. Lal, "Abating climate change and feeding the world through soil carbon sequestration," 444.

40. Srinivasarao et al., "Conservation agriculture and soil carbon sequestration," 482.

41. Lal, "Abating climate change and feeding the world through soil carbon sequestration," 444.

42. Ibid., 450.

43. Harvey et al., "Climate-smart landscapes," 79.

44. IPCC, *Climate Change 2014: Mitigation of Climate Change*, 830–32.

45. Lal, "Abating climate change and feeding the world through soil carbon sequestration," 444.

46. Ibid., 112–13.

47. Ibid., 108–11.

48. Li, *Agro-Ecological Farming Systems in China*, 107.

49. Ibid., 150.

50. Crawford, "Silvoarable systems in Europe," 7.

51. Udawatta and Jose, "Carbon sequestration potential of agroforestry practices in temperate North America," 25.

52. USDA NRCS, *COMET-Planner*, 28.

53. Nair, "Climate change mitigation," 53.

54. Li, *Agro-Ecological Farming Systems in China*, 110–112.

55. Rinaudo, "Farmer managed natural regeneration," 208.

56. Cunningham, "The Farmer-Managed Agroforestry Farming System (FMAFS)," 226.

57. Ibid., 208.

58. Ibid., 226.

59. Rinaudo, "Farmer managed natural regeneration," 208.

60. Garrity et al., "Evergreen agriculture," 209.

61. Ibid., 206.

62. Ibid., 199.

63. Ibid., 202.

64. Boffa, *Agroforestry Parklands in Sub-Saharan Africa*.

65. Plant Resources of Tropical Africa online database.

66. Garrity et al., "Evergreen agriculture," 209.

67. Plant Resources of Tropical Africa online database.

68. Garrity et al., "Evergreen agriculture." 1.

69. Ibid., 200.

70. Ibid., 203.

71. Ibid., 206.

72. Ibid., 201.

74. Neufeldt et al., "Trees on Farms," 4.

74. Ibid., 201–02.

75. Ibid., 2.

76. Garrity et al., "Evergreen agriculture."

77. Neufeldt et al., "Trees on Farms," 17.

78. Ibid., 92–93.

79. Truong et al., *The Vetiver System for Agriculture*, 2–3.

80. Ibid., 33–65.

81. Ibid., 68–76.

82. Ibid., 14.

83. The Vetiver Network International website, www.vetiver.org.

84. Kang et al., *Alley Farming*, 56.

85. Kang et al., *Alley Farming*, 58–60.

86. Li, *Agro-Ecological Farming Systems in China*, 281.

87. Ibid., 283–88.

88. Whitmore and Turner, *Cultivated Landscapes of Middle America on the Eve of Conquest*, 141.

89. Cunningham, "The Farmer-Managed Agroforestry Farming System (FMAFS)."

90. Mindanao Baptist Rural Life Center, *Sloping Agricultural Land Technology*, 3.

91. Legoupil et al., "Conservation agriculture in Southeast Asia," 181.

92. Nair, *Introduction to Agroforestry*, 129.

93. Cunningham, "The Farmer-Managed Agroforestry Farming System (FMAFS)."

94. Albrecht and Kandji, "Carbon sequestration in tropical agroforestry systems," 21.

95. Sileshi et al., "Meta-analysis of maize yield response to woody and herbaceous legumes in sub-Saharan Africa," 4.

96. Plant Resources of Tropical Africa online database.

97. Sileshi et al., "Meta-analysis of maize yield response to woody and herbaceous legumes in sub-Saharan Africa," 1.

98. Williams et al., "Agroforestry in North America and its role in farming systems," 21–22.

99. Udawatta and Jose, "Carbon sequestration potential of agroforestry practices in temperate North America," 22.

100. Nair, "Climate change mitigation," 53.

101. Kort, "Benefits of windbreaks to field and forage crops," 285.

102. Ibid., 171–74.

103. Ibid., 184.

104. Nuberg, "Effect of shelter on temperate crops," 11.

105. Cleugh, "Effects of windbreaks on airflow, microclimates, and crop yields," 76.

106. Ibid., 9–10.

107. Ibid., 7.

108. Baudry, Bunce, and Burel, "Hedgerows," 8–9.

109. Nair, "Climate change mitigation," 53.

110. Ibid., 13.

111. Brownlee, "Grow agriculturally productive buffers."

112. Schoeneberger et al., "Branching out," 133.

113. Schoeneberger et al., "Branching out."

114. Udawatta and Jose, "Carbon sequestration potential of agroforestry practices in temperate North America," 25.

115. Nair, "Climate change mitigation," 53.

116. Li, *Agro-Ecological Farming Systems in China*, 24.

117. Ibid., 105.

118. Crawford, "Silvoarable systems in Europe," 7.

119. Boffa, *Agroforestry Parklands in Sub-Saharan Africa*.

120. Ibid.; Zomer et al., *Trees on Farm*, 3.

121. Luedeling et al., "Carbon sequestration potential of agroforestry systems in Africa," 65.

122. Boffa, *Agroforestry Parklands in Sub-Saharan Africa*.

123. Mertz et al., "Swidden change in Southeast Asia," 261.

124. van Vleit et al., "Trends, drivers, and impacts of changes in swidden cultivation in tropical forest-agriculture frontiers," 2–5.

125. Schmidt-Vogt et al., "An assessment of trends in the extent of swidden in Southeast Asia," 272.

126. Ibid., 259.

127. Fox et al., "Shifting cultivation," 522; 132. Whitmore and Turner, *Cultivated Landscapes of Middle America on the Eve of Conquest*, 78.

128. Bruun et al., "Environmental consequences of the demise of swidden cultivation in Southeast Asia," 383.

129. Padoch and Pinedo-Vasquez, "Saving slash-and-burn to save biodiversity," 551–552.

130. Nair, *An Introduction to Agroforestry*, 75.

131. Ibid., 78.

132. Cubbage, "Global timber production, trade and timberland investments."

133. Li, *Agro-Ecological Farming Systems in China*, 167.

134. White, *Grass, Soil, Hope*, 68–71.

135. Ibid., 77.

136. Finlayson et al., "A bio-economic evaluation of the profitability of adopting subtropical grasses and pasture-cropping on crop-livestock farms," 4.

137. White, *Grass, Soil, Hope*, 76.

138. Ampt and Doornbos, "Benchmark study of innovators," 19.

139. Millar and Badgery, "Pasture cropping," 777.

140. Finlayson et al., "A bio-economic evaluaton of the profitability of adopting subtropical grasses and pasture-cropping on crop-livestock farms," 21.

7. LIVESTOCK SYSTEMS

1. FAO Statistics Division online.

2. Havlík et al., "Climate change mitigation through livestock system transitions," 3709.

3. Steinfeld et al., *Livestock's Long Shadow*, 95.

4. Ibid., 112.

5. Worldwatch Institute, "Global meat production and consumption continue to rise."

6. Wilson, *Livestock Production Systems*, 10.

7. Ibid.

8. Steinfeld et al., *Livestock's Long Shadow*, 118.

9. Ibid., 830–32.

10. IPCC, *Climate Change 2014: Mitigation of Climate Change*, 24.

11. Ibid., 830–32.

12. Although much of that discussion is beyond the scope of this book, I recommend *Livestock's Long Shadow: Environmental Issues and Options* for a grim but informative summary.

13. McKnight, "Want to have a real impact on climate change?"

14. Ibid., 82.

15. Ibid., 95.

16. Ibid., 97.

17. Steinfeld et al., *Livestock's Long Shadow*, 112.

18. Ibid., 120.

19. Carter et al., "Holistic management," 4.

20. FAO, *Review of Evidence on Drylands Pastoral Systems and Climate Change*, 11; IPCC, *Climate Change 2007*, 511.

21. FAO, *Challenges and Opportunities for Carbon Sequestration in Grassland Systems*, 12.

22. Cuartas Cardona et al., "Contribution of intensive silvopastoral systems to animal performance and to adaptation and mitigation of climate change," 12.

23. IPCC, *Climate Change 2007*, 511; Steinfeld et al., *Livestock's Long Shadow*, 119–20.

24. Steinfeld et al., *Livestock's Long Shadow*, 97.

25. Ibid., 107.

26. IPCC, *Climate Change 2007*, 511.

27. Steinfeld et al., *Livestock's Long Shadow*, 121.

28. Ibid., 74.

29. Savory, *How to Fight Desertification and Reverse Climate Change*.

30. Carter et al., "Holistic management," 4.

31. Briske et al., "Commentary," 51.

32. Tennigkeit and Wilkes, *An Assessment of the Potential for Carbon Finance in Rangelands*, 9.

33. Conant, Paustian, and Elliott, "Grassland management and conversion into grassland: Effects on soil carbon," 343.

34. Flannery, *Now or Never*, 88.

35. Ibid., 830–32.

36. Teague et al., "Multi-paddock grazing on rangelands," 712.

37. Soussana et al., "Full accounting of the greenhouse gas (CO_2, N_2O, CH_4) budget of nine European grassland sites," 132.

38. O'Brien, et al., "A case study of the carbon footprint of milk from high-performing confinement and grass-based dairy farms," 1835.

39. Naranjo et al., "Balance de gases de efecto invernadero en sistemas silvopastoriles intensivos con *Leucaena leucocephala* en Colombia."

40. Ibid.

41. IPCC, *Climate Change 2014: Mitigation of Climate Change*, 838.

42. Ibid., 840.

43. FAO, *Review of Evidence on Drylands Pastoral Systems and Climate Change*, 1.

44. Ibid.

45. FAO, *Challenges and Opportunities for Carbon Sequestration in Grassland Systems*, 3.

46. IPCC, *Land Use, Land-Use Change, and Forestry*, 184.

47. Russelle, Entz, and Franzluebbers, "Reconsidering integrated crop–livestock systems in North America," 325.

48. Wilson, *Livestock Production Systems*.

49. Toensmeier, "Integrating livestock in the food forest."

50. Wilkinson, *Nut Grower's Guide*, 40.

51. Toensmeier, "Integrating livestock in the food forest."

52. FAO, *Climate-Smart Agriculture Sourcebook*, 222.

53. IPCC, *Climate Change 2014: Mitigation of Climate Change*, 830–32.

54. FAO, *Climate-Smart Agriculture Sourcebook*, 223.

55. Russelle, Entz, and Franzluebbers, "Reconsidering integrated crop–livestock systems in North America," 327.

56. Ohler, *Modern Coconut Management*, 290–337.

57. FAO, *Challenges and Opportunities for Carbon Sequestration in Grassland Systems*, 11.

58. USDA, Census of Agriculture 2012, 558.

59. Neely and De Leeuw, "Home on the range," 337.

60. Kilani et al., "Al-Hima," 1.

61. FAO, *Challenges and Opportunities for Carbon Sequestration in Grassland Systems*, 11.

62. Flannery, *Now or Never*, 88.

63. Briske et al., "Origin, persistence, and resolution of the rotational grazing debate," 325.

64. IPCC, *Climate Change 2007*, 508.

65. Conant, Paustian, and Elliott, "Grassland management and conversion into grassland: Effects on soil carbon," 343.

66. Tennigkeit and Wilkes, *An Assessment of the Potential for Carbon Finance in Rangelands*, 9.

67. Smith et al., "Greenhouse gas mitigation in agriculture," 795–796; FAO, *Challenges and Opportunities for Carbon Sequestration in Grassland Systems*, 11; IPCC, *Land Use, Land-Use Change, and Forestry*, 184.

68. Briske et al., "Commentary," 51.

69. Ryals and Silver, "Effects of organic matter amendments on net primary productivity and greenhouse gas emissions in annual grasslands," 1.

70. Lal, "Abating climate change and feeding the world through soil carbon sequestration," 447.

71. IPCC, *Climate Change 2014: Mitigation of Climate Change*, 830–32.

72. FAO, *Review of Evidence on Drylands Pastoral Systems and Climate Change*, 14.

73. FAO, *Challenges and Opportunities for Carbon Sequestration in Grassland Systems*, v.

74. Harvey et al., "Climate-smart landscapes," 80.

75. Soussana et al., "Full accounting of the greenhouse gas (CO_2, N_2O, CH_4) budget of nine European grassland sites," 132.

76. Nair, "Climate change mitigation," 47; Zomer et al., *Trees on Farm*, 3; Boffa, *Agroforestry Parklands in Sub-Saharan Africa*; Papanastasis et al., "Traditional agroforestry systems and their evolution in Greece," 95.

77. Montagnini and Nair, "Carbon sequestration," 289.

78. McSherry and Ritchie, "Effects of grazing on grassland soil carbon: a global review," 1353.

79. Ibid., 1347.

80. Ibid., 1355.

81. Henderson et al. "Greenhouse gas mitigation potential of the world's grazing lands: Modeling soil carbon and nitrogen fluxes of mitigation practices," 91.

82. Ibid., 94.

83. Ibid.

84. Ibid., 96.

85. Ibid., 96.

86. See Tables 3.3 and 3.4.

87. Udawatta and Jose, "Carbon sequestration potential of agroforestry practices in temperate North America," 25; Nair, "Climate change mitigation," 53.

88. Nair, "Climate change mitigation," 53.

89. Gutteridge and Shelton, "Animal production potential of agroforestry systems," 9.

90. Hamilton, *Silvopasture*.

91. Schoeneberger et al., "Branching out," 132.

92. Harvey et al., "Climate-smart landscapes," 80.

93. Van Huis et al., *Edible Insects*, 60.

94. International Sericultural Commission, "Statistics."

95. Ibid.

96. Van Huis et al., *Edible Insects*, 22.

97. Ibid., 11.

98. Ibid., 52–53.

99. Ibid.

100. Gutteridge and Shelton, "Animal production potential of agroforestry systems," 13.

101. Boffa, *Agroforestry Parklands in Sub-Saharan Africa*.

102. Dupraz and Newman, "Temperate agroforestry," 202–03.

103. Ibid., 226–28.

104. Gutteridge and Shelton, "Animal production potential of agroforestry systems," 11.

105. Cuartas Cardona et al., "Contribution of intensive silvopastoral systems to animal performance and to adaptation and mitigation of climate change," 7.

106. Ibid.

107. Karki, "Sustainable year-round forage production and grazing/browsing management education program," 128.

108. Cuartas Cardona et al., "Contribution of intensive silvopastoral systems to animal performance and to adaptation and mitigation of climate change," 12.

109. Ibid.

110. Ibid., 9.

111. Ibid., 7–9.

112. Nair, "Climate change mitigation," 53.

113. Sukmana, Abdurachman, and Karama, "Strategies to develop sustainable livestock on marginal land," 55.

114. Irwin and Bratton, "Outdoor living barn," 2.

115. IPCC, *Climate Change 2014: Mitigation of Climate Change*, 830–32.

116. Dupraz and Newman, "Temperate agroforestry," 221.

117. Shepard, *Restoration Agriculture*.

118. Savanna Institute website.

119. Shepard, *Restoration Agriculture*, 153–83.

120. Ibid.

8. Perennial Cropping Systems

1. IPCC, *Climate Change 2014: Mitigation of Climate Change*, 830–32.

2. Ibid.

3. Harvey et al., "Climate-smart landscapes," 80.

4. Montagnini, "Homegardens of Mesoamerica," 43.

5. Nair and Kumar, "Introduction," 2.

6. Ibid., 3.

7. Zomer et al., *Trees on Farm*, 3.

8. Jha et al., "Shade coffee," 4.

9. Wibawa et al., "Rubber based Agroforestry Systems (RAS) as alternative for rubber monoculture system," 2.

10. Manner, "A review of traditional agroforestry in Micronesia," 34.

11. Nair, "Climate change mitigation," 47.

12. FAO Statistics Division online.

13. Elevitch and Wilkinson, *Agroforestry Guides for Pacific Islands*, 103.

14. See chapter 3.

15. Nair, "Climate change mitigation," 53.

16. Clay, *World Agriculture and the Environment*, 73.

17. Ibid.

18. Miccolis et al., "Oil palm and agroforestry systems."

19. FAO Statistics Division online.

20. IPCC, *Land Use, Land-Use Change, and Forestry*, 212.

21. Henley and Yiping, "The climate change challenge and bamboo."

22. See chapter 3 for details.

23. Henley and Yiping, "The climate change challenge and bamboo," 12.

24. Dhamodaran, Gnanaharan, and Pillai, "Bamboo for Pulp and Paper," 32.

25. Hunter and Junqui, "Bamboo Biomass," 4–5.

26. Jansen and Kuiper, "Double green energy from traditional coppice stands in the Netherlands," 401.

27. Jacke and Toensmeier, *Edible Forest Gardens*, vol. 1, 4.

28. Ibid.

29. Clay, *World Agriculture and the Environment*, 309.

30. El Bassam, *Handbook of Bioenergy Crops*, 186.

31. Clay, *World Agriculture and the Environment*, 309.

32. See chapter 3.

33. Nair, "Climate change mitigation," 53.

34. Clay, *World Agriculture and the Environment*, 312–14.

35. Weih, "Intensive short rotation forestry in boreal climates," 1370.
36. See chapter 3.
37. See chapter 3.
38. El Bassam, *Handbook of Bioenergy Crops*, 26.
39. Grogan and Matthews. "A modelling analysis of the potential for soil carbon sequestration under short rotation coppice willow bioenergy plantations," 175–83.
40. El Bassam, *Handbook of Bioenergy Crops*, 26.
41. Rutter, "Woody agriculture."
42. Pasiecznik, "The role of trees in aquaculture systems."
43. Li, *Agro-Ecological Farming Systems in China*, 292.
44. Ibid., 291.
45. Ibid., 292.
46. Ibid., 296.
47. Ibid., 291.
48. Ibid., 296.
49. Whitmore and Turner, *Cultivated Landscapes of Middle America on the Eve of Conquest*, 221.
50. Crossley, "Just beyond the eye," 112.
51. Merlín-Uribe et al., "Environmental and socio-economic sustainability of chinampas (raised beds) in Xochimilco, Mexico City," 216–33.
52. Crossley, "The chinampas of Mexico."
53. Crossley, "Just beyond the eye,"112.
54. Merlín-Uribe et al., "Environmental and socio-economic sustainability of chinampas (raised beds) in Xochimilco, Mexico City," 217.
55. Li, *Agro-Ecological Farming Systems in China*, 299–301.
56. Ibid., 301.
57. Cantini, Gucci, and Sillari, "An alternative to managing olive orchards," 49.
58. Clay, *World Agriculture and the Environment*, 76.
59. Kranemann GmbH equipment website.
60. Sands, Pilgeram, and Morris, "Development and marketing of perennial grains with benefits for human health and nutrition," 211.
61. Bell, "Economics and system applications for perennial grain crops in dryland farming systems in Australia," 178.
62. Wade, "Perennial crops," 10.
63. Van Tassel, personal communication.
64. Cox, Crews, and Jackson, "From genetics and breeding to agronomy to ecology," 163.
65. Soule and Piper, *Farming in Nature's Image*.
66. Cox, Crews, and Jackson, "From genetics and breeding to agronomy to ecology," 167.
67. Bell, "Economics and system applications for perennial grain crops in dryland farming systems in Australia," 178.
68. Rumsey, "Perennial crops and climate change," 14.
69. Zhang et al., "The progression of perennial rice breeding and genetics research in China," 34.

9. Additional Tools

1. Pandey, Gupta, and Anderson, "Rainwater harvesting as an adaptation to climate change," 48.
2. Oweis et al., *Water Harvesting*, 4.
3. Lancaster, *Rainwater Harvesting for Drylands and Beyond*, 61–71.
4. IPCC, *Climate Change 2014: Mitigation of Climate Change*, 830–32.
5. Pandey, Gupta, and Anderson, "Rainwater harvesting as an adaptation to climate change," 46.
6. Fox, Rockström, and Barron, "Risk analysis and economic viability of water harvesting for supplemental irrigation in semi-arid Burkina Faso and Kenya," 250.
7. Beach et al., "Impacts of the ancient Maya on soils and soil erosion in the central Maya Lowlands," 76.
8. Dorren and Rey. "A review of the effect of terracing on erosion," 97.
9. Harvey et al., "Climate-smart landscapes," 80.
10. Ferguson and Lovell, "Permaculture for agroecology," 4.
11. Ibid.
12. IPCC, *Climate Change 2014: Mitigation of Climate Change*, 830–32.
13. Robert Flanagan, "The bright prospect of biochar," 1.
14. Mann, "The real dirt on rainforest fertility," 920.
15. Leach, Fairhead, and Fraser, "Land grabs for biochar?," 2.
16. Ibid., 19.
17. IPCC, *Climate Change 2014: Mitigation of Climate Change*, 830–32.
18. Ibid., 833.
19. Barrow, "Biochar," 21.
20. Gammage, *The Biggest Estate on Earth*, 168, 173.
21. Ibid., 1.
22. Anderson, *Tending the Wild*, 2.
23. Anderson, *Indigenous Uses, Management, and Restoration of Oaks in the Far Western United States*, 14.
24. Mann, *1491*.
25. Ibid.
26. Australian Department of the Environment, "Current states and trends of the land environment."
27. Steinfeld et al., *Livestock's Long Shadow*, 92.
28. Gammage, *The Biggest Estate on Earth*, 218.
29. Tennigkeit and Wilkes, *An Assessment of the Potential for Carbon Finance in Rangelands*, 13; Tropical Savannas CRC, "The West Arnhem Land Fire Abatement Project (WALFA)."

30. Sampson and Knopf, "Prairie conservation in North America," 418.
31. Knapp et al., "The keystone role of bison in North American tallgrass prairie," 39.
32. Williams, "Plains sense," 1.
33. Lal, "Abating climate change and feeding the world through soil carbon sequestration," 443.

10. INTRODUCTION TO SPECIES

1. Hanelt, *Mansfeld's Encyclopedia of Agricultural and Horticultural Crops.*
2. Eswaran, Lal, and Reich, "Land degradation," 2.
3. Pimentel et al., "Annual vs. perennial grain production," 3.
4. Leakey, *Living with the Trees of Life*, 66.
5. Hanelt, *Mansfeld's Encyclopedia of Agricultural and Horticultural Crops.*
6. Leakey, *Living with the Trees of Life*, 66.
7. Ibid., 177.
8. Ibid., 66.
9. Pollan, *The Botany of Desire.*
10. Klein, *This Changes Everything*, 23; IPCC, *Climate Change 2014: Impacts, Adaptation and Vulnerability*, 9.
11. Finlay, *Growing American Rubber*, 74–106.
12. Tokar, "Biofuels and the global food crisis," 128–30.
13. IPCC, *Climate Change 2014: Impacts, Adaptation and Vulnerability*, 515.
14. Murray and Jesseup, "Breeding and genetics of perennial maize," 109.
15. Lombard and Leakey, "Protecting the rights of farmers and communities while securing long term market access for producers of non-timber forest products," 235.
16. Wikipedia, "Plant breeders rights."
17. Rai, "India–US fight on basmati rice is mostly settled."
18. Lombard and Leakey, "Protecting the rights of farmers and communities while securing long term market access for producers of non-timber forest products," 236.
19. Ibid.
20. Ibid., 245–46.
21. Leakey, *Living with the Trees of Life*, 73.
22. Ibid., 76.
23. Ibid., 96.
24. Ibid., 11.
25. Ibid., xiii.
26. Ibid., 72.
27. Ibid., 75.
28. Ibid., 78.
29. Ibid., 72.
30. Ibid., 67.
31. Ibid., 76.

32. DeHaan et al., "Current efforts to develop perennial wheat and domesticate *Thinopyrum intermedium* as a perennial grain," 78.
33. Van Tassel, personal communication.
34. Eswaran et al., "Global land resources and population-supporting capacity," 4.
35. Van Tassel, personal communication.
36. Ibid.
37. Ibid.
38. Van Tassel, personal communication.
39. Ibid.
40. Rutter, Wiegrefe, and Rutter-Daywater, *Growing Hybrid Hazelnuts*, 200.
41. Ibid., 201–03.
42. Ibid., 207–08.
43. Rutter, personal communication.
44. Rutter, Wiegrefe, and Rutter-Daywater, *Growing Hybrid Hazelnuts*, 208.
45. Rutter, "Hybrid swarms and hickory-pecans," 7–9.
46. Tokar, "The GMO threat to food sovereignty," 10.
47. Ibid., 4.
48. Ibid., 3.
49. Tokar, "The GMO threat to food sovereignty," 9.
50. Union of Concerned Scientists, *Failure to Yield*, 2.
51. Denison, *Darwinian Agriculture.*
52. Davis et al., "Don't judge species on their origins," 153–54.
53. Felger, personal communication.
54. Van Tassel, personal communication.
55. Global Invasive Species Database online.
56. Davis et al., "Don't judge species on their origins," 153–154.
57. Ibid.
58. IPCC, *Climate Change 2014: Impacts, Adaptation and Vulnerability*, 12.
59. Lynas, *Six Degrees*, 176–77.
60. Marris, *Rambunctious Garden*, 73–95.
61. Van Noordwijk, "Agroforestry solutions for buffering climate variability and adapting to change," 226.
62. Hobbs, Higgs, and Hall, "Introduction."
63. Mascaro et al., "Origins of the novel ecosystems concept"; Perring and Ellis, "The extent of novel ecosystems."
64. Jackson, "Perspective."
65. Mascaro, "Perspective."
66. Hallett, "Towards a conceptual framework for novel ecosystems"; Davis et al., "Don't judge species on their origins," 153–54.
67. Hallett, "Towards a conceptual framework for novel ecosystems"; Jackson, "Perspective"; Davis et al., "Don't judge species on their origins," 153–54.

68. Davis, *Invasion Biology*, 188.

69. Ibid., 190.

11. INTRODUCTION TO PERENNIAL STAPLE CROPS

1. FAO Statistics Division online.

2. Goren-Ibor et al., "Nuts, nut cracking, and pitted stones at Gesher Benut Ya'aqov, Israel."

3. Koeppel, *Banana*, 18.

4. Leakey *Living with the Trees of Life*, xiii.

5. IPCC, *Climate Change 2014: Mitigation of Climate Change*, 830–32.

6. Pasiecznik et al., *The Prosopis juliflora–Prosopis pallida Complex*, 8; Whitmore and Turner, *Cultivated Landscapes of Middle America on the Eve of Conquest*, 76–111; Janick and Paull, *Encyclopedia of Fruits and Nuts*.

7. Chapman and Watson, "The archaic period and the flotation revolution; Wendy Hodgson, *Food Plants of the Sonoran Desert*; Pearlstein et al., "Nipa (*Distichlis palmeri*)."

8. Brandt et al., *Enset: The "Tree Against Hunger"*; Plant Resources of Tropical Africa.

9. Janick and Paull, *The Encyclopedia of Fruits and Nuts*; Conedera et al., "The cultivation of *Castanea sativa* in Europe."

10. House and Harwood, *Australian Dry-Zone Acacias for Human Food*; Gammage, *The Biggest Estate on Earth*.

11. Elevitch, *Traditional Trees of Pacific Islands*.

12. FAO Statistics Division online.

13. Janick and Paull, *The Encyclopedia of Fruits and Nuts*, 94–95.

14. Smith, *Tree Crops*, 18.

15. FAO Statistics Division online.

16. Gammage, *The Biggest Estate on Earth*.

17. Pogna et al., "Evaluation of nine perennial wheat derivatives grown in Italy," 56.

18. FAO, *Perennial Crops for Food Security*.

19. DeHaan et al., "Current efforts to develop perennial wheat and domesticate *Thinopyrum intermedium* as a perennial grain," 73.

20. FAO, *Perennial Crops for Food Security*, 55.

21. Eswaran et al., "Global land resources and population-supporting capacity," 6.

22. Murray and Jesseup, "Breeding and genetics of perennial maize," 103; Pimentel et al., "Annual vs. perennial grain production," 3.

23. One ton per hectare is equivalent to 892 pounds per acre.

24. Kiptot and Franzel, *Gender and Agroforestry in Africa*, 54; see individual plant profiles for details.

25. Smith, "Acorns."

26. Wilkinson, *Nut Grower's Guide*.

27. Altieri, "Strengthening resilience of farming systems," 58.

28. Wilkinson, *Nut Grower's Guide*.

12. BASIC STARCH CROPS

1. FAO Statistics Division online.

2. Elevitch, *Traditional Trees of Pacific Islands*, 90.

3. Ibid., 88.

4. Janick and Paull, *The Encyclopedia of Fruits and Nuts*.

5. Elevitch, *Traditional Trees of Pacific Islands*, 98.

6. Ibid.

7. Nair, "Whither homegardens?," 359.

8. FAO Statistics Division online.

9. Koeppel, *Banana*, 41.

10. FAO Statistics Division online.

11. Koeppel, *Banana*, 18.

12. Elevitch, *Traditional Trees of Pacific Islands*, 534.

13. Ibid., 537.

14. Ibid., 533.

15. Plant Resources of Southeast Asia online.

16. Janick and Paull, *The Encyclopedia of Fruits and Nuts*.

17. FAO Statistics Division online.

18. Nair, "Whither homegardens?," 359.

19. Elevitch, *Traditional Trees of Pacific Islands*, 536.

20. Elevitch, *Traditional Trees of Pacific Islands*, 559–560.

21. Koeppel, *Banana*, 128.

22. Ibid., 185–242.

23. Martin, *Tropical Yams and Their Potential*, 1.

24. Hanelt, *Mansfeld's Encyclopedia of Agricultural and Horticultural Crops*.

25. Martin, *Tropical Yams and Their Potential*, 16–17.

26. Ibid., 13; Flach and Rumawas, *Plant Resources of Southeast Asia, Plants Yielding Non-Seed Cabohydrates*.

27. Martin, *Tropical Yams and Their Potential*, 1–2.

28. Ibid., 19.

29. Mulualem and Mohammed, "Genetic variability and association among yield and yield related traits in aerial yam (*Dioscorea bulbifera*) accessions at southwestern Ethiopia."

30. Elevitch, *Traditional Trees of Pacific Islands*, 492.

31. Ibid., 492.

32. Ibid., 510.

33. Ibid., 502–08.

34. Flach, *Sago Palm*, 59.

35. Elevitch, *Traditional Trees of Pacific Islands*, 502.

36. Ibid., 510.

37. Brandt, *The "Tree Against Hunger"*, 1; 7.

38. Ibid., v.

39. Elevitch, *Traditional Trees of Pacific Islands*, 552.

40. Brandt, *The "Tree Against Hunger"*, v.

41. Ibid., 41.

42. National Research Council, *Lost Crops of Africa*, vol. 2: *Vegetables*, 183–84.

43. Romero, personal communication.

44. Hanelt, *Mansfeld's Encyclopedia of Agricultural and Horticultural Crops*.

45. Hodgeson, *Food Plants of the Sonoran Desert*, 14–42.

46. Ibid.

47. Global Invasive Species Database online.

13. BALANCED CARBOHYDRATE CROPS

1. FAO Statistics Division online.

2. Smith, *Tree Crops*, 126–55.

3. FAO Statistics Division online.

4. Duke, *Handbook of Nuts*, 80–92.

5. Rutter, personal communication. Burbank trees observed at Luther Burbank Home and Gardens, Sebastopol CA, October 2012.

6. Rutter, personal communication.

7. Wilkinson, *Nut Grower's Guide*, 123.

8. Wilkinson, *Nut Grower's Guide*, 120–22.

9. Li, *Agro-Ecological Farming Systems in China*, 106–07.

10. World Agroforestry Centre, *Participatory Agroforestry Development in DPR Korea*, 152.

11. Rico-Gray and Garcia-Franco, "The Maya and the vegetation of the Yucatan Peninsula," 135–37.

12. Ibid.; Janick and Paull, *The Encyclopedia of Fruits and Nuts*, 492.

13. Janick and Paull, *The Encyclopedia of Fruits and Nuts*, 492.

14. Ibid.

15. Ibid., 479–80.

16. Nair, "Whither homegardens?," 359.

17. Elevitch, *Traditional Trees of Pacific Islands*, 107.

18. Smith, *Tree Crops*, 159.

19. Boffa, *Agroforestry Parklands in Sub-Saharan Africa*.

20. Dupraz and Newman, "Temperate agroforestry in Europe," 203.

21. Hanelt, *Mansfeld's Encyclopedia of Agricultural and Horticultural Crops*.

22. Smith, *Tree Crops*, 158.

23. Janick and Paull, *The Encyclopedia of Fruits and Nuts*, 407.

24. Dupraz and Newman, "Temperate agroforestry in Europe," 202.

25. Hanelt, *Mansfeld's Encyclopedia of Agricultural and Horticultural Crops*.

26. Anderson, *Tending the Wild*.

27. North or South Korea unspecified. Janick and Paull, *The Encyclopedia of Fruits and Nuts*, 407.

28. Smith, personal communication.

29. Hanelt, *Mansfeld's Encyclopedia of Agricultural and Horticultural Crops*.

30. Anderson, *Indigenous Uses, Management, and Restoration of Oaks of the Far Western United States*.

31. Oikos Tree Crops catalog 2014.

32. Fulbright, *A Guide to Nut Tree Culture in North America*, 233.

33. Asmus, personal communication.

34. Fulbright, *A Guide to Nut Tree Culture in North America*, 232.

35. Ciesla, *Non-Wood Forest Products from Temperate Broad-Leaved Trees*.

36. Smith, "Acorns."

37. Anderson, *Indigenous Uses, Management, and Restoration of Oaks of the Far Western United States*, 14.

38. Li, *Agro-Ecological Farming Systems in China*, 106–07.

39. Elevitch, *Traditional Trees of Pacific Islands*, 408.

40. Ibid., 418; Duke, *Handbook of Nuts*, 175.

41. Elevitch, *Traditional Trees of Pacific Islands*, 418.

42. Ibid., 421.

43. Ibid., 407–23.

44. Ibid., 422.

45. Plant Resources of Tropical Africa online database.

46. Ibid.

47. Ibid.

48. Ibid.

49. Ibid.

50. Ibid.

51. Sands, Pilgeram, and Morris, "Development and marketing of perennial grains with benefits for human health and nutrition," 208–09.

52. Zhang et al., "The progression of perennial rice breeding and genetics research in China," 27–38.

53. Ibid., 34.

54. Land Institute, "Perennial rice in tropical China produces grain competitively," 27.

55. Bird, "Perennial rice: In search of a greener, hardier staple crop."

56. Pimentel et al., "Annual vs. perennial grain production."

57. Pearlstein et al., "Nipa (*Distichlis palmeri*)," 60.

58. Felger, personal communication.

59. Ibid.

60. Ibid.

61. Paterson et al., "Viewpoint," 91.

62. Ibid.

63. Van Tassel, personal communication.

64. Paterson et al., "Viewpoint," 95.

65. Bell, "Economics and system applications for perennial grain crops in dryland farming systems in Australia," 167.

66. Pimentel et al., "Annual vs. perennial grain production," 4.

67. Pogna et al., "Evaluation of nine perennial wheat derivatives grown in Italy," 56.

68. Cox et al., "Prospects for developing perennial grain crops," 655.

69. Wade, "Perennial crops," 10; Larkin and Newell, "Perennial wheat breeding," 40.

70. Pimentel et al., "Annual vs. perennial grain production," 6.

71. Wade, "Perennial crops," 10.

72. Pogna et al., "Evaluation of nine perennial wheat derivatives grown in Italy," 41; Land Institute, "The long, hard challenge of wheat," 19.

73. DeHaan et al., "Current efforts to develop perennial wheat and domesticate *Thinopyrum intermedium* as a perennial grain," 77.

74. Murray and Jesseup, "Breeding and genetics of perennial maize," 106.

75. Van Tassel, personal communication.

76. Murray and Jesseup, "Breeding and genetics of perennial maize," 109.

77. Ibid.

78. Van Tassel, personal communication.

79. Zhang et al., "The progression of perennial rice breeding and genetics research in China," 28.

80. Land Institute, "A grain seen as nearest the perennial goal," 21.

81. Jaikumar et al., "Agronomic assessment of perennial wheat and perennial rye as cereal crops," 1.

82. Sands, Pilgeram, and Morris, "Development and marketing of perennial grains with benefits for human health and nutrition," 211.

83. Van Tassel, personal communication.

84. Pogna et al., "Evaluation of nine perennial wheat derivatives grown in Italy," 72.

85. DeHaan et al., "Current efforts to develop perennial wheat and domesticate *Thinopyrum intermedium* as a perennial grain," 81.

86. Ibid., 73.

87. Ibid., 78.

88. Ibid.

89. Plant Resources of Southeast Asia online.

90. Norman, Pearson, and Searle, *The Ecology of Tropical Food Crops*, 164.

91. Low, *Wild Food Plants of Australia*, 159.

92. Smith and Smith, *Grow Your Own Bushfoods*, 87–88.

93. These include *Astrebla squarrosa* and *Panicum decompositum*. See Bill Gammage's *The Biggest Estate on Earth* for details.

94. Sands, Pilgeram, and Morris, "Development and marketing of perennial grains with benefits for human health and nutrition," 211.

95. Janick and Paull, *The Encyclopedia of Fruits and Nuts*.

96. Ibid., 95.

97. Ibid., 94–95.

98. Ibid., 96.

99. Clement, "The potential use of the pejibaye palm in agroforestry systems," 203–04.

100. Ibid., 204.

101. Mora-Urpí, Weber, and Clement, *Peach Palm, Bactris gasipaes* Kunth, 17.

102. Clement, "The potential use of the pejibaye palm in agroforestry systems," 209.

103. Geesing, Felker, and Bingham, "Influence of mesquite (*Prosopis glandulosa*) on soil nitrogen and carbon development," 176.

104. Pasiecznik, *The* Prosopis juliflora–Prosopis pallida *Complex*, 59–61.

105. USDA Plants Database online.

106. Pasiecznik, *The* Prosopis juliflora–Prosopis pallida *Complex*, 59–61.

107. Williams et al., "Agroforestry in North America and its role in farming systems," 30.

108. Pasiecznik, *The* Prosopis juliflora–Prosopis pallida *Complex*, 19–28.

109. Ibid., 85.

110. Ibid., 130.

111. Felger and Logan, "Mesquite," 4.

112. Pasiecznik, *The* Prosopis juliflora–Prosopis pallida *Complex*, 73–84.

113. Ibid., 98, 99.

114. Ibid., 130.

115. Ibid., 5.

116. Felger and Logan, "Mesquite," 4.

117. Pasiecznik, *The* Prosopis juliflora–Prosopis pallida *Complex*, 92–93.

118. Ibid., 130.

119. Ibid., 120–21.

120. Ibid., 1.

121. Ibid., 6.

122. Ibid., 123.

123. Ibid., 120–21.

124. Boffa, *Agroforestry Parklands in Sub-Saharan Africa*, accessed April 21, 2014.

125. Pasiecznik, *The* Prosopis juliflora–Prosopis pallida *Complex*, 1.

126. Habletzul, personal communication.

127. Pasiecznik, *The* Prosopis juliflora–Prosopis pallida *Complex*, 115.

128. Ibid., 133.

129. Ibid., 127.

130. Felger, personal communication.

131. Pasiecznik, *The* Prosopis juliflora–Prosopis pallida *Complex*, 135.

132. Janick and Paull, *The Encyclopedia of Fruits and Nuts*, 389.

133. FAO Statistics Division online.

134. Janick and Paull, *The Encyclopedia of Fruits and Nuts*, 387–88.

135. Solowey, *Growing Bread on Trees*, 193.

136. Janick and Paull, *The Encyclopedia of Fruits and Nuts*, 387.

137. Ibid., 389.

138. FAO Statistics Division online.

139. Janick and Paull, *The Encyclopedia of Fruits and Nuts*, 387.

140. USDA Plants Database online.

141. Smith, *Tree Crops*, 65–80.

142. Seibert et al., "Fuel and chemical co-production from tree crops," 54.

143. Facciola, *Cornucopia II*, 68.

144. Ibid.

145. Dupraz, Newmann, and Gordon, "Temperate agroforestry," 223–25.

146. National Research Council, *Firewood Crops*, vol. 2, 36–37.

147. Dupraz, Newman, and Gordon, "Temperate agroforestry," 223–25.

148. Ibid., 225.

149. Gold, "Honeylocust (*Gleditsia triacanthos*)."

150. Seibert et al., "Fuel and chemical co-production from tree crops," 54.

151. Bryan, *Leguminous Trees with Edible Beans*, 209.

14. Protein Crops

1. FAO Statistics Division online.

2. Florez et al., "*Erythrina edulis*."

3. Winrock International, *Nitrogen Fixing Trees Highlights*.

4. Duarte, *Guia para el cultivo y aprovechamiento del Chachafruto o balú*, 14.

5. Florez et al., "*Erythrina edulis*."

6. Ibid.

7. Duarte, *Guia para el cultivo y aprovechamiento del Chachafruto o balú*, 40–52.

8. FAO Statistics Division online.

9. Daniel and Ong, "Perennial pigeonpea."

10. Plant Resources of Tropical Africa online.

11. Ibid.

12. Daniel and Ong, "Perennial pigeonpea."

13. Hanelt, *Mansfeld's Encyclopedia of Agricultural and Horticultural Crops*.

14. Nair, "Whither homegardens?," 359.

15. FAO Statistics Division online.

16. FAO Statistics Division online.

17. Plant Resources of Tropical Africa online.

18. Ibid.

19. Ibid.

20. Ibid.

21. Ibid.

22. Ibid.

23. Ibid.

24. Ibid.

25. Wink and van Wyk, *Mind-Altering and Poisonous Plants of the World*, 352.

26. Udediebe and Carlini, "Questions and answers to edibility problems of the *Canavalia ensiformis* seeds."

27. House and Harwood, *Australian Dry-Zone Acacias for Human Food*.

28. Ibid.

29. Ibid.

30. Ibid.

31. Francis, *Wattle We Eat for Dinner*, 19.

32. Facciola, *Cornucopia II*, 103.

33. Cram, "Breeding and genetics of *Caragana*," 400–01.

34. Plant Resources of Tropical Africa online.

35. Janick and Paull, *The Encyclopedia of Fruits and Nuts*, 507–509; Duke, *Handbook of Nuts*, 287.

36. Danforth and Noren, *Congo Native Fruits*, 37; Janick and Paull, *The Encyclopedia of Fruits and Nuts*, 509.

37. Janick and Paull, *The Encyclopedia of Fruits and Nuts*, 509.

38. Danforth and Noren, *Congo Native Fruits*, 37.

39. Ibid.

40. Etoamaihe and Ndubueze, "Development and performance evaluation of a dehulling machine for African breadfruit (*Treculia africana*)," 312.

41. Janick and Paull, *The Encyclopedia of Fruits and Nuts*, 509.

42. Hanelt, *Mansfeld's Encyclopedia of Agricultural and Horticultural Crops*.

43. Ibid.

44. Galuppo, "Documantacao do uso y valoriza de olea de piquia (*Caryocar villosum*) e do leite do amapa-doce (*Brosimum parinarioides*) para a comunidade de Piquituba," 8.

45. Ibid., 63.

46. Davys et al., "Leaf concentrate and other benefits of leaf fractionation," 338.

47. Ibid., 353.

48. Ibid., 355; Pirie, *Leaf Protein and Its By-Products in Human and Animal Nutrition*, 47.

49. Dale, "Protein Feeds Coproduction in Biomass Conversion to Fuels and Chemicals," 219–224.

50. IPCC, *Climate Change 2014: Mitigation of Climate Change*, 832.

51. Pirie, *Leaf Protein and Its By-Products in Human and Animal Nutrition*, 41.
52. Davys et al., "Leaf concentrate and other benefits of leaf fractionation," 354.
53. Ibid., 361–62.
54. Ibid., 355.
55. National Research Council, *Lost Crops of Africa*, vol. 2: *Vegetables*, 256.
56. Ibid., 247.
57. Ibid., 256.
58. Ibid., 261.
59. Elevitch, *Specialty Crops for Pacific Islands*, 337.
60. Telek and Graham, *Leaf Protein Concentrates*, 52–111.
61. Hanelt, *Mansfeld's Encyclopedia of Agricultural and Horticultural Crops*.
62. World Vegetable Center, Vegetable Nutrition Database online.
63. Berkelaar, "Chaya: The spinach tree," 176.

15. PROTEIN-OIL CROPS

1. FAO Statistics Division online.
2. Wilkinson, *Nut Grower's Guide*, 81.
3. FAO Statistics Division online.
4. Amy Westervelt, "Is it fair to blame almond farmers for California's drought?"
5. Janick and Paull, *The Encyclopedia of Fruits and Nuts*, 707.
6. Ibid.; Duke, *Handbook of Nuts*, 250.
7. University of California, *Sample Costs to Establish an Almond Orchard and Produce Almonds*, 5.
8. FAO Statistics Division online.
9. Wilkinson, *Nut Grower's Guide*, 84.
10. Hanelt, *Mansfeld's Encyclopedia of Agricultural and Horticultural Crops*.
11. Duke, *Handbook of Nuts*, 194–97.
12. FAO Statistics Division online.
13. Duke, *Handbook of Nuts*, 194–97.
14. Ibid.
15. Ibid.
16. FAO Statistics Division online.
17. Wilkinson, *Nut Grower's Guide*, 194.
18. Li, *Agro-Ecological Farming Systems in China*, 106–07.
19. World Agroforestry Centre, *Participatory Agroforestry Development in DPR Korea*, 146–50.
20. Duke, *Handbook of Nuts*, 184–85.
21. Hanelt, *Mansfeld's Encyclopedia of Agricultural and Horticultural Crops*.
22. National Research Council, *Lost Crops of the Incas*, 323.
23. Duke, *Handbook of Nuts*, 190–93.
24. Hanelt, *Mansfeld's Encyclopedia of Agricultural and Horticultural Crops*.

25. FAO Statistics Division online.
26. Duke, *Handbook of Nuts*, 119; Janick and Paull, *The Encyclopedia of Fruits and Nuts*, 162.
27. Wilkinson, *Nut Grower's Guide*, 129.
28. Rutter, Weigrefe, and Rutter-Daywater, *Growing Hybrid Hazelnuts*, 77.
29. Wilkinson, *Nut Grower's Guide*, 134.
30. Rutter, Weigrefe, and Rutter-Daywater, *Growing Hybrid Hazelnuts*, 157–60.
31. Li, *Agro-Ecological Farming Systems in China*, 106–07.
32. Hanelt, *Mansfeld's Encyclopedia of Agricultural and Horticultural Crops*.
33. Ibid.
34. Ibid.
35. Ibid.
36. Goren-Ibor et al., "Nuts, nut cracking, and pitted stones at Gesher Benut Ya'aqov, Israel," 2455.
37. Wilkinson, *Nut Grower's Guide*, 177.
38. FAO Statistics Division online.
39. Duke, *Handbook of Nuts*, 240; Janick and Paull, *The Encyclopedia of Fruits and Nuts*, 21.
40. Duke, *Handbook of Nuts*, 240.
41. Wilkinson, *Nut Grower's Guide*, 185.
42. FAO Statistics Division online.
43. Wilkinson, *Nut Grower's Guide*, 184.
44. Li, *Agro-Ecological Farming Systems in China*, 106–07.
45. Crawford, "Nut pines."
46. Ibid.
47. Ibid.
48. World Agroforestry Centre, *Participatory Agroforestry Development in DPR Korea*, 21.
49. Crawford, "Nut pines."
50. World Agroforestry Centre, *Participatory Agroforestry Development in DPR Korea*, 21.
51. Duke, *Handbook of Nuts*, 237.
52. Crawford, "Nut pines"; Gymnosperm Database online.
53. World Agroforestry Centre, *Participatory Agroforestry Development in DPR Korea*, 141; Yao, Qi, and Yin, "Biodiesel production from *Xanthoceras sorbifolia* in China," 57–65.
54. World Agroforestry Centre, *Participatory Agroforestry Development in DPR Korea*, 141.
55. Ibid.
56. Ibid.
57. Li, *Agro-Ecological Farming Systems in China*, 106–07.
58. Duke, *Handbook of Nuts*, 19.
59. FAO Statistics Division online.
60. Duke, *Handbook of Nuts*, 20.
61. Wink and van Wyk, *Mind-Altering and Poisonous Plants of the World*, 343.

62. Hanelt, *Mansfeld's Encyclopedia of Agricultural and Horticultural Crops.*
63. Duke, *Handbook of Nuts*, 21.
64. FAO Statistics Division online.
65. Wilkinson, *Nut Grower's Guide*, 107.
66. Nair, "Whither homegardens?," 359
67. de Camargo et al., "How old are large Brazil-nut trees (*Bertholletia excelsa*) in the Amazon?," 389–91.
68. Ibid.
69. Janick and Paull, *The Encyclopedia of Fruits and Nuts*, 451.
70. Ibid., 450.
71. Maués, "Reproductive phenology and pollination of the Brazil nut tree (*Bertholletia excelsa* Humb. & Bonpl. Lecythidaceae) in Eastern Amazonia," 245–54.
72. Smith et al.,"Agroforestry developments and potential in the Brazilian Amazon," 251–63; Beer and Muschler, "Multistrata agroforestry systems with perennial crops."
73. Beer and Muschler, "Multistrata agroforestry systems with perennial crops."
74. Duke, *Handbook of Nuts*, 258.
75. Lee, *Man the Hunter*, 33.
76. Janick and Paull, *The Encyclopedia of Fruits and Nuts*, 374.
77. Plant Resources of Tropical Africa online.
78. Ibid.
79. Ibid.
80. Janick and Paull, *The Encyclopedia of Fruits and Nuts*, 374; Hanelt, *Mansfeld's Encyclopedia of Agricultural and Horticultural Crops.*
81. Duke, *Handbook of Nuts*, 78.
82. Ibid.
83. Janick and Paull, *The Encyclopedia of Fruits and Nuts*, 366.
84. Duke, *Handbook of Nuts*, 78.
85. Nair, "Whither homegardens?," 359.
86. National Research Council, *The Winged Bean*, 13.
87. Ibid., 22.
88. Ibid., 16; French, *Food Plants of Papua New Guinea*, 54.
89. National Research Council, *Lost Crops of Africa*, vol. 2: *Vegetables*, 207.
90. Plant Resources of Tropical Africa online.
91. Plant Resources of Tropical Africa online; National Research Council, *Lost Crops of Africa*, vol. 2: *Vegetables*, 214.
92. National Research Council, *Lost Crops of Africa*, vol. 2: *Vegetables*, 207.
93. Bryan, *Leguminous Trees with Edible Beans*, 156.
94. Plant Resources of Tropical Africa online.
95. Ibid.
96. Danforth and Noren, *Congo Native Fruits*, 67.
97. Plant Resources of Tropical Africa online.
98. Ibid.
99. Bryan, *Leguminous Trees with Edible Beans*, 164.
100. Ma, "Evaluation of fertile lines derived from the hybridization of Glycine max and G. tomentella," 1.
101. Plant Resources of Tropical Africa online.
102. FAO Statistics Division online.
103. Plant Resources of Tropical Africa online.
104. Ibid.
105. Ibid.
106. Ibid.
107. Ibid.
108. Price, personal communication.
109. Duke, *Handbook of Energy Crops.*
110. Giwa, Abdullah, and Adam, "Investigating 'Egusi' (*Citrullus colocynthis* L.) seed oil as potential biodiesel feedstock."
111. Duke, *Handbook of Energy Crops*; Giwa, Abdullah, and Adam, "Investigating 'Egusi' (*Citrullus colocynthis* L.) seed oil as potential biodiesel feedstock."
112. Duke, *Handbook of Energy Crops* online.
113. Giwa, Abdullah, and Adam, "Investigating 'Egusi' (*Citrullus colocynthis* L.) seed oil as potential biodiesel feedstock."
114. Duke, *Handbook of Energy Crops.*
115. Hanelt, *Mansfeld's Encyclopedia of Agricultural and Horticultural Crops.*
116. El Bassam, *Handbook of Bioenergy Crops*, 137.
117. Ibid.
118. Lancaster, personal communication.
119. Duke, *Handbook of Energy Crops*, under profile for *Citrullus colocynthis*; El Bassam, *Handbook of Bioenergy Crops*, 137.
120. El Bassam, *Handbook of Bioenergy Crops*, 137.
121. Ibid.
122. Hanelt, *Mansfeld's Encyclopedia of Agricultural and Horticultural Crops.*
123. Van Wyk and Gerike, *People's Plants*, 34.
124. Plant Resources of Tropical Africa online.
125. Van Wyk and Gerike, *People's Plants*, 34.
126. Plant Resources of Tropical Africa online.
127. Ibid.
128. Van Wyk and Gerike, *People's Plants*, 34.
129. Plant Resources of Tropical Africa online.
130. Van Wyk and Gerike, *People's Plants*, 34.
131. Duke, *Handbook of Nuts*, 169.
132. Van Tassel et al., "Evaluating perennial candidates for domestication," 115.
133. FAO Statistics Division online.
134. Duke, *Handbook of Nuts*, 169; Plant Resources of Tropical Africa online.
135. Plant Resources of Tropical Africa online.
136. Van Tassel et al., "Evaluating perennial candidates for domestication," 115.

137. Ibid.

138. Ibid., 125.

139. Ibid., 115.

140. Ibid.

141. Ibid., 125.

142. National Research Council, *Lost Crops of Africa*, vol. 2: *Vegetables*, 296.

143. Ibid., 297.

144. Plant Resources of Tropical Africa online.

145. Ibid.

146. National Research Council, *Lost Crops of Africa*, vol. 2: *Vegetables*, 294.

147. Ibid., 293.

148. Ibid., 289–90.

149. Plant Resources of Tropical Africa online.

150. National Research Council, *Lost Crops of Africa*, vol. 2: *Vegetables*, 289.

151. National Research Council, *Lost Crops of Africa*, vol. 3: *Fruits*, 74.

152. Ibid., 74.

153. Leakey, *Living with the Trees of Life*, 73.

154. Plant Resources of Tropical Africa online.

155. Ibid.

156. National Research Council, *Lost Crops of Africa*, vol. 3: *Fruits*, 69.

157. Ibid., 73.

158. Leakey, *Living with the Trees of Life*, 90.

159. National Research Council, *Lost Crops of Africa*, vol. 3: *Fruits*, 65.

160. Plant Resources of Tropical Africa online.

161. Ibid.

162. Wink and van Wyk, *Mind-Altering and Poisonous Plants of the World*, 66.

163. Morton, *Fruits of Warm Climates*, 271.

164. Ekué et al., "Uses, traditional management, perception of variation and preferences in ackee (*Blighia sapida* KD Koenig) fruit traits in Benin."

165. Janick and Paull, *The Encyclopedia of Fruits and Nuts*, 793; Plant Resources of Tropical Africa online.

166. Morton, *Fruits of Warm Climates*, 271.

167. Plant Resources of Tropical Africa online.

168. Ekué et al., "Uses, traditional management, perception of variation and preferences in ackee (*Blighia sapida* KD Koenig) fruit traits in Benin."

16. EDIBLE OIL CROPS

1. FAO Statistics Division online.

2. Van Tassel et al., "Evaluating perennial candidates for domestication: Lessons from wild sunflower relatives," 112.

3. Ibid.

4. Janick and Paull, *The Encyclopedia of Fruits and Nuts*, 107–09.

5. Hanelt, *Mansfeld's Encyclopedia of Agricultural and Horticultural Crops*.

6. FAO Statistics Division online.

7. Elevitch, *Traditional Trees of Pacific Islands*, 292.

8. Duke, *Handbook of Nuts*, 101; Ohler, *Modern Coconut Management*, 24.

9. Duke, *Handbook of Nuts*, 101.

10. Elevitch, *Traditional Trees of Pacific Islands*, 292–96.

11. Ohler, *Modern Coconut Management*, 6.

12. FAO Statistics Division online.

13. Ohler, *Modern Coconut Management*, 245.

14. Elevitch, *Traditional Trees of Pacific Islands*, 282.

15. Janick and Paull, *The Encyclopedia of Fruits and Nuts*, 108.

16. Nair, "Whither homegardens?," 359.

17. Elevitch, *Traditional Trees of Pacific Islands*, 291–92.

18. Ibid.

19. Ohler, *Modern Coconut Management*, 263.

20. Ibid., 265.

21. Ibid., 267.

22. Ibid., 272.

23. Ibid., 287.

24. Ibid., 289.

25. Ibid., 317.

26. Duke, *Handbook of Nuts*, 70.

27. Fulbright, *A Guide to Nut Tree Culture in North America*, 120.

28. Janick and Paull, *The Encyclopedia of Fruits and Nuts*, 421.

29. Wilkinson, *Nut Grower's Guide*, 171.

30. Ibid.

31. Ibid., 170.

32. Williams et al., "Agroforestry in North America and its role in farming systems," 36.

33. Li, *Agro-Ecological Farming Systems in China*, 106–07.

34. Rutter, "Hybrid swarms and hickory-pecans."

35. Hanelt, *Mansfeld's Encyclopedia of Agricultural and Horticultural Crops*.

36. Ibid.

37. Ibid.

38. Ibid.

39. Wilkinson, *Nut Grower's Guide*, 142.

40. Hanelt, *Mansfeld's Encyclopedia of Agricultural and Horticultural Crops*.

41. Elevitch, *Specialty Crops for Pacific Islands*, 300.

42. Duke, *Handbook of Nuts*, 207; Janick and Paull, *The Encyclopedia of Fruits and Nuts*, 602.

43. Wilkinson, *Nut Grower's Guide*, 145.

44. Elevitch, *Specialty Crops for Pacific Islands*, 308.
45. Duke, *Handbook of Nuts*, 209.
46. Elevitch, *Specialty Crops for Pacific Islands*, 308.
47. Ibid., 302–05.
48. National Research Council, *Lost Crops of Africa*, vol. 2: *Vegetables*, 303.
49. Ibid., 303–21.
50. FAO Statistics Division online.
51. Plant Resources of Tropical Africa online.
52. Ibid.
53. National Research Council, *Lost Crops of Africa*, vol. 2: *Vegetables*, 315.
54. FAO Statistics Division online.
55. National Research Council, *Lost Crops of Africa*, vol. 2: *Vegetables*, 307.
56. Ibid., 315.
57. Boffa, *Agroforestry Parklands in Sub-Saharan Africa*.
58. National Research Council, *Lost Crops of Africa*, vol. 2: *Vegetables*, 316.
59. FAO Statistics Division online.
60. *Economist*, "The other oil spill."
61. Magdoff, "Twenty-first century land grabs," 7; *Economist*, "The other oil spill."
62. Hanelt, *Mansfeld's Encyclopedia of Agricultural and Horticultural Crops*; Duke, *Handbook of Nuts*, 150.
63. Janick and Paull, *The Encyclopedia of Fruits and Nuts*, 118.
64. Danforth and Noren, *100 Tropical Fruits, Nuts, and Spices for the Central African Home Garden*, 147.
65. Hanelt, *Mansfeld's Encyclopedia of Agricultural and Horticultural Crops*.
66. Duke, *Handbook of Nuts*, 150.
67. Janick and Paull, *The Encyclopedia of Fruits and Nuts*, 118.
68. Miccolis et al., "Oil palm and agroforestry systems."
69. Ibid.
70. Morton, *Fruits of Warm Climates*, 92–96.
71. FAO Statistics Division online.
72. Plant Resources of Southeast Asia online database.
73. Morton, *Fruits of Warm Climates*, 101.
74. Plant Resources of Southeast Asia online database.
75. Human, "Oil as a byproduct of the avocado," 5.
76. Nair, "Whither homegardens?," 359
77. Plant Resources of Tropical Africa online.
78. Hanelt, *Mansfeld's Encyclopedia of Agricultural and Horticultural Crops*.
79. Olive Tree Growers, "Olive trees in Florida."
80. FAO Statistics Division online.
81. Janick and Paull, *The Encyclopedia of Fruits and Nuts*, 566.
82. Plant Resources of Tropical Africa online.
83. Crawford, "Silvoarable systems in Europe," 7.
84. Cantini, Gucci, and Sillari, "An alternative to managing olive orchards," 4.9.
85. Janick and Paull, *The Encyclopedia of Fruits and Nuts*, 339–343.
86. Ibid., 340.
87. Ibid., 342.
88. Li, *Agro-Ecological Farming Systems in China*, 106–07.
89. World Agroforestry Centre, *Participatory Agroforestry Development in DPR Korea*, 137.
90. Lu, "Combating desertification with seabuckthorn," 297.
91. Ibid., 298.
92. Duke, *Handbook of Nuts*, 75–77.
93. Shanley, *Fruit Trees and Useful Plants in Amazonian Life*, 109–19.
94. Hanelt, *Mansfeld's Encyclopedia of Agricultural and Horticultural Crops*.
95. Shanley, *Fruit Trees and Useful Plants in Amazonian Life*, 109.
96. Ibid., 112.
97. Ibid.; Duke, *Handbook of Nuts*, 75–77; Janick and Paull, *The Encyclopedia of Fruits and Nuts*, 249.
98. Shanley, *Fruit Trees and Useful Plants in Amazonian Life*, 112.
99. Duke, *Handbook of Nuts*, 75–77.
100. Shanley, *Fruit Trees and Useful Plants in Amazonian Life*, 113.
101. Ibid., 118.
102. Ibid., 112.
103. Duke, *Handbook of Nuts*, 73–77.
104. Whitmore and Turner, *Cultivated Landscapes of Middle America on the Eve of Conquest*, 70.
105. Hanelt, *Mansfeld's Encyclopedia of Agricultural and Horticultural Crops*.
106. Elevitch, *Specialty Crops for Pacific Islands*, 94.
107. Janick and Paull, *The Encyclopedia of Fruits and Nuts*, 885.
108. Whitmore and Turner, *Cultivated Landscapes of Middle America on the Eve of Conquest*.
109. Zomer, *Trees on Farm*, 3.
110. Elevitch, *Specialty Crops for Pacific Islands*, 84.
111. Elevitch and Wilkinson, *Agroforestry Guides for Pacific Islands*, 105.
112. Hanelt, *Mansfeld's Encyclopedia of Agricultural and Horticultural Crops*.
113. World Agroforestry Centre, *Participatory Agroforestry Development in DPR Korea*, 139.
114. Ibid.
115. FAO Statistics Division online.
116. "Western Front" now known to be *B. napus*, not *B. oleracea*. Toensmeier, *Perennial Vegetables*, 106.
117. Hanelt, *Mansfeld's Encyclopedia of Agricultural and Horticultural Crops*.

118. Van Tassel, personal communication.
119. World Agroforestry Centre, *Participatory Agroforestry Development in DPR Korea*, 139.
120. Solowey, "Upscaling the experimental planting of *Argania spinosa* at Kibbutz Ketura, Israel," 257–262.
121. Ibid.
122. Ibid.
123. Foti et al., "Possible alternative utilization of *Cynara* spp.," 219–28.
124. Ibid.
125. Global Invasive Species Database online.

17. SUGAR CROPS

1. Finlay, *Growing American Rubber*, 172.
2. FAO Statistics Division online.
3. Janke, "Stevia"; Schnelle, "Stevia."
4. FAO Statistics Division online.
5. Ibid.
6. Duke, *Handbook of Energy Crops*.
7. Dong et al., "A nitrogen-fixing endophyte of sugarcane stems," 1139.
8. Calvin, *Fuel Oils from Higher Plants*, 148.
9. Takamizawa, Anderson, and Singh, "Ethanol from lignocellulosic crops," 112.
10. Clay, *World Agriculture and the Environment*, 166.
11. Santiago et al., "Sugarcane," 96.
12. Tokar, "Biofuels and the global food crisis," 129.
13. Duke, *Handbook of Energy Crops*.
14. Feedipedia online.
15. Duke, *Handbook of Energy Crops*.
16. FAO Statistics Division online.
17. Flach and Rumawas, *Plant Resources of Southeast Asia*, vol. 9, 143.
18. Ibid., 147.
19. Duke, *Handbook of Energy Crops*.
20. Clay, *World Agriculture and the Environment*, 160.
21. Duke, *Handbook of Energy Crops*.
22. Clay, *World Agriculture and the Environment*, 170.
23. Ibid., 161.
24. Nair, "Whither homegardens?," 359.
25. Santiago et al., "Sugarcane," 81–84.
26. Nickerson, "A jolt for the science behind harvesting maple sap."
27. Staats, Krasny, and Campbell, "History of the Sugar Maple Tree Improvement Program: Uihlein Sugar Maple Field Station."
28. McKentley, "The story of the sweet-sap silver maple."
29. Westover, "Changing climate may substantially alter maple syrup production."

30. Nickerson, "A jolt for the science behind harvesting maple sap."
31. Re, "The effective communication of agricultural R&D output in the UK beet sugar industry"; *Farmers Guardian*, "Attention to detail essential in meeting beet yield targets."
32. Janick and Paull, *The Encyclopedia of Fruits and Nuts*, 138.
33. Ibid., 142.
34. Morton, *Fruits of Warm Climates*, 8.
35. Janick and Paull, *The Encyclopedia of Fruits and Nuts*, 138.
36. Morton, *Fruits of Warm Climates*, 11.
37. El Bassam, *Handbook of Bioenergy Crops*, 179.
38. Janick and Paull, *The Encyclopedia of Fruits and Nuts*, 138.
39. Morton, *Fruits of Warm Climates*, 9.
40. Encyclopedia Brittanica online; Lawton, *Establishing a Food Forest* DVD.
41. Flach and Rumawas, *Plant Resources of Southeast Asia*, vol. 9, 54.
42. Ibid., 54–55.
43. Ibid., 58.
44. Ibid., 56–57.
45. Ibid., 58.
46. Martini et al., "Sugar palm (*Arenga pinnata* (Wurmb) Merr.) for livelihoods and biodiversity conservation in the orangutan habitat of Batang Toru, North Sumatra, Indonesia," 3.
47. Flach and Rumawas, *Plant Resources of Southeast Asia*, vol. 9, 56.
48. Martini et al., "Sugar palm (*Arenga pinnata* (Wurmb) Merr.) for livelihoods and biodiversity conservation in the orangutan habitat of Batang Toru, North Sumatra, Indonesia," 1.
49. Flach and Rumawas, *Plant Resources of Southeast Asia*, vol. 9, 58.

18. INDUSTRIAL CROPS: MATERIALS, CHEMICALS, AND ENERGY

1. Hanelt, *Mansfeld's Encyclopedia of Agricultural and Horticultural Crops*.
2. Clark and Deswarte, *The Biorefinery Concept*, 1–3.
3. Farrelly, *The Book of Bamboo*, 13.
4. Clark and Deswarte, *The Biorefinery Concept*, 4–5.
5. US Energy Information Administration, "How much oil is used to make plastic?"
6. Thielen, *Bioplastics*, 79.
7. Honary and Singh, "Industrial oil types and uses," 157.
8. Singh, "Overview of industrial crops," 3.
9. Smith, *Tree Crops*, 292–93.
10. Seibert et al., "Fuel and chemical co-production from tree crops," 54.
11. Smith and Perino, "Osage orange (*Maclura pomifera*)," 35.

12. Moser et al., "Preparation of fatty acid methyl esters from Osage orange (*Maclura pomifera*) oil and evaluation as biodiesel," 1870.

13. Ibid., 1869.

14. Smith and Perino, "Osage orange (*Maclura pomifera*)," 35.

15. Ibid., 35–36.

16. Seibert et al., "Fuel and chemical co-production from tree crops," 54.

17. Smith and Perino, "Osage orange (*Maclura pomifera*)," 35.

18. Seibert et al., "Fuel and chemical co-production from tree crops," 54.

19. Moser et al., "Preparation of fatty acid methyl esters from Osage orange (*Maclura pomifera*) oil and evaluation as biodiesel," 1870.

20. Seibert et al., "Fuel and chemical co-production from tree crops," 54.

21. Smith and Perino, "Osage orange (*Maclura pomifera*)," 25.

22. Drieling and Mussig, "Grades and standards," 51–63.

23. Turley, "The chemical value of biomass," 21.

24. IPCC, *Climate Change 2014: Mitigation of Climate Change*, 829.

25. IPCC, *Climate Change 2007*, 501.

26. Leffler, *Petroleum Refining in Nontechnical Language*, 16–18.

27. Fetter, personal communication.

28. For example, see Hardy, "The bio-based economy," 11–15.

29. Hochschild, *King Leopold's Ghost*, 3.

30. Tokar, "Biofuels and the global food crisis," 129.

31. Magdoff, "Twenty-first-century land grabs," 5.

32. Magdoff and Foster, *What Every Environmentalist Needs to Know About Capitalism*, 64.

33. Magdoff, "Twenty-first-century land grabs," 7.

34. Tokar, "Biofuels and the global food crisis," 122.

35. Oxfam International, *Another Inconvenient Truth*.

36. International Panel for Sustainable Resource Management, *Towards Sustainable Production and Use of Resources*, 9.

37. Tokar, "Biofuels and the global food crisis," 126–27.

38. Ibid., 127.

39. Ibid., 124.

40. McKibben, *Eaarth*, 27.

41. Tokar, "Biofuels and the global food crisis," 129–32.

42. International Panel for Sustainable Resource Management, *Towards Sustainable Production and Use of Resources*, 16.

43. Dale, "Protein feeds coproduction in biomass conversion to fuels and chemicals," 219.

44. Anderson and Bows, "Beyond 'dangerous' climate change," 41.

45. Gore, *Our Choice*, 247.

46. Bogdanski et al., *Making Integrated Food–Energy Systems Work for People and Climate*, v.

47. Ibid., 5–6.

48. Ibid., v.

49. Ruan et al., "Size matters."

50. Jacobson and Delucchi, "A plan to power 100 percent of the planet with renewables."

51. Budischak et al., "Cost-minimized combinations of wind power, solar power, and electrochemical storage, powering the grid up to 99.9 percent of the time," 60.

52. Frederiks, personal communication.

53. International Panel for Sustainable Resource Management, *Towards Sustainable Production and Use of Resources*, 9.

54. Frederiks, personal communication.

55. Thielen, *Bioplastics*, 79.

56. Clark and Deswarte, *The Biorefinery Concept*, 7.

57. Ibid., 8.

58. Gore, *Our Choice*, 124.

59. International Panel for Sustainable Resource Management, *Towards Sustainable Production and Use of Resources*, 7.

19. Biomass Crops

1. Turley, "The chemical value of biomass," 32.

2. Koutinas et al., "Production of chemicals from biomass," 77–90.

3. Frederiks, personal communication.

4. Farrelly, *The Book of Bamboo*, 13.

5. Henley and Yiping, "The climate change challenge and bamboo," 12.

6. Yiping et al., *Bamboo and Climate Change Mitigation*, 32–33.

7. Ibid., v.

8. Kuehl and Yiping, *Carbon Off-Setting with Bamboo*, 21.

9. Janssen, *Designing and Building with Bamboo*, 77.

10. Henley and Yiping, "The climate change challenge and bamboo," 18.

11. Dhamodaran, Gnanaharan, and Pillai, "Bamboo for Pulp and Paper," 32.

12. Yiping et al., *Bamboo and Climate Change Mitigation*, 34–36.

13. Li, *Agro-Ecological Farming Systems in China*, 106–07.

14. Qungen and Qingmei, "Bamboo agroforestry in China," 17–21.

15. Chen et al., "Bioenergy industry status and prospects," 21–35.

16. Krawczyk, personal communication.

17. National Academy of Sciences, *Firewood Crops*, 1–9.

18. Romero, personal communcation; Blank, personal communication.

19. National Academy of Sciences, *Firewood Crops*, 24.

20. El Bassam, *Handbook of Bioenergy Crops*, 186.

21. FAO Regional Office for Asia and the Pacific, *Proceedings*.

22. Global Invasive Species Database online.

23. Babe, "Use of introduced trees," 242–44.

24. National Academy of Sciences, *Firewood Crops*, 126–28.

25. Ibid., 24.

26. Babe, "Use of introduced trees," 242–44.
27. Speedy and Pugliese, *Legume Trees and Other Fodder Trees as Protein Sources for Livestock.*
28. Ibid.
29. Tropical Forages Interactive Selection Tool online.
30. Figures given in cubic meters per year, converted to tons per hectare per year as follows. Determined that high-quality firewoods average about 3,800 pounds per cord, or 1.7 tons. A cord is 128 cubic feet, or about 3.6 cubic meters. So 1.7 tons divided by 3.6 cubic meters comes out pretty close to 0.5 ton per cubic meters. Based on Kuhns and Schmidt, "Heating with wood."
31. Nair, *Introduction to Agroforestry.*
32. Global Invasive Species Database online.
33. Gymnosperm Database online.
34. Crawford, "Coast redwood for timber," 33.
35. Dolman, personal communication.
36. Crawford, "Coast redwood for timber," 32.
37. Ibid., 31.
38. Dimitriou and Aronsson, "Willows for energy and phytoremediation in Sweden."
39. Jacke and Toensmeier, *Edible Forest Gardens*, vol. 1, 5.
40. Isebrands et al., "Environmental applications of poplars and willows," 258–62.
41. Hanley, "Willow," 259–72.
42. Dillen et al., "Poplar," 283.
43. Dillen et al., "Poplar."
44. Li, *Agro-Ecological Farming Systems in China*, 106–07.
45. Shim, "Domestication and Improvement of Rattan," 19–23.
46. Hanelt, *Mansfeld's Encyclopedia of Agricultural and Horticultural Crops.*
47. Kennedy, "Turning wood into bones."
48. Fried, "Tropical forests forever?," 204–34.
49. Chen et al., "Bioenergy industry status and prospects," 27.
50. Hanelt, *Mansfeld's Encyclopedia of Agricultural and Horticultural Crops.*
51. Thakur et al., "Eulaliopsis binata."
52. Kim et al., "Developing miscanthus for bioenergy," 302.
53. Ibid., 309–13.
54. Takamizawa, Anderson, and Singh, "Ethanol from lignocellulosic crops," 112.

20. INDUSTRIAL STARCH CROPS

1. Turley, "The chemical value of biomass," 32.
2. Vaca-Garcia, "Biomaterials," 125.
3. Thielen, *Bioplastics*, 6.
4. Ibid., 79.
5. Ibid., 16–19.
6. Stevens, *Green Plastics*, 115.
7. Ibid., 104.
8. Thielen, *Bioplastics*, 7–9.
9. Ibid., 76.
10. Ibid., 7–9.
11. Ibid., 21–44.
12. Open Source Ecology online.
13. Thielen, *Bioplastics*, 79.
14. Ibid. 79.
15. Turley, "The chemical value of biomass," 32–34.
16. Ibid.
17. Ibid.
18. Finlay, *Growing American Rubber*, 172.
19. Ibid.
20. Ibid.
21. See chapter 22.
22. Finlay, *Growing American Rubber*, 172.
23. Ibid.
24. Duke, *Handbook of Nuts*, 26–29.
25. Wink and van Wyk, *Mind-Altering and Poisonous Plants of the World*, 338–419.
26. Duke, *Handbook of Nuts*, 133.
27. Wink and van Wyk, *Mind-Altering and Poisonous Plants of the World*, 44.
28. Duke, *Handbook of Nuts*, 235.
29. Ibid., 234–35.
30. Ibid., 235.

21. INDUSTRIAL OIL CROPS

1. Honary and Singh, "Industrial oil types and uses," 157.
2. Moser et al., "Preparation of fatty acid methyl esters from Osage orange (*Maclura pomifera*) oil and evaluation as biodiesel," 1869.
3. Plant Resources of Tropical Africa database online.
4. FAO Statistics Division online.
5. El Bassam, *Handbook of Bioenergy Crops*, 150.
6. Global Invasive Species Database online.
7. Plant Resources of Tropical Africa database online.
8. Ibid.
9. FAO Statistics Division online.
10. Plant Resources of Tropical Africa database online.
11. Hanelt, *Mansfeld's Encyclopedia of Agricultural and Horticultural Crops.*
12. FAO Statistics Division online.
13. Global Invasive Species Database, accessed April 9, 2014.
14. Duke, *Handbook of Nuts*, 262–65.
15. Ibid., 264.
16. FAO Statistics Division online.
17. Duke, *Handbook of Nuts*, 264.
18. Ibid., 264.

19. Li, *Agro-Ecological Farming Systems in China*, 106–07.
20. Ibid., 182.
21. Elevitch, *Traditional Trees of Pacific Islands*, 42.
22. Ibid., 49.
23. Duke, *Handbook of Nuts*, 13.
24. Global Invasive Species Database online.
25. Duke, *Handbook of Nuts*, 8.
26. FAO Statistics Division online.
27. Duke, *Handbook of Nuts*, 14–15.
28. Hanelt, *Mansfeld's Encyclopedia of Agricultural and Horticultural Crops*.
29. Tokar, "Biofuels and the global food crisis," 128.
30. Ibid., 128–30.
31. Hanelt, *Mansfeld's Encyclopedia of Agricultural and Horticultural Crops*.
32. Sharma and Kuman, "*Jatropha curcas*," 202
33. Duke, *Handbook of Nuts*, 179.
34. FACT Foundation, *The Jatropha Handbook*, 28–30.
35. Hanelt, *Mansfeld's Encyclopedia of Agricultural and Horticultural Crops*.
36. Ibid.
37. Duke, *Handbook of Nuts*, 272–75.
38. El Bassam, *Handbook of Bioenergy Crops*, 216, 95; Duke, *Handbook of Nuts*, 274–75.
39. Borlaug, *Jojoba*, 36.
40. Ibid., 30.
41. Kazakoff, Gresshoff, and Scott, "*Pongamia pinnata*," 236–41.
42. Ibid., 240.
43. Ibid., 252.
44. Ibid., 241.
45. Hanelt, *Mansfeld's Encyclopedia of Agricultural and Horticultural Crops*.
46. Kesari and Rangan, "Development of *Pongamia pinnata* as an alternative biofuel crop," 132.
47. Hanelt, *Mansfeld's Encyclopedia of Agricultural and Horticultural Crops*.

22. Hydrocarbon Crops

1. Langenheim, *Plant Resins*, 197–253.
2. See the discussion in chapter 18.
3. Nishimura and Calvin, "Essential oil of *Eucalptus globulus* in California," 432.
4. Calvin, "Petroleum plantations for fuel and materials," 533–38.
5. Kalita, "Hydrocarbon plant," 457.
6. Calvin, "Petroleum plantations for fuel and materials," 533–38.
7. Duke, *Handbook of Energy Crops*, profile of *Euphorbia lathyris*.
8. Buchanan et al., "Hydrocarbon- and rubber-producing crops," 131–45.
9. Kalita, "Hydrocarbon plant," 456–69.
10. Singh and Singh, "Natural rubber," 358–83.
11. Calvin, "Petroleum plantations for fuel and materials," 533–38.
12. Leffler, *Petroleum Refining in Nontechnical Language*, 167–74.
13. Fini et al., "Chemical characterization of biobinder from swine manure," 1506–13.
14. Singh and Singh, "Natural rubber," 377–78.
15. Langenheim, *Plant Resins*, 307–29.
16. Honary and Singh, "Industrial oil types and uses," 157–82.
17. Calvin, "Petroleum plantations for fuel and materials," 533–38.
18. Ibid.
19. Langenheim, *Plant Resins*, 334–39.
20. Loke, Mesa, and Franken, "*Euphorbia tirucalli* bioenergy manual," 55–56.
21. Langenheim, *Plant Resins*, 336.
22. Kalita, "Hydrocarbon plant," 465.
23. Calvin, "Petroleum plantations for fuel and materials," 537.
24. Duke, *Handbook of Energy Crops*, profile of *Euphorbia tirucalli*.
25. Calvin, *Fuel Oils from Higher Plants*, 153.
26. Ibid., 151; Dollar Times Financial Calculator online, www.dollartimes.com.
27. oil-price.net online.
28. Singh and Singh, "Natural rubber," 358.
29. Finlay, *Growing American Rubber*, 3.
30. Singh and Singh, "Natural rubber," 376.
31. Mann, *1493*, 253–78.
32. Hochschild, *King Leopold's Ghost*, 3.
33. Mann, *1493*, 253–78.
34. See table 22.2.
35. Singh and Singh, "Natural rubber," 377–78.
36. See table 22.2.
37. Finlay, *Growing American Rubber*, 171–72.
38. Ibid., 74–106.
39. FAO Statistics Division online.
40. Finlay, *Growing American Rubber*, 3.
41. Mann, *1493*, 253–78.
42. Singh and Singh, "Natural rubber," 367.
43. FAO Statistics Division online.
44. Singh and Singh, "Natural rubber," 366.
45. Wibawa et al., "Rubber based Agroforestry Systems (RAS) as alternatives for rubber monoculture system," 2.
46. Li, *Agro-Ecological Farming Systems in China*, 164–65.
47. Singh and Singh, "Natural rubber," 365.
48. Finlay, *Growing American Rubber*, 23.

49. Ibid., 154–55.

50. Ray et al., "Guayule," 399.

51. Ibid., 400–02.

52. Ibid., 392.

53. Ibid., 392.

54. Duke, *Handbook of Energy Crops*.

55. Suszkiw, "Milkweed."

56. 13 t/ha biomass @ 5.7% latex for undomesticated plants; from Duke, *Handbook of Energy Crops*, profile of *Asclepias incarnata*.

57. Loke, Mesa, and Franken, "*Euphorbia tirucalli* bioenergy manual," 51–52.

58. Ibid., 4.

59. 2200 liters; a barrel of oil has 159 liters and weighs 138kg. Ibid., 56.

60. Ibid., 49.

61. Ibid., 55.

62. Ibid., 31.

63. Maxwell, Wiatr, and Fay. "Energy potential of leafy spurge (*Euphorbia esula*)," 150–56.

64. Calvin, *Fuel Oils from Higher Plants*, 151.

65. Langenheim, *Plant Resins*, 24.

66. Ibid., 24–25.

67. FAO, "Turpentine from pine resin," 1.

68. Langenheim, *Plant Resins*, 257–455.

69. Ibid., 405–06.

70. Ibid., 463.

71. Ibid., 394–95.

72. Outland, *Tapping the Pines*, 122–61.

73. Langenheim, *Plant Resins*, 406–07.

74. Stanley, "Demystifying the tragedy of the commons."

75. Langenheim, *Plant Resins*, 469–70.

76. Calvin, "Petroleum plantations for fuel and materials," 533–38.

77. Langenheim, *Plant Resins*, 393–95.

78. Glanville, "Queensland farmers invest in diesel-producing trees."

79. Shanley, *Fruit Trees and Useful Plants in Amazonian Life*, 71.

80. Ibid., 72.

81. Duke, *Handbook of Energy Crops*.

82. FAO, "Turpentine from pine resin."

83. Weisman, *Gaviotas*.

84. FAO, "Turpentine from pine resin."

85. Ibid.

86. Hanelt, *Mansfeld's Encyclopedia of Agricultural and Horticultural Crops*.

87. FAO, "Turpentine from pine resin."

88. Langenheim, *Plant Resins*, 453.

89. Gymnosperm Database online.

90. Langenheim, *Plant Resins*, 404–06.

91. Duke, *Handbook of Energy Crops*, profile of *Copaifera langsdorffii*.

92. Calvin, *Fuel Oils from Higher Plants*, 155.

93. Ibid.

23. FIBER CROPS

1. Hanninen and Hughes, "Historical, contemporary, and future applications," 385.

2. The Fiber Year, *The Fiber Year 2013*, 5.

3. Piotrowski and Carus, "Natural fibers in technical applications," 75.

4. FAO Statistics Division online.

5. Piotrowski and Carus, "Natural fibers in technical applications," 75.

6. Clay, *World Agriculture and the Environment*, 292–93.

7. Ibid., 294.

8. Thevs et al., "*Apocynum venetum* L. and *Apocynum pictum* Schrenk (Apocynaceae) as multi-functional and multi-service plant species in Central Asia," 159–67.

9. Clay, *World Agriculture and the Environment*, 285.

10. Beckert, *Empire of Cotton*, xix.

11. Goltenboth and Muhlbauer, "Abacá," 163–81.

12. Leson, Harding, and Dippon, "Natural fibers in geotextiles for soil protection and erosion control," 521.

13. Hanninen and Hughes, "Historical, contemporary, and future applications," 521.

14. Neubauer, "Insulation materials based on natural fibers," 481–503.

15. Jayasekara and Amarasinghe, "Coir," 198.

16. Clay, *World Agriculture and the Environment*, 285.

17. Hanelt, *Mansfeld's Encyclopedia of Agricultural and Horticultural Crops*.

18. Plant Resources of Tropical Africa online.

19. Ibid.; Percival and Kohel, "Distribution, collection, and evaluation of *Gossypium*."

20. Percival and Kohel, "Distribution, collection, and evaluation of *Gossypium*."

21. FAO Statistics Division online.

22. Hanelt, *Mansfeld's Encyclopedia of Agricultural and Horticultural Crops*.

23. Ibid..

24. Plant Resources of Tropical Africa online.

25. Piotrowski and Carus, "Natural fibers in technical applications," 75.

26. Jayasekara and Amarasinghe, "Coir," 198.

27. Ibid., 197–219.

28. Jarman, *Plant Fibre Processing*, 3.

29. 1,000 coconut husks yield 80kg of coir. 18–24t/ha of nuts is 108,000 to 144,000 nuts/ha (at 3,000 nuts per ton), yielding 8.6–10.7 tons of coir per hectare per year. Calculated from Duke, *Handbook of Nuts*, 105.

30. FAO Statistics Division online.

31. Plant Resources of Tropical Africa online.

32. Hanelt, *Mansfeld's Encyclopedia of Agricultural and Horticultural Crops*.

33. Bacci et al., "Fiber yields and quality of fiber nettle (*Urtica dioica* L.) cultivated in Italy," 480–84.

34. Hanelt, *Mansfeld's Encyclopedia of Agricultural and Horticultural Crops*.

35. Ibid.

36. Yields 5 to 6 t/ha at 10% fiber. Calculated from Thevs et al., "*Apocynum venetum* L. and *Apocynum pictum* Schrenk (Apocynaceae) as multi-functional and multi-service plant species in Central Asia," 159–67.

37. Duke, *Handbook of Energy Crops*.

38. Ibid.

39. FAO Statistics Division online.

40. Plant Resources of Tropical Africa online.

41. Anandjiwala and John, "Sisal," 181–97.

42. Plant Resources of Tropical Africa online.

43. Anandjiwala and John, "Sisal," 183.

44. Hanelt, *Mansfeld's Encyclopedia of Agricultural and Horticultural Crops*.

45. Plant Resources of Tropical Africa online.

46. Ibid.

47. Ibid.

48. FAO Statistics Division online.

49. Goltenboth and Muhlbauer, "Abacá," 163–81.

50. Plant Resources of Tropical Africa online.

51. Goltenboth and Muhlbauer, "Abacá," 163–81.

24. OTHER INDUSTRIAL USES

1. Burgess and Green, *Harvesting Color*, vii.

2. Buchanan, *A Weaver's Garden*, 105–08.

3. Ibid.

4. Plant Resources of Tropical Africa online.

5. Hanelt, *Mansfeld's Encyclopedia of Agricultural and Horticultural Crops*.

6. Varela, "Cork and the cork oak system."

7. Ibid.

8. Pereira, *Cork*, 154.

9. Varela, "Cork and the cork oak system."

10. Plant Resources of Southeast Asia online.

11. Hanelt, *Mansfeld's Encyclopedia of Agricultural and Horticultural Crops*.

12. Casadeli and Chikamai, "Gums, resins, and waxes," 412.

13. Ibid.

14. Ibid.

15. Ibid.

16. Ibid., 411–12.

17. Ibid., 427.

18. Ibid., 429.

19. FAO, "Gums, Resins, and Latexes of Plant Origin."

20. Ibid.

21. Ibid.

22. Plant Resources of Tropical Africa online.

23. ECHO, "Neem."

24. Ibid.

25. Ibid.

26. El Bassam, *Handbook of Bioenergy Crops*, 252.

27. ECHO, "Neem."

28. Hanelt, *Mansfeld's Encyclopedia of Agricultural and Horticultural Crops*.

29. Van Wyk and Wink, *Medicinal Plants of the World*, 7.

30. Ibid., 158.

31. Barlow, *The Ghosts of Evolution*, 140.

32. Ibid., 147.

33. Ibid., 139.

34. Van Wyk and Wink, *Medicinal Plants of the World*, 158.

35. Duke, *Handbook of Nuts*, 164.

36. Ibid., 165.

37. Facciola, *Cornucopia II*, 115.

38. Hanelt, *Mansfeld's Encyclopedia of Agricultural and Horticultural Crops*.

39. Ibid.

25. A THREE-POINT PLAN TO SCALE UP CARBON FARMING

1. Mbow et al., "Agroforestry solutions to address food security and climate change challenges in Africa," 63.

2. Project Drawdown online.

3. Klein, *This Changes Everything*, 23.

4. IPCC, *Climate Change 2014: Impacts, Adaptation and Vulnerability*, 9.

5. Anderson and Bows, "Beyond 'dangerous' climate change."

6. Klein, *This Changes Everything*, 55.

7. Gore, *An Inconvenient Truth*.

8. Klein, *This Changes Everything*, 56.

9. Ibid., 25.

10. Confino, "Al Gore."

11. Lal, "Abating climate change and feeding the world through soil carbon sequestration," 453.

12. IPCC, *Climate Change 2007*, 499.

13. Leakey, "Twelve principles for better food and more food from mature perennial agroecosystems," 284.

14. Rosset and Benjamin, *The Greening of the Revolution*, 5.
15. Chan and Roach, *Unfinished Puzzle*, xii.
16. Rosset and Benjamin, *The Greening of the Revolution*, 3.
17. Ibid., 16.
18. Ibid., 18.
19. Ibid., 4.
20. Chan and Roach, *Unfinished Puzzle*, 13.
21. Rosset and Benjamin, *The Greening of the Revolution*, 32; Chan and Roach, *Unfinished Puzzle*, 15.
22. Chan and Roach, *Unfinished Puzzle*, 16.
23. Ibid., 15.
24. What other national leader would say, "The production of these essential proteins and calories per hectare, and the cost (in terms of energy and CO_2 emissions) of each crop, should be in the manuals of policy makers all over the world. Having this information is perhaps just as important as knowing how to read and write." Chan and Roach, *Unfinished Puzzle*, 70, 73.
25. Ibid., 99.
26. Ibid., 6–8.
27. Ibid., 73.
28. Ibid., 43.
29. Ibid., 27–35.
30. Ibid., 76.
31. Ibid., 91.
32. Ibid., 76–81.
33. Ibid., 22.
34. Ibid., 82.
35. FAO Statistics Division online.
36. Chan and Roach, *Unfinished Puzzle*, 91.
37. Rosset and Benjamin, *The Greening of the Revolution*, 54.
38. Chan and Roach, *Unfinished Puzzle*, 12.
39. Rosset and Benjamin, *The Greening of the Revolution*, 29.
40. Chan and Roach, *Unfinished Puzzle*, 12.
41. Ibid., 85.
42. Ibid., 19.
43. Ibid., 69.
44. Rosset and Benjamin, *The Greening of the Revolution*, 26.
45. Chan and Roach, *Unfinished Puzzle*, 100.

26. SUPPORT FARMERS AND FARMING ORGANIZATIONS TO MAKE THE TRANSITION

1. Gliessman, *Agroecology*, 277.
2. McCarthy et al., "Climate-smart agriculture."
3. De Pinto, Ringler, and Magalhaes, "Economic challenges facing agricultural access to carbon markets," 63.
4. McCarthy et al., "Climate-smart agriculture."
5. Ibid.
6. Ibid.
7. Gliessman, *Agroecology*, 277.
8. McCarthy et al., "Climate-smart agriculture."
9. Pattanayak et al., "Taking stock of agroforestry adoption studies."
10. World Agroforestry Centre, *Climate-Smart Landscapes*, xviii.
11. Ibid., 4.
12. Ibid., xxxii.
13. Ibid., 397.
14. Leakey, *Living with the Trees of Life*, 178.
15. Wollenberg et al., *Climate Change Mitigation and Agriculture*, 217–83.
16. Colomb et al., "Review of GHG calculators in agriculture and forestry sectors," 29.
17. Nair, "Methodological challenges in estimating carbon sequestration potential of agroforestry systems," 3–16.
18. Seebauer et al., "Carbon accounting for smallholder agricultural soil carbon projects," 264.
19. Climate Yardstick online; Cool Farm Tool online.

27. EFFECTIVELY FINANCE CARBON FARMING

1. Hansen et al., "Target atmospheric CO_2," 16–17.
2. Falconer and Stadelmann, "What is climate finance?," 1.
3. Climate Policy Initiative, *The Global Landscape of Climate Finance*, ii–iv.
4. Ibid.
5. Ibid., 14–17.
6. Ibid.
7. GRAIN, "Food, climate change, and healthy soils," 19–20.
8. Climate Focus, "The geographical distribution of climate finance for agriculture."
9. Climate Policy Initiative, *The Global Landscape of Climate Finance*, 14–17.
10. Climate Focus, "The geographical distribution of climate finance for agriculture."
11. http://en.wikipedia.org/wiki/Clean_Development_Mechanism.
12. IPCC, *Climate Change 2014: Mitigation of Climate Change*, 864.
13. Ibid., 838.
14. Klein, *This Changes Everything*, 225.
15. Havemann, "Financing mitigation in smallholder agricultural systems," 131–32.
16. Kuehl and Yiping, *Carbon Off-Setting with Bamboo*, 21; Kaonga et al., *Agroforestry for Biodiversity and Ecosystem Services.*
17. De Pinto, Ringler, and Magalhaes, "Economic challenges facing agricultural access to carbon markets," 62.
18. Shames, Buck, and Scherr, "Reducing costs and improving benefits in smallholder agriculture carbon projects," 70.

19. Baroudy and Hooda, "Sustainable land management and carbon finance," 129.

20. Carbon Market Watch, "Joint implementation"; Schneider, "Is the CDM fulfilling its environmental and sustainable development objectives?," 15.

21. Magdoff and Foster, *What Every Environmentalist Needs to Know About Capitalism*, 118.

22. Rainforest Alliance, "SAN Climate Module."

23. Havemann, "Financing mitigation in smallholder agricultural systems," 159–69.

24. Pattanayak et al., "Taking stock of agroforestry adoption studies."

25. Werneck, "The potential for microfinance as a channel for carbon payments."

26. Grant Space, "What is program-related investment?"

27. Heifer International, "The foundation of transformation."

28. IPCC, *Climate Change 2014: Mitigation of Climate Change*, 26.

29. Klein, *This Changes Everything*, 379.

30. IPCC, *Climate Change 2014: Mitigation of Climate Change*, 817.

31. Hansen et al., "Target atmospheric CO_2," 16.

32. Klein, *This Changes Everything*, 114.

33. Gilbertson and Reyes, "Carbon trading," 10.

34. Ibid., 11.

35. Ibid., 11.

36. Klein, *This Changes Everything*, 225.

37. Magdoff and Foster, *What Every Environmentalist Needs to Know About Capitalism*, 118.

38. Gilbertson and Reyes, "Carbon trading," 11.

39. IPCC, *Climate Change 2014: Mitigation of Climate Change*, 28.

40. Ibid.

41. Klein, *This Changes Everything*, 112.

42. Magdoff and Foster, *What Every Environmentalist Needs to Know About Capitalism*, 126.

43. Leakey, *Living with the Trees of Life*, 135.

28. Remove National and International Policy Barriers

1. GRAIN, "Food, climate change, and healthy soils,"19.

2. FAO, *Climate-Smart Agriculture Sourcebook*, 353.

3. Gore, *Our Choice*, 15–16; Flannery, *Now or Never*, 65; IPCC, *Climate Change 2014: Impacts, Adaptation, and Vulnerability*, 1104.

4. IPCC, *Climate Change 2014: Impacts, Adaptation, and Vulnerability*, 1110.

5. IPCC, *Climate Change 2014: Mitigation of Climate Change*, 25.

6. Klein, *This Changes Everything*, 414.

7. Leakey, *Living with the Trees of Life*, 70.

8. Bala et al., "Combined climate and carbon-cycle effects of large-scale deforestation."

9. Leakey, *Living with the Trees of Life*, 136.

10. Ibid., 137.

11. Ching, "The importance of international trade, trade rules, and market structures," 252.

12. Ibid., 260.

13. Ibid., 252.

14. Feyder, "Agriculture," 11.

15. Ishii-Eiteman, "Democratizing control of agriculture to meet the needs of the twenty first century," 65–67.

16. Ibid., 19.

17. Müller, "A critical analysis of commodity and food price speculation," 293.

18. Place et al., "Improved policies for facilitating the adoption of agroforestry,"121.

19. Kuehl and Yiping, *Carbon Off-Setting with Bamboo*, 21.

20. USDA National Agroforestry Center website; Nair, *An Introduction to Agroforestry*, 42–45.

21. Kiptot and Franzel, *Gender and Agroforestry in Africa*, 2; Kiptot, Franzel, and Degrande, "Gender, agroforestry and food security in Africa," 104.

22. IPCC, *Climate Change 2014: Mitigation of Climate Change*, 855.

23. Ibid., 857.

24. Leach, Fairhead, and Fraser, "Land grabs for biochar?," 19.

25. Parfitt and Barthel, "Food waste reduction," 91.

26. Reij, "Adapting to climate change and improving household food security in Africa through agroforestry," 201.

27. IPCC, *Climate Change 2014: Mitigation of Climate Change*, 867.

28. Climate Policy Initiative, *The Global Landscape of Climate Finance*, ii–iv.

29. Van Tassel, personal communication.

30. IPCC, *Climate Change 2014: Mitigation of Climate Change*, 869.

31. Australian Carbon Farming Initiative Regulations 2011.

32. Australian Department of the Environment, "Carbon farming initiative project transition into the emissions reduction fund."

33. Lal, "Abating climate change and feeding the world through soil carbon sequestration," 453.

34. White, *Grass, Soil, Hope*, 220.

35. De Pinto, Ringler, and Magalhaes, "Economic challenges facing agricultural access to carbon markets," 62.

36. Steele, personal communication.

37. Chambers, personal communication.

38. IPCC, *Climate Change 2014: Mitigation of Climate Change*, 867.

39. Ibid., 863.

40. Klein, *This Changes Everything*, 134.

41. Ho, "Sustainable agriculture and off-grid renewable energy," 72.

42. McKibben, *Eaarth*, 167–68.

43. GRAIN, *Hungry for Land*, 3–9.

44. Ibid., 13.

45. Kaplan, Ifejika-Speranza, and Sholz, "Promoting resilient agriculture in sub-Saharan Africa as a major priority in climate-change adaptation," 43.

46. Kaonga et al., *Agroforestry for Biodiversity and Ecosystem Services*.

47. Ibid., 119.

48. Kaplan, Ifejika-Speranza, and Sholz, "Promoting resilient agriculture in sub-Saharan Africa as a major priority in climate-change adaptation," 43.

49. IPCC, *Climate Change 2014: Mitigation of Climate Change*, 859; 28.

50. Klein, *This Changes Everything*, 88.

51. Magdoff and Foster, *What Every Environmentalist Needs to Know About Capitalism*, 43.

52. Ibid., 39–40.

53. Klein, *This Changes Everything*, 114.

54. Magdoff and Foster, *What Every Environmentalist Needs to Know About Capitalism*, 32.

55. Anderson and Bows, "Beyond 'dangerous' climate change."

56. Klein, *This Changes Everything*, 88.

57. Ibid., 21.

58. Ibid., 91.

59. FAO, *Climate-Smart Agriculture Sourcebook*, 548.

60. Global Alliance for Climate-Smart Agriculture, "GACSA Action Plan."

61. Neufeldt et al., "Beyond climate-smart agriculture."

62. La Via Campesina, "Unmasking climate smart agriculture," 1.

63. Institute for Agriculture and Trade Policy, "Corporations represent at UN Trade Summit," 1.

64. La Via Campesina, "Unmasking climate smart agriculture," 1.

Appendix A: Global Species Matrix

1. The Plant List: A Working List of All Plant Species, http://www.theplantlist.org.

2. Global Invasive Species Database, http://www.issg.org/database/welcome.

3. Dave's Garden, www.davesgarden.com.

Appendix D: Changes in Latin Names

1. www.theplantlist.org.

BIBLIOGRAPHY

Abraham, Binju, et al. *The System of Crop Intensification: Agro-ecological Innovations for Improving Agricultural Production, Food Security, and Resilience to Climate Change.* Ithaca, N.Y.: Sri International Network and Resources Center and Technical Centre for Agricultural and Rural Cooperation, 2014.

Adams, J. M. "Estimates of preanthropogenic carbon storage in global ecosystem types." http://www.esd.ornl.gov/projects/qen/carbon3.html.

Agroecology in Action website. "What is Agroecology?" http://agroeco.org.

AGROOF. *Agroforesterie: Produire Autremont* DVD, 2009.

Albrecht, Alain, and Serigne T. Kandji. "Carbon sequestration in tropical agroforestry systems." *Agriculture, Ecosystems & Environment* 99, no. 1 (2003): 15–27.

Altieri, Miguel. "Strengthening resilience of farming systems: A prerequisite for sustainable agricultural production." In *Wake Up Before It Is Too Late: Make Agriculture Truly Sustainable Now for Food Security in a Changing Climate.* Geneva: UNCTAD, 2013.

Altieri, Miguel A. "Agroecology, small farms, and food sovereignty." *Monthly Review* 61, no. 3 (2009): 102–13.

Ampt, Peter, and Sarah Doornbos. "Benchmark study of innovators." A Communities in Landscapes project. Draft report, September 2011. http://sydney.edu.au/agriculture/documents/2011/reports/Ampt_CiL_BM_CombinedReportSept2011DRAFT.pdf.

Anandjiwala, Rajesh D., and Maya John. "Sisal: Cultivation, processing and products." In Jörg Müssig, ed., *Industrial Applications of Natural Fibres: Structure, Properties and Technical Applications*, pp. 181–95. Chichester, UK: John Wiley, 2010.

Anderegg, William R. L., James W. Prall, Jacob Harold, and Stephen H. Schneider. "Expert credibility in climate change." *Proceedings of the National Academy of Sciences* 107, no. 27 (2010): 12107–09.

Anderson, Kat. *Indigenous Uses, Management, and Restoration of Oaks of the Far Western United States.* Technical Note. Washington, DC: USDA Natural Resources Conservation Service, 2007.

———. *Tending the Wild: Native American Knowledge and the Management of California's Natural Resources.* Berkeley: University of California Press, 2005.

Anderson, Kevin. "Climate change going beyond dangerous: Brutal numbers and tenuous hope." *Development Dialogue* 61 (2012): 16–39.

Anderson, Kevin, and Alice Bows. "Beyond 'dangerous' climate change: Emission scenarios for a new world." *Philosophical Transactions of the Royal Society A: Mathematical, Physical and Engineering Sciences* 369, no. 1934 (2011): 20–44.

Angers, D. A., and N. S. Eriksen-Hamel. "Full-inversion tillage and organic carbon distribution in soil profiles: A meta-analysis." *Soil Science Society of America Journal* 72, no. 5 (2008): 1370–74.

Australian Carbon Farming Initiative Regulations 2011. http://www.comlaw.gov.au/Series/F2011L02583.

Australian Department of the Environment. "Carbon farming initiative project transition into the emissions reduction fund." http://www.environment.gov.au/climate-change/emissions-reduction-fund/carbon-farming-initiative-project-transition.

———. "Current states and trends of the land environment." http://www.environment.gov.au/science/soe/2011-report/5-land/2-state-and-trends/2-2-soil.

Aylott, Matthew J., E. Casella, I. Tubby, N. R. Street, P. Smith, and Gail Taylor. "Yield and spatial supply of bioenergy poplar and willow short-rotation coppice in the UK." *New Phytologist* 178, no. 2 (2008): 358–70.

Babe, Rodney. "Use of introduced trees: A positive experience with eucalyptus." In *Agricultural Options for the Poor.* Fort Myers, Fla.: ECHO, 2014.

Bacci, L., S. Baronti, S. Predieri, and N. Di Virgilio. "Fiber yield and quality of fiber nettle (*Urtica dioica* L.) cultivated in Italy." *Industrial Crops and Products* 29, no. 2 (2009): 480–84.

Badgley, Catherine, Jeremy Moghtader, Eileen Quintero, Emily Zakem, M. Jahi Chappell, Katia Aviles-Vazquez, Andrea Samulon, and Ivette Perfecto. "Organic agriculture and the global food supply." *Renewable Agriculture and Food Systems* 22, no. 2 (2007): 86–108.

Bala, Govindasamy, K. Caldeira, M. Wickett, T. J. Phillips, D. B. Lobell, C. Delire, and A. Mirin. "Combined climate and carbon-cycle effects of large-scale deforestation." *Proceedings of the National Academy of Sciences* 104, no. 16 (2007): 6550–55.

Barlow, Connie C. *The Ghosts of Evolution: Nonsensical Fruit, Missing Partners, and Other Ecological Anachronisms.* New York: Basic Books, 2000.

Baroudy, Ellysar, and Neeta Hooda. "Sustainable land management and carbon finance: The experience of the biocarbon fund." In Eva Wollenberg et al., eds., *Climate Change Mitigation and Agriculture.* London: Earthscan, 2012.

Barringer, Felicity. "Water sources for almonds in California may run dry." *New York Times,* December 7, 2014.

Barrow, C. J. "Biochar: Potential for countering land degradation and for improving agriculture." *Applied Geography* 34 (2012): 21–28.

Baudry, J., R. G. H. Bunce, and F. Burel. "Hedgerows: An international perspective on their origin, function and management." *Journal of Environmental Management* 60, no. 1 (2000): 7–22.

Beach, Timothy, Nicholas Dunning, Sheryl Luzzadder-Beach, Duncan E. Cook, and Jon Lohse. "Impacts of the ancient Maya on soils and soil erosion in the central Maya Lowlands." *Catena* 65, no. 2 (2006): 166–78.

Beckert, Sven. *Empire of Cotton: A Global History.* New York: Knopf, 2014.

Beer, John, and Reinhold Muschler. "Multistrata agroforestry systems with perennial crops: Selected papers from international symposium held at CATIE, Turrialba, Costa Rica." *Agroforestry Systems* 53, no. 2 (2001).

Bell, Lindsay W. "Economics and system applications for perennial grain crops in dryland farming systems in Australia." In *Perennial Crops for Food Security: Proceedings of the FAO Expert Workshop.* Rome: FAO, 2013.

Bemis, W. P., L. D. Curtis, C. W. Weber, and J. Berry. "The feral buffalo gourd, *Cucurbita foetidissima.*" *Economic Botany* 32, no. 1 (1978): 87–95.

Bemis, W. P., L. D. Curtis, C. W. Weber, J. W. Berry, and J. M. Nelson. *The Buffalo Gourd* (Cucurbita foetidissima) *HBK: A Potential Crop for the Production of Protein, Oil, and Starch on Arid Lands.* Office of Agriculture, Technical Assistance Bureau, Agency for International Development, 1975.

Berkelaar, Dawn. "Chaya: The spinach tree." In *Agricultural Options for the Poor,* Fort Myers, Fla.: ECHO, 2014.

Berkhout, Ezra, Dominic Glover, and Arie Kuyvenhoven. "On-farm impact of the System of Rice Intensification (SRI): Evidence and knowledge gaps."*Agricultural Systems* 132 (2015): 157–66.

Betts, Richard A. "Offset of the potential carbon sink from boreal forestation by decreases in surface albedo." *Nature* 408, no. 6809 (2000): 187–90.

Bindraban, Prem S. "Agroecological intelligence needed to prepare agriculture for climate change." *Combating Climate Change: An Agricultural Perspective* (2013): 89.

Bird, Winifred. "Perennial rice: In search of a greener, hardier staple crop." http://e360.yale.edu/feature/perennial_rice_in_search_of_a_greener_hardier_staple_crop/2853.

Boffa, Jean-Marc. *Agroforestry Parklands in Sub-Saharan Africa.* Rome: FAO, 1999.

Bogdanski, Anne, Olivier Dubois, Craig Jamieson, and Rainer Krell. *Making Integrated Food–Energy Systems Work for People and Climate: An Overview.* Food and Agriculture Organization of the United Nations (FAO), 2011.

Borlaug, Norman, et al. *Jojoba: New Crop for Arid Lands, New Raw Material for Industry.* Washington, D.C.: National Academy Press, 1985.

Brakas, Shushan Ghirmai, and Jens B. Aune. "Biomass and carbon accumulation in land use systems of Claveria, the Philippines." In *Carbon Sequestration Potential of Agroforestry Systems,* pp. 163–175. Dordrecht, Neth., Springer, 2011.

Brandt, Steven A. et al. *The "Tree Against Hunger": Enset-Based Agricultural System in Ethiopia.* Washington, D.C.: American Association for the Advancement of Science, 1997.

Briske, D. D., Nathan F. Sayre, Lynn Huntsinger, Maria Fernandez-Gimenez, Bob Budd, and J. D. Derner. "Origin, persistence, and resolution of the rotational grazing debate: Integrating human dimensions into rangeland research." *Rangeland Ecology & Management* 64, no. 4 (2011): 325–34.

Briske, David D., Andrew J. Ash, Justin D. Derner, and Lynn Huntsinger. "Commentary: A critical assessment of the policy endorsement for holistic management." *Agricultural Systems* 125 (2014): 50–53.

Brown, Christopher. *The Global Outlook for Future Wood Supply from Forest Plantations.* Rome: FAO, Forestry Policy and Planning Division, 2000.

Brownlee, Elizabeth. "Grow agriculturally productive buffers." Burlington: University of Vermont, 2013.

Bruun, Thilde Bech, Andreas de Neergaard, Deborah Lawrence, and Alan D. Ziegler. "Environmental consequences of the demise in swidden cultivation in Southeast Asia: Carbon storage and soil quality." *Human Ecology* 37, no. 3 (2009): 375–88.

Bryan, James Alfred. *Leguminous Trees with Edible Beans, with Indications of a Rhizobial Symbiosis in Non-Nodulating Legumes.* New Haven, Conn.: Yale University, 1995.

Buchanan, R. A., Irene M. Cull, F. H. Otey, and C. R. Russell. "Hydrocarbon- and rubber-producing crops." *Economic Botany* 32, no. 2 (1978): 146–53.

Buchanan, Rita. *A Weaver's Garden.* Loveland, Colo.: Interweave Press, 1987.

Budischak, Cory, DeAnna Sewell, Heather Thomson, Leon Mach, Dana E. Veron, and Willett Kempton. "Cost-minimized combinations of wind power, solar power and electrochemical storage, powering the grid up to 99.9 percent of the time." *Journal of Power Sources* 225 (2013): 60–74.

Burgess, Rebecca, and Paige Green. *Harvesting Color: How to Find Plants and Make Natural Dyes.* New York: Artisan, 2011.

Burney, Jennifer A., Steven J. Davis, and David B. Lobell. "Greenhouse gas mitigation by agricultural intensification." *Proceedings of the National Academy of Sciences* 107, no. 26 (2010): 12052–57.

Calvin, Melvin. *Fuel Oils from Higher Plants.* No. LBL-19374. Berkeley, Calif.: Lawrence Berkeley Laboratory, 1985.

———. "Petroleum plantations for fuel and materials." *Bioscience* 29, no. 9 (1979): 533–38.

Cantini, Claudio, Riccardo Gucci, and Balilla Sillari. "An alternative to managing olive orchards: The coppice system." *Hortechnology* 8, no. 3 (July–September 1998): 409.

Carbon Market Watch. "Joint implementation: CDM's little brother grew up to be big and nasty." http://carbon marketwatch.org/joint-implementation-cdms-little-brother -grew-up-to-be-big-and-nasty.

Carter, John, Allison Jones, Mary O'Brien, Jonathan Ratner, and George Wuerthner. "Holistic management: Misinformation on the science of grazed ecosystems." *International Journal of Biodiversity* 2014 (2014).

Casadeli, Enrico, and Ben Chikamai. "Gums, resins, and waxes." In Bharat P. Singh, ed., *Industrial Crops and Uses*, p. 411. Wallingford, UK: CABI, 2010.

Ceccolini, Lorenzo. "The homegardens of Soqotra Island, Yemen: An example of agroforestry approach to multiple land-use in an isolated location." *Agroforestry Systems* 56, no. 2 (2002): 107–15.

Chan, May Ling, and Eduardo Francisco Roach. *Unfinished Puzzle: Cuban Agriculture: The Challenges, Lessons and Opportunities.* New York: Perseus Books, 2013.

Chapman, Jefferson, and Patty Jo Watson. "The archaic period and the flotation revolution." In C. Margaret Scarry, ed., *Foraging and Farming in the Eastern Woodlands.* Gainesville: University of Florida, 1993.

Chen, Wan, Wang, Cheng, Liu, Lin, Ruan, and Singh. "Bioenergy industry status and prospects." In Bharat P. Singh, ed., *Industrial Crops and Uses*, p. 21. Wallingford, UK: CABI, 2010.

Cherrett, Ian. "Quesungual agroforestry." *The Overstory* 259. http://www.agroforestry.net/the-overstory.

Chi, Kelly Rae, James MacGregor, and Richard King. *Fair Miles: Recharting the Food Miles Map.* London: IIED, 2009.

Ching, Lim Li. "The importance of international trade, trade rules, and market structures." in *Wake Up Before It Is Too Late: Make Agriculture Truly Sustainable Now for Food Security in a Changing Climate.* Geneva: UNCTAD, 2013.

Chu, Jianmin, Xinqiao Xu, and Yinglong Zhang. "Production and properties of biodiesel produced from *Amygdalus pedunculata* Pall." *Bioresource Technology* 134 (2013): 374–76.

Ciesla, William M. *Non-Wood Forest Products from Temperate Broad-Leaved Trees.* Rome: FAO, 2002.

Clark, James H., and Fabien EI Deswarte. *The Biorefinery Concept: An Integrated Approach.* Chichester, UK: John Wiley & Sons, 2008.

Clay, Jason W. *World Agriculture and the Environment: A Commodity-by-Commodity Guide to Impacts and Practices.* Washington, D.C.: Island Press, 2004.

Clement, Charles. "The potential use of the pejibaye palm in agroforestry systems." In *Agroforestry Systems* 7, no. 3 (1989): 201–12.

Cleugh, H. A. "Effects of windbreaks on airflow, microclimates and crop yields." *Agroforestry Systems* 41, no. 1 (1998): 55–84.

Climate Policy Initiative. *The Global Landscape of Climate Finance 2013.* CPI Report, 2013, access March 9, 2015, http:// climatepolicyinitiative.org/wpcontent/uploads/2013/10 /The-Global-Landscape-of-Climate-Finance-2013.pdf.

Climate Focus. "The geographical distribution of climate finance for agriculture." http://www.climatefocus.com/sites/default /files/the_geographical_distribution_of_climate_finance _for_agriculture.0.pdf.

Climate Yardstick. http://www.climateyardstick.eu.

Colomb, Vincent, et al. "Review of GHG calculators in agriculture and forestry sectors: A guideline for appropriate choice and use of landscape based tools" (2012). http://www.fao.org /fileadmin/templates/ex_act/pdf/ADEME/Review_existing GHGtool_VF_UK4.pdf.

Conant, Richard T., Keith Paustian, and Edward T. Elliott. "Grass-land management and conversion into grassland: Effects on soil carbon." *Ecological Applications* 11, no. 2 (2001): 343–55.

Conedera, M., et al. "The cultivation of *Castanea sativa* in Europe, from its origin to its diffusion on a continental scale." *Vegetation History and Archaeobotany* 13 (2004): 161–79.

Confino, Joe. "Al Gore: Oil companies use our atmosphere as an open sewer." *Guardian* (January 21, 2015).

Cool Farm Tool. http://www.coolfarmtool.org.

Cooper, P. J. M., et al. "Agroforestry and the mitigation of land degradation in the humid and sub-humid tropics of Africa." *Experimental Agriculture*, 32, no. 3 (1996): 235–90.

Costanza, Robert, Rudolf De Groot, Paul Sutton, Sander van der Ploeg, Sharolyn J. Anderson, Ida Kubiszewski, Stephen Farber, and R. Kerry Turner. "Changes in the global value of ecosystem services." *Global Environmental Change* 26 (2014): 152–58.

Cox, Stan, Tim Crews, and Wes Jackson. "From genetics and breeding to agronomy to ecology." In *Perennial Crops for Food*

Security: Proceedings of the FAO Expert Workshop. Rome: FAO, 2013.

Cox, Thomas, et al. "Prospects for developing perennial grain crops." *BioScience* 56, no. 8 (August 2006): 649–59.

Cram, W. H. "Breeding and genetics of *Caragana*." *Forestry Chronicle* 45, no. 6 (1969): 400–01.

Crawford, Martin. "Coast redwood for timber." *Agroforestry News* 17, no. 3.

———. "Cork and cork oaks." *Agroforestry News* 11, no. 2: 9–10.

———. *Creating a Forest Garden: Working with Nature to Grow Edible Crops.* Totnes, UK: Green Books, 2010.

———. "Nut pines." *Agroforestry News* 3, no. 1.

———. "Silvoarable systems in Europe." *Agroforestry News* 15, no. 3.

Crossley, Philip. "The Chinampas of Mexico." http://chinampas .info.

Crossley, Philip L. "Just beyond the eye: Floating gardens in Aztec Mexico." *Historical Geography* 32 (2004): 111–35.

Cuartas Cardona, César A., Juan F. Naranjo Ramírez, Ariel M. Tarazona Morales, Enrique Murgueitio Restrepo, Julián D. Chará Orozco, Juan Ku Vera, Francisco J. Solorio Sánchez, Martha X. Flores Estrada, Baldomero Solorio Sánchez, and Rolando Barahona Rosales. "Contribution of intensive silvo-pastoral systems to animal performance and to adaptation and mitigation of climate change." *Revista Colombiana de Ciencias Pecuarias* 27, no. 2 (2014): 76–94.

Cubbage, Frederick. "Global timber production, trade, and timber-land investments." (Speech presented at the Forest Research Assessment Consortium, Research Triangle Park, NC, October 2008.) http://www4.ncsu.edu/~bobabt/CUBBAGE _TIMBER_PRODUCTION_INVESTMENTS_V7.pdf.

Cubbage, Frederick, Sadharga Koesbandana, Patricio Mac Donagh, Rafael Rubilar, Gustavo Balmelli, Virginia Morales Olmos, Rafael De La Torre, et al. "Global timber investments, wood costs, regulation, and risk." *Biomass and Bioenergy* 34, no. 12 (2010): 1667–78.

Cunningham, Peter. "The Farmer-Managed Agroforestry Farming System (FMAFS)." In *Agricultural Options for the Poor: A Handbook for Those Who Serve Them.* Fort Myers, Fla.: ECHO, 2013.

Czako, Mihaly, and Laszlo Marton. "Subtropical and tropical reeds for biomass." In Nigel Halford and Angela Karp, eds., *Energy Crops.* Cambridge: Royal Society of Chemistry, 2011.

Dale, Bruce, et. al., "Protein feeds coproduction in biomass conversion to fuel and chemicals," *Biofuels, Bioproducts and Biorefining* 3, no. 2 (2009): 219–230.

———. *Congo Native Fruits: Twenty-Five of the Best.* Self-published, 1997.

Danforth, Roy M., and Paul D. Noren. *100 Tropical Fruits, Nuts, and Spices for the Central African Home Garden.* Self-published, 2014.

Daniel, J. N., and Ong, C. K. (1990). "Perennial pigeonpea: A multipurpose species for agroforestry systems." *Agroforestry Systems* 10 (1990): 113–29.

Dar, William D., and C. L. Laxmipathi Gowda. "Declining agricultural productivity and global food security." *Journal of Crop Improvement* 27, no. 2 (2013): 242–54.

Dave's Garden. http://davesgarden.com.

Davis, Mark A. *Invasion Biology.* Oxford: Oxford University Press, 2009.

Davis, Mark A., Matthew K. Chew, Richard J. Hobbs, Ariel E. Lugo, John J. Ewel, Geerat J. Vermeij, James H. Brown, et al. "Don't judge species on their origins." *Nature* 474, no. 7350 (2011): 153–54.

Davys, M. N. G., et al. "Leaf concentrate and other benefits of leaf fractionation." In Brian Thompson and Leslie Amoroso, eds., *Combating Micronutrient Deficiencies: Food-Based Approaches.* Rome: FAO, 2011.

de Camargo, P. B., R. de P. Salomão, S. Trumbore, and L. A. Marti-nelli. "How old are large Brazil-nut trees (*Bertholletia excelsa*) in the Amazon?" *Scientia Agricola* 51, no. 2 (1994): 389–91.

DeHaan, et al. "Current efforts to develop perennial wheat and domesticate *Thinopyrum intermedium* as a perennial grain." In *Perennial Crops for Food Security: Proceedings of the FAO Expert Workshop.* Rome: FAO, 2013.

Denison, R. Ford. *Darwinian Agriculture: How Understanding Evolution Can Improve Agriculture.* Princeton, N.J.: Princeton University Press, 2012.

De Pinto, Alessandro, Claudia Ringler, and Marilia Magalhaes. "Economic challenges facing agricultural access to carbon markets." In Eva Wollenberg et al., eds., *Climate Change Mitigation and Agriculture.* London: Earthscan, 2012.

de Schutter, Olivier. "Agroecology: A solution to the crises of food systems and climate change." In *Wake Up Before It Is Too Late: Make Agriculture Truly Sustainable Now for Food Security in a Changing Climate.* Geneva: UNCTAD, 2013.

Dhamodaran, T. K., R. Gnanaharan, and K. Sankara Pillai. "Bamboo for pulp and paper." Kerala Forest Research Institute, India (2003).

Dillen, S. Y., et al. "Poplar." In Nigel Halford and Angela Karp, eds., *Energy Crops.* Cambridge: Royal Society of Chemistry, 2011.

Dimitriou, I., and P. Aronsson. "Willows for energy and phytore-mediation in Sweden." *UNASYLVA-FAO* 56, no. 2 (2005): 47.

Dixon, R. K. "Agroforestry systems: Sources of sinks of green-house gases?" *Agroforestry Systems* 31, no. 2 (1995): 99–116.

Dollar Times Financial Calculator online. www.dollartimes.com, accessed April 15, 2014.

Dong, Zhongmin, Martin J. Canny, Margaret E. McCully, Maria Regla Roboredo, Clemente Fernandez Cabadilla, Eduardo Ortega, and Rosita Rodes. "A nitrogen-fixing endophyte of sugarcane stems (a new role for the apoplast)." *Plant Physiology* 105, no. 4 (1994): 1139–47.

Donovan, Peter. "The calculation: Soil organic matter needed to bring down atmospheric carbon." http://www.soilcarbon coalition.org/calculation.

Dorren, Luuk, and Freddy Rey. "A review of the effect of terracing on erosion." A Soil Conservation and Protection for Europe (SCAPE) project (2004).

Dov, Y. Ben, et al. "Introduction of drought and salt tolerant plants for landscaping." In *Combating Desertification with Plants*. New York: Kluwer Academic/Plenum Publishers, 2001.

Drieling, Axel, and Jorg Mussig. "Grades and standards." In *Industrial Application of Natural Fibres: Structure, Properties, and Technical Applications*. Chichester, UK: Wiley, 2010.

Duany, Andres, and Duany Plater-Zyberk & Co. *Garden Cities: Theory & Practice of Agrarian Urbanism*. UK: Prince's Foundation for the Built Environment, 2011.

Duarte, Luis Enrique Acero. *Guia para el cultivo y aprovechamiento del Chachafruto o balú* (Erythrina edulis). Bogotá, Colombia: Convenio Andres Bello, 2002.

Duke, James. *Handbook of Energy Crops* online. https://www.hort.purdue.edu/newcrop/duke_energy/dukeindex.html.

———. *Handbook of Nuts*. Boca Raton, Fla.: CRC Press, 2001.

Dupraz, C., and S. M. Newman. "Temperate agroforestry in Europe." In *Temperate Agroforestry Systems*. Wallingford, UK: CABI, 1997.

Dupraz, C., S. M. Newman, and A. M. Gordon. "Temperate agroforestry: the European way." *Temperate agroforestry systems*. (1997): 181–236.

ECHO. "Neem: Indian lilac." Fort Myers, FL: ECHO, 2006.

Economist. "The other oil spill." http://www.economist.com /node/16423833?story_id=16423833, accessed February 23, 2015.

Egbe, E. A., and P. T. Tabot. "Carbon sequestration in eight woody non-timber forest species and their economic potentials in southwestern Cameroon."*Applied Ecology and Environmental Research* 9, no. 4 (2011): 369–85.

eHALOPH Database online. http://www.sussex.ac.uk/affiliates/ halophytes/index.php.

Ekué, Marius R. M., Brice Sinsin, Oscar Eyog-Matig, and Reiner Finkeldey. "Uses, traditional management, perception of variation and preferences in ackee (*Blighia sapida* KD Koenig) fruit traits in Benin: Implications for domestication and conservation." *Journal of Ethnobiology and Ethnomedicine* 6, no. 12 (2010): 1–14.

El Bassam, Nasir. *Handbook of Bioenergy Crops: A Complete Reference to Species, Development and Applications*. London: Earthscan, 2010.

Elevitch, Craig R., ed. *Specialty Crops for Pacific Islands*. Holualoa, Hawaii: Permanent Agriculture Resources, 2011.

———. *Traditional Trees of Pacific Islands: Their Culture, Environment, and Use*. Holualoa, Hawaii: Permanent Agriculture Resources, 2006.

Elevitch, Craig R., and Kim M. Wilkinson, eds. *Agroforestry Guides for Pacific Islands*. Holualoa, Hawaii: Permanent Agriculture Resources, 2000.

Encyclopedia Brittanica online. http://www.britannica.com.

Epicos Business Directory. http://www.epicos.com/EPCompany ProfileWeb/GeneralInformation.aspx?id=18322.

Eswaran, Hari, Fred Beinroth, and Paul Reich. "Global land resources and population-supporting capacity." *American Journal of Alternative Agriculture* 14, no. 3 (1999): 129–36.

Eswaran, H., R. Lal, and P. F. Reich. "Land degradation: An overview." *Responses to Land Degradation: Proceedings of the Second International Conference on Land Degradation and Desertification, Khon Kaen, Thailand*, pp. 20–35. New Delhi: Oxford Press, 2001.

Etoamaihe, U. J., and K. C. Ndubueze. "Development and performance evaluation of a dehulling machine for African breadfruit (*Treculia africana*)." *Journal of Engineering and Applied Sciences* 5, no. 4 (2010): 312–15.

Evangelista, Regine. "The smallholder farmers' secret weapon: Trees and their big roles in climate change adaptation." http:// dialogues.cgiar.org/blog/smallholder-farmers-secret-weapon.

Ewel, John J. "Designing agricultural ecosystems for the humid tropics." *Annual Review of Ecology and Systematics* (1986): 245–71.

Facciola, Stephen. *Cornucopia II: A Sourcebook of Edible Plants*. Vista, Calif.: Kampong Publications, 1990.

FACT Foundation. *The Jatropha Handbook: From Cultivation to Application*. FACT Foundation, 2010.

Fagg, C. W. "*Faidherbia albida*: Inverted phenology supports dryzone agroforestry." *NFT Highlights*, NFTA 95-01 (1995).

Falconer, Angela, and Martin Stadelmann. "What is climate finance?: Definitions to improve tracking and scale up climate finance." http://climatepolicyinitiative.org/wp-content /uploads/2014/09/Climate-Finance-Brief-Definitions-to -Improve-Tracking-and-Scale-Up.pdf.

FAO. *Challenges and Opportunities for Carbon Sequestration in Grassland Systems*. Rome: FAO, 2010.

FAO. *Climate-Smart Agriculture Sourcebook*. Rome: FAO, 2013.

FAO. *Gums, resins and latexes of plant origin*. Non-Wood Forest Products 6. Rome: FAO, 1995.

FAO. Perennial Crops for Food Security: Proceedings of the FAO Expert Workshop. Rome: FAO, 2013.

FAO. *Review of Evidence on Drylands Pastoral Systems and Climate Change*. Rome: FAO, 2009.

FAO. "Turpentine from pine resin." In *Flavors and Fragrances of Plant Origin*. Rome: FAO, 1995.

FAO Regional Office for Asia and the Pacific. *Proceedings: Regional Expert Consultation on Eucalyptus, Bangkok*. Rome: FAO, 1993.

FAO Statistics Division. http:// http://faostat3.fao.org.

Farmers Guardian online. "Attention to detail essential in meeting beet yield targets." 2010.

Farooq, Muhammad, and Kadambot, Siddique. "Conservation agriculture: Concepts, brief history, and impacts on agricultural systems." In Muhammad Farooq and Kadambot Siddique, eds., *Conservation Agriculture*, pp. 3–20. New York: Springer, 2015.

Farrelly, David. *The Book of Bamboo*. San Francisco: Sierra Club Books, 1984.

Feedipedia: Animal Feed Resources Information System. http:// www.feedipedia.org.

Felger, Richard, and Neil Logan. "Mesquite: Food for the world." In *Eat Mesquite!: A Cookbook*. Tucson, Ariz.: Green Press Initiative, 2011.

Ferguson, Rafter. "Confronting the context: Permaculture and capitalism." http://liberationecology.org/2013/09/11 /confronting-context-permaculture-capitalism.

Ferguson, Rafter Sass, and Sarah Taylor Lovell. "Permaculture for agroecology: Design, movement, practice, and worldview. A review." *Agronomy for Sustainable Development* 34, no. 2 (2014): 251–74.

Feyder, Jean. "Agriculture: A unique sector in economic, ecological, and social terms." In *Wake Up Before It Is Too Late: Make Agriculture Truly Sustainable Now for Food Security in a Changing Climate*. Geneva: UNCTAD, 2013.

Fiber Year, The. *The Fiber Year 2013: World Survey on Textiles & Nonwovens*, Frankfurt, (2013).

Fini, Elham H., Eric W. Kalberer, Abolghasem Shahbazi, Mufeed Basti, Zhanping You, Hasan Ozer, and Qazi Aurangzeb. "Chemical characterization of biobinder from swine manure: Sustainable modifier for asphalt binder." *Journal of Materials in Civil Engineering* 23, no. 11 (2011): 1506–13.

Finlay, Mark R. *Growing American Rubber: Strategic Plants and the Politics of National Security*. New Brunswick, N.J.: Rutgers University Press, 2009.

Finlayson, J. D., R. A. Lawes, Tess Metcalf, M. J. Robertson, David Ferris, and M. A. Ewing. "A bio-economic evaluation of the profitability of adopting subtropical grasses and pasture-cropping on crop–livestock farms." *Agricultural Systems* 106, no. 1 (2012): 102–12.

Flach, M., and F. Rumawas, eds. *Plant Resources of Southeast Asia*, vol. 9: *Plants Yielding Non-Seed Carbohydrates*. Leiden, Netherlands: Backhuys Publishers, 1996.

Flach, Michael. *Sago Palm* (Metroxylon sagu). Rome: International Plant Genetic Resources Institute, 1997.

Flanagan, Robert. "The bright prospect of biochar." *Nature Reports: Climate Change*. http://www.nature.com/climate /2009/0906/full/climate.2009.48.html.

Flannery, Tim F. *Now or Never: Why We Must Act Now to End Climate Change and Create a Sustainable Future*. New York: Atlantic Monthly Press, 2009.

Florez, Jaime-Eduardo Munoz, et al. "*Erythrina edulis*, and Andean giant bean for human consumption." *Grain Legumes* 22 (1998): 26–27.

Foti, S., G. Mauromicale, S. A. Raccuia, B. Fallico, F. Fanella, and E. Maccarone. "Possible alternative utilization of *Cynara* spp.: I. Biomass, grain yield and chemical composition of grain." *Industrial Crops and Products* 10, no. 3 (1999): 219–28.

Fox, Jefferson, Dao Minh Truong, A. Terry Rambo, Nghiem Phuong Tuyen, and Stephen Leisz. "Shifting cultivation: A new old paradigm for managing tropical forests." *BioScience* 50, no. 6 (2000): 521–28.

Fox, P., J. Rockström, and J. Barron. "Risk analysis and economic viability of water harvesting for supplemental irrigation in semi-arid Burkina Faso and Kenya." *Agricultural Systems* 83, no. 3 (2005): 231–50.

Francis, Pope. "Encyclical Letter Laudato Si' of the Holy Father Francis on Care for Our Common Home." http://w2.vatican .va/content/francesco/en/encyclicals/documents/papa -francesco_20150524_enciclica-laudato-si.html.

Francis, Rob, ed. *Wattle We Eat for Dinner: Proceedings of the Wattle We Eat for Dinner Workshop on Australian Acacias for Food Security*. World Vision Australia, 2011.

French, Bruce. *Food Plants of Papua New Guinea*. http://food plantsinternational.com/free-resources-for-download/png.

Fried, Stephanie G. "Tropical forests forever? A contextual ecology of Bentian rattan agroforestry systems." *People, Plants, and Justice: The Politics of Nature Conservation* (2000): 203–33.

Fulbright, Dennis W. *A Guide to Nut Tree Culture in North America*, 3rd ed. East Lansing, Mich.: Northern Nut Growers Association, 2003.

Galuppo, Silvio Carla. "Documantacao do uso y valoriza de olea de piquia (*Caryocar villosum*) e do leite do amapa-doce (*Brosimum parinarioides*) para a comunidade de Piquituba." Floresta Nacional de Tapajos.

Gammage, Bill. *The Biggest Estate on Earth: How Aborigines Made Australia*. Crows Nest, NSW, Australia: Allen & Unwin, 2011.

Garrity, Dennis Philip, Festus K. Akinnifesi, Oluyede C. Ajayi, Sileshi G. Weldesemayat, Jeremias G. Mowo, Antoine Kalinganire, Mahamane Larwanou, and Jules Bayala. "Evergreen agriculture: A robust approach to sustainable food security in Africa." *Food Security* 2, no. 3 (2010): 197–214.

Geesing, Dieter, Peter Felker, and Ralph L. Bingham. "Influence of mesquite (*Prosopis glandulosa*) on soil nitrogen and carbon development: Implications for global carbon sequestration." *Journal of Arid Environments* 46, no. 2 (2000): 176.

Geraldo, Daniel, Pedro Jose Correia, Jose Filipe, and Luis Nunes. "Carob-tree as CO_2 sink in the carbon market." *Advances in Climate Changes, Global Warming, Biological Problems and Natural Hazards* (2010).

Geyer, W. A., "Biomass production in the Central Great Plains of the USA under various coppice regimes," *Biomass and Bioenergy* 30 (2006): 778–83.

Gilbertson, Tamra, and Oscar Reyes. "Carbon trading: How it works and why it fails." *Soundings* 45, no. 45 (2010).

Giwa, Solomon, Luqman Chuah Abdullah, and Nor Mariah Adam. "Investigating 'Egusi' (*Citrullus colocynthis* L.) seed oil as potential biodiesel feedstock." *Energies* 3, no. 4 (2010): 607–18.

Glanville, Brigid. "Queensland farmers invest in diesel-producing trees." *ABC News*, March 28, 2008.

Gliessman, Stephen R. *Agroecology: The Ecology of Sustainable Food Systems*, 3rd ed. Boca Raton, Fla.: CRC Press, 2015.

Global Alliance for Climate-Smart Agriculture. "GACSA Action Plan." http://www.un.org/climatechange/summit/wp-content/uploads/sites/2/2014/09/AGRICULTURE-Action-Plan.pdf.

Global Climate Change Student Guide. "The climate system." http://web.archive.org/web/20071121192518/http://www.ace.mmu.ac.uk/resources/gcc/1-3-3.html.

Global Invasive Species Database online. http://www.issg.org.

Gold, Michael. "Honeylocust (*Gleditsia triacanthos*), a multi-purpose tree for the temperate zone." *International Tree Crops Journal* 7, no. 4 (1993).

Goldenberg, Suzanne. "Al Gore says use of geo-engineering to head off climate disaster is insane." *Guardian*, January 15, 2014.

Goltenboth, Friedhelm, and Werner Muhlbauer. "Abacá: Cultivation, extraction and processing." In Jörg Müssig, ed., *Industrial Applications of Natural Fibres: Structure, Properties and Technical Applications*, pp. 163–79. Chichester, UK: John Wiley, 2010.

Gore, Al. *An Inconvenient Truth: The Planetary Emergency of Global Warming and What We Can Do About It*. New York: Rodale Press, 2006.

Gore, Al. *Our Choice: A Plan to Solve the Climate Crisis*. Emmaus, Pa.: Rodale, 2009.

Goren-Ibor, Naama, et al. "Nuts, nut cracking, and pitted stones at Gesher Benut Ya'aqov, Israel." *Proceedings of the National Academy of Sciences*, February 19, 2002.

GRAIN. "Food, climate change, and healthy soils: The forgotten link." In *Wake Up Before It Is Too Late: Make Agriculture Truly Sustainable Now for Food Security in a Changing Climate*. Geneva: UNCTAD, 2013.

———. *Hungry for Land: Small Farmers Feed the World with Less than a Quarter of All Farmland*. http://www.grain.org/article/entries/4929-hungry-for-land-small-farmers-feed-the-world-with-less-than-a-quarter-of-all-farmland.

Grant Space. "What is program-related investment?" http://grantspace.org/tools/knowledge-base/Grantmakers/pris.

Grobe, Hannes. "Albedo-e hg.png." Alfred Wegener Institute for Polar and Marine Research, Bremerhaven, Germany. http://commons.wikimedia.org/wiki/File:Albedo-e_hg.png.

Grogan, P., and R. Matthews. "A modelling analysis of the potential for soil carbon sequestration under short rotation coppice willow bioenergy plantations." *Soil Use and Management* 18, no. 3 (2002): 175–83.

Gutteridge, R. C., and H. M. Shelton. "Animal production potential of agroforestry systems." *ACIAR Proceedings*, Australian Centre for International Agricultural Research, 1994.

Gymnosperm Database online. http://www.conifers.org.

Hallett, Lauren M. "Towards a conceptual framework for novel ecosystems." In Richard J. Hobbs, Eric S. Higgs, and Carol M. Hall, eds., *Novel Ecosystems: Intervening in the New Ecological World Order*. Oxford, UK: John Wiley, 2013.

Hamilton, Jim. *Silvopasture: Establishment and Management Principles for Pine Forests in the Southeastern United States*. Lincoln: University of Nebraska, 2008.

Hanelt, Peter. *Mansfeld's Encyclopedia of Agricultural and Horticultural Crops*. Berlin: Springer, 2001.

Hanley, S. J. "Willow." In Nigel Halford and Angela Karp, eds., *Energy Crops*. Cambridge: Royal Society of Chemistry, 2011.

Hänninen, T., and Mark Hughes. "Historical, contemporary and future applications." In Jörg Müssig, ed., *Industrial Applications of Natural Fibres: Structure, Properties and Technical Applications*, pp. 385–95. Chichester, UK: John Wiley, 2010.

Hansen, J., M. Sato, P. Hearty, R. Ruedy, M. Kelley, V. Masson-Delmotte, G. Russell, G. Tselioudis, J. Cao, E. Rignot, I. Velicogna, E. Kandiano, K. von Schuckmann, P. Kharecha, A. N. Legrande, M. Bauer, and K. –W. Lo. "Ice melt, sea level rise and superstorms: Evidence from paleoclimate data, climate modeling, and modern observations that 2°C global warming is highly dangerous." *Atmospheric Chemistry and Physics* 15, issue 14 (2015): 20059-20179. doi:10.5194/acpd-15-20059-2015.

Hansen, James, Makiko Sato, Pushker Kharecha, David Beerling, Robert Berner, Valerie Masson-Delmotte, Mark Pagani, Maureen Raymo, Dana L. Royer, and James C. Zachos. "Target atmospheric CO_2: Where should humanity aim?" arXiv preprint arXiv:0804.1126, 2008.

Hardy, Ralph W. F. "The bio-based economy." In *Trends in New Crops and New Uses*, pp. 11–16. Alexandria, Va.: ASHS Press, 2002.

Harvey, Celia A., Mario Chacón, Camila I. Donatti, Eva Garen, Lee Hannah, Angela Andrade, Lucio Bede, et al. "Climate-smart landscapes: Opportunities and challenges for integrating adaptation and mitigation in tropical agriculture." *Conservation Letters* 7, no. 2 (2014): 77–90.

Havemann, Tanja. "Financing mitigation in smallholder agricultural systems: Issues and opportunities." In Eva Wollenberg et al., eds., *Climate Change Mitigation and Agriculture*. London: Earthscan, 2012.

Havlík, Petr, Hugo Valin, Mario Herrero, Michael Obersteiner, Erwin Schmid, Mariana C. Rufino, Aline Mosnier, et al. "Climate change mitigation through livestock system transitions." *Proceedings of the National Academy of Sciences* 111, no. 10 (2014): 3709–14.

Heifer International. "The foundation of transformation." http://www.heifer.org/ending-hunger/the-heifer-way/mission-and-cornerstones.html#cornerstone-passing.

Henderson, Benjamin, Pierre Gerber, Tom Hilinski, Alessandra Falcucci, Dennis Ojima, Mirella Salvatore, and Richard Conant. "Greenhouse gas mitigation potential of the world's grazing lands: Modeling soil carbon and nitrogen fluxes of mitigation practices." *Agriculture, Ecosystems and Environment* 207 (2015): 91-100. doi:10.1016/j.agee.2015.03.029.

Henley, G., and L. Yiping. "The climate change challenge and bamboo: Mitigation and adaptation." International Network for Bamboo and Rattan, Beijing, 2009.

Hepperly, Paul, Don Lotter, Christine Ziegler Ulsh, Rita Seidel, and Carolyn Reider. "Compost, manure and synthetic fertilizer influences crop yields, soil properties, nitrate leaching and crop nutrient content." *Compost Science & Utilization* 17, no. 2 (2009): 117–26.

Ho, Mae-Wan. "Sustainable agriculture and off-grid renewable energy." In *Wake Up Before It Is Too Late: Make Agriculture Truly Sustainable Now for Food Security in a Changing Climate*. Geneva: UNCTAD, 2013.

Hobbs, Richard J., Eric S. Higgs, and Carol M. Hall. "Introduction: Why novel ecosystems." In Richard J. Hobbs, Eric S. Higgs, and Carol M. Hall, eds., *Novel Ecosystems: Intervening in the New Ecological World Order*. Oxford, UK: John Wiley, 2013.

Hochschild, Adam. *King Leopold's Ghost: A Story of Greed, Terror, and Heroism in Colonial Africa*. Boston: Houghton Mifflin, 1999.

Hodgson, Wendy C. *Food Plants of the Sonoran Desert*. Tucson: University of Arizona Press, 2001.

Hoffmann, Ulrich. "Agriculture at the crossroads: Assuring food security in developing countries under the challenges of global warming." In *Wake Up Before It Is Too Late: Make Agriculture Truly Sustainable Now for Food Security in a Changing Climate*. Geneva: UNCTAD, 2013.

Hoggan, James, and Richard D. Littlemore. *Climate Cover-Up: The Crusade to Deny Global Warming*. Vancouver: Greystone Books, 2009.

Högy, Petra, and Andreas Fangmeier. "Yield and yield quality of major cereals under climate change." In *Wake Up Before It Is Too Late: Make Agriculture Truly Sustainable Now for Food Security in a Changing Climate*. Geneva: UNCTAD, 2013.

Honary, Lou, and B. P. Singh. "Industrial oil types and uses." In Bharat P. Singh, ed., *Industrial Crops and Uses*, p. 157. Wallingford, UK: CABI, 2010.

Hongwei, Tan, Zhou Liuqiang, Xie Rulin, and Huang Meifu. "Attaining high yield and high quality banana production in Guangxi." *Better Crops* 88, no. 4 (2004): 22–24.

House, A. P. N., and C. E. Harwood. *Australian Dry-Zone Acacias for Human Food: Proceedings of a Workshop Held at Glen Helen, Northern Territory, Australia, 7–10 August 1991*. Reprinted with Additional Material as Appendix 6. Canberra: CSIRO Division of Forestry, Australian Tree Seed Centre, 1994.

Human, T. R. "Oil as a byproduct of the avocado." *South African Avocado Growers' Association* 10 (1987): 163–164.

Hunter, I. R., and Wu Junqi. "Bamboo biomass." *International Network for Bamboo and Rattan (INBAR)* 11 (2002).

Huo, Yongkang. "Mulberry cultivation and utilization in China." *Mulberry for Animal Production*, FAO Animal Production and Health Paper 147 (2002): 11–43.

IFOAM. "Definition of organic agriculture." http://www.ifoam.bio/en/organic-landmarks/definition-organic-agriculture.

———. *The World of Organic Agriculture: Statistics and Emerging Trends 2014*. FiBL and IFOAM, 2014.

Institute for Agriculture and Trade Policy. "Corporations represent at UN Climate Summit." http://www.iatp.org/blog/201409/corporations-represent-at-un-climate-summit.

International Panel for Sustainable Resource Management. *Towards Sustainable Production and Use of Resources: Assessing Biofuels (Executive Summary)*. UNEP, 2009.

International Sericultural Commission. "Statistics." http://inserco.org/en/statistics.

IPCC. *Climate Change 2007: Mitigation of Climate Change: Contribution of Working Group III to the Fourth Assessment Report of the Intergovernmental Panel on Climate Change*. Cambridge, UK: Cambridge University Press, 2007.

———. *Climate Change 2013: The Physical Science Basis: Contribution of Working Group I to the Fifth Assessment Report of the Intergovernmental Panel on Climate Change*. Cambridge, UK: Cambridge University Press, 2013.

———. *Climate Change 2014: Impacts, Adaptation, and Vulnerability. Part A: Global and Sectoral Aspects: Contribution of Working Group II to the Fifth Assessment Report of the Intergovernmental Panel on Climate Change.* Cambridge, UK: Cambridge University Press, 2014.

———. *Climate Change 2014: Mitigation of Climate Change: Contribution of Working Group II to the Fifth Assessment Report of the Intergovernmental Panel on Climate Change.* Cambridge, UK: Cambridge University Press, 2014.

———. *Contribution of Working Group II to the Fourth Assessment Report of the Intergovernmental Panel on Climate Change, Summary for Policymakers.* Cambridge, UK: Cambridge University Press, 2007.

———. *Land Use, Land-Use Change, and Forestry: A Special Report of the Intergovernmental Panel on Climate Change.* Cambridge, UK: Cambridge University Press, 2000.

Irwin, Kris, and Jerry Bratton. "Outdoor living barn: A specialized windbreak." Lincoln, Neb.: National Agroforestry Center, USDA Forest Service, 1996.

Isebrands, J. G., P. Aronsson, M. Carlson, R. Ceulemans, M. Coleman, N. Dickinson, J. Dimitriou, et al. "Environmental applications of poplars and willows." In J. G. Isebrands and J. Richardson, eds., *Poplars and Willows: Trees for Society and the Environment*, p. 258. Wallingford, UK, and Boston: CABI, 2014.

Ishii-Eiteman, Marcia. "Democratizing control of agriculture to meet the needs of the twenty first century." In *Wake Up Before It Is Too Late: Make Agriculture Truly Sustainable Now for Food Security in a Changing Climate.* Geneva: UNCTAD, 2013.

IUCN. *Red List of Threatened Species.* http://www.iucnredlist.org.

Jacke, Dave, and Eric Toensmeier. *Edible Forest Gardens.* 2 vols. White River Junction, VT: Chelsea Green, 2005.

Jackson, Robert B., Esteban G. Jobbágy, Roni Avissar, Somnath Baidya Roy, Damian J. Barrett, Charles W. Cook, Kathleen A. Farley, David C. Le Maitre, Bruce A. McCarl, and Brian C. Murray. "Trading water for carbon with biological carbon sequestration." *Science* 310, no. 5756 (2005): 1944–47.

Jackson, Stephen T. "Perspective: Ecological novelty is not new." In Richard J. Hobbs, Eric S. Higgs, and Carol M. Hall, eds., *Novel Ecosystems: Intervening in the New Ecological World Order.* Oxford, UK: John Wiley, 2013.

Jacobson, Mark Z., and Mark A. Delucchi. "A plan to power 100 percent of the planet with renewables." *Scientific American* 26 (2009).

Jaikumar, N. S., et al. "Agronomic assessment of perennial wheat and perennial rye as cereal crops." *Agronomy Journal* 104, no. 6 (2012): 1716–26.

Jamnadass, Ramni, et al. "The benefits of agroforestry systems for food and nutritional security." World Agroforestry Centre, 2013.

Janick, Jules, and Robert Paull, eds. *The Encyclopedia of Fruits and Nuts.* Cambridge, UK: CABI, 2008.

Janke, Rhonda. "Stevia: A grower's guide." Kansas State University, 2004.

Jansen, Patrick, and Leen Kuiper. "Double green energy from traditional coppice stands in the Netherlands." *Biomass and Bioenergy* 26, no. 4 (2004): 401–02.

Janssen, Jules J. A. *Designing and building with bamboo.* China: International Network for Bamboo and Rattan, 2000.

Jarman, Cyril, *Plant Fibre Processing: A Handbook*, London, Intermediate Technology Publications, 1998.

Jayasekara, Chitrangani, and Nalinie Amarasinghe. "Coir—coconut cultivation, extraction and processing of coir." In Jörg Müssig, ed., *Industrial Applications of Natural Fibres: Structure, Properties and Technical Applications*, pp. 197–217. Chichester, UK: John Wiley, 2010.

Jha, Shalene, Christopher M. Bacon, Stacy M. Philpott, V. Ernesto Méndez, Peter Läderach, and Robert A. Rice. "Shade coffee: Update on a disappearing refuge for biodiversity." *BioScience* 64, no. 5 (2014): 416–28.

Jones, Christine. "Liquid carbon pathway unrecognized." http://www.amazingcarbon.com/PDF/JONES-LiquidCarbon Pathway(AFJ-July08).pdf.

Jordan, N., G. Boody, W. Broussard, J. D. Glover, D. Keeney, B. H. McCown, G. McIsaac, et al. "Sustainable development of the agricultural bio-economy." *Science* 316, no. 5831 (2007): 1570.

Jose, Shibu. "Agroforestry for ecosystem services and environmental benefits: An overview." *Agroforestry Systems* 76, no. 1 (2009): 1–10.

Kalita, Dipul. "Hydrocarbon plant: New source of energy for future." *Renewable and Sustainable Energy Reviews* 12, no. 2 (2008): 455–71.

Kang, B. T., A. N. Atta-Krah, and L. Reynolds. *Alley Farming.* Basingstoke, UK: Macmillan Education, 1999.

Kaonga, Martin Leckson, et al. *Agroforestry for Biodiversity and Ecosystem Services: Service and Practice.* Rijeka, Croatia: InTech, 2012.

Kaplan, Marcus, Chinwe Ifejika-Speranza, and Imme Sholz. "Promoting resilient agriculture in sub-Saharan Africa as a major priority in climate-change adaptation." In *Wake Up Before It Is Too Late: Make Agriculture Truly Sustainable Now for Food Security in a Changing Climate.* Geneva: UNCTAD, 2013.

Kapsoot, Daniel. "Agroforestry in India: New national policy sets the bar high." *Guardian*, February 17, 2014.

Karki, Uma. "Sustainable year-round forage production and grazing/browsing management education program." In *2014 ADSA-ASAS-CSAS Joint Annual Meeting.* Tuskegee 2014.

Kazakoff, Stephen, Peter Gresshoff, and Paul Scott. "*Pongamia pinnata*, a sustainable feedstock for biodiesel production." In

Nigel Halford and Angela Karp, eds., *Energy Crops*. Cambridge: Royal Society of Chemistry, 2011.

Keller, David P., Ellias Y. Feng, and Andreas Oschlies. "Potential climate engineering effectiveness and side effects during a high carbon dioxide–emission scenario." *Nature Communications* 5 (2014).

Kennedy, Duncan. "Turning wood into bones." *BBC News*, January 8, 2010.

Kesari, Vigya, and Latha Rangan. "Development of *Pongamia pinnata* as an alternative biofuel crop: Current status and scope of plantations in India." *Journal of Crop Science and Biotechnology* 13, no. 3 (2010): 127–37.

Khorramdel, Surur, Alireza Koocheki, Mehdi Nassiri Mahallati, Reza Khorasani, and Reza Ghorbani. "Evaluation of carbon sequestration potential in corn fields with different management systems." *Soil and Tillage Research* 133 (2013): 25–31.

Kilani, Hala, Assaad Serhal, and Othman Llewlyn. "Al-Hima: A way of life." Amman, Jordan: IUCN West Asia Regional Office, 2007.

Kim, Do-Soon, Lin S. Huang, Richard Flavell, Steve A. Renvoize, Joerg M. Greef, Tsai Wen Hsu, Astley Hastings, et al. "Developing miscanthus for bioenergy." In Nigel Halford and Angela Karp, eds., *Energy Crops*. Cambridge: Royal Society of Chemistry, 2011.

Kime, Lynne. "Diversification of your operation, why?" Penn State Extension. http://extension.psu.edu/business/farm/resources/publications/diversification-of-your-operation-why.

Kiptot, Evelyne, and Steven Franzel. *Gender and Agroforestry in Africa: Are Women Participating?* Nairobi: World Agroforestry Centre, 2011.

Kiptot, Evelyne, Steven Franzel, and Ann Degrande. "Gender, agroforestry and food security in Africa." *Current Opinion in Environmental Sustainability* 6 (2014): 104–09.

Klein, Naomi. *This Changes Everything: Capitalism Versus the Climate*. London: Allen Lane, 2014.

———. "Why #BlackLivesMatter should transform the climate debate." http://www.thenation.com/article/192801/what-does-blacklivesmatter-have-do-climate-change.

Knapp, Alan K., John M. Blair, John M. Briggs, Scott L. Collins, David C. Hartnett, Loretta C. Johnson, and E. Gene Towne. "The keystone role of bison in North American tallgrass prairie." *BioScience* 49, no. 1 (1999): 39–50.

Koeppel, Dan. *Banana: The Fate of the Fruit That Changed the World*. New York: Hudson Street Press, 2008.

Kort, John. "9. Benefits of windbreaks to field and forage crops." *Agriculture, Ecosystems & Environment* 22 (1988): 165–90.

Koutinas, Apostolis A., C. Du, R. H. Wang, and Colin Webb. "Production of chemicals from biomass." In James H. Clark and Fabien Deswarte, eds., *Introduction to Chemicals from Biomass*, pp. 77–101. Chichester, UK: John Wiley, 2008.

Kranemann GmbH equipment website. www.kranemann.org/eng/seabuckthorn.html.

Kuehl, Yannick, and Lou Yiping. *Carbon Off-Setting with Bamboo*. Beijing: INBAR, 2012.

Kuhns, Michael, and Tom Schmidt. *Heating with Wood: Species, Characteristics and Volumes*. Lincoln: University of Nebraska Extension, 1988.

Kumar, B. M. "Carbon sequestration potential of tropical homegardens." In *Tropical Homegardens*. Dordrecht, Neth.: Springer, 2006.

Kumar, B. M., and P. K. R. Nair. "Introduction." In *Tropical Homegardens*. Dordrecht, Neth.: Springer, 2006.

Kumar, Rajeew, Sharad Pandey, and Apurv Pandey. "Plant roots and carbon sequestration." *Current Science* 91, no. 7 (2006): 885.

La Via Campesina. 2014. "Unmasking climate smart agriculture." http://viacampesina.org/en/index.php/main-issues-main-menu-27/sustainable-peasants-agriculture-mainmenu-42/1670-un-masking-climate-smart-agriculture.

Lal, Rattan. "Abating climate change and feeding the world through soil carbon sequestration." In *Soil as World Heritage*, pp. 443–57. Amsterdam: Springer, 2014.

———. "Intensive agriculture and the soil carbon pool." *Journal of Crop Improvement* 27, no. 6 (2013): 735–51.

———. "Managing soils and ecosystems for mitigating anthropogenic carbon emissions and advancing global food security." *BioScience* 60, no. 9 (2010): 708–21.

Lamade, Emmanuelle, and Jean-Pierre Bouillet. "Carbon storage and global change: the role of oil palm." *Oléagineux, corps gras, lipides* 12, no. 2 (2005): 154–60.

Lancaster, Brad. *Rainwater Harvesting for Drylands and Beyond*. Tucson, Ariz.: Rainsource Press, 2008.

Land Institute. "A grain seen as nearest the perennial goal: Rye." *Land Report* 110 (Fall 2014).

———. "The long, hard challenge of wheat." *Land Report* 110 (Fall 2014).

———. "Perennial rice in tropical China produces grain competitively." *Land Report* 110 (Fall 2014).

Langenheim, Jean H. *Plant Resins: Chemistry, Evolution, Ecology, and Ethnobotany*. Portland, Ore.: Timber Press, 2003.

Larkin, Philip, and Matthew Newell. "Perennial wheat breeding: Current germplasm and a way forward for breeding and global cooperation." In *Perennial Crops for Food Security: Proceedings of the FAO Expert Workshop*. Rome: FAO, 2013.

Las Cañadas website. http://www.bosquedeniebla.com.mx.

Lawton, Geoff. *Establishing a Food Forest* DVD, Permaculture Research Institute of Australia, 2008.

Leach, Melissa, James Fairhead, and James Fraser. "Land grabs for biochar? Narratives and counter narratives in Africa's emerging biogenic carbon sequestration economy." In *International Conference on Global Land Grabbing, Sussex, England*. 2011.

Leakey, Roger. "Twelve principles for better food and more food from mature perennial agroecosystems." In *Perennial Crops for Food Security: Proceedings of the FAO Expert Workshop*. Rome: FAO, 2013.

Leakey, Roger R. B. "Addressing the causes of land degradation, food/nutritional insecurity and poverty: A new approach to agricultural intensification in the tropics and subtropics." In *Trade and Environment Review 2013: Wake Up Before It Is Too Late: Make Agriculture Truly Sustainable Now for Food Security in a Changing Climate*, pp. 192–98. Geneva: UN Publications, 2013.

———. *Living with the Trees of Life: Towards the Transformation of Tropical Agriculture*. Wallingford, UK: CABI, 2012.

Lee, Richard B. *Man the Hunter*. Chicago: Aldine, 1969.

Leffler, William L. *Petroleum Refining in Nontechnical Language*, 4th ed. Tulsa, Okla.: PennWell, 2008.

Lefroy, E. C. "Perennial farming systems." In Tim Barlow and Roberta Thorburn, eds., *Balancing Conservation and Production in Grassy Landscapes*, p. 170. Canberra: Environment Australia, 1999.

Legoupil, Jean-Claude, Pascal Lienhard, and Anonh Khamhoung. "Conservation agriculture in Southeast Asia." In *Conservation Agriculture*, pp. 285–310. Amsterdam: Springer International Publishing, 2015.

Lemus, R., and R. Lal. "Bioenergy crops and carbon sequestration." *Critical Reviews in Plant Sciences* 24, no. 1 (2005): 1–21.

Leson, Gero, Michael V. Harding, and Klaus Dippon. "Natural fibres in geotextiles for soil protection and erosion control." In Jörg Müssig, ed., *Industrial Applications of Natural Fibres: Structure, Properties and Technical Applications*, pp. 509–22. Chichester, UK: John Wiley, 2010.

Leu, Andre. "Managing climate change with soil organic matter in organic production systems." In *Wake Up Before It Is Too Late: Make Agriculture Truly Sustainable Now for Food Security in a Changing Climate*. Geneva: UNCTAD, 2013.

Leung, W. T. W. and M. Flores. *Food Composition Table for Use in Latin America*. Washington, D.C., National Institutes of Health, 1961.

Li, Wenhua. *Agro-Ecological Farming Systems in China*. Paris: UNESCO, 2001.

Liguori, Giorgia, Giovanni Gugliuzza, and Paolo Inglese. "Evaluating carbon fluxes in orange orchards in relation to planting density." *Journal of Agricultural Science* 147, no. 6 (2009): 637–45.

Loke, John, Luz Adriana Mesa, and Ywe Jan Franken. "*Euphorbia tirucalli* bioenergy manual: Feedstock production, bioenergy conversion, applications, economics." Wageningen, Neth.: FACT Foundation, 2011.

Lombard, C., and R. Leakey. "Protecting the rights of farmers and communities while securing long term market access for producers of non-timber forest products: Experience in southern Africa." In *Forests, Trees and Livelihoods* 19 (2010): 235–49.

Low, Tim. *Wild Food Plants of Australia*. North Ryde, NSW, Australia: Angus & Robertson Publishers, 1988.

Lu, Rongsen. "Combating desertification with seabuckthorn." In *Combating Desertification with Plants*, pp. 291–99. Amsterdam: Springer, 2001.

Luedeling, Eike, Gudeta Sileshi, Tracy Beedy, and Johannes Dietz. "Carbon sequestration potential of agroforestry systems in Africa." In *Carbon Sequestration Potential of Agroforestry Systems*, pp. 61–83. Amsterdam: Springer, 2011.

Luske, Boki, and Joris van der Kamp. "Carbon sequestration potential of reclaimed desert soils in Egypt." Louis Bolk Instituut and Soil & More International, 2009.

Lynas, Mark. *Six Degrees: Our Future on a Hotter Planet*. London: Fourth Estate, 2007.

Ma, Justin Man-yin. "Evaluation of fertile lines derived from the hybridization of *Glycine max* and *G. tomentella*." PhD dissertation, University of Illinois at Urbana-Champaign, 2011.

Magdoff, Fred. "Twenty-first-century land grabs: Accumulation by agricultural dispossession." *Centre for Research on Globalization (Global Research)* 4 (2013).

Magdoff, Fred, and John Bellamy Foster. *What Every Environmentalist Needs to Know About Capitalism: A Citizen's Guide to Capitalism and the Environment*. New York: Monthly Review Press, 2011.

Mann, C. C. "The real dirt on rainforest fertility." *Science* 297, no. 5583 (August 2002): 920–23.

Mann, Charles C. *1491: New Revelations of the Americas Before Columbus*. New York: Knopf, 2005.

———. *1493: Uncovering the New World Columbus Created*. New York: Knopf, 2011.

Manner, Harley I. "A review of traditional agroforestry in Micronesia." USDA Forest Service Gen. Tech. Rep. PSW-GTR-140 (1993), pp. 32–36.

Marin Carbon Project. "What is carbon farming?" http://www.marincarbonproject.org/what-is-carbon-farming.

Marris, Emma. *Rambunctious Garden: Saving Nature in a Post-Wild World*. New York: Bloomsbury, 2011.

Martellozzo, F., J. S. Landry, D. Plouffe, V. Seufert, P. Rowhani, and N. Ramankutty. "Urban agriculture: A global analysis of the space constraint to meet urban vegetable demand." *Environmental Research Letters* 9, no. 6 (2014): 064025.

Martin, Franklin W. *Tropical Yams and Their Potential*. Washington, D.C.: USDA Agricultural Research Service in cooperation with USAID, 1974.

Martinez, Steve. *Local Food Systems: Concepts, Impacts, and Issues*. Washington, D.C.: USDA Economic Research Service, 2010.

Martini, Endri, James M. Roshetko, Meine van Noordwijk, Arif Rahmanulloh, Elok Mulyoutami, Laxman Joshi, and Suseno Budidarsono. "Sugar palm (*Arenga pinnata* (Wurmb) Merr.) for livelihoods and biodiversity conservation in the orangutan habitat of Batang Toru, North Sumatra, Indonesia: Mixed prospects for domestication." *Agroforestry Systems* 86, no. 3 (2012): 401–17.

Mascaro, Joseph. "Perspective: From rivets to rivers." In Richard J. Hobbs, Eric S. Higgs, and Carol M. Hall, eds., *Novel Ecosystems: Intervening in the New Ecological World Order*. Oxford: John Wiley, 2013.

Mascaro, Joseph, et al. "Origins of the novel ecosystems concept." In Richard J. Hobbs, Eric S. Higgs, and Carol M. Hall, eds., *Novel Ecosystems: Intervening in the New Ecological World Order*. Oxford: John Wiley, 2013.

Maués, Márcia Motta. "Reproductive phenology and pollination of the Brazil nut tree (*Bertholletia excelsa* Humb. & Bonpl. Lecythidaceae) in Eastern Amazonia." *Pollinating Bees: The Conservation Link Between Agriculture and Nature*. Brazilia: Ministry of Environment, 2002: 245–54.

Maxwell, Bruce D., Stanley M. Wiatr, and Peter K. Fay. "Energy potential of leafy spurge (*Euphorbia esula*)." *Economic Botany* 39, no. 2 (1985): 150–56.

Mbow, Cheikh, Meine Van Noordwijk, Eike Luedeling, Henry Neufeldt, Peter A. Minang, and Godwin Kowero. "Agroforestry solutions to address food security and climate change challenges in Africa." *Current Opinion in Environmental Sustainability* 6 (2014): 61–67.

McAdam, J. H., P. J. Burgess, A. R. Graves, A. Rigueiro-Rodríguez, and M. R. Mosquera-Losada. "Classifications and functions of agroforestry systems in Europe." In *Agroforestry in Europe*, pp. 21–41. Amsterdam: Springer, 2009.

McCarthy, Nancy, et al. "Evaluating synergies and trade-offs among food security, development, and climate change." In Eva Wollenberg et al., eds., *Climate Change Mitigation and Agriculture*. London: Earthscan, 2012.

McCarthy, Nancy, Leslie Lipper, and Giacomo Branca. "Climate-smart agriculture: Smallholder adoption and implications for climate change adaptation and mitigation." *Mitigation of Climate Change in Agriculture Series 4*. Rome: FAO, 2011.

McKentley, Bill. "The story of the sweet-sap silver maple." St. Lawrence Nursery catalog, 2015.

McKibben, Bill. *Eaarth*. New York: Time Books/Henry Holt, 2010.

McKnight, Travis. "Want to have a real impact on climate change?" *Guardian*, August 4, 2014.

McSherry, Megan, and Mark Ritchie. "Effects of grazing on grassland soil carbon: A global review." *Global Change Biology* 19 (2013): 1347-1357. doi:10.1111/gcb.12144.

Mehra, Rekha, and Mary Hill Rojas. *Women, Food Security and Agriculture in a Global Marketplace*. Washington, D.C.: International Center for Research on Women (ICRW), 2008.

Merlín-Uribe, Yair, Carlos E. González-Esquivel, Armando Contreras-Hernández, Luis Zambrano, Patricia Moreno-Casasola, and Marta Astier. "Environmental and socio-economic sustainability of chinampas (raised beds) in Xochimilco, Mexico City." *International Journal of Agricultural Sustainability* 11, no. 3 (2013): 216–33.

Merriam-Webster Dictionary online. www.merriam-webster.com.

Mertz, Ole, Christine Padoch, Jefferson Fox, Rob A. Cramb, Stephen J. Leisz, Nguyen Thanh Lam, and Tran Duc Vien. "Swidden change in Southeast Asia: Understanding causes and consequences." *Human Ecology* 37, no. 3 (2009): 259–64.

Miccolis, Andrew, et al. "Oil palm and agroforestry systems: Coupling yields with environmental services, an experiment in the Brazilian Amazon." http://www.slideshare.net/agroforestry /session-66-oil-palm-agroforestry-systems-brazilian-amazon.

Millar, G. D., and W. B. Badgery. "Pasture cropping: A new approach to integrate crop and livestock farming systems." *Animal Production Science* 49, no. 10 (2009): 777–87.

Miller, Ethan. "Other Economies Are Possible!" http://www. dollarsandsense.org/archives/2006/0706emiller.html.

Mindanao Baptist Rural Life Center. *Sloping Agricultural Land Technology: How to Farm Hilly Land Without Losing Soil*. Fort Myers, Fla.: ECHO, 2012.

Mollison, B. C., and Reny Mia Slay. *Introduction to Permaculture*, 2nd ed. Tyalgum, Australia: Tagari Publications, 1994.

Monbiot, George. "After Capitalism." http://www.monbiot.com /2012/08/28/after-capitalism.

Montagnini, Florencia. "Homegardens of Mesoamerica: Biodiversity, food security, and nutrient management." In *Tropical Homegardens*, pp. 61–84. Amsterdam: Springer, 2006.

Montagnini, Florencia, and P. K. R. Nair. "Carbon sequestration: An underexploited environmental benefit of agroforestry systems." *Agroforestry Systems* 61, no. 1–3 (2004): 281–95.

Mora-Urpí, Jorge, John C. Weber, and Charles R. Clement. *Peach Palm*, Bactris gasipaes *Kunth*. Rome: IPGRI, 1997.

Morton, Julia Frances. *Fruits of Warm Climates*. Miami: J. F. Morton, 1987.

Moser, Bryan R., Fred J. Eller, Brent H. Tisserat, and Alan Gravett. "Preparation of fatty acid methyl esters from Osage orange (*Maclura pomifera*) oil and evaluation as biodiesel." *Energy & Fuels* 25, no. 4 (2011): 1869–77.

Mosquera-Losada, M. R., Dirk Freese, and A. Rigueiro-Rodríguez. "Carbon sequestration in European agroforestry systems." In *Carbon Sequestration Potential of Agroforestry Systems*, pp. 43–59. Amsterdam: Springer, 2011.

Mosquera-Losada, M. R., J. H. McAdam, R. Romero-Franco, J. J. Santiago-Freijanes, and A. Rigueiro-Rodríguez. "Definitions and components of agroforestry practices in Europe." In *Agroforestry in Europe*, pp. 3–19. Amsterdam: Springer, 2009.

Moyer, Jeffrey. *Organic No-Till Farming: Advancing No-Till Agriculture—Crops, Soils, Equipment*. Austin, Tex.: Acres USA, 2011.

Mtaita, T. A., B. K. Manqwiro, and A. N. Mphúru. "The role of horticulture plants in combating desertification." In *Combating Desertification with Plants*, pp. 33–43. Amsterdam: Springer, 2001.

Müller, Dirk. "A critical analysis of commodity and food price speculation." In *Wake Up Before It Is Too Late: Make Agriculture Truly Sustainable Now for Food Security in a Changing Climate*. Geneva: UNCTAD, 2013.

Mulualem, Tewodros, and Hussein Mohammed. "Genetic variability and association among yield and yield related traits in aerial yam (*Dioscorea bulbifera*) accessions at Southwestern Ethiopia." *Journal of Natural Sciences Research* 2, no. 9 (2012).

Mundler, Patrick, and Lucas Rumpus. "The energy efficiency of local food systems: A comparison between different modes of distribution." *Food Policy* 37, no. 6 (2012): 609–15.

Murphy, Tim, Graham Jones, Jerry Vanclay, and Kevin Glencross. "Preliminary carbon sequestration modelling for the Australian macadamia industry." *Agroforestry Systems* 87, no. 3 (2013): 689–98.

Murray, Seth C., and Russell W. Jesseup. "Breeding and genetics of perennial maize: Progress, opportunities, and challenges." In *Perennial Crops for Food Security: Proceedings of the FAO Expert Workshop*. Rome: FAO, 2013.

Nair, P. K. "The coming of age of agroforestry." *Journal of the Science of Food and Agriculture* 87, no. 9 (2007): 1613–19.

———. "Whither homegardens?" In *Tropical Homegardens*. Dordrecht, Neth.: Springer, 2006.

Nair, P. K., and B. M. Kumar. "Introduction." In *Tropical Homegardens*. Dordrecht, Neth.: Springer, 2006.

Nair, P. K. R. "Climate change mitigation: A low-hanging fruit of agroforestry." In *Agroforestry: The Future of Global Land Use*, pp. 31–67. Amsterdam: Springer, 2012.

———. *An Introduction to Agroforestry*. Dordrecht, Neth.: Kluwer Academic Publishers in cooperation with International Centre for Research in Agroforestry, 1993.

Nair, P. K. Ramachandran. "Methodological challenges in estimating carbon sequestration potential of agroforestry systems." In *Carbon Sequestration Potential of Agroforestry Systems*, pp. 3–16. Amsterdam: Springer, 2011.

Nair, P. K. Ramachandran, Vimala D. Nair, B. Mohan Kumar, and Julia M. Showalter. "Carbon sequestration in agroforestry systems." *Advances in agronomy* 108 (2010): 237–307.

Naranjo, Juan Fernando, César Augusto Cuartas, Enrique Murgueitio, Julián Chará, and Rolando Barahona. "Balance de gases de efecto invernadero en sistemas silvopastoriles intensivos con *Leucaena leucocephala* en Colombia." *Development* 24 (2012): 8.

Nath, Arun Jyoti, Rattan Lal, and Ashesh Kumar Das. "Managing woody bamboos for carbon farming and carbon trading." *Global Ecology and Conservation* 3 (2015): 654–63.

National Academy of Sciences. *Firewood Crops: Shrub and Tree Species for Energy Production*. Washington, D.C.: National Academy of Sciences, 1980.

National Corn Growers Association. *2012 World of Corn Statistics Book*. http://www.ncga.com/upload/files/documents/pdf/2012_woc_metric.pdf.

National Research Council. *Firewood Crops: Shrub and Tree Species for Energy Production*, vol. 2: *Report of an Ad Hoc Panel of the Advisory Committee on Technology Innovation, Board on Science and Technology for International Development, Office of International Affairs*. Washington, D.C.: National Academy Press, 1983.

———. *Lost Crops of Africa*, vol. 1: *Grains*. Washington, D.C.: National Academy Press, 1996.

———. *Lost Crops of Africa*, vol. 2: *Vegetables*. Washington, D.C.: National Academy Press, 2006.

———. *Lost Crops of Africa*, vol. 3: *Fruits*. Washington, D.C.: National Academy Press, 2008.

———. *Lost Crops of the Incas: Little-Known Plants of the Andes with Promise for Worldwide Cultivation*. Washington, D.C.: National Academy Press, 1989.

———. *The Winged Bean, a High-Protein Crop for the Tropics: Report of an Ad Hoc Panel of the Advisory Committee on Technology Innovation, Board on Science and Technology for International Development, Commission on International Relations*, 1978 ed. Washington, D.C.: National Academy of Sciences, 1975.

Neely, Constance, and Jan De Leeuw. "Home on the range: The contribution of rangeland management to climate change mitigation." In Eva Wollenberg et al., eds., *Climate Change Mitigation and Agriculture*. London: Earthscan, 2012.

Nelson, Gerald C., et. al., *Food Security, Farming, and Climate Change to 2050: Scenarios, Results, Policy Options*. Washington, D.C.: International Food Policy Research Institute, 2010.

Neubauer, Franz. "Insulation materials based on natural fibers." In Jörg Müssig, ed., *Industrial Applications of Natural Fibres:*

Structure, Properties and Technical Applications. Chichester, UK: John Wiley, 2010.

Neufeldt, Henry, Molly Jahn, Bruce M. Campbell, John R. Beddington, Fabrice DeClerck, Alessandro De Pinto, Jay Gulledge, et al. "Beyond climate-smart agriculture: Toward safe operating spaces for global food systems." *Agriculture & Food Security* 2, no. 1 (2013): 12.

Neufeldt, Henry, Andreas Wilkes, R. J. Zomer, J. Xu, E. Nang'ole, C. Munster, and Frank Place. "Trees on farms: Tackling the triple challenges of mitigation, adaptation and food security." *World Agroforestry Centre Policy Brief* 7 (2009): 2.

New, Mark, Diana Liverman, Heike Schroder, and Kevin Anderson. "Four degrees and beyond: The potential for a global temperature increase of four degrees and its implications." *Philosophical Transactions of the Royal Society of London A: Mathematical, Physical and Engineering Sciences* 369, no. 1934 (2011): 6–19.

Newell, Joshua P., and Robert O. Vos. "Accounting for forest carbon pool dynamics in product carbon footprints: Challenges and opportunities." *Environmental Impact Assessment Review* 37 (2012): 23–36.

Nickerson, Colin. "A jolt for the science behind harvesting maple sap." *Boston Globe*, March 2, 2014.

Nishimura, Hiroyuki, and Melvin Calvin. "Essential oil of *Eucalyptus globulus* in California." *Journal of Agricultural and Food Chemistry* 27, no. 2 (1979): 432–35.

Nobrega, Camila. "Solidarity economy: Finding a new way out of poverty." *Guardian*, October 9, 2013.

Norman, Michael John Thornley, Craig John Pearson, and P. G. E. Searle. *The Ecology of Tropical Food Crops*, 2nd ed. Cambridge, UK: Cambridge University Press, 1995.

Nuberg, I. K. "Effect of shelter on temperate crops: A review to define research for Australian conditions." *Agroforestry Systems* 41, no. 1 (1998): 3–34.

O'Barr, Scott. *Alternative Crops for Drylands: Proactively Adapting to Climate Change and Water Shortages*. Santa Barbara, Calif.: Amaigabe Press, 2013.

O'Brien, D., J. L. Capper, P. C. Garnsworthy, C. Grainger, and L. Shalloo. "A case study of the carbon footprint of milk from high-performing confinement and grass-based dairy farms." *Journal of Dairy Science* 97, no. 3 (2014): 1835–51.

Ohler, J. G., ed. *Modern Coconut Management: Palm Cultivation and Products*. London: Intermediate Technology Publications, 1999.

Oikos Tree Crops online catalog 2014.

OilPrice.net. http://www.oil-price.net.

Olive Tree Growers. "Olive trees in Florida—A brief history." olivetreegrowers.com, accessed March 5, 2014.

Open Source Ecology website. opensourceecology.org, accessed April 14, 2014.

Outland, Robert B. *Tapping the Pines: The Naval Stores Industry in the American South*. Baton Rouge: Louisiana State University Press, 2004.

Oweis, Theib, Dieter Prinz, and Ahmed Hachum. *Water Harvesting: Indigenous Knowledge for the Future of the Drier Environments*. Beirut: ICARDA, 2001.

Oxfam International. "Another inconvenient truth: How biofuel policies are deepening poverty and accelerating climate change." *Oxfam Policy and Practice: Climate Change and Resilience* 4, no. 2 (2008): 1–58.

———. *Suffering the Science: Climate Change, People, and Poverty*. Oxford, UK: Oxfam International, 2009.

Padoch, Christine, and Miguel Pinedo-Vasquez. "Saving slash-and-burn to save biodiversity." *Biotropica* 42, no. 5 (2010): 550–52.

Pandey, Deep Narayan, Anil K. Gupta, and David M. Anderson. "Rainwater harvesting as an adaptation to climate change." *Current Science* 85, no. 1 (2003): 46–59.

Papanastasis, V. P., K. Mantzanas, O. Dini-Papanastasi, and I. Ispikoudis. "Traditional agroforestry systems and their evolution in Greece." In *Agroforestry in Europe*, pp. 89–109. Amsterdam: Springer, 2009.

Parfitt, Julian, and Mark Barthel. "Food waste reduction: A global imperative." In *Wake Up Before It Is Too Late: Make Agriculture Truly Sustainable Now for Food Security in a Changing Climate*. Geneva: UNCTAD, 2013.

Parrotta, John A. "Productivity, nutrient cycling, and succession in single- and mixed-species plantations of *Casuarina equisetifolia*, *Eucalyptus robusta*, and *Leucaena leucocephala* in Puerto Rico." *Forest Ecology and Management* 124, no. 1 (1999): 45–77.

Pasiecznik, N. M., et al. *The* Prosopis juliflora–Prosopis pallida *Complex: A Monograph*. Coventry, UK: HDRA, 2001.

Pasiecznik, Nick. "The role of trees in aquaculture systems." *Overstory* 174 (2006).

Pasternak, Dov. "Combating poverty with plants." In *Combating Desertification with Plants*, pp. 17–30. Amsterdam: Springer, 2001.

Paterson, Andrew, et al. "Viewpoint: Multiple-harvest sorghums toward improved food security." In *Perennial Crops for Food Security: Proceedings of the FAO Expert Workshop*. Rome: FAO, 2013.

Pattanayak, Subhrendu K., D. Evan Mercer, Erin Sills, and Jui-Chen Yang. "Taking stock of agroforestry adoption studies." *Agroforestry Systems* 57, no. 3 (2003): 173–86.

Pearlstein, S. L., et al. "Nipa (*Distichlis palmeri*): A perennial grain crop for saltwater irrigation." *Journal of Arid Environments* 82 (July 2012): 60–70.

Pelletier, Nathan, Eric Audsley, Sonja Brodt, Tara Garnett, Patrik Henriksson, Alissa Kendall, Klaas Jan Kramer, David Murphy, Thomas Nemecek, and Max Troell. "Energy intensity of agriculture and food systems." *Annual Review of Environment and Resources* 36 (November 2011): 223–46.

Percival, A. Edward, and Russell J. Kohel. "Distribution, collection, and evaluation of *Gossypium*." *Advances in Agronomy* 44 (1990): 225–56.

Pereira, Helena. *Cork: Biology, Production and Uses*. Amsterdam: Elsevier, 2007.

Perring, Michael P., and Erle C. Ellis. "The extent of novel ecosystems: Long in time and broad in space." In Richard J. Hobbs, Eric S. Higgs, and Carol M. Hall, eds., *Novel Ecosystems: Intervening in the New Ecological World Order*. Oxford, UK: John Wiley, 2013.

Pimentel, David, David Cerasale, Rose C. Stanley, Rachel Perlman, Elise M. Newman, Lincoln C. Brent, Amanda Mullan, and Debbie Tai-I. Chang. "Annual vs. perennial grain production." *Agriculture, Ecosystems & Environment* 161 (2012): 1–9.

Piotrowski, Stephan, and Michael Carus. "Natural fibers in technical applications: Market and trends." In Jörg Müssig, ed., *Industrial Applications of Natural Fibres: Structure, Properties and Technical Applications*. Chichester, UK: John Wiley, 2010.

Pirie, N. W. *Leaf Protein and Its By-Products in Human and Animal Nutrition*, 2nd ed. Cambridge, UK: Cambridge University Press, 1987.

Place, Frank, Oluyede C. Ajayi, Emmanuel Torquebiau, Guillermo Detlefsen, Michelle Gauthier, and Gérard Buttoud. "Improved Policies for Facilitating the Adoption of Agroforestry" in *Agroforestry for Biodiversity and Ecosystem Services - Science and Practice*, Dr. Martin Kaonga, ed. doi:10.5772/34524, Available from http://www.intechopen.com/books/agroforestry-for-biodiversity-and-ecosystem-services-science-and-practice/improved-policies-for-facilitating-the-adoption-of-agroforestry.

Plain, Ron. "Annual cattle inventory summary." Columbia: University of Missouri, 2014.

Plant List. http://www.theplantlist.org.

Plant Resources of Southeast Asia online database. http://proseanet.org/prosea/eprosea.php.

Plant Resources of Tropical Africa online database. http://www.prota4u.org.

Pogna, Norberto E., et al. "Evaluation of nine perennial wheat derivatives grown in Italy." In *Perennial Crops for Food Security: Proceedings of the FAO Expert Workshop*. Rome: FAO, 2013.

Pollan, Michael. *The Botany of Desire: A Plant's-Eye View of the World*. New York: Random House, 2001.

Ponisio, Lauren C., Leithen K. M'Gonigle, Kevi C. Mace, Jenny Palomino, Perry de Valpine, and Claire Kremen. "Diversification practices reduce organic to conventional yield gap." *Proceedings of the Royal Society of London B: Biological Sciences* 282, no. 1799 (2015): 20141396.

Powlson, David S., Clare M. Stirling, M. L. Jat, Bruno G. Gerard, Cheryl A. Palm, Pedro A. Sanchez, and Kenneth G. Cassman. "Limited potential of no-till agriculture for climate change mitigation." *Nature Climate Change* 4, no. 8 (2014): 678–83.

Pretty, Jules N., Andrew D. Noble, Deborah Bossio, John Dixon, Rachel E. Hine, Frits W. T. Penning de Vries, and James IL. Morison. "Resource-conserving agriculture increases yields in developing countries." *Environmental Science & Technology* 40, no. 4 (2006): 1114–19.

Project Drawdown. http://www.drawdown.org.

Quinkenstein, Ansgar, Christian Böhm, Eduardo da Silva Matos, Dirk Freese, and Reinhard F. Hüttl. "Assessing the carbon sequestration in short rotation coppices of *Robinia pseudoacacia* L. on marginal sites in northeast Germany." In *Carbon Sequestration Potential of Agroforestry Systems*, pp. 201–16. Amsterdam: Springer 2011.

Qungen, Cao, and Zhang Qingmei. "Bamboo agroforestry in China." *Agroforestry News* 7, no. 2: 17 (1999).

Rai, Sarathi. "India–US fight on basmati rice is mostly settled." *New York Times*, August 25, 2001.

Rainforest Alliance. "SAN Climate Module." http://www.rainforest-alliance.org/work/climate/climate-smart-agriculture/san-climate-module.

Rao, A. N., V. Ramanatha Rao, and John Trevor Williams, eds. *Priority Species of Bamboo and Rattan*. International Plant Genetic Resources Institute (IPGRI) and International Network for Bamboo and Rattan (INBAR), 1998.

Ray, Dennis T., Michael A. Foster, Terry A. Coffelt, and Colleen McMahan. "Guayule: Culture, breeding and rubber production." In Bharat P. Singh, ed., *Industrial Crops and Uses*, pp. 384–410. Wallingford, UK: CABI, 2010.

Re, Limb. "The effective communication of agricultural R&D output in the UK beet sugar industry." *Proceedings of the South African Sugar Technological Association* 81 (2008): 107–15.

Reij, Chris. "Adapting to climate change and improving household food security in Africa through agroforestry: Some lessons from the Sahel." In *Wake Up Before It Is Too Late: Make Agriculture Truly Sustainable Now for Food Security in a Changing Climate*. Geneva: UNCTAD, 2013.

Renting, H., W. A. H. Rossing, J. C. J. Groot, J. D. Van der Ploeg, C. Laurent, D. Perraud, Derk Jan Stobbelaar, and M. K. Van Ittersum. "Exploring multifunctional agriculture: A review of conceptual approaches and prospects for an integrative transitional framework." *Journal of Environmental Management* 90 (2009): S112–23.

Rico-Gray, Victor, and Jose Garcia-Franco. "The Maya and the vegetation of the Yucatan Peninsula." *Journal of Ethnobiology* 11, no. 1: 135–42.

Rinaudo, Tony. "Farmer managed natural regeneration: Exceptional impact of a novel approach to reforestation in sub-Saharan Africa." In *Agricultural Options for the Poor*. Fort Myers, Fla.: ECHO, 2014.

Rodale Institute. *Regenerative Organic Agriculture and Climate Change: A Down-to-Earth Solution to Global Warming*. Kutztown, Pa.: Rodale Institute, 2014.

Roni, Phil, Karrie Hanson, and Tim Beechie. "Global review of the physical and biological effectiveness of stream habitat rehabilitation techniques." *North American Journal of Fisheries Management* 28, no. 3 (2008): 856–90.

Rosset, Peter. and Medea Benjamin, eds. *The Greening of the Revolution: Cuba's Experiment with Organic Agriculture*. Minneapolis: Ocean Press, 2002.

Ruan, Roger R., Paul Chen, Richard Hemmingsen, Vance Morey, and Doug Tiffany. "Size matters: Small distributed biomass energy production systems for economic viability." *International Journal of Agricultural and Biological Engineering* 1, no. 1 (2008): 64–68.

Rumpel, Cornelia, and Ingrid Kögel-Knabner. "Deep soil organic matter: A key but poorly understood component of terrestrial C cycle." *Plant and Soil* 338, no. 1–2 (2011): 143–58.

Rumsey, Brian. "Perennial crops and climate change: Longer, livelier roots should restore carbon from atmosphere to soil." *Land Report* 109 (Summer 2014).

Russelle, Michael P., Martin H. Entz, and Alan J. Franzluebbers. "Reconsidering integrated crop–livestock systems in North America." *Agronomy Journal* 99, no. 2 (2007): 325–34.

Rutter, Phil. "Hybrid swarms and hickory-pecans." *Permaculture Activist* 95 (Spring 2014): 1.

———. "Woody agriculture: Increased carbon fixation and co-production of food and fuel." In *Annual Report of the Northern Nut Growers Association*, 1989.

Rutter, Philip, Susan Wiegrefe, and Brandon Rutter-Daywater. *Growing Hybrid Hazelnuts: The New Resilient Crop for a Changing Climate*. White River Junction, Vt.: Chelsea Green, 2015.

Ryals, Rebecca, and Whendee L. Silver. "Effects of organic matter amendments on net primary productivity and greenhouse gas emissions in annual grasslands." *Ecological Applications* 23, no. 1 (2013): 46–59.

Sackle, Sackey Augustina, and Kwaw Emmanuel. "Nutritional and sensory analysis of *Parkia biglobosa* (Dawadawa) based cookies." *Journal of Food and Nutrition Sciences* 1, no. 4 (2013): 43–49. doi: 10.11648/j.jfns.20130104.13.

Samson, Fred, and Fritz Knopf. "Prairie conservation in North America." *BioScience* (1994): 418–21.

Sands, David, Alice Pilgeram, and Cindy Morris. "Development and marketing of perennial grains with benefits for human health and nutrition." In *Perennial Crops for Food Security: Proceedings of the FAO Expert Workshop*. Rome: FAO, 2013.

Santiago, A. D., et al. "Sugarcane." In Nigel Halford and Angela Karp, eds., *Energy Crops*. Cambridge, UK: Royal Society of Chemistry, 2011.

Savanna Institute website, www.savannainstitute.org.

Savory, Allan. *How to Fight Desertification and Reverse Climate Change*. TED Talk transcript. http://www.ted.com/talks/allan_savory_how_to_green_the_world_s_deserts_and_reverse_climate_change/transcript?language=en.

Schmidt-Vogt, Dietrich, Stephen J. Leisz, Ole Mertz, Andreas Heinimann, Thiha Thiha, Peter Messerli, Michael Epprecht, et al. "An assessment of trends in the extent of swidden in Southeast Asia." *Human Ecology* 37, no. 3 (2009): 269–80.

Schneider, Lambert. "Is the CDM fulfilling its environmental and sustainable development objectives? An evaluation of the CDM and options for improvement." *Öko-Institut for Applied Ecology, Berlin* 248 (2007): 1685–703.

Schnelle, Rebecca. "Stevia." University of Kentucky, 2010.

Schoeneberger, Michele, Gary Bentrup, Henry de Gooijer, Raju Soolanayakanahally, Tom Sauer, James Brandle, Xinhua Zhou, and Dean Current. "Branching out: Agroforestry as a climate change mitigation and adaptation tool for agriculture." *Journal of Soil and Water Conservation* 67, no. 5 (2012): 128A–36A.

Schröter, Matthias, Emma H. Zanden, Alexander P. E. Oudenhoven, Roy P. Remme, Hector M. Serna-Chavez, Rudolf S. Groot, and Paul Opdam. "Ecosystem services as a contested concept: A synthesis of critique and counter-arguments." *Conservation Letters* 7, no. 6 (2014): 514–23.

Schroth, Götz, Sammya Agra D'Angelo, Wenceslau Geraldes Teixeira, Daniel Haag, and Reinhard Lieberei. "Conversion of secondary forest into agroforestry and monoculture plantations in Amazonia: Consequences for biomass, litter and soil carbon stocks after 7 years." *Forest Ecology and Management* 163, no. 1 (2002): 131–50.

Seebauer, Matthias, et al. "Carbon accounting for smallholder agricultural soil carbon projects." In Eva Wollenberg et al., eds., *Climate Change Mitigation and Agriculture*. London: Earthscan, 2012.

Seebert-Elverfeldt, Christina, and Marja-Lisa Tapio-Bistrom. "Agricultural mitigation approaches for smallholders." In Eva Wollenberg et al., eds., *Climate Change Mitigation and Agriculture*. London: Earthscan, 2012.

Seibert, Michael, et al. "Fuel and chemical co-production from tree crops." *Biomass* 9, no. 1 (1986): 49–66.

Shames, Seth, Louise E. Buck, and Sar Scherr. "Reducing costs and improving benefits in smallholder agriculture carbon projects." In Eva Wollenberg et al., eds., *Climate Change Mitigation and Agriculture*. London: Earthscan, 2012.

Shanley, Patricia. *Fruit Trees and Useful Plants in Amazonian Life*, revised English ed. Rome: FAO, 2011.

Sharma, Satyawati, and Ashwani Kuman. "*Jatropha curcas*: A source of energy and other applications." In Nigel Halford and Angela Karp, eds., *Energy Crops*. Cambridge, UK: Royal Society of Chemistry, 2011.

Shepard, Mark. *Restoration Agriculture: Real-World Permaculture for Farmers*. Austin, Tex.: Acres USA, 2013.

Shim, P. S. "Domestication and improvement of rattan." INBAR Working Paper No. 5, 1995.

Shiva, Vandana. *Monocultures of the Mind: Perspectives on Biodiversity and Biotechnology*. London: Zed Books, 1993.

Sidibe, M., and J. T. Williams. *Baobab (Adansonia digitata)*. Southampton, UK: International Centre for Underutilized Crops, 2002.

Sileshi, Gudeta, Festus K. Akinnifesi, Oluyede C. Ajayi, and Frank Place. "Meta-analysis of maize yield response to woody and herbaceous legumes in sub-Saharan Africa." *Plant and Soil* 307, no. 1–2 (2008): 1–19.

Singh, B. "Overview of industrial crops." In Bharat P. Singh, ed., *Industrial Crops and Uses*, pp. 1–20. Wallingford, UK: CABI, 2010.

Singh, Hari P., and Bharat P. Singh. "Natural rubber." In Bharat P. Singh, ed., *Industrial Crops and Uses*. Wallingford, UK: CABI, 2010.

Singh, Kamal Kishor. *Neem: A Treatise*. New Delhi and Bangalore: I. K. International, 2009.

Skeptical Science website. www.skepticalscience.com.

Smith, J. Russell. *Tree Crops: A Permanent Agriculture*. New York: Harper & Row, 1978.

Smith, Jeffrey L., and Janice V. Perino. "Osage orange (*Maclura pomifera*): History and economic uses." *Economic Botany* 35, no. 1 (1981): 24–41.

Smith, Keith, and Irene Smith. *Grow Your Own Bushfoods*. Sydney: New Holland Publishers, 1999.

Smith, Lincoln. "Acorns: The most overlooked food in America." http://www.foodday.org/acorns_the_most_overlooked_food_in_america.

Smith, Nigel J. H., T. J. Fik, P. de T. Alvim, Italo C. Falesi, and E. A. S. Serrao. "Agroforestry developments and potential in the Brazilian Amazon." *Land Degradation & Development* 6, no. 4 (1995): 251–63.

Smith, Pete, Daniel Martino, Zucong Cai, Daniel Gwary, Henry Janzen, Pushpam Kumar, Bruce McCarl, et al. "Greenhouse gas mitigation in agriculture." *Philosophical Transactions of the Royal Society B: Biological Sciences* 363, no. 1492 (2008): 789–813.

Sofo, Adriano, Vitale Nuzzo, Assunta Maria Palese, Cristos Xiloyannis, Giuseppe Celano, Paul Zukowskyj, and Bartolomeo Dichio. "Net CO_2 storage in Mediterranean olive and peach orchards." *Scientia horticulturae* 107, no. 1 (2005): 17–24.

Solowey, Elaine. *Growing Bread on Trees: The Case for Perennial Agriculture*. Lexington, Mass.: Biblio Books, 2010.

Solowey, Elaine M. "Upscaling the experimental planting of *Argania spinosa* at Kibbutz Ketura, Israel." In *Combating Desertification with Plants*, pp. 257–261. Amsterdam: Springer US, 2001.

Soule, Judith D., and Jon K. Piper. *Farming in Nature's Image: An Ecological Approach to Agriculture*. Washington, D.C.: Island Press, 1992.

Soussana, J. F., V. Allard, Kim Pilegaard, Per Ambus, C. Amman, C. Campbell, E. Ceschia, et al. "Full accounting of the greenhouse gas (CO_2, N_2O, CH_4) budget of nine European grassland sites." *Agriculture, Ecosystems & Environment* 121, no. 1 (2007): 121–34.

Speedy, Andrew, and Pierre-Luc Pugliese. *Legume Trees and Other Fodder Trees as Protein Sources for Livestock*. Rome: FAO, 1992.

Spurr, Dan. "Interview with Joel Salatin." http://rockymountaingardening.com/blogs/editors-blog/2013/06/interview-with-joel-salatin.

Srinivasarao, Ch., Rattan Lal, Sumanta Kundu, and Pravin B. Thakur. "Conservation agriculture and soil carbon sequestration." In *Conservation Agriculture*, pp. 479–524. Amsterdam: Springer, 2015.

Staats, Lewis, M. E. Krasny, and C. Campbell. "History of the Sugar Maple Tree Improvement Program: Uihlein Sugar Maple Field Station." http://maple.dnr.cornell.edu/Ext/history_tree_imp.htm.

Stanley, Denise. "Demystifying the tragedy of the commons: The resin tappers of Honduras." *Grassroots Development: Journal of the Inter-American Foundation* (1991).

Steinfeld, Henning, Pierre Gerber, Tom Wassenaar, Vincent Castel, Mauricio Rosales, and Cees De Haan. *Livestock's Long Shadow*. Rome: FAO, 2006.

Stevens, E. S. *Green Plastics: An Introduction to the New Science of Biodegradable Plastics*. Princeton, N.J.: Princeton University Press, 2002.

Stewart, Omer Call, and Henry T. Lewis. *Forgotten Fires: Native Americans and the Transient Wilderness*. Norman: University of Oklahoma Press, 2002.

Sukmana, S., A. Abdurachman, and A. Syarifuddin Karama. "Strategies to develop sustainable livestock on marginal land." *ACIAR Proceedings*, Australian Centre for International Agricultural Research, 1994.

Sundriyal, Manju, and R. C. Sundriyal. "Wild edible plants of the Sikkim Himalaya: Nutritive values of selected species." *Economic Botany* 55, no. 3 (2001): 377–90.

Suszkiw, Jan. "Milkweed: From floss to fun in the sun." *Agricultural Research* 1 (2009).

Tachieva, Galina. *Sprawl Repair Manual*. Washington, D.C.: Island Press, 2010.

Takamizawa, Kazuhiro, William Anderson, and Hari P. Singh. "Ethanol from lignocellulosic crops." In Bharat P. Singh, ed., *Industrial Crops and Uses*, pp. 104–39. Wallingford, UK: CABI, 2010.

Teague, Richard et al. "Multi-paddock grazing on rangelands: Why the perceptual dichotomy between research results and rancher experience?" *Journal of Environmental Management* 128 (2013): 699-717.

Telek, Lehel and Horace D. Graham. *Leaf Protein Concentrates*. Westport, Conn.: AVI, 1983.

Tennigkeit, Timm, and Andreas Wilkes. *An Assessment of the Potential for Carbon Finance in Rangelands*. Nairobi, World Agroforestry Centre, 2008.

Thakur, V. K., M. K. Thakur, and R. Gupta. "*Eulaliopsis binata*: Utilization of waste biomass in green composites." *Green Composites from Natural Resources* (2013): 125–30.

Than, Ker. "Causes of California drought linked to climate change, Stanford scientists say." http://news.stanford.edu/news/2014/september/drought-climate-change-092914.html.

Thebo, A. L., P. Drechsel, and E. F. Lambin. "Global assessment of urban and peri-urban agriculture: Irrigated and rainfed croplands." *Environmental Research Letters* 9, no. 11 (2014): 114002.

Thevs, N., S. Zerbe, Y. Kyosev, A. Rozi, B. Tang, N. Abdusalih, and Z. Novitskiy. "*Apocynum venetum* L. and *Apocynum pictum* Schrenk (Apocynaceae) as multi-functional and multi-service plant species in Central Asia: A review on biology, ecology, and utilization." *Journal of Applied Botany and Food Quality* 85, no. 2 (2013): 159.

Thielen, Michael. *Bioplastics: Basics, Applications, Markets*, 1st ed. Mönchengladbach, Germany: Polymedia, 2012.

Toensmeier, Eric. "Integrating livestock in the food forest." http://www.perennialsolutions.org/livestock-integration-reducing-labor-and-fossil-fuel-inputs.

———. *Perennial Vegetables: From Artichokes to 'Zuiki' Taro, a Gardener's Guide to Over 100 Delicious, Easy-to-Grow Edibles*, White River Junction, Vt.: Chelsea Green, 2007.

Tokar, Brian. "Biofuels and the global food crisis." In Fred Magdoff, *Agriculture and Food in Crisis: Conflict, Resistance, and Renewal*. New York: Monthly Review Press, 2010.

———. "The GMO threat to food sovereignty: Science, resistance and transformation." In William D. Schanbacher, ed., *The Global Food System: Issues and Solutions*. Santa Barbara, Calif.: ABC-CLIO, 2014.

———. *Toward Climate Justice: Perspectives on the Climate Crisis and Social Change*. Porsgrunn, Norway: New Compass Press, 2014.

Tropical Forages Interactive Selection Tool online. http://www.tropicalforages.info.

Tropical Savannas CRC. "The West Arnhem Land Fire Abatement Project (WALFA)." http://savanna.cdu.edu.au/information/arnhem_fire_project.html.

Truong, P. N., and Tran Tan Van. *The Vetiver System for Agriculture*. Leesburg, Va.: Vetiver Network International, 2008.

Tsujimoto, Yasuhiro, Takeshi Horie, Hamon Randriamihary, Tatsuhiko Shiraiwa, and Koki Homma. "Soil management: The key factors for higher productivity in the fields utilizing the System of Rice Intensification (SRI) in the central highland of Madagascar." *Agricultural Systems* 100, no. 1 (2009): 61–71.

Turley, David B. "The chemical value of biomass." In James H. Clark and Fabien Deswarte, eds., *Introduction to Chemicals from Biomass*, pp. 21–46. Chichester, UK: John Wiley, 2008.

Udawatta, Ranjith P., and Shibu Jose. "Carbon sequestration potential of agroforestry practices in temperate North America." In *Carbon Sequestration Potential of Agroforestry Systems*, pp. 17–42. Amsterdam: Springer, 2011.

Udediebe, A. B. I., and Carlini, C. R. (2000). "Questions and answers to edibility problems of the *Canavalia ensiformis* seeds: A review." *Animal Feed Science and Technology* 74, no. 2: 95–106.

UNCAD, *Wake up before it is too late: Make agriculture truly sustainable now for food security in a changing climate*. Switzerland: UNCAD, 2013.

Union of Concerned Scientists. *Failure to Yield: Biotechnology's Broken Promises*. Cambridge, UK: Union of Concerned Scientists, 2009.

University of California. *Sample Costs to Establish a Date Palm Orchard and Produce Dates in the Coachella Valley*. Davis: University of California, 2006.

———. *Sample Costs to Establish a Pecan Orchard and Produce Pecans*. Davis: University of California, 2005.

———. *Sample Costs to Establish a Walnut Orchard and Produce Walnuts*. Davis: University of California, 2013.

———. *Sample Costs to Establish an Almond Orchard and Produce Almonds*. Davis: University of California, 2011.

———. *Sample Costs to Establish an Orchard and Produce Almonds*. Davis: University of California, 2011.

———. *Sample Costs to Establish and Produce Pistachios*. Davis: University of California, 2008.

———. *Sample Costs to Produce Blackeye Beans*. Davis: University of California, 2008.

———. *Sample Costs to Produce Grain Corn*. Davis: University of California, 2008.

———. *Sample Costs to Produce Rice*. Davis: University of California, 2012.

———. *Sample Costs to Produce Sunflowers*. Davis: University of California, 2011.

Uphoff, Norman. "Agroecological alternatives: Capitalising on existing genetic potentials." *Journal of Development Studies* 43, no. 1 (2007): 218–36.

USDA, Census of Agriculture 2012, table 43: "Selected practices 2012."

USDA National Agrofroestry Center website. http://nac.unl.edu/#about.

USDA NRCS. *COMET-Planner.* http://cometfarm.nrel.colostate.edu/

———. *Soil Quality Indicators.* http://www.nrcs.usda.gov/wps/portal/nrcs/detail/soils/health/assessment/?cid=stelprdb1237387.

USDA Nutrient Database. http://ndb.nal.usda.gov.

USDA Plants Database. http://plants.usda.gov, accessed December 1, 2014.

US Energy Information Administration. "How much oil is used to make plastic?" http://www.eia.gov/tools/faqs/faq.cfm?id=34&t=6.

Vaca-Garcia, Carlos. "Biomaterials." In James H. Clark and Fabien Deswarte, eds., *Introduction to Chemicals from Biomass*, pp. 77–101. Chichester, UK: John Wiley, 2008.

van Huis, A., J. Van Itterbeeck, H. Klunder, E. Mertens, A. Halloran, G. Muir, and P. Vantomme. *Edible Insects: Future Prospects for Food and Feed Security.* Rome: FAO, 2014, p. 1.

van Noordwijk, Meine, Jules Bayala, Kurniatun Hairiah, Betha Lusiana, Catherine Muthuri, N. Khasanah, and R. Mulia. "Agroforestry solutions for buffering climate variability and adapting to change." *Climate Change Impact and Adaptation in Agricultural Systems*, pp. 216–232. Wallingford, UK: CABI, 2014.

Van Tassel et al. "Evaluating perennial candidates for domestication: Lessons from wild sunflower relatives." In *Perennial Crops for Food Security: Proceedings of the FAO Expert Workshop.* Rome: FAO, 2013.

Van Vliet, Nathalie, Ole Mertz, Andreas Heinimann, Tobias Langanke, Unai Pascual, Birgit Schmook, Cristina Adams, et al. "Trends, drivers and impacts of changes in swidden cultivation in tropical forest-agriculture frontiers: A global assessment." *Global Environmental Change* 22, no. 2 (2012): 418–29.

van Wyk, Ben-Erik, and Nigel Gerike. *People's Plants: A Guide to Useful Plants of Southern Africa.* Arcadia, South Africa: Briza Publications, 2000.

van Wyk, Ben-Erik, and Michael Wink. *Medicinal Plants of the World.* Portland, Ore.: Timber Press, 2004.

Vanek, Francis M., and Largus T. Angenent. *Sustainable Transportation Systems Engineering.* New York: McGraw-Hill, 2014.

Varela, M. C. "Cork and the cork oak system." *Unasylva* 50, no. 197: 42–44.

Verchot, Louis. "On forests' role in climate, *New York Times* op-ed gets it wrong." http://blog.cifor.org/24311/on-forests-role-in-climate-new-york-times-op-ed-gets-it-wrong#.VPuqwinC3XA.

Vetiver Network International website. www.vetiver.org.

Vidal, John. "India's rice revolution." *Guardian*, February 16, 2013.

Vogel, Kenneth P., et al. "Switchgrass." In Nigel Halford and Angela Karp, eds., *Energy Crops.* Cambridge, UK: Royal Society of Chemistry, 2011.

Vujević, Predrag, Nada Vahčić, Bernardica Milinović, Tvrtko Jelačić, and Dunja Halapija. "Pomological traits and proximate chemical composition of hazelnut (*Corylus avellana* L.) varieties grown in Croatia." *African Journal of Agricultural Research* 5, no. 15 (2010): 2023–29.

Wade, Len J. "Perennial crops: Needs, perceptions, essentials." In *Perennial Crops for Food Security: Proceedings of the FAO Expert Workshop.* Rome: FAO, 2013.

Weih, Martin. "Intensive short rotation forestry in boreal climates: Present and future perspectives." *Canadian Journal of Forest Research* 34, no. 7 (2004): 1369–78.

Weisman, Alan. *Gaviotas: A Village to Reinvent the World.* White River Junction, Vt.: Chelsea Green, 1998.

Werneck, Fred. "The potential for microfinance as a channel for carbon payments." In Eva Wollenberg et al., eds., *Climate Change Mitigation and Agriculture.* London: Earthscan, 2012.

Westervelt, Amy. "Is it fair to blame almond farmers for California's drought?" http://www.theguardian.com/sustainable-business/2015/apr/13/is-it-fair-to-blame-almond-farmers-for-californias-drought.

Westover, Robert. "Changing climate may substantially alter maple syrup production." USDA blog online, September 11, 2012.

White, Courtney. *Grass, Soil, Hope: A Journey Through Carbon Country.* White River Junction, Vt.: Chelsea Green, 2013.

Whitmore, Thomas M., and B. L. Turner. *Cultivated Landscapes of Middle America on the Eve of Conquest.* Oxford, UK: Oxford University Press, 2001.

Wibawa, Gede, Laxman Joshi, Meine Van Noordwijk, and Eric André Penot. "Rubber based Agroforestry Systems (RAS) as alternatives for rubber monoculture system." IRRDB annual conference, 2006, Ho Chi Minh City, Vietnam.

Wilkinson, Jennifer. *Nut Grower's Guide: The Complete Handbook for Producers and Hobbyists.* Melbourne, Australia: CSIRO, 2005.

Williams, Florence. "Plains sense: Frank and Deborah Popper's 'Buffalo Commons' is creeping toward reality." *High Country News* 15 (2001).

Williams, P. A., A. M. Gordon, H. E. Garrett, L. Buck, and S. M. Newman. "Agroforestry in North America and its role in farming systems." In *Temperate Agroforestry Systems*, pp. 9–84. Wallingford, UK: CABI, 1997.

Williams-Linera, Guadaloupe. *El Bosque de Niebla del Centro de Veracruz: Ecología, Historia y Destino en Tiempos de Fragmentacion y Cambio Climático*. Xalapa, Mexico: Instituto de Ecología, AC, 2007.

Wilson, R. T. *Livestock Production Systems*. London: Macmillan Education in cooperation with the CTA, 1995.

Wink, Michael, and Ben-Erik van Wyk. *Mind-Altering and Poisonous Plants of the World*. Portland, Ore.: Timber Press, 2008.

Winrock International. *Nitrogen Fixing Trees Highlights*. http://www.nzdl.org/gsdlmod?e=d-00000-00---off-0hdl--00-0---
-0-10-0---0---0direct-10---4-------0-1l--11-en-50---20-help--
-00-0-1-00-0-0-11-1-1utfZz-8-00&cl=CL1.12&d=HASH011e-c19a37bb817d72319187.23&x=1.

Wojtkowski, Paul A. *The Theory and Practice of Agroforestry Design: A Comprehensive Study of the Theories, Concepts and Conventions That Underlie the Successful Use of Agroforestry*. Enfield, N.H.: Science Publishers, 1998.

Wollenberg, Eva, et al., eds. *Climate Change Mitigation and Agriculture*. London: Earthscan, 2012.

World Agroforestry Centre. *Climate-Smart Landscapes: Multifunctionality in Practice*. Nairobi: World Agrofrestry Centre, 2015.

———. *Participatory Agroforestry Development in DPR Korea*. Kunming, China: World Agroforestry Center, 2011.

———. *Treesilience: An Assessment of the Resilience Provided by Trees in the Drylands of Eastern Africa*. Nairobi: World Agroforestry Centre, 2014.

World Health Organization. "Food Security." http://www.who.int/trade/glossary/story028/en.

World Vegetable Center. Vegetable Nutrition Database online. http://avrdcnutrition.gtdtestsite.comoj.com/nutrition.

Worldwatch Institute. "Global meat production and consumption continue to rise." http://www.worldwatch.org/global-meat-production-and-consumption-continue-rise-1.

———. "Leading the fight for food sovereignty: An interview with La Via Campesina's Dena Hoff." http://www.worldwatch.org/node/6514.

Wu, Ting, Yi Wang, Changjiang Yu, Rawee Chiarawipa, Xinzhong Zhang, Zhenhai Han, and Lianhai Wu. "Carbon sequestration by fruit trees: Chinese apple orchards as an example." *PloS one* 7, no. 6 (2012): e38883.

Yao, Zeng-Yu, Jian-Hua Qi, and Li-Ming Yin. "Biodiesel production from *Xanthoceras sorbifolia* in China: Opportunities and challenges." *Renewable and Sustainable Energy Reviews* 24 (2013): 57–65.

Yiping, Lou, et al. *Bamboo and Climate Change Mitigation*. Beijing: International Network for Bamboo and Rattan, 2010.

Young, Anthony. *Agroforestry for Soil Conservation*. Wallingford, UK: CABI, 2012.

Zhang, Shila, et al. "The progression of perennial rice breeding and genetics research in China." In *Perennial Crops for Food Security: Proceedings of the FAO Expert Workshop*. Rome: FAO, 2013.

Zimmer, Carl. "Putting a price tag on nature's' defenses." *New York Times*, June 5, 2014.

Zomer, Robert J., Deborah A. Bossio, Antonio Trabucco, Li Yuanjie, Diwan C. Gupta, and Virenda P. Singh. *Trees and Water: Smallholder Agroforestry on Irrigated Lands in Northern India*. Colombo, Sri Lanka: International Water Management Institute, 2007.

Zomer, Robert J., Antonio Trabucco, Richard Coe, and Frank Place. *Trees on Farm: Analysis of Global Extent and Geographical Patterns of Agroforestry*. Nairobi: World Agroforestry Centre, 2009.

INDEX

Note: Page numbers in *italics* refer to photographs and figures; page numbers followed by *t* refer to tables.

About the Author

Eric Toensmeier is the award-winning author of *Paradise Lot* and *Perennial Vegetables*, and the coauthor of *Edible Forest Gardens*. Eric is an appointed lecturer at Yale University, a Senior Fellow with Project Drawdown, and an international trainer. He presents in English and Spanish throughout the United States, Canada, Mexico, Guatemala, and the Caribbean. Eric has studied useful perennial plants and their roles in agroforestry systems for over two decades and cultivates about 300 species in his urban garden. His writing can be viewed online at perennialsolutions.org.

Photo by Rob Deza

About the Foreword Author

Dr. Hans Herren is an internationally recognized scientist who lived for twenty-seven years in Kenya, Benin, and Nigeria and worked across Africa in agriculture, health, and environmental research and capacity development. He is the recipient of numerous awards that recognize his distinguished and continuing achievements in original research for sustainable development, among them The Right Livelihood Award, which he received in 2013, and The World Food Prize, which he was awarded in 1995. He is also a member of the National Academy of Sciences and the Third World Academy of Sciences, and served as a co-chair of the International Assessment of Agricultural Science and Technology for Development (IAASTD). Dr. Herren currently serves as president and CEO of the Millennium Institute.

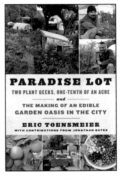